2nd Transdisciplinary Research on Environmental Problems in Southeast Asia (TREPSEA 2016)

IOP Conference Series: Earth and Environmental Science
Volume 71

Bandung, Indonesia
20 – 22 September 2016

ISBN: 978-1-5108-4264-9
ISSN: 1755-1307

Printed from e-media with permission by:

Curran Associates, Inc.
57 Morehouse Lane
Red Hook, NY 12571

proceedings.com

Some format issues inherent in the e-media version may also appear in this print version.

Printed by Curran Associates, Inc. (2017)

For permission requests, please contact the Institute of Physics
at the address below.

Institute of Physics
Dirac House, Temple Back
Bristol BS1 6BE UK

Phone: 44 1 17 929 7481
Fax: 44 1 17 920 0979

techtracking@iop.org

Additional copies of this publication are available from:

Curran Associates, Inc.
57 Morehouse Lane
Red Hook, NY 12571 USA
Phone: 845-758-0400
Fax: 845-758-2633
Email: curran@proceedings.com
Web: www.proceedings.com

Table of contents

Volume 71

2nd Transdisciplinary Research on Environmental Problems in Southeast Asia

20–22 September 2016, Bandung, Indonesia

Accepted papers received: 7 June 2017
Published online: 20 June 2017

Preface

Disaster Mitigation

Measure and Improvement to Urban Environmental Problem

Sustainable Development and Environmental Preservation

General Session

IOP Conf. Series: Earth and Environmental Science **71** (2017) 011001 doi:10.1088/1755-1315/71/1/011001

Preface

The Transdisciplinary Research on Environmental Problem in Southeastern Asia (TREPSEA) international conference was started in 2011 by the 1st International Seminar of Environmental Geoscience (ISEGA) to provide a forum for its members to discuss and share their latest research. In 2014, the conference name was changed to 1st TREPSEA and it took place in Swiss Belinn Hotel, Makassar, Indonesia on 4-5 September 2014. The last but not the least conference is the current 2nd TREPSEA, which was held in the Papandayan Hotel Bandung, West Java with the great theme of "Disaster and Environmental". Our community is leading the way in studies to understand how to solve the environmental problems in Southeast Asia using Transdiciplinary approaches.

The 2nd TREPSEA collected 63 papers handed by colleagues from university, researcher and professional. All papers are divided into 4 main themes: Disaster mitigation, Measure and Improvement to Urban Environmental problem, Sustainable development and Environmental Preservation, and General. From 63 abstracts the committee selected 34 manuscripts to handle as Post Proceeding TREPSEA.

The chairman would like to express deep appreciation to the 3 universities, Bandung Government city for support to this event, and 2 conference sponsor companies. The 3 universities are Bandung Institute of Technology (ITB), Ehime University (EU), and State of Gorontalo University (UNG). The 2 companies include Medco Energy and PT. LAPI ITB.

The transdisciplinary research is defined as research efforts conducted by researchers from different disciplines and non-academic stakeholders working jointly to create new conceptual, theoretical, methodological, and translational innovations. The stakeholders are funder, government and development organizations, businesses and industries, civil societies (inhabitants, NGO's etc), and media for completion of the environmental problems. We look forward to cooperate with all of you to produce a deep, thoughtful set of works that can guide our activities in the future.

Chairman
Prof. Dr Emmy Suparka

Acknowledgement to Sponsor of TREPSEA 2016

On behalf of committee members and participant, the chairman would like to express deep appreciation to the sponsor companies that have helped us to keep down the cost of TREPSEA 2016 for all participants.

1. Medco Energy
2. PT LAPI ITB

Acknowledgement to Reviewers of TREPSEA 2016

The chairman and editors of TREPSEA would like to express their sincere gratitude to the following reviewers for assessing manuscripts in TREPSEA 2016. They are

Asep Saepuloh
Hisanari Sugawara
Hyodo Satoshi
Indra Gunawan
Idham Andri Kurniawan
Kenji Okazaki
Nguyen Thi Hoang Ha
Nishiuchi Hiroaki
Alfend Rudyawan
Very Susanto
Nurcahyo Indra Basuki
Nobuhiko Matsumura
Mirzam Abdurrachman
Masayuki Sakakibara
Motoko Shimagami
Shusaku Yamazaki
Hiromitsu Yamagishi
Tomohiro Tsuji

IOP Conf. Series: Earth and Environmental Science **71** (2017) 011001 doi:10.1088/1755-1315/71/1/011001

List of Committee Members

Chairman

General Chair	: Prof. Dr Emmy Suparka	ITB
Vice Chair	: Prof. Masayuki Sakakibara, Ph.D	EU
General Committee	: Prof. Dr. Dwia Aries Tina Pulubuhu, M.A	UNHAS
	Prof. Dr. Hasanudin Z. Abidin	ITB
	Prof. Dr. Ir. Herry Suhardiyanto, M.Sc	IPB
	Prof. Dr. Masrurah Mokhtar	UMI
	Prof. Dr. Suwarno Hadisusanto, M.Sc	UGM
	Prof. Dr. Syamsu Qamar Badu, M.Pd	UNG
Scientific Committee	: Prof. Budu	UNHAS
	Dr. Edy Hartulistiyoso	IPB
	Dr. Eng. Imam Achmad Sadisun	ITB
	Prof. Mai Trong Nhuan	VNU
	Dr. rer.nat. Mohammad Jahja	UNG
	Dr. Retno Peni Sancayaningsih, M Sc	UGM
	Prof. Toshio Yoshii	EU
Local Committee	: Dr Eng Mirzam Abdurrachman	ITB
	Dr. IGB Eddy Sucipta	ITB
	Dr. Eng Asep Saepuloh	ITB
	Dr. Purnama Sendjaja	PSG
	Dr. Supartoyo	PVMBG
	Arif Susanto M.T	ITB
	Dr Idham Andri Kurniawan	EU/ITB
	Robby Ginanjar S.T	ITB
	Jomae Kyoko	EU

SPONSORED BY:
THE 2nd INTERNATIONAL CONFERENCE OF
TREPSEA 2016
MEDCOENERGI
PAPANDAYAN HOTEL, BANDUNG, WEST JAVA, INDONESIA
SUPPORTED BY:

IOP Conf. Series: Earth and Environmental Science **71** (2017) 011001 doi:10.1088/1755-1315/71/1/011001

IOP Conf. Series: Earth and Environmental Science **71** (2017) 011001 doi:10.1088/1755-1315/71/1/011001

ICHAS

IOP Conf. Series: Earth and Environmental Science 71 (2017) 011001 doi:10.1088/1755-1315/71/1/011001

Peer review statement

All papers published in this volume of *IOP Conference Series: Earth and Environmental Science* have been peer reviewed through processes administered by the proceedings Editors. Reviews were conducted by expert referees to the professional and scientific standards expected of a proceedings journal published by IOP Publishing.

Tsunami Evidence in South Coast Java, Case Study: Tsunami Deposit along South Coast of Cilacap

Yan Rizal[1], Aswan[1], Yahdi Zaim[1], Wahyu Dwijo Santoso[1], Nur Rochim[1], Daryono[2], Suci Dewi Anugrah[2], Wijayanto[2], Indra Gunawan[2], Tatok Yatimantoro[2], Hidayanti[2], Resti Herdiyani Rahayu[2], Priyobudi[2]

[1] Geology Department, Institute of Technology Bandung (ITB), Jl. Ganesha 10 Bandung, Indonesia (40132).
[2] Tsunami Mitigation Division - Meteorological Climatological and Geophysical Agency (BMKG), Jakarta- Indonesia
Email: yan@gc.itb.ac.id

Abstract. Cilacap Area is situated in coastal area of Southern Java and directly affected by tsunami hazard in 2006. This event was triggered by active subduction in Java Trench which active since long time ago. To detect tsunami and active tectonic in Southern Java, paleo-tsunami study is performed which is targeted paleo-tsunami deposit older than fifty years ago. During 2011 - 2016, 16 locations which suspected as paleo-tsunami location were visited and the test-pits were performed to obtain characteristic and stratigraphy of paleo-tsunami layers. Paleo-tsunami layer was identified by the presence of light-sand in the upper part of paleo-soil, liquefaction fine grain sandstone, and many rip-up clast of mudstone. The systematic samples were taken and analysis (micro-fauna, grainsize and dating analysis). Micro-fauna result shows that paleo-tsunami layer consist of benthonic foraminifera assemblages from different bathymetry and mixing in one layer. Moreover, grainsize shows random grain distribution which characterized as turbulence and strong wave deposit. Paleo-tsunami layers in Cilacap area are correlated using paleo-soil as marker. There are three paleo-tsunami layers and the distribution can be identified as PS-A, PS-B and PS-C. The samples which were taken in Glempang Pasir layer are being dated using Pb – Zn (Lead-Zinc) method. The result of Pb - Zn (Lead-Zinc) dating shows that PS-A was deposited in 139 years ago, PS-B in 21 years ago, and PS C in 10 years ago. This result indicates that PS -1 occurred in 1883 earthquake activity while PS B formed in 1982 earthquake and PS-C was formed by 2006 earthquake. For ongoing research, the older paleo-tsunami layers were determined in the Gua Nagaraja, close to Selok location and 6 layers of Paleo-tsunami suspect found which shown a similar characteristic with the layers from another location. The three layers deeper approximately have an older age than another location in Cilacap.
Keywords: Cilacap; Dating; Earthquake; Paleotsunami layers; Tsunami.

1. Introduction

Tsunami hazard is the most dangerous in coastal hazard and give the dramatic human and infrastructure losses [1,2]. Therefore, geologic research has focused on modern tsunami and paleo-tsunami, including characteristic, causes, and evidence, in recent years. 75% of the Indonesian coastline is tsunami warning area [3]. NOAA (National Oceanic and Atmospheric Administration) data shows that tsunami killed more than 470. 000 lives worldwide during the last 450 years tsunamis. One of the locations which affected by tsunami and had the good record of tsunami sediment is Cilacap Area, Central Java.

Cilacap Area is situated in coastal area of Southern Java and directly affected by tsunami hazard in 2006. This event was triggered by active subduction in Java Trench which active not only in 2006 but also much older than it. Several record of tsunami and earthquake hazard has been recorded and reported by previous study, such as [4-6].

Newcamb and Mc Can [4] successfully reported that many paleo - tsunami had been detected in Southern Java on 4 January 4th, 1840, October 20th, 1859, and September 11th, 1921. Yudhicara et al described and dated two key beds which were suspected as paleo-tsunami deposit [5]. However, two key beds give the age in 2002 and 2007 and the beds are not correlated with paleo-tsunami deposit.

Catalogue which was developed by BMKG [6] reported many earthquake and tsunami related with Java trench activity in Central Java Area. Sixteen times tsunami and earthquake data have been collected, with the oldest earthquake were performed in 1416 (from NOAA reference) and the youngest was happened in 2006. The 2006 tsunami injured 600 people death and 15 people lost [7].

This study was focused for characterized and determine the age of paleo-tsunami deposit by observe, collect, and describe all of tsunami and paleo-tsunami deposit in Cilacap area. Moreover, regional correlation of paleo-tsunami deposit in Cilacap area is revealed, attempt to correlate paleo-tsunami deposit time by time, and help to prove and reconstruct previous paleo-tsunami reported before [6].

2. Material and Method

The study site is located in Cilacap areas, which have $\pm$105 km coastlines [8]. The research was conducted as joint research program between Institut Teknologi Bandung (ITB) with Meteorology, Climatology, and Geophysics Agency (BMKG in Indonesia) in 2015. Fifteen locations suspected as paleo-tsunami location were visited and examined based on image and topographic map analysis (Figure 1). All of the locations were confirmed as paleo-swale and back of ridge which give the good preservation of paleo-tsunami deposit. To get the precision location, a GPS system collaborated with topography measurement using total station was executed and corrected with sea level.

The identification of paleo-tsunami deposit was performed with made trenching area in the suspect location. After that, the samples were taken by systematically using grab and coring method. Laboratory analysis was examined in this study including micro-fauna, grainsize, and dating analysis.

Micro fauna analysis was examined to identify the mixing and transporting of marine fauna from different environment. Besides that, micro fauna analysis can be used to know how deep the fossils were transported to the land caused by tsunami. Paleo-tsunami samples were washed and soaked in the H_2O_2. After that, the samples were dried using oven, and ready to determine. The samples were observed in Paleontology Laboratory, Institut Teknologi Bandung, using reflected-light microscope Nikon SMZ 1500 with 16X multiplies.

Grainsize analysis was applied using - 4.5 multiplies using oven, and ready to determine. The samples show grain size distribution of paleo-tsunami deposit. The 100 gram samples of paleo-tsunami were dried and the grains coarse were sieved out of each sample. The samples must be weighed before and after sieving to get the precision calculation of the weight percentage of the coarse fraction. The standard of the samples was that the different of weight before and after sieving must be less than one gram. For the analysis, grain size computation and plotting were performed using Program of [9]. This analysis was executed in Sedimentology Laboratory, Institut Teknologi Bandung.

Dating analysis for the paleo-tsunami deposit was conducted using ^{210}Pb method. The age can be known by counting ^{210}Pb content inside the samples [10]. The samples of ^{210}Pb dating were taken by systematically with 5 cm interval from top to bottom. This analysis performed in the Dating

Laboratory, Indonesian Atomic Agency. For the age measurement, the formula ^{210}Pb decay followed Olfield and Appleby method [11];

$$t_{1/2} = 22.3 \text{ y } (t = 32.2 \text{ y})$$

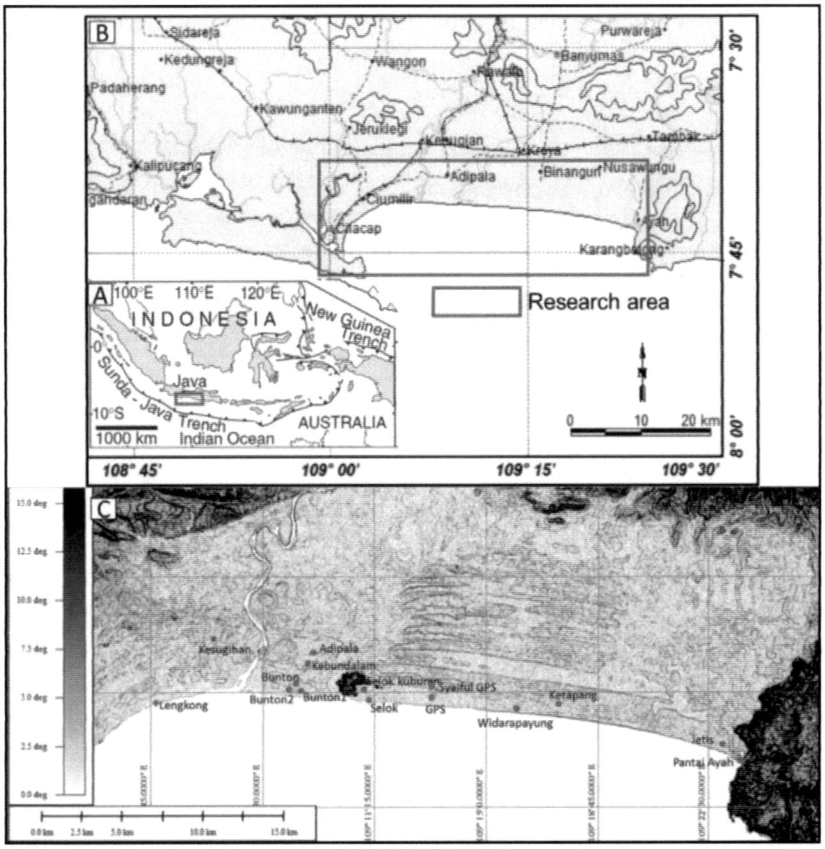

Figure 1 Location of research area. (A). Plate tectonic framework of Java [12]. Cilacap area (marked with red box) faces with Java Trench as earthquake trigger. (B). Administrative and road access map of research area (map from [5]). (C). Location of paleo-tsunami suspect (marked with red dot) overlay with satellite image (Image from USGS [13]).

3. Result and Discussion

3.1. Paleo-tsunami Deposit Characteristic

Many of sedimentary signatures were used to examine and separate paleo-tsunami deposit from another sand beach deposit. Paleo-tsunami deposit has special characteristic. Physical signatures of paleo-tsunami deposit can be distinguished by its colors, bedding contact, sedimentary structure, grain size, and the compositions. The signatures of paleo-tsunami deposit refer to previous study in several areas. Horton et al. [14] described the erosional contact as signatures in US Golf Coast. Another study from [15] in Florida Coast gives the similar result with [14]. Another characteristic of tsunami deposit is rip-up clast, which consist of intra-clast or reworked material. This characteristic was observed by [16] in Thailand and [15] in Florida Coast. The light sand in the upper part of paleo-soil is another signature of paleo-tsunami deposit. Cochran et al. [17] dug the paleo-tsunami deposit in Takui River, New Zealand, and found this character as paleo-tsunami signatures.

In Cilacap area, fifteen locations were visited and the result showed that nine locations contain paleo-tsunami deposits and six locations other indicate without paleo-tsunami deposit. Paleo-tsunami deposit can be found in paleo-swale and back of paleo-ridge area, with elevation from 5 meters to 18 meters mean sea level. Topography measurement shows that the location of paleo-tsunami deposit has 250 meters from the beach as the nearest location (Selok location) and 3 kilometers from the beach as the farthest location (Kesugihan location). The summary of paleo-tsunami layers in Cilacap area can be seen in Table 1.

Table 1 Summary of paleo-tsunami layers in Cilacap Area

Location	**Coordinate**	**The amount of Paleo-tsunami layer**	**Paleo-tsunami signatures**
Lengkong	07° 41’ 36.96” E 109° 03’54.396” N	2	Erosional contact, on-lap topography, liquefaction structure.
Kesugihan	07° 39’37.20” E 109°05’52.332” N	2	Light brown sand, fine to medium grain size.
Bunton	07° 41’ 2. 724” E 109° 08’ 43.908 N	2	Light gray sand, erosional contact.
Bunton 1	07° 41’15.288” E 109° 08’47.796” N	None	
Bunton 2	07°41’10.2012” E 109°08’38.904” N	None	
Adipala	07° 39’ 56.736” E 109° 09’7.344” N	None	
Kebundalam	07° 40’ 18.012” E 109°08’55.068” N	3	Light gray sand, fine to medium grain size.
Selok Kuburan	07° 41’ 9.816”E 109°10’51.384” N	3	Light gray sand, erosional contact.
Selok	07° 41’ 24.54” E 109°10’57.252” N	2	Light gray sand, medium grain size, erosional contact.
Syaiful GPS	07° 41’ 4.848’ E 109°12’58.568” N	None	
Glempang Pasir (GPS)	07° 41’18/024” E 109°12’56/376” N	3	Rip up clast, light gray sand, paleosoil, erosional contact.
Widara payung	07° 41’ 49/092 “ E 109° 16’3.144” N	1	Light gray sand, paleosoil, erosional contact.
Ketapang	07° 41’ 48.984” E 109°17’30.984” N	None	
Jetis	07° 43’ 4.834” E 109°22’53.148” N	None	
Pantai Ayah	07° 43’31.656” E 109°23’40.704” N	3	Light gray sand, medium to coarse grain size, fragmented.

The ideal profile of paleo-tsunami deposit in Cilacap area can be observed in Glempang Pasir (GPS) location (see Figure 1) by trenching method. Compared with beach sand, tsunami material has lighter color than beach sand, more enriched in fine grained sand and mix with silt and shale material, and rip-up clast is found in several place. Beach sediment in Cilacap Area is characterized by black sand sediments. The beach sediment is consisted of iron sand layer, medium grain size, well sorted, and without clay material as a product of traction flow during deposition.

Three layers could be examined as paleo-tsunami candidates, namely PS A, PS B, and PS C (Figure 2). PS- A description shows light gray sand, very fine – fine grain, with carbon fragment in the bottom with 5 cm thickness. PS-B, the younger paleo-tsunami candidate than PS-A, consist of sandy clay, light gray, and paleo-soil in the bottom part. The thickness of PS B has about 20 cm, with paleo-soil around 3 cm. At the upper part of trenching, the thickest paleo-tsunami candidate, namely PS-C, was deposited with 50 cm thickness. PS C has light sand, medium – coarse sand, rip-up clay, and erosional contact in bottom part.

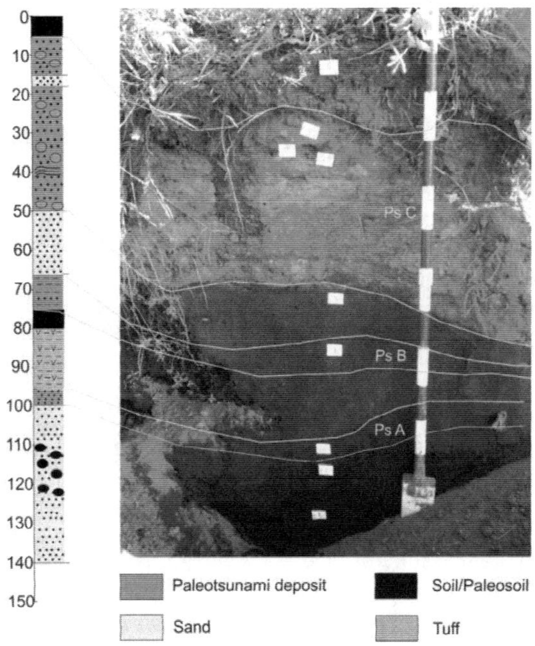

Figure 2 Profile of Paleo-tsunami suspect in Glempang Pasir (GPS) location.

Tsunami energy erodes the sediment from deeper marine and deposits it in the shallower environment. To detect how far the material was transported, micro-fauna analysis was performed as indicator. Previous research by [18] in Thailand indicated that the transportation of the large volumes of coastal material and seaward material during the tsunami was followed by relocation of mangrove foraminifera. Another research by [19] observed in 2004 tsunami deposit in Thailand and Malaysia, tsunami layers consist of a mixing layer of fauna from of freshwater, brackish, and marine species. Therefore, paleo-tsunami layers can be concluded by sediments which contain from rework and transported fossils (Figure 3).

Observation of paleo-tsunami layer in Cilacap area indicates the similar result. The faunas which live in the neritic environment, *Oolina* sp., *Bolivina* sp., *Lagena* sp., and *Amphistegina* sp., mix with *Elphidium* sp. and *Ammonia* sp. which adapt in the litoral environment (Figure 3). Moreover, many foraminifer shells are broken and abraded which can be formed due to high energy condition.

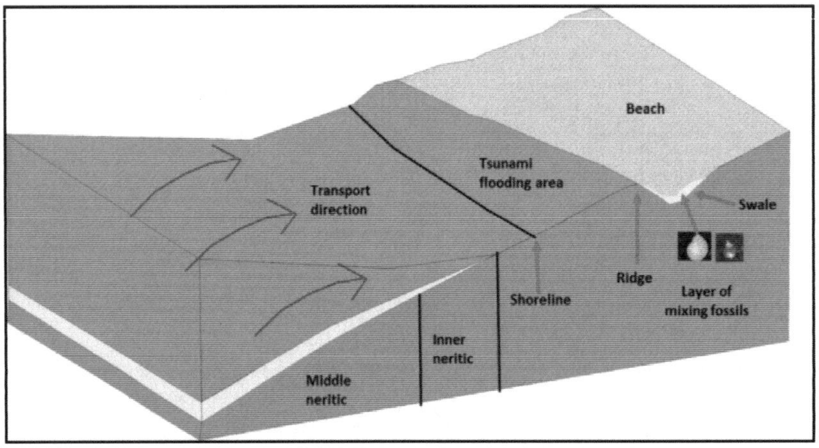

Figure 3 Illustration of transport fossil mechanism during tsunami activity. Paleo-tsunami was characterized by layer of mixing faunas and deposited in swale area.

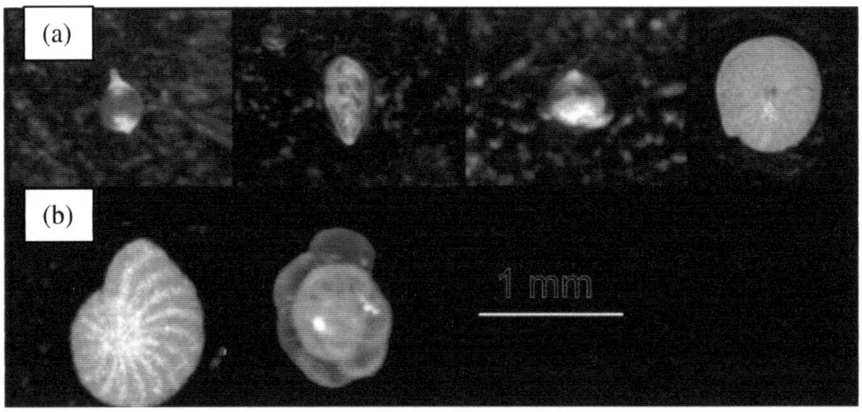

Figure 4 Micro-fauna from paleo-tsunami layers in Cilacap Area. (a) Microfossil from middle neritic environment (left to right); *Lagena* sp., *Bolivina* sp., *Oolina* sp., *Amphistegina* sp. (b) Micro-fauna from litoral environment (left to right); *Elphidium sp.* and *Ammonia sp.*

Grain size analysis of paleo-tsunami deposits performs fine – medium sandy sediment with rip-up clast as paleo-tsunami characteristic in Cilacap Area. Mean grain size of paleo-tsunami deposit is vary and shows fining–up succession and decreasing grain size from 2.63ϕ to 2.44 ϕ. Relative dispersion has 0.474 and indicates excellent homogeneity and well-sorted sand. The distribution shows very finely skewed (around 0.1 – 0.3). All distribution reveal meso-kurtosis (kurtosis from 2.5 – 3.7). The distribution curves show the unimodal curve with another small peak at the end (Figure 5).

The unimodal grain-size distribution curves are interpreted as a re-deposition of the material from one source. The primary source of the mud fractions as the rip-up material is the re-deposited marine mud. The fact that the sediment source came from the marine deposit is revealed by the presence of neritic benthic foraminifera fossils (Figure 4) among which littoral benthic fossil. This indication is concluded as bottom erosion was developed during tsunami which carried out neritic foraminifera and settled out in the tsunami deposits [20].

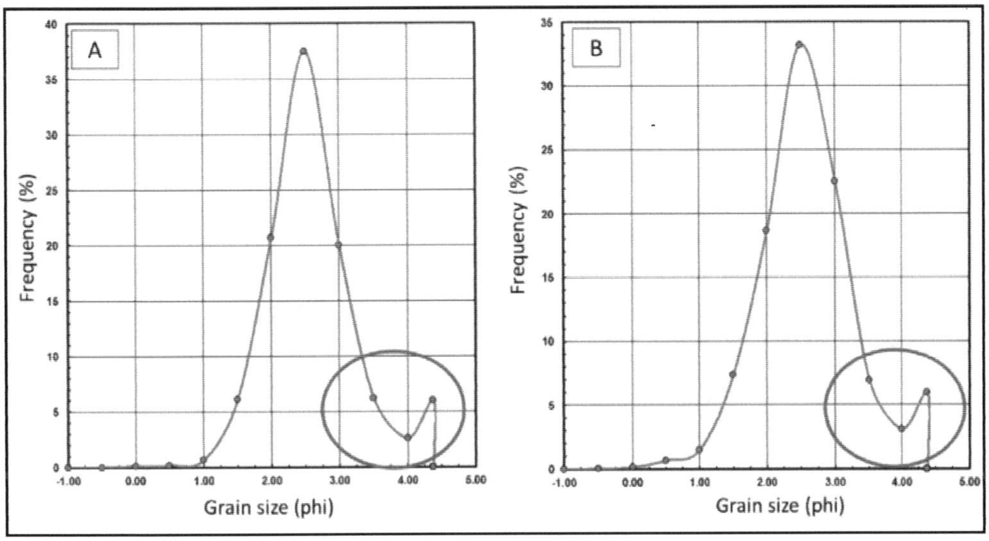

Figure 5 Distribution curves of paleotsunami deposit from Cilacap Area; (A) Glempang Pasir (GPS) location and (B) Selok. Small peak is observed at the end of curve (blue circle).

Yudhicara et al. [5] was analyzed the 2002 and 2007 storm deposit from Bunton location, Cilacap (Figure 6) and it became the good comparison between storm and paleo-tsunami deposit in Cilacap Area. Yudichara et al. [5] reported that storm deposits showed the coarse sand material, with coarsening upward features, poorly sorted material, without rip-up clast. This layers were not found fossil contain. The distribution curve showed unimodal peak with small peak at the start of distribution (Figure 7). Compared to storm deposit, the tsunami deposits in Cilacap Area exhibit the different characteristic which mentioned before. Tsunami deposit show fine sand – clay, rich fossils contain, and rip – up material.

Figure 6 The storm deposit from Bunton location, Cilacap [5].

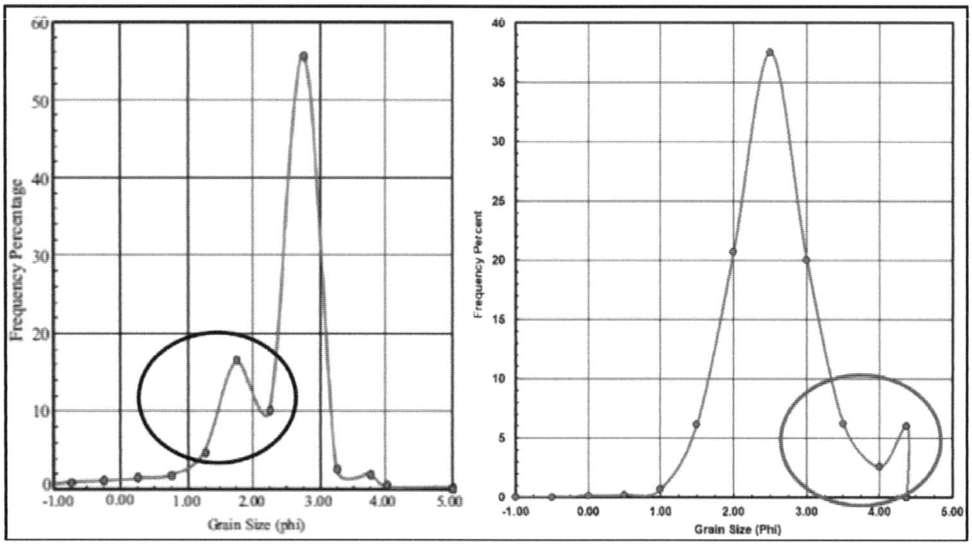

Figure 7 Distribution curve comparison between storm deposit (left) and tsunami deposit (right). The storm deposit has coarser material than tsunami deposit and small peak at the start (black circle). However, tsunami deposit has finer material and small peak at the end (blue circle). The picture of storm deposited was cited from [5].

3.2. Dating and the age of Paleo-tsunami deposits

Accurate and precision dating analysis is very important to reconstruct history and frequency of paleo-tsunami in Cilacap Area. The samples from Glempang Pasir (GPS) location were chosen for ^{210}Pb dating because the outcrop in this location showed clearly paleo-tsunami signatures and easy to distinguish with another beach sand deposit which was deposited below paleo-tsunami sediments. The result of ^{210}Pb dating can be seen in Table 2.

Table 2 Result of ^{210}Pb dating from Glempang Pasir (GPS) location.

Paleo-tsunami layer	Paleo-tsunami signatures	The age of sediment (years ago)
PS-C	Rip up clast, erosional contact.	10.6 ~ 2006
PS-B	light gray sandy clay, paleosoil,	29.6 ~ 1982
PS-A	Light gray sand	138.9 ~ 1883

^{210}Pb dating represents the conclusive result of paleo-tsunami history in the Cilacap Area. Three paleo-tsunami candidates give the strong correlation with earthquake event in Southern Java. The result of PS-A around 138.9 years ago correlates with 1883 earthquake which reported by BMKG catalogue [6]. Based on this report, 1883 earthquake triggered tsunami hazard which the source of earthquake and tsunami came from java trench movement in Southern Java (BMKG, 2010; source data from NOAA). Unfortunately, the catalogue was not written the magnitude of earthquake and the amount of fatalities which caused by tsunami and earthquake.

PS-B layer shows 29.6 years ago. This sediment was deposited as a product of earthquake activity in 1982, which near with ^{210}Pb calculation. BMKG [6] reported that 1982 earthquake was magnitude

5.4 Richter Scales and resulted 10 meters tsunami waves. Total of damage and injuries were not reported in BMKG [6] (source data from NOAA).

PS-C layer indicates 10.6 years ago as evidence of 2006 tsunami. The source of earthquake was Java Trench, with magnitude 7.7 Richter Scales. The damage was reported total 664 peoples death in Southern Java [6]. Rosyidie [7] counted the effect of tsunami resulted 600 people death and 15 people lost. Tsunami 2006 covered not only Cilacap Area, but also Ciamis, Pangandaran, Cipatujah, Tasikmalaya. Almost all of public facilities and recreational area were collapse [6]. The documentation of dangerous tsunami 2006 in the electricity power plant of Cilacap was recorded by Putra and can be seen in Figure 8.

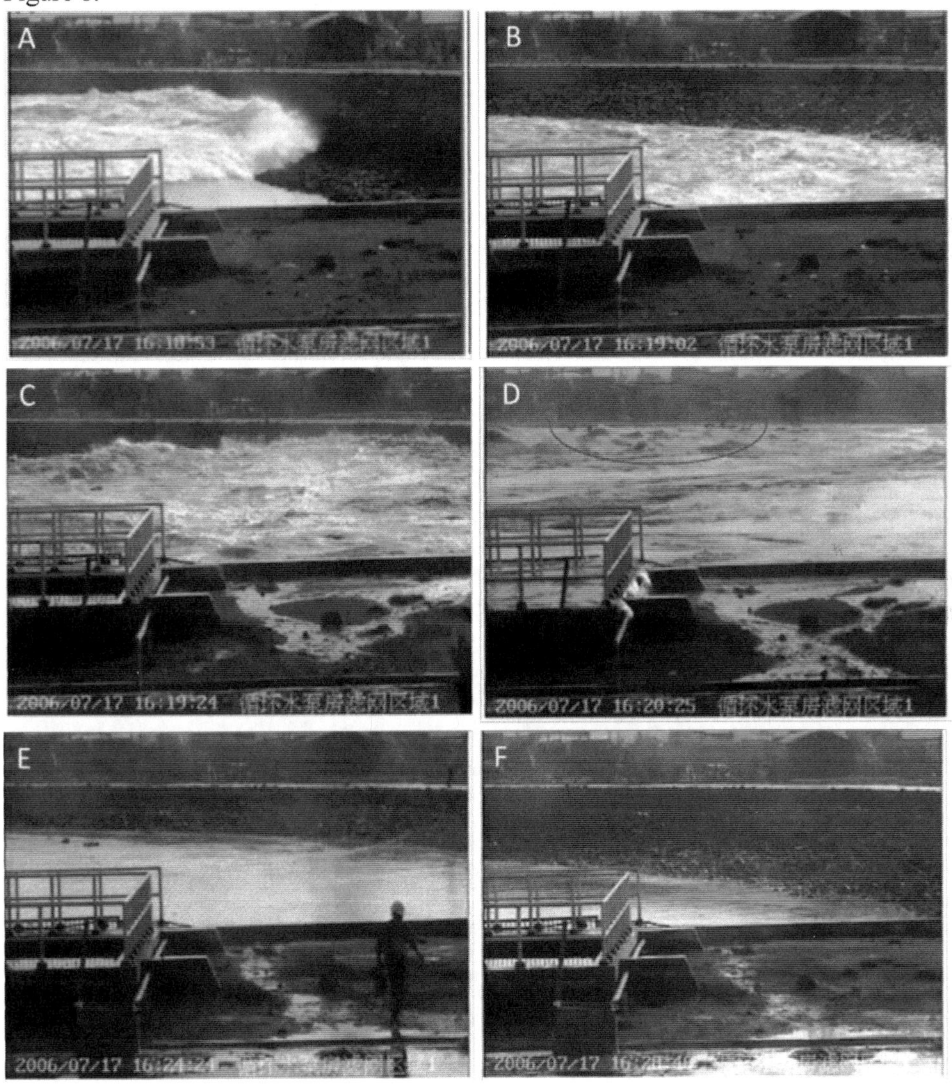

Figure 8 Effect of 2006 tsunami in the electricity power plant of Cilacap. (A). Tsunami started to flood at 16:18:53. (B). First wave of tsunami at 16:19:02 filled power plant area. (C) The second wave came at 16:19:24. (D). Highest waves at 16:20:25. Power plant workers were brought by tsunami waves (red circle). (E) The tsunami waves started down at 16:24:24. (F) Normal condition of sea level at 16:28:40 LT. All of photographs were taken by Putra and cited from [3].

3.3. Regional correlation

Paleo-tsunami correlation is developed to establish stratigraphic continuity of target paleo-tsunami layers in Cilacap Area. Paleo-tsunami event is correlated based on: a)clearly observation of the deposits in the field, b) number of paleo-tsunami layers, c) morphology of paleo-tsunami sites. Glempang Pasir (GPS) location was chosen as the key profile because this section has complete laboratory analysis, including micro-fauna, grainsize, and dating, and clearly observed in the field.

Three paleo-tsunami layers which were recognized in Glempang Pasir (GPS) location is correlated into paleo-tsunami deposits from another location in Cilacap Area (Figure 9). PS- A layer from GPS location continue into Kebundalam, Kesugihan, and Lengkong location. PS-A layer is missing in the Bunton, Selok, and Widarapayung because the sites were not dug deeper. PS-A has similar thickness in all of location, and it indicates that beach morphology when the tsunami happened has flat morphology. 1883 earthquake which triggered tsunami brought and uniformly distributed along Cilacap Coastline.

PS-B layer was deposited with the maximum thickness in Kesugihan Location, around 15 cm. This thickness represents that Kesugihan Location when tsunami destroyed in 1994 was a depression and the other location located in higher morphology than Kesugihan.

PS – C layer, a product of 2006 earthquake, deposited along Glempang Pasir – Selok – Bunton for the maximum thickness. In higher location, Kesugihan, Kebundalam, Widarapayung, and Ketapang, tsunami deposits deposited with the thinner thickness. Erosion made tsunami deposits preserved in weather condition and reduce the thickness of deposits.

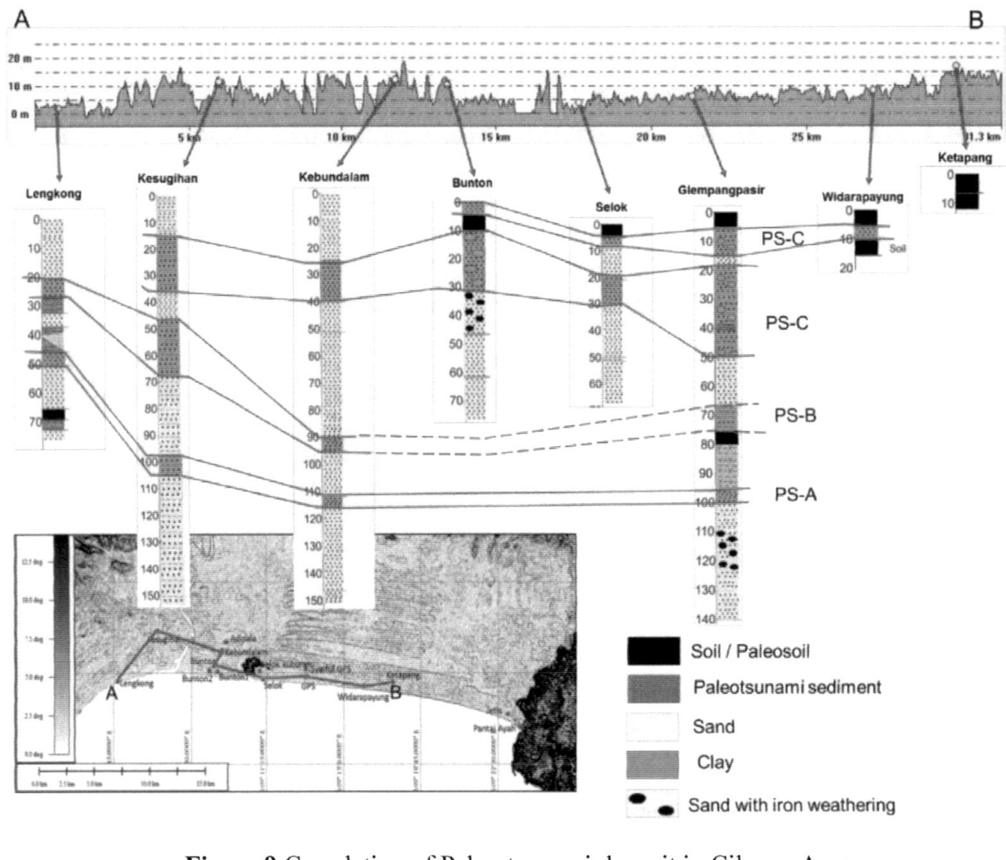

Figure 9 Correlation of Paleo-tsunami deposit in Cilacap Area

4. Conclusion

Three paleo-tsunami deposits in Cilacap Area were clearly identified, namely PS-A, PS-B, PS-C. PS-A has light gray sand, very fine – fine grain, with carbon fragment in the bottom with 5 cm thickness. PS-B observation shows sandy clay, light gray, and paleo-soil in the bottom part. PS - C can be distinguished by light sand, medium – coarse sand, rip-up clay, and erosional contact in bottom part.

The result of micropaleontology represents the mixing layer of faunas as paleo-tsunami indicator. The fossils which live in the neritic environment, *Oolina* sp., *Bolivina* sp., *Lagena* sp., and *Amphistegina* sp., mix with *Elphidium* sp. and *Ammonia* sp. which adapt in the litoral environment. Grain size analysis revealed fine – medium sandy sediment, rip-up clast, and the unimodal curve as paleo-tsunami characteristic in Cilacap Area. The unimodal grain-size distribution curves perform as a re-deposition of the material from one source.

The dating analysis shows the strong correlation between paleo-tsunami and earthquake event which the source was triggered by java trench activity. PS-A was deposited as a product of 1883 earthquake. PS – B was triggered by 1982 earthquake with magnitude 5.4 Richter Scales. PS – C, the youngest deposit in Cilacap Area, formed as evidence of 2006 tsunami with magnitude 7.7 Richter Scales.

5. Acknowledgement

The authors thank to BMKG (Meteorology, Climatology, and Geophysics Agency) for funding this study. All of BMKG's crews who support us during field and laboratory study. We say thank you to Bang Acong, our driver, who help us and give the full support in the field. This research was funded by BMKG and a part of paleo-tsunami joint research between Institut Teknologi Bandung (ITB) and BMKG (Meteorology, Climatology, and Geophysics Agency).

6. References

[1] Morton, R. A., Gelfenbaum, G., and Jaffe, B.E., 2007, Physical criteria for distinguishing sandy tsunami and storm deposits using modern examples, Sedimentary Geology, doi: 10.1016/j.sedgeo.2007.01.003

[2] Engel, M. and Bruckner, H., 2011, The identification of palaeo-tsunami deposits a major challenge in coastal sedimentary research, Prozesse, Zusammenhänge und Auswirkungen Coastline Reports 17, Koln.

[3] Lavigne, F., Gomez, C., Giffo, M., Wassmer, P., Hoebreck, C., Mardiatno, D., Prioyono, J., and Paris, R., 2007, Field observations of the 17 July 2006 Tsunami in Java, Nat. Hazards Earth Syst. Sci., 7, 177–183.

[4] Newcomb, K.R. and McCann, W.R., 1987, Seismic and tectonics of the Sunda Arc, Journal of Geophysical Research, 92(B1), p.421-439.

[5] Yudhicara, Zaim, Y., Rizal, Y., Aswan, Triyono, R., Setiyono, U., and Hartanto, D., 2013, Characteristics of Paleo-tsunami Sediments, A Case Study in Cilacap and Pangandaran *Coastal Areas, Jawa, Indonesia*, Indonesian Journal of Geology, Vol. 8, 163 – 175.

[6] BMKG (Meteorology, Climatology, and Geophysics Agency), 2010, Tsunami catalogue in Indonesia, BMKG Internal Publication, Jakarta (Text in Indonesia).

[7] Rosyidie, A., 2006, The impacts of disarter in coastal area: Learning from Aceh Tsunami, Journal of City Area Planning, Vol. 17 No. 3 (Text in Indonesia).

[8] Mahendra, A. M., Armono, H.A., Sambodho, K., 2013, Study of public resilience in Cilacap Coast against tsunami disaster, ITS Journal of Technic, Vol. 2 No.1 (Text in Indonesia).

[9] Balsillie, J. H., Donoghue, J. F., Butler, K. M., and Koch, J. L., 2002, Plotting equation for Gaussian, percentiles and a spreadsheet program for generating probability plots. Journal of Sedimentary Research, 72 (6), p.929-933.

[10] Boer, W., Van den Bergh, G. D., De Haas, H., De Stigter, H. C., Gieles, R., & Van Weering, T.

C. 2006. Validation of accumulation rates in Teluk Banten (Indonesia) from commonly applied 210 Pb models, using the 1883 Krakatau tephra as time marker. *Marine Geology*, *227*(3), 263-277.

[11] Oldfield,F. and Appleby, P.G., 1984, Empirical testing of 210Pb dating models. In: E.Y. Haworth and J.G. Lund (eds.), Lake Sediments and Environmental History, Leicester Univ. Press, 93-124.

[12] Hall, R., 2002, Cenozoic geological and plate tectonic evolution of SE Asia and the SW Pacific. Computer-based reconstructions, model and animations. Journal of Asian Earth Sciences 20, 353–431.

[13] USGS (United States if Geological Survey), 2015, Satellite image of Java with 30 m accuracy, downloaded on October 5th, 2015

[14] Horton, B.P., Rossi, V., and Hawkes, A.D., 2009, The Sedimentary record of the 2005 hurricane season from the Mississippi and Alabama coastlines. In: Quaternary International 195: 15–30.

[15] Wang, P., and Horwitz, M.H., 2007, Erosional and depositional characteristics of regional overwash deposits caused by multiple hurricanes, In: Sedimentology 54: 545–564.

[16] Srisutam, C., and Wagner, F.F., 2010, Tsunami sediment characteristics at the Thai Andaman coast. In: Pure and Applied Geophysics 167: 215–232.

[17] Cochran, U.A., Clarck, K.J., and Strong, D.T., 2014, Paleo-tsunami investigations in the Southland Region, GNS Science Concultancy, 163 p.

[18] Sugawara, D., Minoura, K., Nemoto, N., Tsukawaki, S., Goto, K., and Imamura, F., 2009, Foraminiferal evidence of submarine sediment transport and deposition by backwash during the 2004 Indian Ocean tsunami, Island Arc 18, 513–525.

[19] Sawai, Y., Jankaew, K., Martin, M.E., Prendergast, A., Choowong, M., Charoentitirat, T., 2009, Diatom assemblages in tsunami deposits associated with the 2004 Indian Ocean tsunami at Phra Thong Island, Thailand. Mar. Micropaleontol. 73 (1–2), 70–79.

[20] Razjigaeva, N.G., Ganzey, L.A., Grebennikova, T.A., Ivanova, E.D., Kharlamov, A.A., Laistrenko, V. M., Arslanov, K.A., and Chernov, S.B., 2014, The Tohoku Tsunami of 11 March 2011: The key event to understanding tsunami sedimentation on the Coasts of Closed Bays of the Lesser Kuril Islands, Pure Appl. Geophys., doi 10.1007/s00024-014-0794-y.

Tephra Fallout Hazard Assessment for VEI5 Plinian Eruption at Kuju Volcano, Japan, Using TEPHRA2

Tomohiro Tsuji[1], Michiharu Ikeda[1], Hiroshi Kishimoto[2], Koji Fujita[2], Naoki Nishizaka[3] and Kozo Onishi[3]

[1] Shikoku Research Institute Inc, 2109-8,Yashimanisi-machi, Takamatsu, 761-0192, Japan.
[2] Asia Air Survey Co. Ltd., Manpukuji, Asao-ku, Kawasaki, 215-0004, Japan.
[3] Shikoku Electric Power Co. Inc, 2-5, Marunouchi, Takamatsu, 760-8573, Japan.
Email: t-tsuji@ssken.co.jp

Abstract. Tephra fallout has a potential impact on engineered structures and systems at nuclear power plants. We provide the first report estimating potential accumulations of tephra fallout as big as VEI5 eruption from Kuju Volcano and calculated hazard curves at the Ikata Power Plant, using the TEPHRA2 computer program. We reconstructed the eruptive parameters of Kj-P1 tephra fallout deposit based on geological survey and literature review. A series of parameter studies were carried out to determine the best values of empirical parameters, such as diffusion coefficient and the fall time threshold. Based on such a reconstruction, we represent probabilistic analyses which assess the variation in meteorological condition, using wind profiles extracted from a 22 year long wind dataset. The obtained hazard curves and probability maps of tephra fallout associated to a Plinian eruption were used to discuss the exceeding probability at the site and the implications of such a severe eruption scenario.
Keywords: TEPHRA2; tephra fallout; simulation; Kuju Volcano; hazard assessment; probabilistic study.

1. Introduction

Nuclear power plants require extreme care to assure safety. The Volcanic Impact Assessment Guide for Nuclear Power Plants published by the Nuclear Regulation Authority [1] requires power plants to identify the volcanos that could affect them and take into consideration tephra fallout that could potentially occur due to eruptions during the operating life of the plants. In addition to literature reviews, geological surveys and volcanological surveys, the use of numerical simulation to estimate the volume of tephra fallout on power generation facilities has been increasing in recent years [2].

Although Shikoku has no active volcanoes, a large-scale eruption of a volcano located upwind on Kyushu could still affect various power generation facilities located in Shikoku via tephra fallout carried by the prevailing westerlies. Several active volcanoes are sitting on the volcanic front on the central to eastern regions of Kyushu Island (Figure 1). Kuju Volcano in particular among these active volcanoes has a history of major eruptions, and its eruption rates in the last 10,000 years have been high [3]; it is also recognized as one of the volcanoes that is most likely to affect the power plant in the future due to its location relative to the power plant [4]. The occurrence of tephra fallout from the major Kuju Volcano eruption 54,000 years ago has been confirmed in Shikoku [5, 6] (Figure 1b).

Although there is no record to indicate that the volcanic ejecta from Kuju Volcano eruption has fallen on the site of the Ikata Power Plant in the past, we must assume that tephra fallout could occur depending on atmospheric conditions at the time of an eruption, particularly wind direction. The use of calculations to simulate advection, dispersal and deposition of tephra (e.g. TEPHRA2 [7], Fall3D [8, 9], HAZMAP [10], PUFF [11]) is an effective technique for forecasting the impact of tephra fallout from volcanic eruptions.

Several probabilistic assessment methods are available for forecasting the impact of tephra fallout during a future large-scale eruption [7, 12, 13]. A probabilistic assessment can be used to assess events that occur infrequently with a range of impacts that varies significantly depending on atmospheric conditions. If records of long-term meteorological observations are available, rare atmospheric conditions can be included in the assessment.

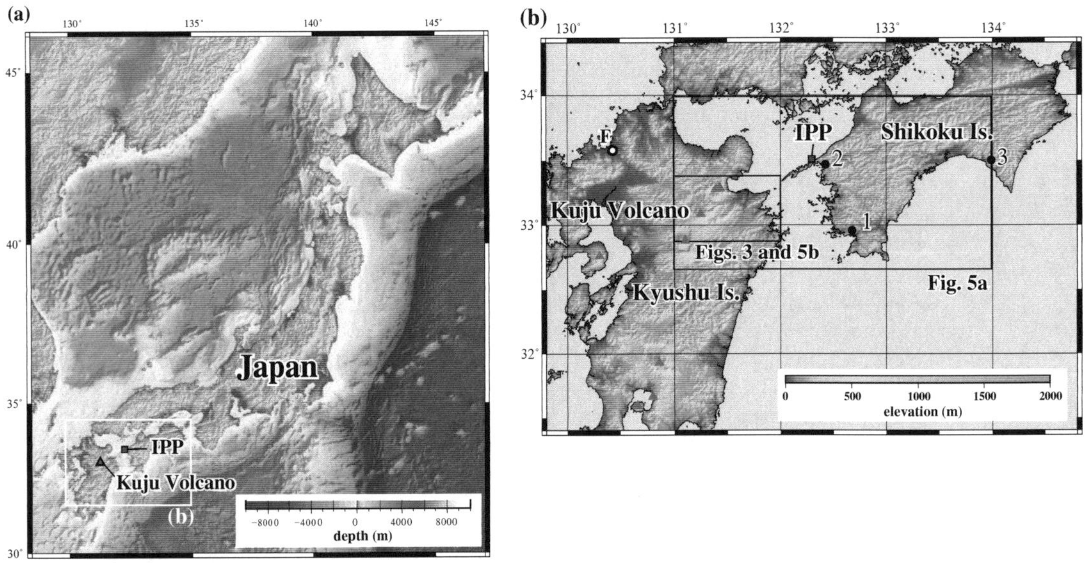

Figure 1. Maps of Japanese archipelago (a), and Kyushu and Shikoku Islands (b) showing localities of Kuju Volcano, Ikata Power Plant (IPP) and Fukuoka observation site (F). 1, 2 and 3 in (b) indicates localities of previous report of Kj-P1 tephras (Kumahara and Nagaoka [5], our unpublished data and Matsu'ura [6]). Red triangles represent active volcanoes. Area of Figs. 3, 5a and 5b are shown in (b).).

In order to probabilistically assess the volume of tephra fallout at the Ikata Power Plant in the event the Kuju Volcano erupts in a scale similar to the largest historical eruption that caused the Handa pyroclastic flow [3], we carried out a parameter study and probabilistic examination using a geological survey and tephra fallout simulation (TEPHRA2). In this paper, we present an outline of the air fall pumice from the Kuju eruption, followed by an estimation of the key eruption parameters (i.e., eruption volume, grain size and eruption column height) based on the results of our geological survey and existing literature. We then estimate the diffusion coefficient and fall time threshold by comparing the tephra fallout distribution between the results of our parameter study using TEPHRA2 and actual data. Based on the results, we develop an exceeding probability curve using wind data for 22 years [14] and a probability maps using wind data for a single year; the probabilistic assessment is discussed.

2. Outline of Kj-P1 fallout deposit

Kuju Volcano, which is located in central eastern Kyushu, has had three major VEI5-class [15] eruptions in the past [3]. The largest-scale eruption in recent time was the Handa pyroclastic eruption [3] about 54,000 years ago [16, 17]. A series of eruptions released, from the bottom layer, Kuju-D volcanic ash (Kj-D; 0.41 km^3), Handa pyroclastic flow (Kj-Hd; 5 km^3) and Kuju air fall pumice (Kj-P1) [18]. The volume of Kj-P1 is estimated to be 2.03 km^3 [19] or 6.2 km^3 [18].

Figure 2. Occurrence of Kj-P1 air-fall pumice and Kj-D air-fall ash at about 19 km east from vent.

Kj-P1 consists mainly of dacitic pumice and lithic fragments. The pumice is rich in phenocryst, and contains heavy minerals such as hornblende, orthopyroxene and biotite. Near Kuju Volcano (within 25 km from the vent), it is accompanied by Kj-D at a lower level (Figure 2), presenting an unstratified, inversely graded structure. The axis of the distribution of layer thickness of Kj-P1 is primarily oriented eastward, with thin and wide distribution to the west (upwind) and north sides of the vent. The Kuju tephra is preserved in sediment in southwestern Shikoku about 150km away from the volcano [5], and its presence is confirmed as crypto-tephra in Yahatahama city 110 km away (unpublished data) as well as in the southeastern area of Shikoku about 280 km away [6] (Figure 1b).

Table 1. Input parameters of Kj-P1 tephra fallout deposit for analysis of TEPHRA2. TGSD; total grain size distribution.

Parameters	Case-1	Case-2	Case-3	References
Eruption Parameters				
Locality of the vent				
Easting		709900.1		[20]
Northing		3663051.9		
Elevation (m)		1971		
Erupted volume (km^3)		2.54		Field data
Plume height (km)		25		Field data, [21]
TGSD				
Md phi		1.5		Field data, [22]
Sigma phi		3.0		
Grain parameters				
Diffusion coefficient (m^2/s)	200-10000	10000	10000	comparison with TEPHRA2 and field data
Fall time threshold (s)	200 -3000	1000	1000	
Dencity (kg/m^3)				
Pumice		1000		[7]
Lithic		2600		
Wind parameters				
Wind speed	Feb. 2010	1988-2010	2010	[14]
Wind direction	Feb. 2010 +8 degree	1988-2010	2010	

3. Methodology

We used TEPHRA2, which has a proven track record in tephra fallout simulation, for the analysis in this study. TEPHRA2 is a simulation program based on an advective diffusion model [7]. TEPHRA2 can calculate the distribution of deposits by providing appropriate initial parameters [23]. It is suitable for parameter studies and probabilistic assessments as it can readily and rapidly perform the calculations. The analytical program is downloadable from the site at the University of South Florida (http://www.cas.usf.edu/~cconnor/vg@usf/tephra.html). Four types of parameters are required for calculation, namely, eruption, atmospheric, grain and grid (Table 1). For the atmospheric parameters, we used the Yearbook of Aerological Observation [14] (Table 1). For the grid parameters, we developed a 2-km grid mesh using digital maps available from the Geospatial Information Authority of Japan (50-m DEM) and the National Institute of Advanced Industrial Science and Technology (1-km DEM). Calculations with TEPHRA2 produce tephra fallout per unit area (kg/m^2) and its grain composition. In our study, we express this as layer thickness (cm) per unit area, assuming that the density of the pyroclastic fall is 1000 kg/m^3.

4. Reconstruction of Eruptive Parameters of Kj-P1 Fallout Deposit

The important eruption parameters that determine the characteristics of an eruption consist mainly of coordinates of the vent, eruption volume, grain size, and eruption column height. For the coordinates of the vent, we referred to the coordinates of Nakadake given in the catalogue published by the Japan Meteorological Agency [20] (Table 1). As the eruption volume and grain size have a significant impact on the results, they should be estimated accurately by a geological survey. However, the grain size and eruption column height of Kj-P1 have not been examined in the past. In this study, therefore, we carried out a geological survey with respect to Kj-P1 and estimated the eruption volume, grain size and eruption column height of the eruption.

4.1. Eruptive Volume

In this study, we successfully reconfirmed the outcrops that had been identified by Nagaoka & Okuno [18], and expanded the dataset by adding new exposed outcrops (Figure 3). We set the survey points more densely along the principal axis of distribution that was important for examination of layer thickness and grain size (e.g. one collection point per 2 km^2) in order to collect a larger volume of data. Based on the collected data, we then examined the eruption volume by methods used for calculation of eruption volume including the contour method, the Hayakawa Method [24] and the Weibull Method [25].

4.1.1. Conter Method

We added reports for three areas in Shikoku to the results of our survey in Kyushu, and drew an isopach map by partitioning layer thickness (cm) into bins of >200, 200–100, 100–50, 50–25, 25–10, 10–5, 5–2, and 2–1 cm. We calculated the eruption volume to be 1.95 km^3 by totaling the values obtained by multiplying each of the layer thicknesses by its area. Note that we did not estimate the volume of the layer less than 1-cm thick with this method.

4.1.2. Weibull Method

The Weibull Method approximates the relationship between layer thickness T (cm) and area of isopach A (km^2; $A=x^2$) by the following formula:

$$T = \theta(x/\lambda)^{k-2}e^\wedge(x/\lambda)^k$$

Where λ is the attenuation length of deposit layer thickness (km), and k and θ the scale of layer thickness (cm; $\theta = eT(\lambda)$). The value of e is 2.718 (Euler/Napier constant) and k is a dimensionless shape parameter. According to this formula, the eruption volume of tephra V (km^3) is calculated by the following formula:

$$V = \int TdA = 2\theta\lambda^2/k$$

where λ, θ and k are determined by the method of least squares with the measured values. The Weibull Method requires the use of a specified range of layer thickness data. The eruption volume was estimated to be 1.85 km^3 using 4 isopachs of the layer thicknesses from 500 cm to 50 cm, 3.26 km^3 using 5 isopachs including layer thicknesses to 25 cm, and 2.25 km^3 using 6 isopachs including layer thicknesses to 10 cm.

4.1.3. Hayakawa Method

The Hayakawa Method expresses the eruption volume V (km^3), layer thickness T (cm) and area of isopach A (km^2) by the following formula:

$$V = 12.2TA$$

The Hayakawa Method requires the use of a specified layer thickness. In order to increase the accuracy of the eruption volume, we used the layer thickness of 100 cm, which had the highest data density and closed contour lines. Using the layer thickness of 100 cm, the eruption volume was estimated to be 2.54 km^3.

4.1.4. Eruptive Volume for Analysis

Using the above three techniques, we estimated the eruption volume to be between 1.95 and 3.26 km^3. We then examined the eruption volume to be used in our analysis by comparing these results. As the estimate by the contour method does not include areas with less than 1cm-thick layer, the actual eruption volume is presumed to be more than 1.96 km^3. The Weibull Method produced different results depending on the range used although the variations are not great. The average of the three estimates produced by the Weibull Method was 2.45 km^3. The eruption volume of 2.54 km^3 as calculated by the Hayakawa Method using the layer thickness of 100 cm is comparable to the average of three values calculated by the Weibull Method. It was also consistent with the result of the contour method that indicated the volume to be more than 1.95 km^3. Based on the results of the above examination, we set the eruption volume to be used in our analysis to be 2.54 km^3 (Table 1). This value is close to 2.03 km^3 estimated by Suto et al. [19] but considerably smaller than 6.2 km^3 set by Nagaoka and Okuno [18]. The reason for this difference may be due to the estimates made by Nagaoka and Okuno [18] for distant sites, particularly Shikoku and eastward, being too large. In a case, such as Kj-P1, in which the layer thickness data at distant sites is scarce, we believe that the use of techniques that allow estimation based on data for data-rich areas improves the reliability of eruption volume estimates.

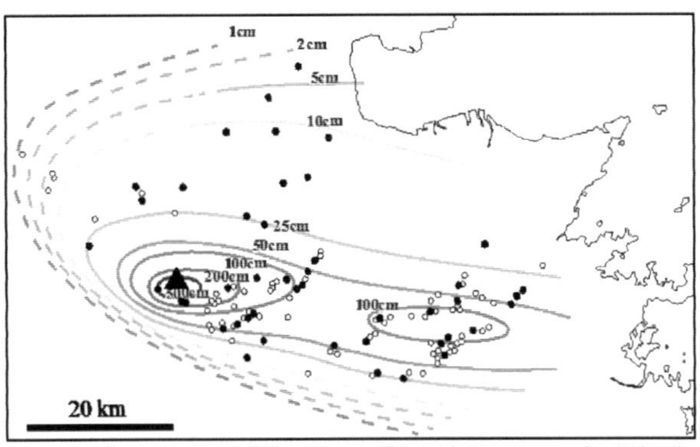

Figure 3. Isopach map and localities of outcrops for Kj-P1 deposit. Black and open circles are localities of our unpublished data and data compiled data from [18]. Black triangle indicates the crater of Kuju Volcano.

4.2. Total grain-size distribution

It is difficult to accurately estimate the grain size in the whole tephra (i.e., total grain size distribution; TGSD) from old and weathered pyroclastic fallout, such as Kj-P1, due to the conditions of outcrops and the effects of weathering [2]. Accordingly, we collected samples from outcrops that were relatively unweathered and therefore measurable, analyzed their grain sizes using a sieve method and a laser diffraction/scattering grain size distribution measuring apparatus (Nikkiso Microtrac Ver. 10.4.2) in order to obtain the median grain size (Mdφ) ($\varphi = -\log_2 D$, D = diameter (mm)) as well as the sortedness (σφ) [26]. We then estimated the median grain size and dispersal of Kj-P1 by comparing this value against the data for tephra for which TGSD had been examined in detail. As a product of an eruption that had the eruption volume, magma composition, quantity of crystals, layer thickness distribution and grain size that were similar to Kj-P1, we selected Ta-b pumice fallout from Tarumae Volcano [22]. Ta-b is a crystal-rich andesitic pumice fallout having a scale of eruption and layer thickness distribution that are similar to Kj-P1. According to Suzuki et al. [22], the eruption volume of Ta-b was 1.96 km^3 with average grain size Mdφ of –0.32φ and sortedness of 3.2φ [22].

4.2.1. Median grain-size

Figure 4 presents the grain size data collected from the area located between about 5 km to 50 km from the vent along the principal axis of the Kj-P1 distribution. The median grain size of Kj-P1 (Mdφ) at 10 km from the vent was about –2φ and at 40 km 0φ, indicating that it became finer rapidly with increasing distance from the vent (Figure 4). This is because coarse-grained particles fall out first as the volcanic eruption column travels through the atmosphere. A logarithmic curve well approximates the changes in the grain size of Kj-P1 with distance (Figure 4).

Ta-b contains a number of fall units; one of these, Ta-b_8, is the most coarse-grained fall unit and its TGSD is well understood [22]. The plot of the median grain size of Ta-b_8 distributed downwind from the vent along the principal axis of distribution is presented in Figure 4. Changes in the median grain size of Ta-b_8 with distance from the vent can be approximated by a logarithmic curve in the same way as for Kj-P1 (Figure 4).

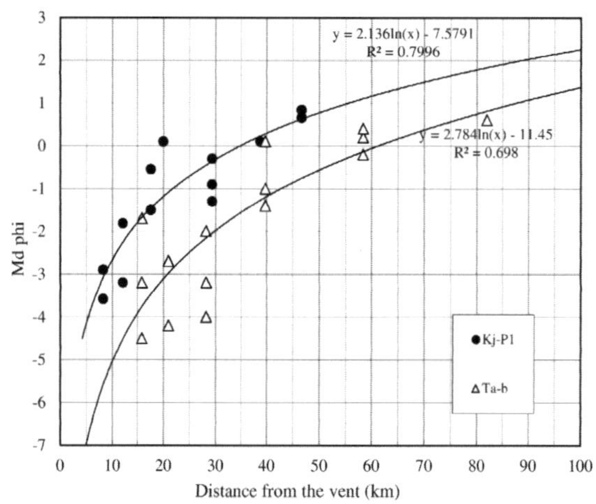

Figure 4. Plots of average grain size Mdφ of Kj-P1 (black circle) and Ta-b (white triangle) [22] distributed downwind along the axes of distribution from vent.

Next, we estimated the grain size distribution for the entire Kj-P1 by reading the differences in grain diameters of Ta-b_8 and Kj-P1from Figure 4. Generally, the TGSD of the tephra is significantly affected by the grain size of the deposit near the vent [27]. This stems from the fact that a majority of tephra is deposited in the vicinity of the vent. Furthermore, Suzuki et al. [22] pointed out that the layer thickness of Ta-b_8 became exponentially thinner with distance from the vent. Accordingly, we identified the median grain size at distances of 5, 10, 20, 40, and 80 km from the vent so that the degree of the effect on grain size farther away from the vent was exponentially smaller. Differences at each distance were, for example, 2.85φ at 5 km, 1.95φ at 20 km, 1.05φ at 80 km (Figure 4). In addition, Suzuki et al. [22] estimated the tephra fallout that was deposited 100 km or farther away from the vent to be approximately 20% of the total eruption volume. In order to consider this, we extrapolated the approximate curves in Fig. 4 to 160 km from the vent, and found that the difference was approximately 0.95φ. We adopted this value as representative of the differences in the grain size in the areas 100 km or farther away from the vent. The above examination indicates that the grain size of Kj-P1 was 0.95 to 2.85φ smaller than that of Ta-b at the same distance from the vent. In addition, we averaged the values at these distances and estimated that the grain size of Kj-P1 was 1.77φ smaller than that of Ta-b_8. Since the median grain size of the entire Ta-b is –0.32φ [22], we estimated Mdφ for Kj-P1 to be about 1.45φ. As a result, we used the value of 1.5φ for the median grain size in our analysis (Table 1). Since the typical grain size distribution in tephra generally has a median grain size of –1 to 4φ or smaller [2], the result of our examination is consistent with this range.

4.2.2. Sortness (Standard Deviation)
The approximate curves of Kj-P1 and Ta-b_8 presented in Figure 4 show similar gradients, which indicates that both are similar to each other in terms of the trends for change in grain diameter with distance and the width of the grain diameter distribution. In this study, we used this similarity in determining the sortedness, an indicator for the width of grain size distribution [26], and set the sortedness of Kj-P1 to be 3φ, which was similar to 3.2φ for Ta-b (Table 1). The result of our examination is consistent with the sortedness of tephra from Plinian eruptions being 2 to 3φ [2], as is generally believed.

4.2.3. Column height
With respect to eruption column height for which there are no observation records, a model that reconstructs the data from grain size has been proposed by Carey and Sparks [21]. In this study, we measured the downwind range and the crosswind range [21] for the grain size of the tephra deposit according to their eruption column model. We measured long and short axes of five grains taken from the largest grains of pumice and lithic fragments observed in outcrops, and determined the largest grain diameter by averaging their values. The downwind ranges of pumice with the largest grain diameters of 4 and 2 cm were 23.0 and 51.3 km and the crosswind ranges for the pumice grains were 8.7 and 12.8 km, respectively. The downwind ranges of lithic fragments with diameters of 3.2 and 1.6 cm were 21 and 39 km and the crosswind ranges of the lithic fragments were 5.1 and 9 km, respectively. By comparing these results with the model map by Carey and Sparks [21], the eruption column height of Kj-P1 was estimated to be 22 to 28 km. The eruption column heights estimated by varying the grain size were relatively consistent. Accordingly, we selected a column height of 25 ± 3 km for our analysis (Table 1). This estimate is consistent with the eruption column height of a VEI5-class eruption, which is the scale of the Kj-P1 eruption, being 20 to 35 km, as stated by Newhall and Self [15]. The effect of changes in the eruption column heights on thickness distribution of the tephra deposit was examined by Tsuji et al. [4] using values in a range of 20, 25 and 30 km, and was confirmed not to have a major effect on the results.

5. Reconstruction of Grain Parameters

In order to set the grain parameters, we needed data for densities of pumice and lithic fragments, diffusion coefficient, fall time threshold and volcanic eruption column model. For the densities of pumice and lithic fragments, we referred to Bonadonna et al. [7] (Table 1). In the following sections, we examine reasonable values for the diffusion coefficient and fall time threshold by comparing the thickness distribution of the tephra deposit layers that we observed in our field study against the results of our parameter study using TEPHRA2.

5.1. Wind Selection

In order to determine the values of the diffusion coefficient and the fall time threshold for our analytical model, we deterministically selected wind profiles. Tsuji et al. [4] examined the effects of changes in the direction and velocity of wind by month based on the normal values by month, which represented the average of the wind directions and velocity data collected at various observation sites at altitudes between ground level and 30 km [14]. According to the results, tephra fallout is carried northeastward from the vent and distributed in a slightly wide area in summer (July–September) when the wind is weak. In contrast, the tephra fallout is carried eastward in winter (December–February) in a narrow distribution pattern due to strong westerlies [4]. As the analytical results for the winter are comparable to the actual layer thickness distribution of Kj-P1, we used winter wind (February) as the comparative wind profile. In order to reconcile the direction of primary axis of the deposit distribution of the analytical results and the field data, however, we adjusted the direction of the normal February wind profile northward by 8 degrees.

5.2. Diffusion coefficient and fall time threshold

We carried out an analysis by varying the diffusion coefficient between 200 and 10,000 m^2/s and the fall time threshold between 200 and 3000 s in order to find the diffusion coefficient and fall time threshold at which the computed layer thickness (the "computed data") and the observed layer thickness (the "observed data") matched the best. The diffusion coefficient of 10,000 m^2/s and the fall time threshold of 1000 s matched best with the results of our survey (Figs. 5 and 6). Figure 6 presents a comparison of the computed and observed data for diffusion coefficient of 10,000 m^2/s and fall time threshold of 1000 s. The results show a positive correlation between the computed and observed data (Figure 6). The layer thickness was thinner in the range of 5–25 cm for the computed data than the observed data whereas the layers were slightly thicker in the range of 40–130 cm for the computed data (Figure 6). Overall, the deposit layers were slightly thicker in the computed data than in the observed data (Figure 6).

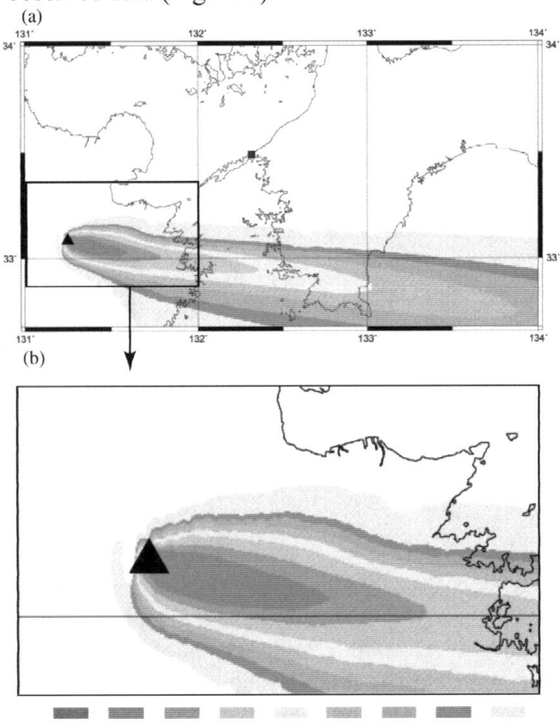

Figure 5. Calculated result with the value of case-1 (diffusion coefficient, 10000; fall time threshold, 1000; Table 1).

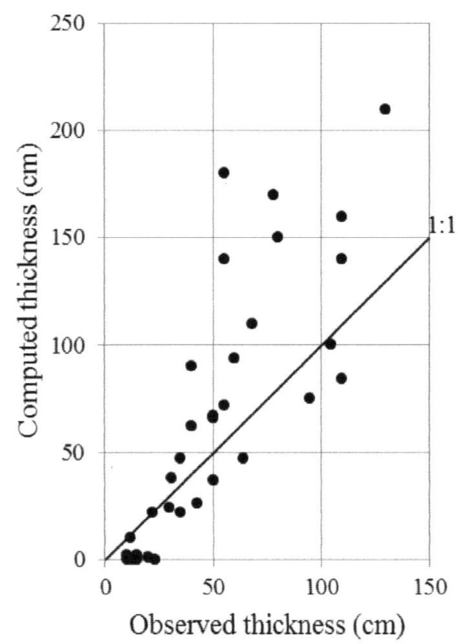

Figure 6. Comparison between observed data and data computed with the same value of Figure 5.

In our parameter study, we could not satisfactorily reproduce the tephra fallout distribution spread thinly on the west (upwind) and north sides of the vent (Figure 3 and 5b). We need to review the possibilities that the eruption rate or the wind direction/velocity changes during the eruption or that the wind direction varies by altitude.

6. Result and Discussion

6.1. Hazard Curves

Based on the eruption and grain parameters (Table 1) obtained by the work described above, we developed a hazard curve for the Ikata Power Plant using daily wind profiles for the 22-year period from 1988 to 2010 compiled by the Japan Meteorological Agency [14] (Case-2 of Table 1). The wind data represent the wind profiles at 9am at the Fukuoka Observation Station for 7976 days between March 1, 1982 and December 31, 2010 (Table 1). Figure 7 presents the exceeding probability curve at Ikata Power Plant. Generally, volcanic ash fall causes poor visibility and difficulty in uphill driving for automobiles at an accumulation of 1 cm, and may pose risk for collapse of structurally weak houses at 10 cm [28, 29]. The results of our present analysis indicated that the probability of ash fall of 10 cm or more at Ikata Power Plant was 4.44% and 5 cm 12.74%. The probability of exceedance with respect to thickness of Kuju Volcano origin ash fall at the Ikata Power Plant is obtained by multiplying each of these values by the probability of eruption.

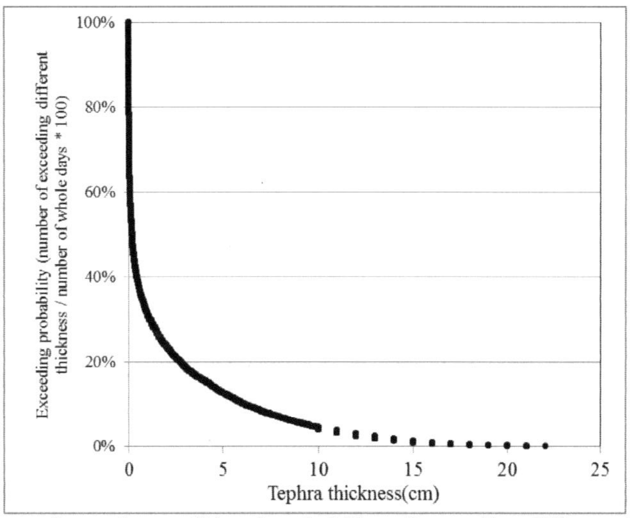

Figure 7. Hazard curve shows the conditional probability of exceeding different thickness of tephra at Ikata Power Plant, given a VEI5 volcanic eruption. The curve was generated from TEPHRA2 output, based on the simulations for 7976 days from March 1, 1982 to December 31, 2010 with the value of case-2 (Table 1).

6.2. Probability Maps

Based on the eruption and grain parameters obtained by the above work (Table 1), we developed a distribution map of the probability of exceedance for ash fall from an eruption of Kj-P1 scale (Figure 8). This map shows that the probability of a 1 cm-thick ash fall during an eruption of Kj-P1 scale is higher than 80% in western area of Usuki city and 40–60% between Yahatahama and Sukumo area in southwestern Shikoku. The probability falls rapidly in the areas further south and north with a probability of about 30% at the Ikata Power Plant site. For 5-cm thick ash fall, the probability is higher than 60% in western area of Usuki city and 10–30% in southwestern Shikoku. As the above results indicate, the probability of thick ash fall from the Kuju Volcano eruption is higher on the east side of the crater. This is a result of the strong influence of the jet stream (westerlies) flow at around an altitude of 10 km.

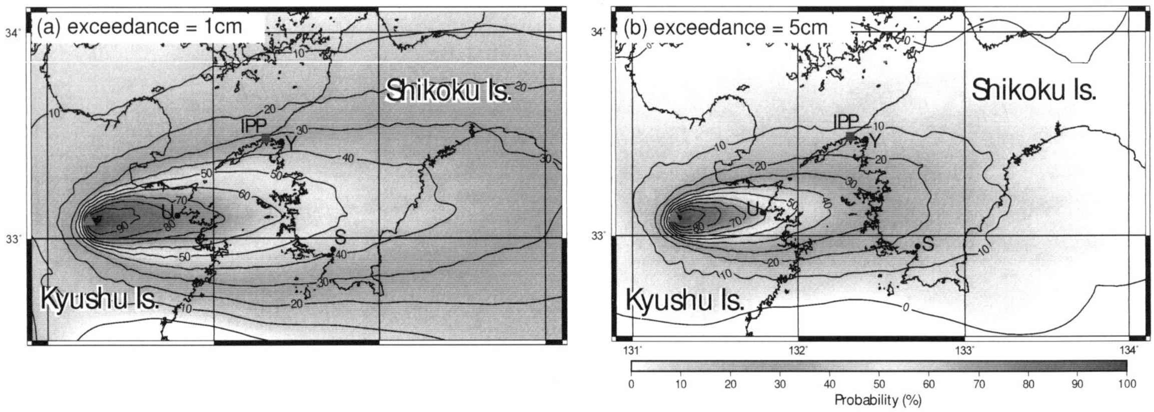

Figure 8. The maps contour the probabilities of tephra accumulations exceeding 1 cm (a) and 5 cm (b), given a VEI5 plinian scenario. The maps were generated from the TEPHRA2 output, based on the simulations for 365 days for 2010 with the value of case-3 (Table 1). IPP: Ikata Power Plant, S: Sukumo city, Y: Yahatahama city, U: Usuki city.

7. Conclusions

In this study, we carried out a geological survey and literature review and estimated the eruption parameters (eruptive volume, TGSD and column height) for a Plinian eruption (Kj-P1) under a VEI5 scenario, which are the most basic as well as the most important data for the assessment of volcanic eruption risks. Furthermore, we estimated the grain parameters (diffution coefficient and fall time threshold) by comparing the tephra fallout distribution between the results of our parameter study using TEPHRA2 and actual data. On the basis of such a reconstruction, we analysed TEPHRA2 using 22-year long record of wind profiles and obtained the results indicating that the probability of ash fall of 10 cm or more at Ikata Power Plant was 4.44% and 5 cm 12.74%. This study also estimate the probability of the range and volume of ash fall during a future large-scale eruption of Kuju Volcano. These findings should contribute to the development of hazard maps not only for Ikata Power Plant but also for regions located downwind from the vent.

8. Acknowledgements

The authors would like to sincerely thank D. Miura, K. Toshida, S. Takeuchi, Y. Hattori, S. Uesawa for their insightful discussions on eruption process, simulation of Kj-P1 eruption and volume calcuration techniques. H. Inakura and K. Mannen provided valuable information about TEPHRA2 technique. M. Okuno provided helpful comments on eruptive process. T. Tsuji also thanks T. Ouchi for correcting English, F. Suzuki for assistance in grain size analysis, TEPHRA2 simulation operation and making figures, T. Yanagihara and S. Yamamoto for their kindful support on grain size analysis. Some of the figures in this paper were made using the Generic Mapping Tools [30].

9. References

[1] Nuclear Regulation Authority, 2013. Volcanic Impact Assessment Guide for Nuclear Power Plants. p. 26 (Japanese).

[2] Volentic, A. C. M., Connor, C. B., Connor, L. J. And Bonadonna, C., 2009. Aspects of volcanic hazard assessment for the Bataan nuclear power plant, Luzon Peninsula, Philippines. In: Connor, C. B. et al. (eds.) Volcanic and Tectonic Hazard Assessment for Nuclear Facilities. Cambridge University Press, 229-256.

[3] Kawanabe, Y., Hoshizumi, H., Ito, J. and Yamazaki, S., 2015. Geological map of Kuju Volcano. 1: 25,000 in scale, National Institute of Advanced Industrial Science and Technology, Geological map of volcanoes, 19 (Japanese with English abstract).

[4] Tsuji, T., Ikeda, M., Kato, S., Nishizaka, N. and Onishi, K., 2015. Simulation analysis for air-fall ash influences on Shikoku Island. Research Report of Shikoku Electoric Power Co. and Shikku Research Institute, 103, 9-21 (Japanese with English Abstract).
[5] Kumahara, Y. and Nagaoka, S., 2002. Identification of the Kuju-Daiichi Tephra Layer, and the Age of the Late Pleistocene Fluvial Terrace along the Matsuda River in Southwestern Shikoku, Japan. The Quaternary Research, 41, 213-219 (Japanese with English abstract).
[6] Matsu'ura, T., 2015. Late Quaternary uplift rate inferred from marine terraces,Muroto Peninsula, southwest Japan: Forearc deformation in an oblique subduction zone. Geomorphology, 234, 133-150.
[7] Bonadonna, C., Connor, C. B., Houghton, B. F., Connor, L., Byrne, M., Laing, A. and Hincks, T. K., 2005. Probabilistic modeling of tephra dispersal: Hazard assessment of a multiphase rhyolitic eruption at Tarawera, New Zealand. Journal of Geophysical Research, 110, B03203, doi:10.1029/2003JB002896.
[8] Costa, A., Macedonio, G. and Folch, A., 2006. A three-dimensional Eulerian model for transport and deposition of volcanic ashes. Earth Planet Sci. Lett., 241, 634–647.
[9] Folch, A., Costa, A. and Macedonio, G., 2009. FALL3D: A computational model for transport and deposition of volcanic ash, Comput. Geosci., 6, 1334–1342.
[10] Macedonio, G., Costa, A. and Longo, A., 2005. A computer model for volcanic ash fallout and assessment of subsequent hazard. Comput. Geosci. 31, 837–845.
[11] Searcy, C., K. Dean, and W. Stringer (1998), PUFF: A high-resolution volcanic ash tracking model, J. Volcanol. Geotherm. Res., 80, 1–16.
[12] Bonasia, R., Capra, L., Costa, A., Macedonio, G. and Saucedo, R., 2011. Tephra fallout hazard assessment for a Plinian eruption scenario at Volcán de Colima (Mexico). Journal of Volcanology and Geothermal Research 203, 12–22.
[13] Bonasia R., Costa, A, Folch, A., Macedonio, G. and Capra, L., 2012. Numerical simulation of tephra transport and deposition of the 1982 El Chichón eruption and implications for hazard assessment., Journal of Volcanology and Geothermal Research, 231-232, 39-49.
[14] Japan Meteorological Agency, 2012.
[15] Newhall, C. G. and Self, S., 1982. The volcanic explosivity index (VEI) an estimate of explosive magnitude for historical volcanism. Journal of Geophysical Research, 87(C2), 1231-1238.
[16] Okuno, M., Nagaoka, S. and Kobayashi, T., 2012. Studies on Eruptive History of Kuju Volcano: A Review. Summaries of Researches Using AMS at Nagoya University, XXIII, 164-170.
[17] Nagaoka, S. and Okuno, M., 2015. Eruptive History of Kuju Volcanic Group, SW Japan. Transactions, Japanese Geomorphological Union, 36, 141-158.
[18] Nagaoka, S. and Okuno, M., 2014. Tephra-staratigraphy of Kuju Volcano in southwest Japan. Monthly Earth, 36, 281-296 (Japanese with English abstract).
[19] Suto, S., Inomata, T., Sasaki, H. and Mukoyama, S., 2007. Data base of the volcanic ash fall distribution map of Japan. Bulletin of Geological Survey of Japan, 58, 261-321 (Japanese with English Abstract).
[20] Japan Meteorological Agency, 2013. National catalogue of the active volcanoes in Japan (The fourth edition).
[21] Carey, S. and Sparks, R. S. J., 1986. Quantitative models of the fallout and dispersal of tephra from volcanic eruption columns. Bull. Volcanol. 48, 109–125.
[22] Suzuki, T., Katsui, Y. and Nakamura, T., 1973. Size Distribution of the Tarumai Ta–b Pumice–fall Deposit. Volcanology, 18, 47-63 (Japanese with English Abstract).
[23] Mannen, K., 2013. Theoretical background and recent progress of the pyroclastic fall simulation code Tephra2 : essence for application in Quaternary research. The Quaternary Research, 52, 173-187 (Japanese with English abstract).
[24] Hayakawa, Y., 1985. Pyroclastic Geology of Towada Volcano. Bulletin of the Earthquake Research Institute, University of Tokyo, 60, 507-592.
[25] Bonadonna, C. and Costa, A., 2012. Estimating the volume of tephra deposits: a new simple

strategy, Geology, 40, 415-418.
[26] Inman, D. L., 1952. Measures for describing the size distribution of sediments. J. Sediment. Petrol., 22, 125-145.
[27] Bonadonna, C., and Houghton, B.F., 2005, Total grain-size distribution and volume of tephra-fall deposits. Bull. Volcanol., 67, 441-456.
[28] Barnard, S., 2003. Potential effects of any future 1886-type eruption from Tarawera volcano on the Bay of Plenty region. Unpublished MSc thesis, University of Canterbury, Christchurch.
[29] Tauranga-WBOP Emergency Management, 2002. Civil Defence Plan 2002. Emergency management Office, Western Bay of Plenty District Council, Tauranga.
[30] Wessel, P. and Smith, W.H.F., 1995, New Version of the Generic Mapping Tools released. EOS, 76, 329.

Comparison landslide-triggering rainfall threshold using satellite data: TRMM and GPM in South Bandung area

Gian Nanda Pratama[1], Rusmawan Suwarman[2,3], I Dewa Gede Agung Junnaedhi[2], Edi Riawan[2], Aan Anugrah[4]

[1] Undergraduate of Department of Meteorology, Bandung Institute of Technology
[2] Research Group of Sains Atmosphere, Bandung Institute of Technology
[3] Department of Magister of Hydrogeology Engineering, Bandung Institute of Technology
[4] Non government organisations, Garda Caah, Majalaya
Email: giannandapratama@gmail.com/giannanda@students.itb.ac.id

Abstract: The number of landslides events in Bandung ranked fourth in Indonesia. The landslide is affected not only by the geographical character, but also by the rainfall intensity and duration. Thus, the information of the threshold of landslide triggering rainfall is important to be obtained. This study is aim to determine the threshold of rainfall in South Bandung using rainfall data provided by TRMM and GPM satellites. The satellite data is corrected by ground-observation data in three sites. The empirical method used in this study performs the relationship between the rainfall intensity and duration to the landslide event and the threshold curve obtained by lowering the regression line intensity-duration (ID) to lowest point rainfall in causing the landslide. The relationship of rainfall ID for landslide threshold is expressed in $I = 2.3\ e^{(-0.03*D)}$ for TRMM data, while $I = 7.20\ e^{(-0.09*D)}$ for GPM data. The result shows that the GPM data better than TRMM data and the analysis of measurement data showed a pattern corresponding to the observed values.
Keywords: Threshold, GPM, rainfall trigger landslide, observations, landslide, TRMM

1. Introduction

Geology Agency, Center for Volcanology and Geological Disaster Mitigation [1], states the potential for ground movement occurs in Bandung District fall into the category of medium to height. That's because region of Southern Bandung have a 235.213 Ha with slope <4° and then 451.913 Ha with slope >4°[2].

Hong [3] used satellite Tropical Rainfall Measuring Mission (TRMM) data to determine for global events landslides-triggering rainfall threshold. Muttaqin [4] evaluate threshold Hong for West Java, but not have been able to cover all historical landslides data, especially in West Java.

Fauziah [5] conduct research determines the intensity-duration for West Java use satellite data Tropical Rainfall Measuring Mission (TRMM) with a coarse scale. Presently, NASA have two satellites rainfall: Tropical Rainfall Measuring Mission (TRMM) with a coarse scale (0.25°x0.25°) and Global Precipitation Measurement (GPM) with a higher scale (0.1°x0.1°). A minimum rainfall threshold is defined as the lowest level below which a process of landslide triggered by rainfall always occurs. The threshold rainfall has two methods, physical and empirical method. The physical model requires a lot of supporting data as it combines spatial analysis, soil type, lithology, morphology, and slope. This information is difficult to be collected accurately and must have a great resource.

Furthermore, empirical method is used by applying analysis historical of rainfall events resulted in the landslides. This empirical threshold can be defined global, regional or local and independent of soil type, lithology, morphology, and land-use condition. The combination of rainfall to determine the threshold triggering landslides are divided into four subcategories: intensity-duration (ID) thresholds, thresholds based on the total event rainfall, rainfall event-duration (ED) thresholds, and rainfall event-intensity (EI) thresholds.

Intensity – duration (ID) thresholds is the most common and are widely used in proposed preceding study [3]. This threshold uses a model known as an ID curve. The general equation of this ID curve is

$$I = c + \alpha D^{\beta} \quad (1)$$

where I is rainfall intensity, D is rainfall duration, and $c \geq 0$. α and β are parameters obtained empirically. Based on previous researches, β value is ranging from -2.0 to -0.19 and α value is ranging from 4.0 to 176.44 [3].

Information rainfall observation in-situ is importance to research. And then, December 2015, KK - Atmospheric Sciences – Faculty of Earth Science and Technology (FITB) ITB cooperation with PT. Reasuransi Maipark Indonesia and Garda Caah for the community service by installing rain gauges at some point in South Bandung area (Figure 1).

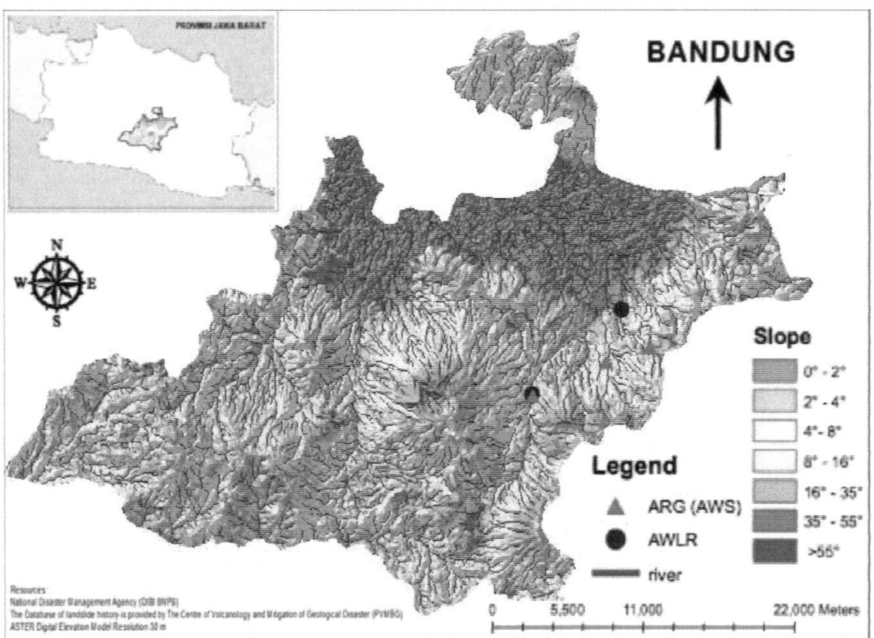

Figure 1. Area study and observations points AWS

Therefore, the area of South Bandung as one of landslides-prone needs to have a rainfall threshold equation. Rainfall Information Satelit Data TRMM has a coarse scale but long history data and then the data GPM which has a detail scale and short history data. This paper, we aim to determine and to compare threshold landslide-triggering rainfall based on the calculation of the intensity-duration of rainfall with empirical methods and analyze the relationship between the rainfall threshold and slopes in South Bandung.

2. Data and Method

2.1. Data

To determine the rainfall threshold, we need historical event landslides data, rainfall data and topographic data to analyze the relationship between threshold and the degree slope:

- The history of landslide is provided by The Centre of Volcanology and Mitigation of Geological Disaster (PVMBG) and National Disaster Management Agency (BNPB) for South Bandung from 2009 to 2016.
- Rainfall intensity is obtained from rainfall TRMM and GPM, which can be downloaded at (ftp://pmm.nasa.gov and ftp://disc2.nascom.nasa.gov). The spatial resolution is 0.25° x 0.25° and 0.10° x 0.10°.
- Rainfall data observations Weather and Climate Prediction Laboratory (WCPL) which can be downloaded at www.weather.meteo.itb.ac.id
- The topography data obtained from Digital Elevation Model (DEM), with a resolution of 30 x 30 m, can be downloaded at *(http://gdem.ersdac.jspacesystems.or.jp/)*

To get a landslide triggered by rainfall (not triggered by other causes, such as earthquake or human interference). Landslide occurrence data obtained from PVMBG and BNPB to do filtering. The number of events recorded landslide is as much as 88 events.

Filtering methods landslide occurrence in this study:

1. Double data case
2. Undefined Coordinate
3. Slope <4°
4. No description of the place
5. Rainless (No rain consecutive 24 hours before the incident)

From a total of 88 landslide that meet the criteria for filtering results by 26% or 23 events, double data case 24%, undefined coordinate 21%, slope <4° 11%, no description of the place 7% and rainless 11% (Figure 2).

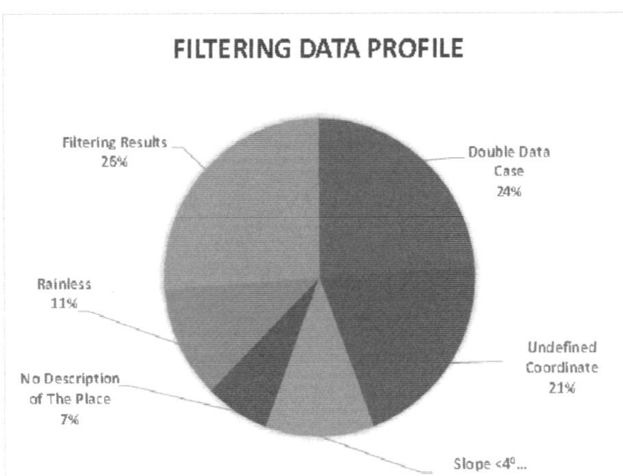

Figure 2. Filtering data landslides

2.2. Method

Method to get a rainfall threshold in this study using empirical method ID because the availability of the data support the use of this method, other than that methods of empirical ID produce better results when adding rainfall data that does not resulted landslides. The method of this study is described on point 2.2.1 *Verification of satellite data,* 2.2.2 *Comparison of thresholds TRMM and GPM,* 2.2.3 *Classification Slopes*:

2.2.1 Verification of satellite data

TRMM, GPM and observations have a different grid and temporal resolutions, then for validation data TRMM and GPM with observational data should be equated temporal resolution and view the observation point in the grid. There are two points observation located in one grid (Figure 3 and Figure 4), the observation data should be used as one entity similar to the method of rainfall region.

$$R = (R1+R2)/n \quad (2)$$

where *R* is rainfall intensity, *R1* observation (1), *R2* observation (2), and n is total stations. Validation data and comparison of data quality in this study using Spearmann Rank Correlation is the ranking method, in this research, ranking ranging from the largest to the smallest.

$$rs = 1 - \frac{6\,\Sigma d^2}{n(n^2-1)} \quad (3)$$

$$z = rs\sqrt{n-1} \quad (4)$$

where *rs* is coefficient correlation spearman, Σd^2 is Total square of the difference between ranking, n is total sample and z is z count. For data < =30 use (equation 3), if data>30 use (equation 4).

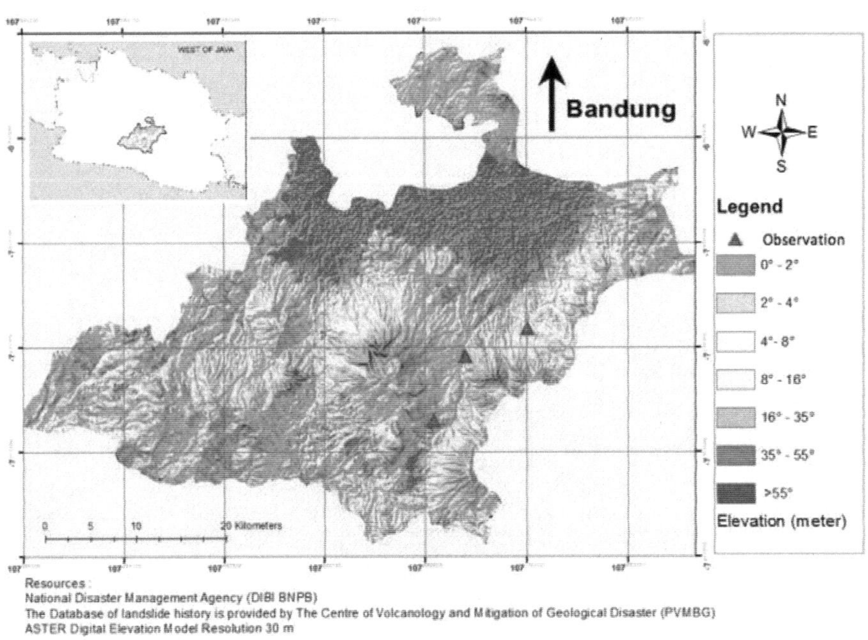

Figure 3. Observation point AWS in grid GPM

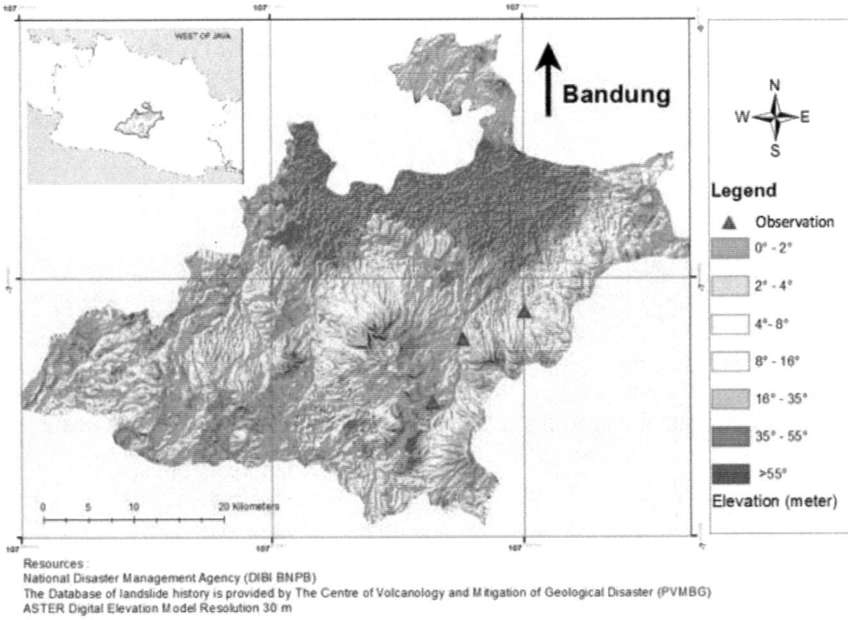

Figure 4. Observation point AWS in grid TRMM

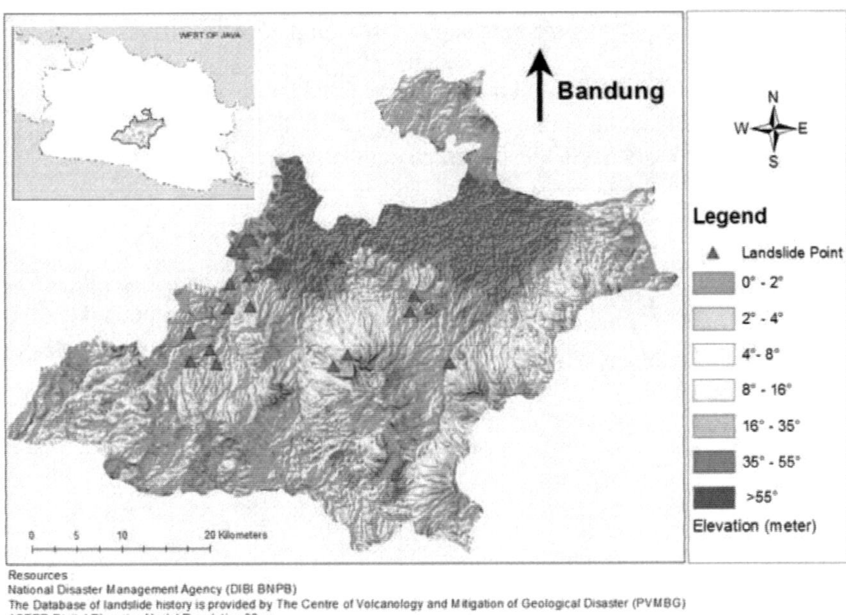

Figure 5. Distribution of Landslides

2.2.2 Comparison of thresholds TRMM and GPM

To compare of thresholds use the regression line of data intensity-duration, most suitable in this method is exponential regression because can describe the landslide events and lowered it to the limited landslide events to get a rainfall thresholds.

2.2.3 Classification Slopes

The slope is one of the factors supporting the occurrence of landslides. From the results of filtering landslide (23 events), classified according to the classes slope Zuidam [6]. Distribution of landslides data that used in this study the results of slope classification and filtering data (Figure 5).

3. Result and Discussion

3.1. Verification of satellite data

By using a confidence interval of 0.02 or 98% can be seen in (Table 1) showed that the Z count > Z table, but the Spearman correlation coefficient GPM is still better than the TRMM although GPM has been reduced resolution.

Table 1. Correlation observations and satellit (mm/3 hrs) with Z count and Z table

	AWS and TRMM	AWS and GPM
Correlation	0.44	0.59
Z Count	2.99	4.06
Z Table	-2.33 and 2.33	-2.33 and 2.33

The results of the analysis showed that of a significant positive relationship between the variables X and Y but GPM more like observational data seen from GPM Spearmann Correlation Coefficient is greater than the TRMM Spearmann Correlation Coefficient.

3.2. Comparison Thresholds TRMM and GPM

From the results of the threshold (Figure 6 and Figure 7) the most suitable is use exponential regression, because it could cover all the historical landslide and shown in (Table 2), showed the value of R-Square of the fitting rather well 0.4.

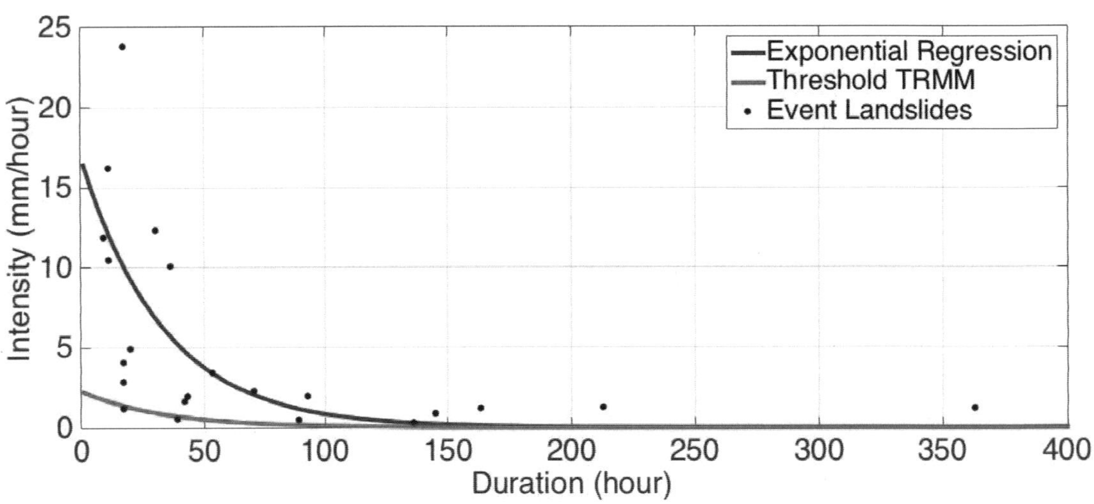

Figure 6. TRMM threshold with exponential regression and event landslides

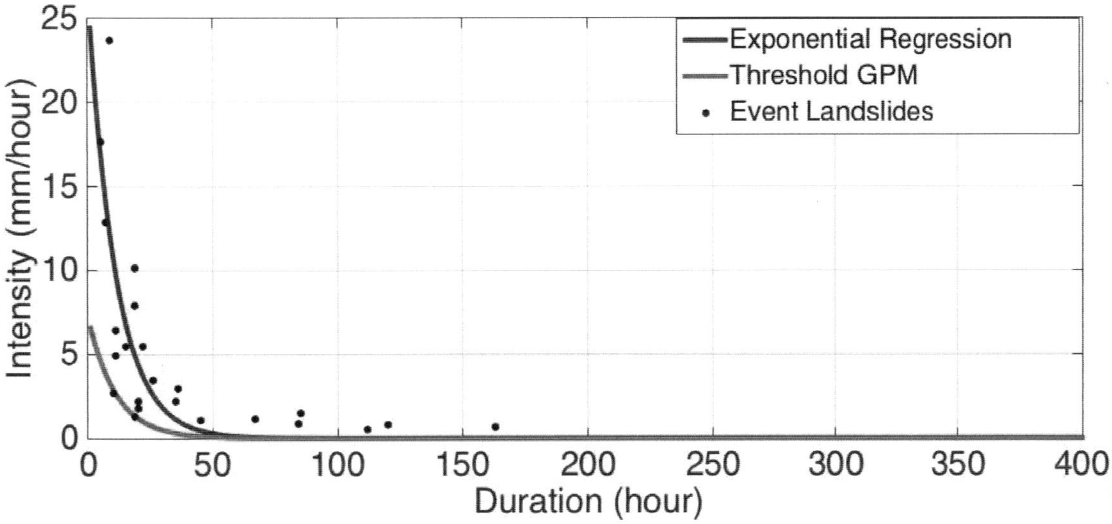

Figure 7. GPM threshold with exponential regression and event landslides

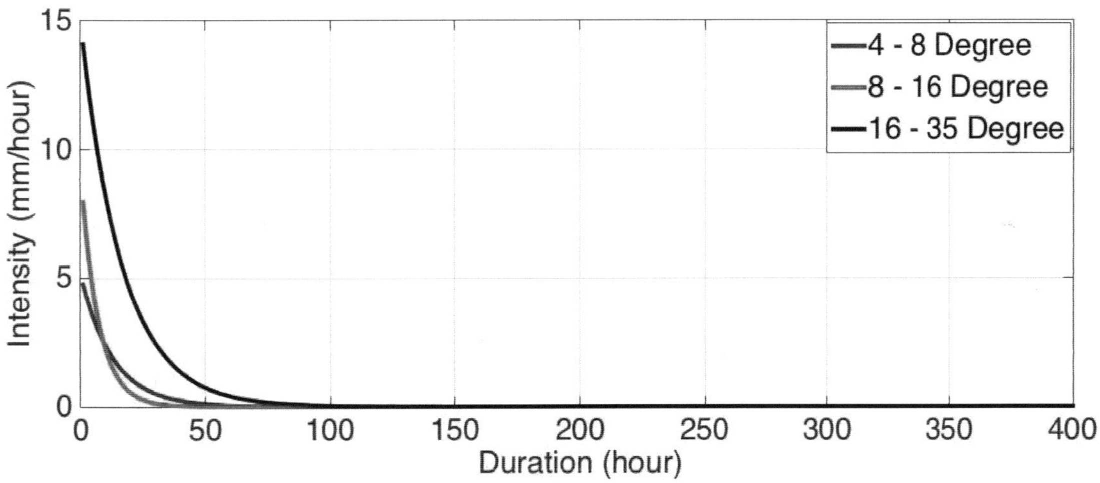

Figure 8. GPM threshold with gradient slope

Table 2. Comparison Threshold TRMM and GPM with R^2 and RMSE

Data	Threshold	R^2	RMSE
GPM	$I=7.2\ e^{(-0.09\ D)}$	0.59	3.9
TRMM	$I=2.3\ e^{(-0.03\ D)}$	0.42	4.8

From the results fitting data (Table 2) RMSE of GPM data smaller than TRMM data is 3.9, it shows that the GPM more similar observation data because it has a small error. Landslide caused by some factor not only by rainfall, can be seen in (Figure 6 and Figure 7) exponential regression line (blue) and threshold (red) produces different distances between TRMM and GPM. The distance between (blue and red) GPM smaller than the TRMM, it shows that the threshold GPM interference from other factors such as lithology, soil type, land cover, and etc was considered almost equal. Therefore, GPM can describe the threshold of rainfall triggers landslide

3.3. Relationship Between Thresholds with Slope

Slope is one of the important factors of the landslide besides rainfall. This sub-chapter explains only GPM data because this is a better threshold to describe landslide event (Figure 8). Slope with a higher than 6 degree is more contrast than the other classes. The possible things explain this condition because of the lower degrees faster to be saturated rather than steeper degrees. The threshold value on every slope is different class of GPM (Table 3).

Table 3. Thresholds value with gradient slope

Slope	Thresholds GPM
4° - 8°	$I=5.14\ e^{(-0,14\ D)}$
8° - 16°	$I=9.4\ e^{(-0.14\ D)}$
16° - 35°	$I=14.95\ e^{(-0.06\ D)}$

4. Conclusion

Based on the analysis of the relationship rainfall intensity and duration on landslides events studied over South Bandung Area, obtained the following conclusions:

- The obtained threshold from 23 landslide event occurrences in South Bandung Area with GPM data is $I= 7.2\ e^{(-0.09\ D)}$ and with TRMM data is $I= 2.3\ e^{(-0.03\ D)}$
- Global Precipitation Measurement (GPM) data more like observational data because correlation GPM is greater than Tropical Rainfall Measuring Mission (TRMM)
- Relationship between slope and threshold landslide-triggering rainfall line shows that the steeper needs to have higher intensity of rainfall and longer duration than flatter slope.

5. Acknowledgements

This research was supported by KK-Atmospheric Sciences FITB-ITB, PT. Reasuransi Maipark Indonesia and Garda Caah for getting rainfall observation data. The Centre of Volcanology and Mitigation of Geological Disaster (PVMBG), National Disaster Management Agency (BNPB) for getting Landslide Inventory Database Indonesia (LIDIA).

6. References

[1] Center for Volcanology and Geological Disaster Mitigation 2008.
[2] Andri N Ardiansyah 2011 Landslide Disaster Risk Areas in Bandung Area. University of Indonesia.
[3] Y. Hong, R.F. Adler, A. Negri, G.J. Huffman 2007 Flood and Landslide Applications of Near Real-time Satellite Rainfall Products. NASA Goddard Space Flight Center.
[4] Imam Muttaqin 2015 Evaluation Threshold I-D Hong 2006 Events for Landslide in West Java. Bandung Institute of Technology.
[5] Selma Nurul Fauziah 2015 The Determination of Rainfall Threshold Triggering Landslide in West Java Using TRMM Satellite Data. Bandung Institute of Technology.

[6] Zuidam, R.A. & Zuidam Cancelado, F.I.,1979 and 1985, Terrain Analysis and Classification Using Areal Photographs, A Geomorphologycal Approach, Netherland, Enschede: ITC.
[7] F. Guzzeti, S. Peruccacci, M. Rossi, and C.P Stark 2007 Rainfall Thresholds for thte initiation of landslides in Central and Southern Europe. Meterol Atmos Phys 98,. 239-267. The Netherlands.
[8] M. D. Syaifullah 2014 Validasi dan Koreksi Data Satelit TRMM pada Tiga Pola Hujan di Indonesia. Badan Meteorologi Klimatologi dan Geofisika. Jakarta.

IOP Conf. Series: Earth and Environmental Science **71** (2017) 012004 doi:10.1088/1755-1315/71/1/012004

Identifying Successive Eruption of Guntur Volcanic Complex Using Magnetic Susceptibility and Polarimetric Synthetic Aperture Radar (PolSAR) Data

Asep Saepuloh and Erwin Bakker

Study Program of Geology Engineering, Faculty of Earth Sciences and Technology, Bandung Institute of Technology (ITB), Indonesia
Corresponding author e-mail address: saepuloh@gc.itb.ac.id

Abstract. Identifying distribution and stratigraphic of volcanic products are important not only for mitigating volcanic hazards, but also to know the characteristics of the successive eruptions. Guntur volcanic complex located in Garut, West Java, Indonesia was selected as study area because of the last eruption took place in 1847 and the volcanic activity has been dormant since then, however its seismicity is still active. During the period of July to October 2009, the hypocentre distribution of volcano tectonic earthquakes is mostly located at western flank of the volcano, beneath Guntur - Gandapura craters at the depth of less than 5 km. This study is aimed to identify distribution and succession of volcanic products based on their magnetic properties and backscattering signal of Polarimetric Synthetic Aperture Radar (PolSAR) data. The polarimetric decomposition method was used to identify the distribution of the volcanic products based on their scattering characteristics. Then, the field measurement using SM-30 magnetic susceptibility meter was performed to confirm the units of volcanic products and interpret their successions. According to the polarimetric decomposition method, we could identify fifteen successive eruptions formed Guntur Volcano Complex and termed as Khuluk and Gumuk in Indonesian standard. The successions were produced Gumuk Windu, Gumuk Malang, Gumuk Pulus, Gumuk Putrri, Khuluk Meungpeuk, Gumuk Cakra, Gumuk Gandapura, Gumuk Putri, Gumuk Gajah, Gumuk Batususun, Khuluk Pasirlaku, Gumuk Agung, Gumuk Picung, Gumuk Pasirmalang, Gumuk Masigit, Khuluk Kabuyutan and Khuluk Guntur. The magnetic susceptibility confirmed that the variations of magnetic susceptibility of rocks at each gumuk agreed with their stratigraphy.

1. Introduction

The phenomenon of an ancient volcanic activities is relatively difficult to be investigated due to intensive erosion and weathering processes. To define a framework to improve early warning and hazard tracking of volcanic eruptions, the physics of volcanoes including what causes them, where they occur, and what hazards they pose must be addressed [1]. Volcanic eruptions could cause a great damage, so it is necessary to monitor periodically in order that losses caused can be reduced [2]. One of the monitoring was done by studying the rocks of the ancient volcanic products including identifying type and distribution of rocks. Surface volcanic rocks identification in active volcano is crucial not only to mitigate volcanic hazards, but also to characterize eruption, urban rehabilitation, and reconstruction especially after eruption [3].

Identifying volcanic rock distributions solely based on field investigation is not effective when applied under tropical condition such as Indonesia because of dense vegetation and intensive weathering. This condition caused a difficulty to find rock outcrops and boundary of volcanic products or geological structures. Therefore, the field observation requires a lot of time and cost. Utilization of active remote sensing technology is a great advantage for mapping under large coverage area regardless time and weather conditions [3].

The change of volcanic rocks in one composite volcano might reflect the occurrence of magmatic evolution presented by variation in petrology and geochemistry of the rocks. The variation may serve as an indicator for characterizing the previous eruptions. Magnetic susceptibility of the rocks represented magnetization of the rocks when applied the magnetic field is a physical parameter to characterize and differentiate rock types [4]. Magnetic properties of rocks reflect the composition of the rocks minerals such as iron highly magnetic and other magnetic materials the weak such as silicates and carbonates. The difference of this magnetic property depends on chemical composition, ratio of iron oxide and petrogenesis of rocks. Magnetic susceptibility of rocks depends on composition of magnetic minerals in rock. Mafic rocks generally have higher magnetic susceptibilities than felsic rocks because mafic rocks are typically more abundant in strongly magnetic minerals such as magnetite [5]. According to the capability of magnetic parameter to characterize the rock type, we raised a combination method between magnetic susceptibility measurement at field with Polarimetric Synthetic Aperture Radar (PolSAR) image interpretations to delineate the boundary of volcanic groups and correlate with their successive eruptions.

This research is aimed to identify successive eruption at the Guntur Volcanic Complex based on magnetic susceptibility and fully polarimetric SAR (PolSAR) data from PALSAR (Phased Array Type L-Band Synthetic Aperture Radar) onboard the Advanced Land Observing Satellite (ALOS). The magnetic susceptibility was used to interpret of the magnetic minerals within volcanic products. The PolSAR data was used to delineate boundary of volcanic groups following the Indonesian standard volcano stratigraphy.

The study area is located at Guntur Volcanic Complex in Garut, West Java, Indonesia (Figure 1). The first historical eruption of the volcano was occurred in 1690 and the last eruption was recorded in the 1847 [6]. The volcano-tectonic earthquake activity recently is still quite high more than 100 events in October 1997, May 1999, November 2002, and June 2005 [7].

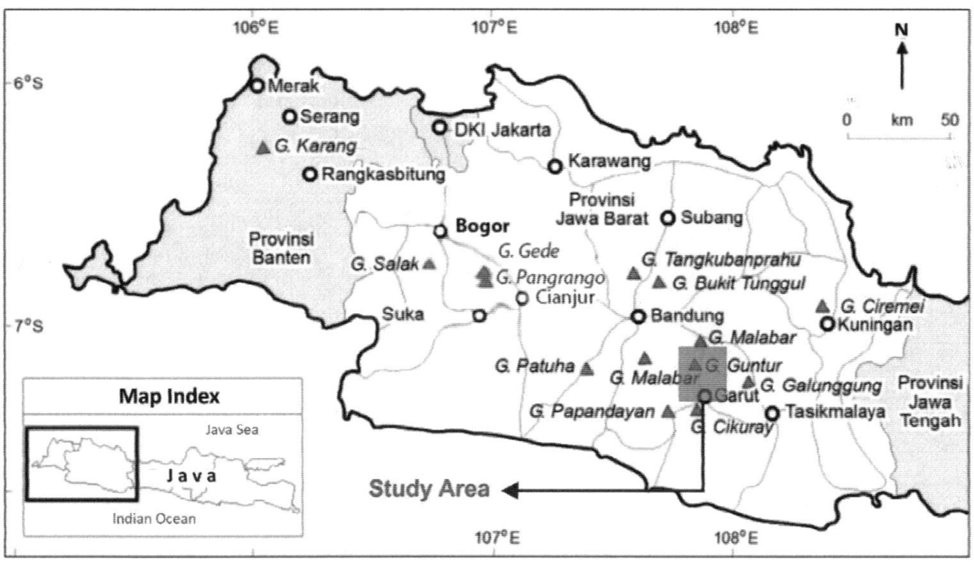

Figure 1. Study area located at Guntur Volcanic Complex, Garut District, West Java, Indonesia presented by pink square.

2. Data and Method

2.1. ALOS PALSAR

ALOS PALSAR is one of the satellites are capable of sending and receiving electromagnetic waves with full polarization. Polarization is an index describing electromagnetic waves propagation and relative orientation of the electric field. The electromagnetic waves propagate vertically and horizontally, they are termed as vertical (V) and horizontal (H) polarizations, respectively. Synthetic aperture radar satellite is usually designed to transmit and receive either horizontally or vertically polarized signals. For example, the Japan Earth Resources Satellite 1 (JERS-1) SAR utilized horizontally polarized signals. However, ALOS PALSAR can transmit and receive both horizontally or vertically polarized signals. Thus, a combined polarization radar image could be HH (horizontal transmitting, horizontal receiving), VV (vertical transmitting, vertical receiving), HV (horizontal transmitting, vertical receiving) or VH (the reverse of HV) as presented by Figure 2. Characteristics of scattering depend on the polarization properties of the target. The different scattering patterns among polarizations can be observed from ALOS PALSAR polarimetric image. PolSAR data of Guntur Volcanic Complex were derived from ALOS PALSAR level 1.1 with a spatial resolution 12.5 m and acquired on May 4, 2011. ALOS PALSAR operates using active microwave sensor with the L-band frequency by minimizing the atmosphere and canopy vegetation effects [8]. The acquired data are fully polarization types HH, HV, VH, and VV.

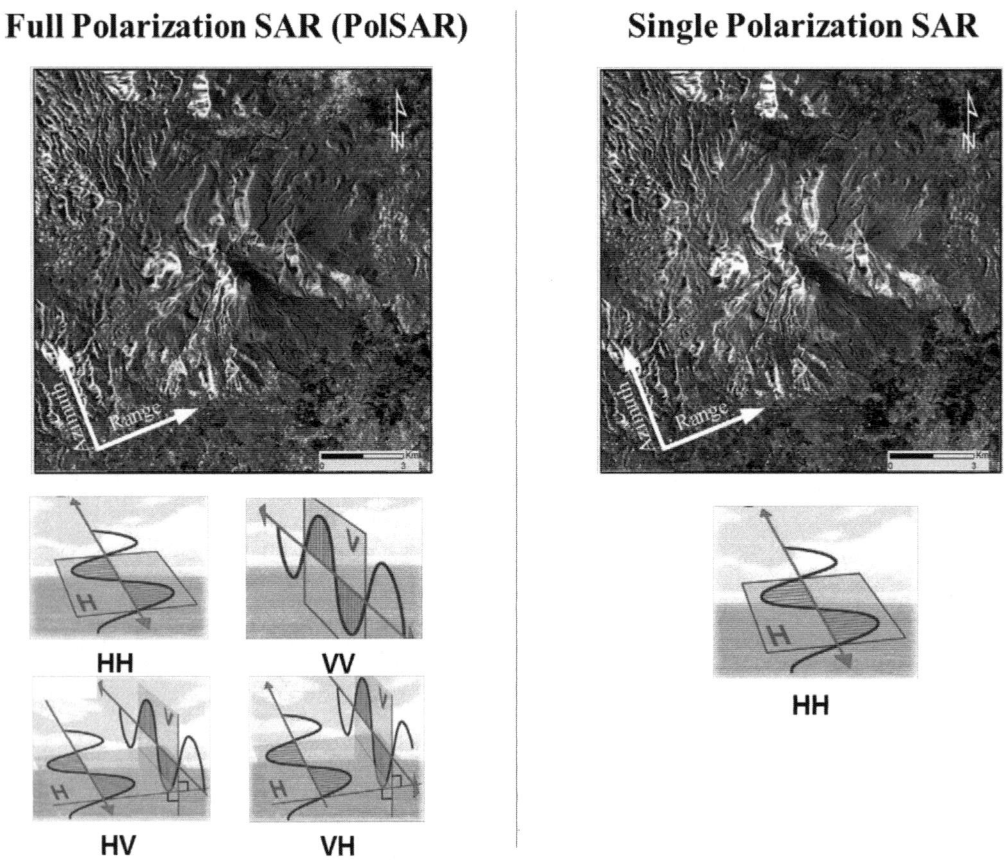

Figure 2. The full (left) and single (right) polarimetric SAR (PolSAR) images and illustration of their electromagnetic propagations.

The ALOS PALSAR data processing on this study is divided into two steps: processing of backscattering intensity and polarimetric decomposition. For processing of backscattering intensity, The ALOS PALSAR data Level 1.1 in Single Look Complex (SLC) were transformed into Multi Look Image (MLI) by splitting into four polarization modes HH, HV, VH, and VV of backscattering intensity. The polarized-backscattering intensity was quantified by a multi-look processing with 1×7 factor to keep the spatial resolution of image along the range and azimuth directions as 28 and 25 m, respectively. Finally, the MLI was corrected geometrically by transforming the slant to ground range coordinate using a simulated DEM derived from the SRTM version 4 with spatial resolution 90 m. Yamaguchi decomposition method was applied to extract backscattering characteristics from object based on polarimetric decomposition [9]. Polarimetric decomposition of ALOS PALSAR data were performed by decomposing a single polarization into matrix scattering [S] 2×2 followed by extraction of matrix coherency [T3]. In horizontal polarization (H) and vertical (V), the matrix scattering [S] could be expressed by:

$$S = \begin{bmatrix} S_{HH} & S_{HV} \\ S_{VH} & S_{VV} \end{bmatrix} \tag{1}$$

Assuming the reciprocal condition $S_{HV} = S_{VH}$, the matrix coherency [T3] could be expressed as follows:

$$T3 = \begin{bmatrix} T_{11} & T_{12} & T_{13} \\ T_{21} & T_{22} & T_{23} \\ T_{31} & T_{32} & T_{33} \end{bmatrix} \tag{2}$$

It should be noted that T_{11}, T_{22} and T_{33} are real-valued terms and that T_{12}, T_{13}, and T_{23} are complex valued. Therefore there are 9 real-valued parameters in the matrix. The matrix [T3] has advantages in the mathematical operations as well as in interpreting physical backscattering phenomenon [10].

To classify the backscattering characteristics of the ground surface, we used the four-component scattering power decomposition of the matrix [T3] by dividing the polarimetric data in pixel area into scattering of surface, double bounce, volume, and helix with 4 expansion sub-matrices as follows:

$$T3 = \langle [T] \rangle = P_s \langle [T] \rangle_{Surface} + P_d \langle [T] \rangle_{Double} + P_v \langle [T] \rangle_{Volume} + P_c \langle [T] \rangle_{Helix} \tag{3}$$

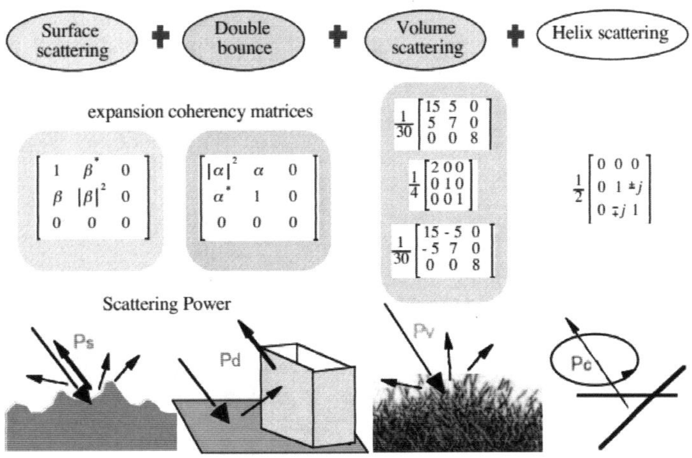

Figure 3. Four mechanisms backscattering characteristics with coherence matrix [9].

Eq. (3) is an extension of the three-component decomposition to deal with non-reflection symmetry scattering case as illustrated by Figure 3. Figure 4 showed the classification steps based on Yamaguchi four component decomposition techniques [11].

In geology, the utilization of the decomposition characteristics of backscattering useful for detecting surface objects such as:

- *Surface scattering* : The scattering mechanism is mainly caused by rough surfaces such as bare soil, volcano summits, agricultural fields;

- *Double bounce scattering* : Right angle structures such as big fragmented boulder size, intrusive body, hill, and river surfaces;
- *Volume scattering*: eruptions centre, brecciate volcanic flow, and intensive fracture zone. Classification of object based on backscattering characteristics obtained from a combination of dual polarization signals, classification of characteristics

2.2. Magnetic Susceptibility
In addition to image processing, also check field and magnetic susceptibility measurement of rocks. Rocks that measured magnetic susceptibility this study focused only on igneous rock from volcanic products. Magnetic susceptibility measurement was performed using Magnetic Susceptibility Meter SM-30 with error measurement up to 1×10^{-7} SI and maximum scale is 0.1 SI. Method of measurement can be performed directly on rock outcrops in the field as well as on samples of rocks, but in this study, measurements carried out directly on the outcrop in the field.

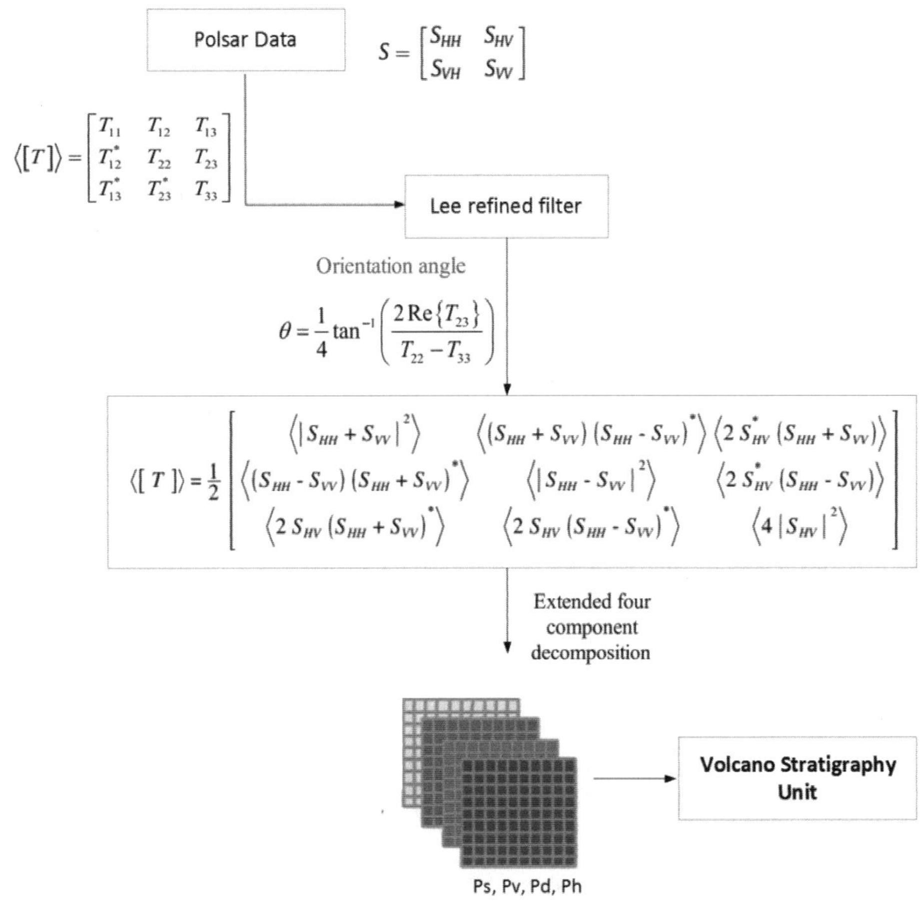

Figure 4. Flowchart of Yamaguchi SAR decomposition method modified from [11].

3. Result
The utilization of L-band SAR data with full polarization is effective for target detection under tropical regions. A technique adopted in this study is visual interpretation based on composite backscattering intensity and decomposition method. The technique was used to identify volcanic groups as well as stratigraphy termed as khuluk and gumuk. Khuluk and gumuk are Indonesian standard volcano stratigraphy to distinguish the groups of volcanic product based on their volcano origin. Khuluk is composed by groups of volcanic rocks from one or more eruption center and formed

large body of volcano. Gumuk is the smallest group of volcanic products originated from an eruption centre or single parasitic eruption [12]. Khuluk could be composed by several gumuks, but not in vice versa.

Based on the visual analysis of tonal pattern, texture, brightness, geomorphology and structure from SAR image, the composite backscattering intensity is useful to identify boundary of the gumuk. The optimum visual contrast was obtained by color composite for R=HH, G= HV, and B=VH (Figure 5A). On the other hand, the decomposition method showed a high detectability for the khuluk of the volcano. Result of the decomposition method is distinguished of backscattering characteristics from object on the surface in the form of surface scattering, double scattering, volume scattering, and helix scattering, that was then in the combined using RGB composite to R=double scattering, G=volume scattering, dan B=surface scattering (Figure 5B). This colour composite showed khuluk volcano with clear boundaries. Boundaries of khuluk volcano from decomposition results are shown with characteristics of the surface scattering (blue). Results analysis of the scattering intensity and decomposition characteristics, then on the location research complex consists of nine Khuluk that is formed by fifteen Gumuk.

A B

Figure 5. (A) Composite RGB backscattering intensity image R=HH, G=HV, B=VH; (B) Composite decomposition polarimetric image R=Double Bounce Scattering (*Pd*), G=Volume Scattering (*Pv*), and B=Surface Scattering.

Magnetic susceptibility rocks characteristics are controlled by mineralogy of rock. The characteristic of the concentration is an indicator of magnetic minerals in rocks. For Guntur Volcanic Complex, magnetic susceptibility values range from 16.17×10^{-3} to 30.2×10^{-3} SI. The stratigraphic of each Khuluk were arranged based on PolSAR image interpretation and magnetic susceptibility measurement at field, except for Khuluk Pasirlaku. The khuluk has a lowest magnetic susceptibility about 19.7×10^{-3} SI, but the stratigraphy is older than Khuluk Gajah with magnetic susceptibility about 21.18×10^{-3} SI. The discrepancy of the magnetic susceptibility at Khuluk Pasirlaku was interpreted from magmatic phenomena during ascend from the deep to the shallow reservoir such as magma differentiation or mixing with surrounding rocks (Table.1). The highest magnetic susceptibility is about 30.2×10^{-3} SI for the Khuluk Guntur. The magnetic susceptibility shows an agreement with the stratigraphy of volcanic groups in general, and also with their absolute age. The magnetic susceptibility of the rocks at the study area increased from the old to the young volcanic products in general (Table.1).

Table 1. Comparison of magnetic susceptibility with their absolute age following [13] of Guntur Volcanic Complex.

Khuluk	Gumuk	Age (Ma)	Susceptibility (× 10^{-3} SI)	Rock Type
Guntur	Guntur	0.05±0.38	30.2	Basalt
Kabuyutan	Kabuyutan		26.76	Basalt
Masigit	Masigit			Andesit
	Pasirmalang		24.53	Andesit
	Picung	0.07±0.02	23.73	Andesit
	Agung	0.08±0.4	22.07	Andesit
Pasirlaku	Pasirlaku		19.7	Andesit
Gajah	Batususun		21.18	Andesit
	Gajah			
Putri	Putri	0.14±0.08	21.03	Andesit
Gandapura	Gandapura	0.33±0.02	20.9	Dacite
	Cakra			
Meungpeuk	Meungpeuk		19.87	Andesit
Windu	Windu			
	Malang			
	Pulus			
	Putrri			

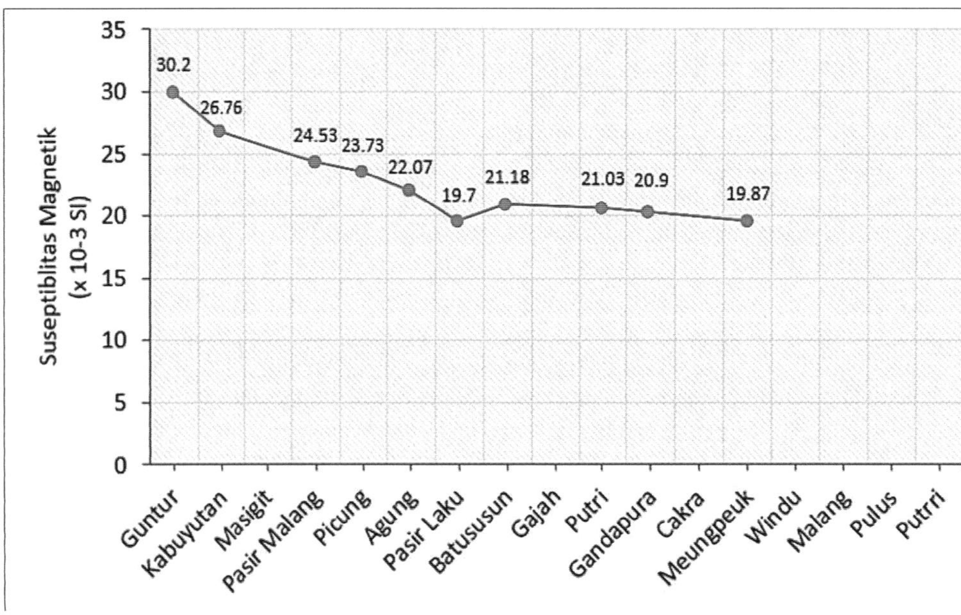

Figure 6. The cross-plot between stratigraphy of Guntur Volcanic Complex and magnetic susceptibility showed an agreement, except at Khuluk Pasirlaku.

4. Discussion

Several Polarimetric SAR processing methods have been widely used in the study of volcanoes and very useful and effective to observe ground surface. Compared with optic remote sensing data, Fully Polarimetric SAR is suitable to detect and classify the characteristics of the geometric of ground surface. As in Figure 7 directed Gumuk Guntur dominated by Surface Scattering (blue portion in Figure 7) that shows the characteristics of smooth surfaces such as soil until gravel sized rocks.

However some places at Khuluk Guntur show the double scattering characteristics the resulting by scattering of blocks boulder rocks sized and of residential building at Guntur volcanic foot (Figure 8A). Double bounce characteristics (red portion in Figure 7) dominated at residential areas that formed by scattering from the building. These characteristic can be used for mitigation of eruption because it can classify the residential areas well, so it is useful for the evacuation of residents and also relocation of affected areas after eruption. While on the khuluk or gumuk volcanoes, characteristic double bounce dominated on Khuluk Pasirlaku that formed by scattering from the blocks of lava flows (Figure 8B). The rocks were mined by local people. Volume scattering originated from dense vegetation at the old volcanic products and presented by green portion in Figure 7. An intensive weathering of the old volcanic products caused decomposition of the rocks into the soil layers. Therefore, the vegetation has growth densely at this zone. These characteristics dominate almost the entire old volcanic products as seen on the Gumuk Picung, Agung, and Masigit (Figure 8C).

Samples of rocks on the Guntur Volcanic Complex show changes the value of the higher magnetic susceptibility toward younger rock as listed in Table 1. Change the value of magnetic susceptibility indicates that the level and stability of magnetisation are dependent on the quantity and composition of the magnetic minerals (which are controlled by the composition of the original liquid magma), their oxidation states and grain sizes [14]. If the change on the composition of magma derived from original magma then it can be interpreted as differentiation occurred and produced different type of rock with different magnetic susceptibility values, which is also reflection of changes in rock type, but this needs further analysis on rock geochemistry.

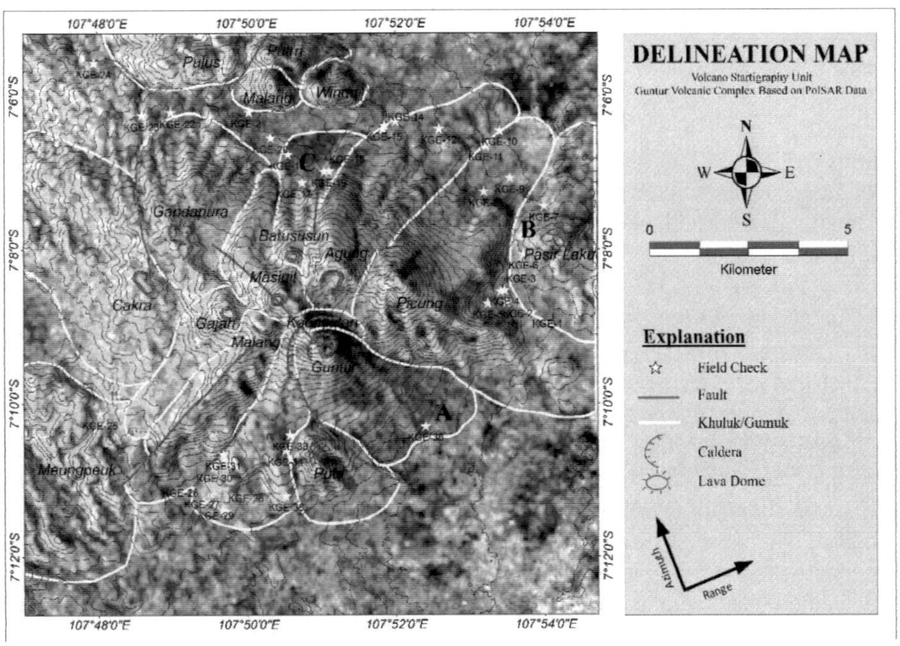

Figure 7. Delineation of volcanic groups at Guntur Volcanic Complex based on PolSAR image interpretation and magnetic susceptibility measurement following geological field observations presented by white lines.

According to [6], rock type of young Guntur Volcano are generally basalt (result of the eruption in 1840), this is confirmed with high magnetic susceptibility that is 30.2×10^{-3} SI and rock sampling when field check. The correlation between the absolute age of rocks and magnetic susceptibility data also show changes of the magnetic susceptibility higher towards the age of younger rocks. As shown in the Gumuk Gandapura, Putri, Agung, Picung and Guntur (Table 1). Therefore, if the data is correlated with magnetic susceptibility, then change the magnetic susceptibility is higher towards younger rocks reflect changes of rocks from acidic to basic. These changes are also visible from rock samples in the field as in the old aged rocks tend to andesitic and basaltic tend towards older.

Figure 8. Field photographs of volcanic products at Guntur Volcanic Complex with their locations presented in Figure 7.

5. Conclusions

Utilization of magnetic susceptibility and Polarimetric SAR (PolSAR) data to the geological field observation is effective to delineate the boundary of volcanic units under tropical area. The colour composites of backscattering characteristics such as PolSAR intensity images and decomposition of the full polarized data were useful for identifying part of the volcanic stratigraphic units in the form of Khuluk and Gumuk. In addition, the geomorphologic and structural features could be used for basis interpretation. The geomorphology and structural features influenced to the interpretation process because the distribution of volcanic products followed the morphology and structure. Yamaguchi decomposition could show characteristics of ground surface such as soil, loose sediment, and gravel. They were identified by surface scattering (soil and loose sediment), double bounce scattering (building and boulder to block sized of rocks), and volume scattering (dense vegetation). Magnetic susceptibility was also useful to confirm the occurrence of the magnetized minerals in the rock type of volcanic complex. The changes of magnetic susceptibility value reflected the changes in mineralogical composition of rocks. We obtained that volcanic products in the same gumuk have similar susceptibility value. Then, the significant different was interpreted that the volcanic rocks were produced by the other gumuk. Successive eruptions of Guntur Volcanic Complex based on PolSAR and magnetic susceptibility data from an old to young volcanic products are as follows: Gumuk Windu, Gumuk Malang, Gumuk Pulus, Gumuk Putrri, Khuluk Meungpeuk, Gumuk Cakra, Gumuk Gandapura, Gumuk Putri, Gumuk Gajah, Gumuk Batususun, Khuluk Pasirlaku, Gumuk Agung, Gumuk Picung, Gumuk Pasirmalang, Gumuk Masigit, Khuluk Kabuyutan and Khuluk Guntur. The causes of the change of magnetic susceptibility with mineral composition of the successive eruptions were aimed to the next step of this study.

6. Acknowledgments

The authors thank to LPPM-ITB and Ristekdikti under IPTEKS research scheme for providing equipment and financial supports to this research. The thanks were also extended to the anonymous reviewer for greatly improving the clarity of the paper.

7. References

[1] Wylie M 2008 *Volcanic activity: Processing of observation and remote sensing data (VAPOR)* (Dublin Institute of Technology)

[2] Suminar W, Saepuloh A and Meilano I 2016 Identifying hazard parameter to develop quantitative and dynamic hazard map of an active volcano in Indonesia *AIP Conference Proceedings* THE 5TH INTERNATIONAL SYMPOSIUM ON EARTHHAZARD AND DISASTER MITIGATION: The Annual Symposium on Earthquake and Related Geohazard Research for Disaster Risk Reduction vol 1730 (AIP Publishing) p 050007

[3] Saepuloh A, Koike K and Omura M 2012 Applying Bayesian Decision Classification to Pi-SAR Polarimetric Data for Detailed Extraction of the Geomorphologic and Structural Features of an Active Volcano *IEEE Geosci. Remote Sens. Lett.* **9** 554–8

[4] Acharya R and Paudyal K 2016 Magnetic Susceptibility as a Tool of Lithological Mapping: A Case Study from Malekhu-Damauli Area of Central Nepal, Lesser Himalaya *Nepal J. Sci. Technol.* **16** 49–58

[5] Carmichael R S 1982 *Magnetic properties of minerals and rocks* vol 2 (CRC Press Boca Raton, FL, USA)

[6] Kusumadinata K 1979 *Catalogue of references on Indonesian volcanoes with eruptions in historical time* (Republik Indonesia, Departemen Pertambangan dan Energi, Direktorat Jenderal Pertambangan Umum, Direktorat Vulkanologi)

[7] Sadikin N 2008 *Volcano-tectonic earthquakes and magma supply system at Guntur volcano, with long-term dormant period* (Kyoto University)

[8] Saepuloh A, Koike K, Urai M and Sumantyo J T S 2015 Identifying Surface Materials on an Active Volcano by Deriving Dielectric Permittivity From Polarimetric SAR Data *IEEE Geosci. Remote Sens. Lett.* **12** 1620–4

[9] Yamaguchi Y, Moriyama T, Ishido M and Yamada H 2005 Four-component scattering model for polarimetric SAR image decomposition *IEEE Trans. Geosci. Remote Sens.* **43** 1699–706

[10] Yamaguchi Y, Sato A, Boerner W-M, Sato R and Yamada H 2011 Four-component scattering power decomposition with rotation of coherency matrix *IEEE Trans. Geosci. Remote Sens.* **49** 2251–8

[11] Xiang D, Tang T, Ban Y, Su Y and Kuang G 2016 Unsupervised polarimetric SAR urban area classification based on model-based decomposition with cross scattering *ISPRS J. Photogramm. Remote Sens.* **116** 86–100

[12] Martodjojo S 1996 Sandi Stratigrafi Indonesia, Komisi Sandi stratigrafi Indonesia *Ikat. Ahli Geol. Indones.*

[13] Purbawinata M A 1990 *Petrology and Geochemistry of the Guntur-Gandapura Volcanic Complex, West Java, Indonesia* (Dunedin, Otago, New Zealand: University of Otago)

[14] Ubangoh R U, Pacca I G, Nyobe J l, Hell J and Ateba B 2005 Petromagnetic characteristics of Cameroon Line volcanic rocks *J. Volcanol. Geotherm. Res.* **142** 225–41

Tidal inundation ("Rob") investigation using time series of high resolution satellite image data and from institu measurements along northern coast of Java (Pantura)

Heri Andreas, Usriyah, Hasanuddin Zainal Abidin, Dina Anggreni Sarsito

Geodesy Research Division Faculty of Earth Science and Technology, Institute of Technology Bandung, LABTEX IXC Jl. Ganesha 10 Bandung 40132 - Indonesia
Email: heri@gd.itb.ac.id; Heriandreas49@gmail.com

Abstract. Tidal inundation (in Javanese they call it "Rob") is now becoming a well known phenomenon along northern coast of Java Indonesia (Pantura). The occurrence of tidal inundation was recognized at least in the early 2000 and even earlier. In the recent years the tidal inundation comes not only at a high tide but even at the regular tide in some area across Pantura. In fact in location such as Pondok Bali, north of Blanakan, north of Pekalongan, north of Semarang and north west of Demak, seems those areas are sinking to the sea through times. Sea level rise and land subsidence are considered as main factors deriving the occurrence of this tidal inundation. We were using time series of high resolution satellite image data and insitu data measurements to mapping the tidal inundation along northern coast of Java. All available data from google data satellite archives (year 2000- recent years) and any available sources being analyze together with field surveys tagging and also from media information. As a result we can see the tidal inundation are taking place in Tanggerang, Jakarta, Bekasi, Cilamaya, Pondok Bali, Blanakan, Indramayu, Cirebon, Brebes, Tegal, Pemalang, Pekalongan, Kendal, Semarang, Demak, Gresik, Surabaya, Sidoarjo and Pasuruan.
Keywords: tidal inundation; time series satellite image data; insitu measurements

1. Introduction

Pantura is a local common name for northern coast of Java Indonesia. It started from Merak in northern part of west of Java province crossed to Banyuwangi in east Java Province. Pantura existed along with the Java Sea. Despite of any features can be found along Pantura, one of them is Picture in figure 1 showing an example where tidal inundation occurs in many places there in the regular times. Tidal inundation (in Javanese they call it "Rob") is now becoming a well known phenomenon along Pantura. The occurrence of tidal inundation was recognized at least in the early 2000 and even earlier. In the recent years the tidal inundation comes not only at a high tide but even at the regular tide in some area along Pantura. In fact in location such as Pondok Bali, north of Blanakan, north of Pekalongan, north of Semarang and north west of Demak, seems those areas are sinking to the sea through times.

Investigating this phenomenon is considered necessary because what is happening in the areas indeed formed a disaster. People in the village of Pondok Bali or Worosari Demak are example of victims that already evacuated since their homes now are permanently inundated by sea water. Soon many would follows. The investigation should include finding the causes. Sea level rise, land subsidence and abrasion are some factors to be considered as the causes [1, 2]. This paper will use time series of high resolution satellite image data, insitu and others data for investigation.

Figure 1. Tidal Inundation located at Pondok Bali Pamanukan Pantura West Java part.

2. Data and methods

We were using time series of high resolution satellite image from all available google data archives (year 2000- recent years) together with any others similar available sources. In this case we identify all of coast and nearby areas along Pantura. We analized the newest until the oldest satellite image available data one by one. The changing of coast line and the colors or tones of inundation areas can be identified from all these data. Bellow we give two example from Pondok Bali and Demak.

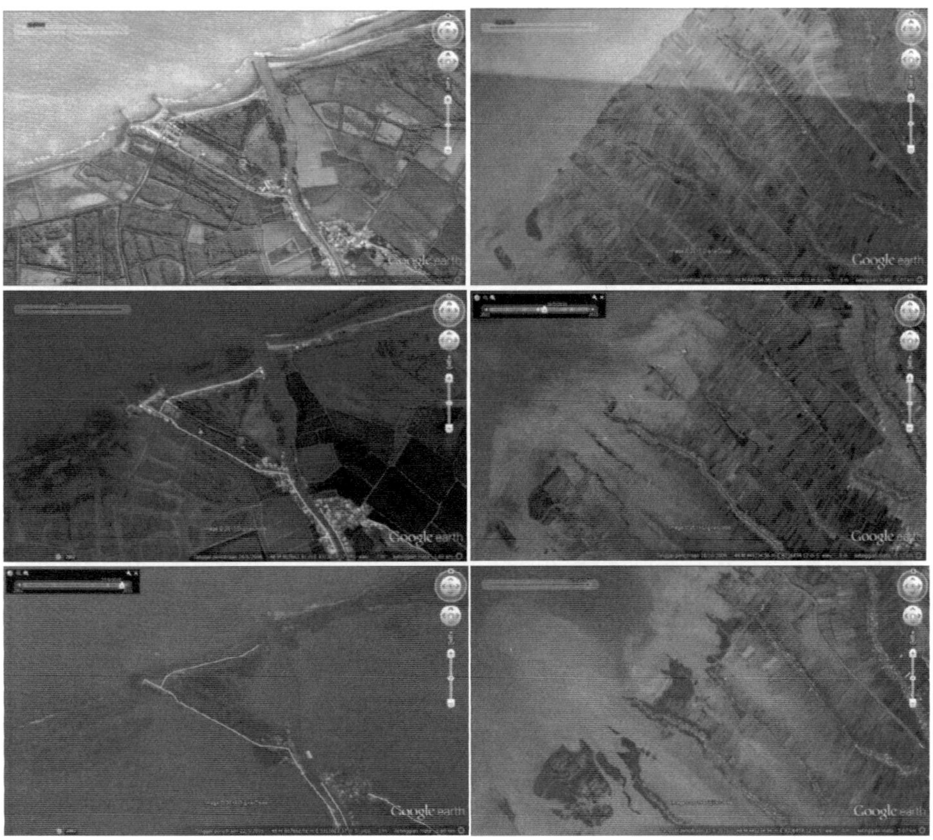

(a) (b)

Figure 2. (a) Series in rows of satellite image data of Pondok Bali Blanakan area, (b) Series in rows of satellite image data of northen Demak area. First picture belong to data acquisition in 2002, second belong to 2006 and the third one belong to 2015. We can see the changing of coast line and the colors of inundation (grey water colors and tones).

From the images, we delineate the old predicted coast line with new probable coast line (far inland of tidal inundation), and then we made polygon area to represent the area that at least being inundated (see figure 3 for examples). To delineate the exact line we have limitations on the image colors and tones interpretation since in the dense housing area or images timing being taken in the low tide, etc. in this case tones or a color of inundation was unclear. In order to make sure the true existence of tidal inundation area, in several places we did insitu measurements. Place that we visited included Tanggerang, Jakarta, Bekasi, Pondok Bali, Pekalongan, Semarang, and Demak. We found the fact that the area being inundated especially the temporary one is much larger compared to what we delineated from image analysis.

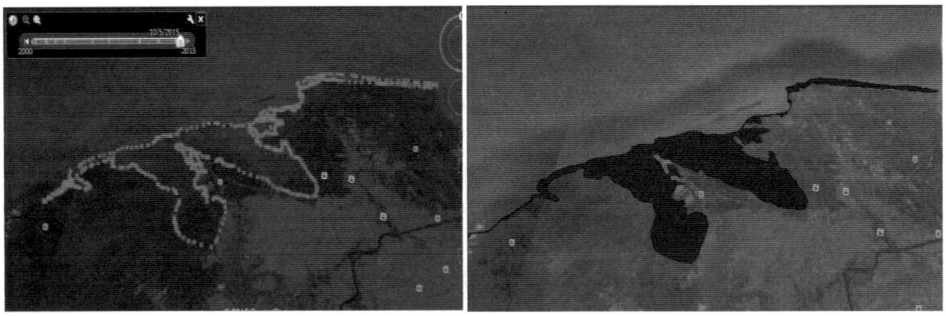

(a) **(b)**

Figure 3. (a) Example delineates the old predicted coast line with far inland of tidal inundation represent new coast line or furthest inundation, (b) example of polygon area represented tidal inundation area taken from delineation the old coast line with far inland or furthest inundation.

Since for some reason we cannot visited all areas of Pantura that expected suffering tidal inundation, we did internet surfed to collect as many information including photo documentations that were reporting tidal inundation around Pantura. Once we found the place we determined the position (coordinates) by using POI (Point Of Interest) of Google map. We noted at least 15 website and we collected more than thirty photo documents. Figure 4 to 5 are example of tidal inundation documentations from internet, each in Salembaran Tanggerang, Muara Bekasi, Tegal and Demak. Here are listed of collected internet site; republika.co.id [7, 8], antarafoto.com [9], regional.kompas.com [10], poskotanews.com [11], menlh.go.id [12], panturanews.com [13], kompasiana.com [14], suaramerdeka.com [15], nasional.news.viva.co.id [16], kabar24.bisnis.com [17], suarasurabaya.net [18], nasional.tempo.co [19], wartabromo.com [20], portalkbr.com [21].

Time series of high resolution satellite image from all available google data archives together with any others similar available source have been mixed up with insitu measurements and internet surfed to create comprehensive spatial information on tidal inundation along Pantura. Each data can be complementary or cross checked each others. Result will be discuss at chapter result and discussion.

For investigation the causes, we processed some data and also refered to some publications e.g. Sinking cities in Indonesia: ALOS PALSAR detects rapid subsidence due to groundwater and gas extraction by Chaussard et al [5], Land subsidence in Jakarta and Semarang by Abidin et al [1-4]. From these information's we can see well land subsidence and sea level rise along Northern Coast of Java or Pantura. For the next research we begin to install new GPS (Global Positioning System) network along Pantura.

Figure 4. Tidal inundation located at Selembaran Tanggerang West Java part of Pantura (left), tidal inundation located at Muara Bekasi West Java part of Pantura (right) [9, 18].

Figure 5. Tidal Inundation located at Northern Tegal Center Java part of Pantura (left), tidal inundation located at Demak Center Java part of Pantura (right) [13, 17].

3. Result and Discussion

From all high resolution satellite image data that we have been collected, processed, and analized, we can see the tidal inundation are taking place in Tanggerang, Jakarta, Bekasi, Pondok Bali, Blanakan, Indramayu, Cirebon, Brebes, Tegal, Pemalang, Pekalongan, Batang, Kendal, Semarang, Demak, Gresik, Surabaya, and Sidoarjo. Figure 6 shows map of tidal inundation in most locations along Pantura as result of above works. We calculate the area being inundated from the polygon. In Tanggerang at least 94.6 hectares are suffering inundation. In Jakarta at least 7.2 hectares are suffering inundation. In Bekasi at least 279.0 hectares are suffering inundation. In Karawang at least 82.0 hectares are suffering inundation. In Subang (Pondok Bali and Blanakan) at least 943.0 hectares are suffering inundation. In Indramayu at least 261.0 hectares are suffering inundation. In Cirebon at least 99.0 hectares are suffering inundation. In Pemalang and Pekalongan at least 21.0 and 13.8 hectares are suffering inundation. In Semarang at least 1157.8 hectares are suffering inundation. In Demak at least 3221.0 hectares are suffering inundation. Table 1 shows complete summary location and each total area being inundate along Pantura. We can see that Demak area is suffering the largest tidal inundation followed by Semarang.

As mentioned earlier, to delineate the exact line from the satellite images, we would have limitation on the image colours and tones interpretation since in the dense housing area or images timing being taken in the low tide, in this case tones or colours of inundation was unclear. It means that the area suffering tidal inundation along Pantura can be much larger. Amplitude analysis on the satellite images, surfing the internet or insitu measurement is a way that might answer larger area being inundated. Indeed we have tried to answer this matter by summarized all places that are informed being inundated from internet site and from insitu measurements. Figure 7 shows map of tidal inundation at most locations along Pantura as result of these works. Table 2 summarize most of the districts and the name of villages being inundated (e.g. in Jakarta, Tanggerang, Bekasi, Subang, Indramayu, Cirebon, Tegal, Pemalang, Semarang, Demak, Gresik, Surabaya, Sidoarjo, and Pasuruan).

In a total, number of village almost reach 100 and can be more in the real measurenments. Figures 8 to 10 shows comparation among satellite image analysis, insitu measurement, and internet info. It is interesting to see in some part are well match, others shows inundation taking place that none can seen from the satellite image.

Figure 6. Tidal Inundation in most locations along Pantura from all high resolution satellite images data being collected and analized.

Table 1. Summary location and each total area being inundated in Pantura from satellite image analysis.

No	District	Area calculated suffering inundation (in Hectare)
1	Tanggerang	**94.6**
2	Jakarta	**7.2**
3	Bekasi	**279.0**
4	Karawang	**82.0**
5	Subang	**943.0**
6	Indramayu	**261.0**
7	Cirebon	**99.0**
8	Brebes	**299.0**
9	Pemalang	**21.0**
10	Pekalongan	**13.8**
11	Kendal	**258.9**
13	Semarang	**1157.8**
13	Demak	**3221.0**

Figure 7. Tidal Inundation in most location along Pantura from insitu measurements and collections of internet information [7-21].

Table 2. Table Summarize most of the districts and the name of villages being inundated from insitu measurements and internet informations [7-21].

No	District	List of villages reported suffering tidal inundation
1	Tangerang and Jakarta	Dadap Muara baru, Sunter agung, Pademangan barat, Muara Angke, Cilincing
2	Bekasi	Hurip jaya, Donggala, Muara angke, Pantai mekar,
3	Subang	Desa Mayangan, Desa Legon wetan, Desa Legon kulon
4	Indramayu and Cirebon	Desa Eretan, Kelurahan Lemahwungkuk
5	Tegal	Desa Tegal sari dan Desa Muarareja, Desa Mintaragen dan Desa Panggung
6	Pemalang	Desa Pedangung, Desa Sususkan, Desa Kebojongan dan Desa Klegen, Desa Jatirejo, Desa Sidokare, Desa Ujung gede dan Desa Losari, Desa Kendal doyong, Desa Kendal rejo, Desa Tebuireng, Desa Widodaren, Desa Pasucen, Desa Panjunan, Desa Sirangkang, Desa Limbangan mojo, Desa Tasik rejo dan Desa Pesantren, Desa Pendowo dan Desa Kelang depok
7	Semarang and Demak	Marina, Tanjung Mas, Kaligawe, Terboyo, Desa Sriwulan, Desa Bendono, Desa Purwasari, Desa Sidogemah, Desa Gemulak, Desa Tugu, Desa Timbulsloko, Desa Suradadi, Desa Sidorejo, dan Desa Banjarsari
8	Gresik and Surabaya	Kelurahan Lumpur, Desa Kalianak, Desa Margomulyo, dan Desa Greges
9	Sidoarjo and Pasuruan	Desa Weru, Desa Kupang, Desa Sedati, Desa Buduran, Desa Sidoarjo, Desa Jabon, Desa Candi, dan Kelurahan Ngemplakrejo

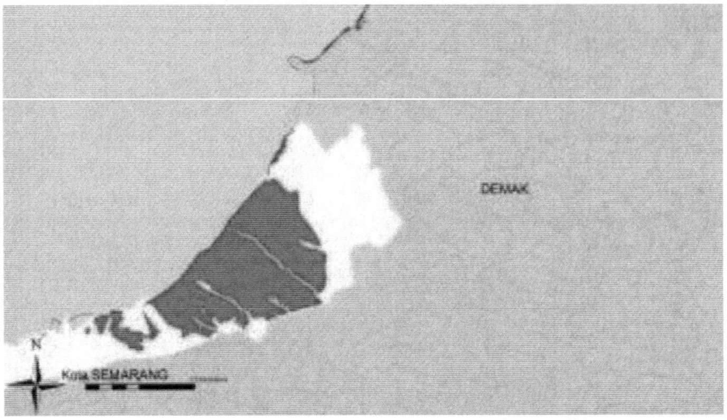

Figure 8. Tidal Inundation location that quite match among satellite image analysis, insitu measurement, and internet informations [14, 17].

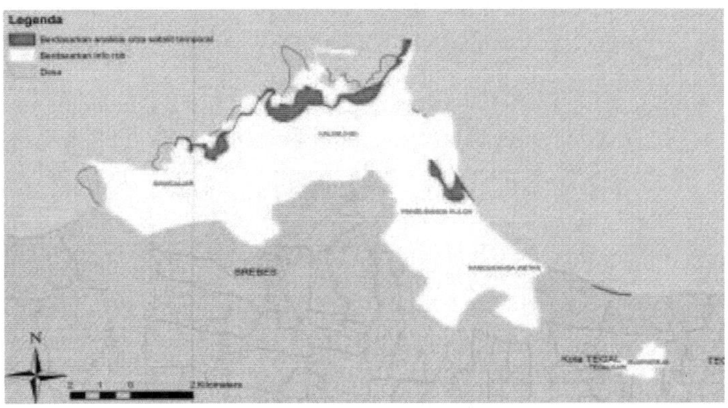

Figure 9. Tidal Inundation location informing larger area base on internet informations [13] compared with satellite image analysis.

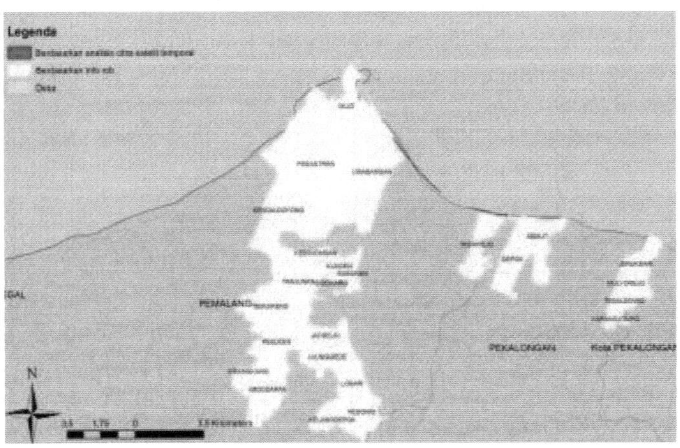

Figure 10. Tidal Inundation location informed only from internet informations [15] and none can seen from satellite image analysis.

Indeed the area suffering tidal inundation along Pantura is something that we cannot neglect. It is already a disaster and maybe sometime in the future it will be worsening and becoming very serious disaster if we do nothing. No doubt we should take an action and started with understands the causes. As mentioned earlier we suspect sea level rise, land subsidence, and abrasion are factors that causes the tidal inundation.

InSAR and GPS results have shown many places along Pantura (e.g. Jakarta, Pondok Bali, Pemalang, Semarang, and Demak) is experiencing land subsidence with yearly rate varying between 1-20 centimeters/years. Figure 11 shows map of land subsidence along Pantura. Figure 12 shows linear trend graphic of land subsidence in Jakarta and Semarang (sample place).

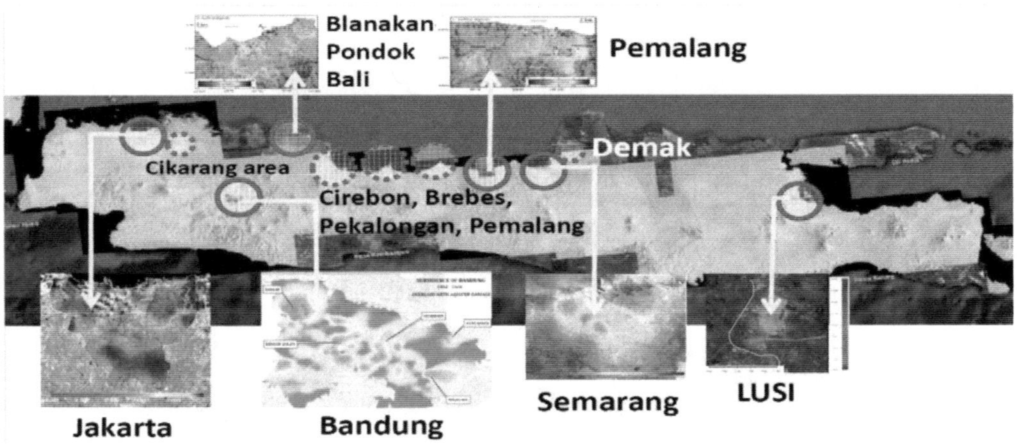

Figure 11. Map of subsidence along Pantura (modified from Chaussard et al [5], Abidin et al [1-4].

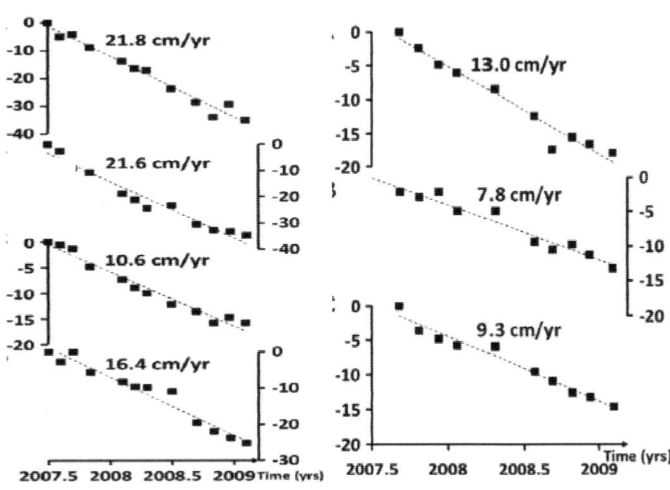

(a) (b)

Figure 12. (a) examples of linear trend graphic of land subsidence in Jakarta (b) examples of linear trend graphic of land subsidence in Semarang (modified from Chaussard et al [5].

Sea level rise along Pantura base on satellite altimetry measurements shows value between 0.1 to 0.6 centimeters/years [6]. This value is less significant compare to subsidence. In this paper we do not yet investigate the abrasion. Nevertheless the tidal inundation considered to be influenced mostly by the land subsidence.

Since the land subsidence is continuing in linier trend, in this case if the linier trend is continuing for many years in the future, then what would happened in future relating tidal inundation is indeed worrying. We make simulation in Jakarta area on how the topography would respond to land subsidence, sea level rise, and the tidal. We used LIDAR data to generate the topography. Result can be seen on figure 13. Around 26.86% of Jakarta maybe inundated by tidal in year 2025, and around 35.61% of Jakarta maybe inundated by the tidal in year 2050.

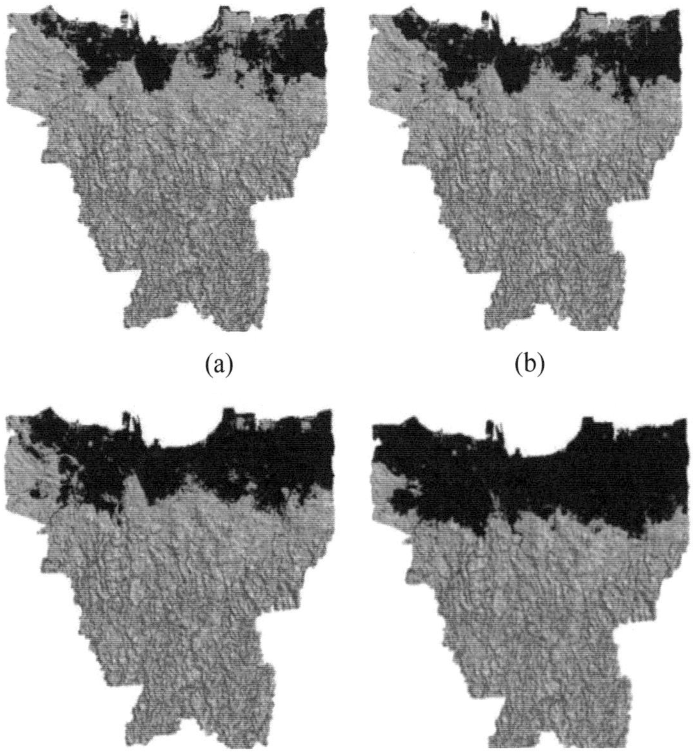

(a) (b)

(c) (d)

Figure 13. (a) Around 15.58% of Jakarta area prone to inundated by tidal in year 2007 (b) Around 18.78% of Jakarta prone to inundated by tidal in year 2012 (c) Around 26.86% of Jakarta maybe inundated by tidal in year 2025 (d) Around 35.61% of Jakarta maybe inundated by the tidal in year 2050.

We can make simulation for whole Pantura if accurate topography data is available. If the model can created, then we can see the future disaster in Pantura from land subsidence, sea level rise, tidal inundation, and any related coastal hazards. Unfortunately not so many people including the government are truly aware of these matters until now. Hopefully awarnes comes sometimes in the future. We are not expecting many others people need to be evacuate as what had happened in Pondok Bali Blanakan and Worosari Demak, and in few unnoticed places.

We already began to install new GPS network along Pantura and working on more for the next research. Until now there are around 10 GPS subsidence points in Tanggerang, 150 GPS in Jakarta, 6 points in Bekasi, 4 points in Karawang, 8 points in Pondok Bali, 7 points in Pekalongan, 40 points in Semarang, and 9 points in Demak (see figure 14).

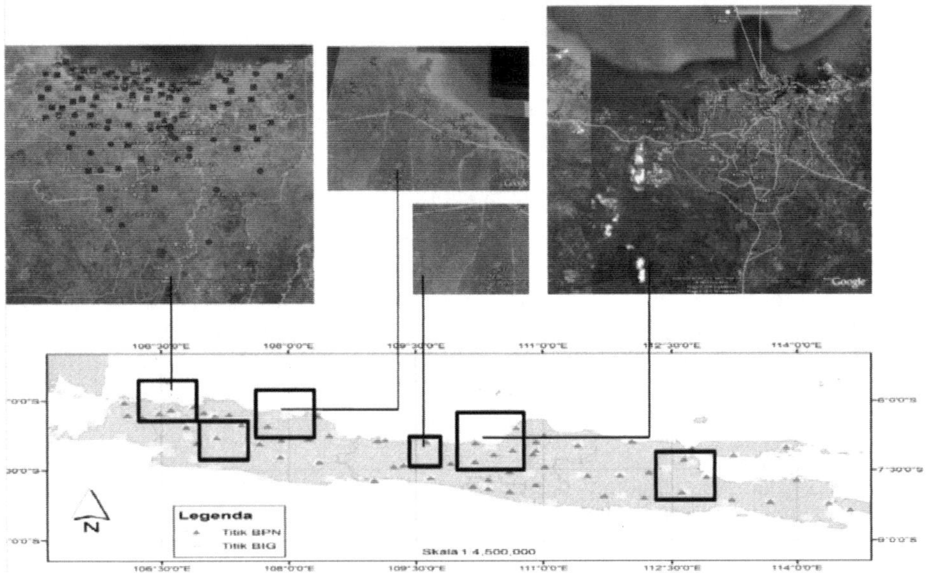

Figure 14. GPS Network along Pantura for monitoring land subsidence in these area.

Since the land subsidence is continuing in linier trend, in this case we should understand why this fact is happing and what the causes are. Our strong hypothesis said that the land subsidence is happening mostly because of effects from overexploitation of ground water, oil and gas. Load from the building, land setting/reclamation, natural compaction and tectonic are complementary or minor causes. Our next research will focus also to prove the strong correlation between land subsidence and ground water, oil and gas exploitation.

4. Conclusions

Investigating tidal inundation phenomenon is considered necessary because what is happening with the tidal inundation indeed formed a disaster in some places along Pantura. People in Pondok Bali or Worosari Demak villages, and probably in others village unnoticed are example of today victim's disaster. They already evacuated to others places since their homes now are permanently inundated by sea water. Soon many would follows we believed.

Tidal inundation considered to be influenced mostly by the land subsidence. In other hand our strong hypothesis said that the land subsidence is happening mostly because of effects from overexploitation ground water, oil and gas. From our investigation we can see the future potential disaster in Pantura from land subsidence, sea level rise, tidal inundation, and any related coastal hazards. Unfortunately not so many people including the government are truly aware of these matters until now. Hopefully awarnes comes sometimes in the future. We are not expecting many people being victims and need to be evacuate.

5. Acknowledgements

Many thanks and appreciation to the students who helped the investigation in the field and especially to the local people for kindly sharing the information and experience relating tidal inundation since indeed most of them are victims now from this ongoing disaster.

6. References

[1] Abidin H, Andreas H, Djaja R, Darmawan D and Gamal M 2008 *Land subsidence characteristics of Jakarta between 1997 and 2005, as estimated using GPS surveys* GPS Solutions, 12(1), 23–32, http://dx.doi.org/10.1007/s10291-007-0061-0

[2] Abidin H, Andreas H, Gumilar I, Sidiq T, Gamal M and Murdohardono D 2010 *Studying land subsidence in Semarang (Indonesia) using geodetic methods* FIG Congress, Facing the Challenges—Building the Capacity, Sydney, Australia
[3] Abidin H, Andreas H, Gumilar I, Fukuda Y, Pohan Y E, and Deguchi T 2011 *Land subsidence of Jakarta (Indonesia) and its relation with urban development* Natural Hazards, 59(3), 1753–1771, http://dx.doi.org/10.1007/s11069-011-9866-9
[4] Abidin H, Andreas H, Gumilar I, Sidiq T P and Fukuda Y 2012 *Land subsidence in coastal city of Semarang (Indonesia): Characteristics, impacts and causes* Geomatics, Natural Hazards and Risk, 1–15, http://dx.doi.org/10.1080/19475705.2012.692336
[5] Chaussard Estelle, Amelung F, Abidin HZ, San Hoon Hong 2013 *Sinking cities in Indonesia: ALOS PALSAR detects rapid subsidence due to groundwater and gas extraction* Remote Sensing of Environment 128 (2013) 150 -161 http://dx.doi.org/10.1016/j.rse.2012.10.015
[6] Nurmaulia S L, Fenoglio-Marc L, Becker M 2010 *Long term sea level change from satellite altimetry and tide gauges in the indonesian region* paper presented at the EGU General Assembly 2010, 2–7 May. Vienna, Austria
[7] Republika.co.id/berita/breakingnews/metropolitan/10/10/15/ Accessed on July 2, 2016
[8] Republika.co.id/berita/nasional/jabodetabeknasional/13/06/14/ Accessed on July 2, 2016
[9] Antarafoto.com/peristiwa/v1448263201 Accessed on July 7, 2016
[10] Regional.kompas.com/read/2011/03/28/13180440/ Accessed on July 7, 2016
[11] Poskotanews.com/2012/06/19 Accessed on July 7, 2016
[12] Menlh.go.id Accessed on July 15, 2016
[13] Panturanews.com/index.php/panturanews/baca/8304/07/05/19/ Accessed on July 15, 2016
[14] Kompasiana.com/ abdulah-sungkar /genangan- rob Accessed on July 15, 2016
[15] Suaramerdeka.com/v1/index.php/read/ news_pantura/ 2013/05/31 Accessed on July 15, 2016
[16] Nasional.news.viva.co.id/news/read/21633 Accessed on August 3, 2016
[17] Kabar24.bisnis.com/read/20140814/78/250020/ Accessed on August 3, 2016
[18] Suarasurabaya.net/fokus/115/2013/119688 Accessed on August 3, 2016
[19] Nasional.tempo.co/read/news/2011/03/13/18031968 Accessed on August 3, 2016
[20] Wartabromo.com/2014/07/13/ Accessed on August 10, 2016
[21] Portalkbr.com/hermawan_arifianto/05-2013/ Accessed on August 10, 2016

Morphometric analysis of relative tectonic activity in the Baturagung Mountain, Central Java, Indonesia

Rahmi Mulyasari[1], Budi Brahmantyo[1] and Supartoyo[2]

[1] Faculty of Earth Science and Technology, Bandung Technology Institute (ITB), Bandung, Indonesia
[2] Center for Vulcanology and Geological Hazard Mitigation, Geological Agency, Bandung, Indonesia
Email: rahmimulyasari@gmail.com

Abstract. Special Region of Yogyakarta and Klaten district, Central Java is one of areas in Indonesia that is prone to earthquake caused by subduction in Indian Ocean and active fault in land. The earthquake sources from active fault probable from Opak and other faults located in Baturagung Mountain. Active faults controlling landform development in tectonically active regions, and it has significantly affected fluvial systems and mountain – front landscapes in the Baturagung Mountain. To assess tectonic activities in the area used quantitative analysis (morphometric). Morphometric analysis consists of 5 parameters geomorphic indices: drainage basin asymmetry (AF), hypsometric curve and integral (Hc and Hi), stream length gradient (SL) index, basin shape index (Bs), and mountain-front sinuosity (Smf). These indices were combined to yield the relative tectonic activity index (RTAI) using geographic information systems (GIS). The result found that RTAI in the study area are divided into three classes: Class 2 (high 0.6% of the watershed area (1.32 km^2)); Class 3 (moderate 58.9% (122.1 km^2)); and Class 4 (low 40.4% (83.75 km^2)). All of morphometric analysis generally indicates this area more influenced by tectonics than erosion. The results are consistent with geomorphological observations.
Keywords: Baturagung Mountain; geomorphic indices; morphometric; active fault.

1. Introduction

Tectonic geomorphology is defined as the study of landforms produced by tectonic processes, or the application of geomorphic principles to the solution of tectonic problems [1]. The quantitative measurement of landscape is based on the calculation of geomorphic indices using topographic maps, aerial photographs and field work. The results of several indices can be combined in order to highlight tectonic activity and to provide an assessment of a relative degree of tectonic activity in an area [1].

Baturagung Mountain is located in Central Java area. This area is one of areas in Indonesia vulnerable to earthquake from subduction zone in Indian Ocean and active fault in land. One of the destructive earthquake sources is from active fault. Destructive earthquake events occurred in this area caused by active fault on 1867, 2006 and 2010 [2].

Opak fault in SE – NW direction inferred as a source of earthquake in this area. In addition there are other faults located in mountain – front zone North of Baturagung which is expected to contribute to the southern Klaten damage [3]. In general, faults that caused earthquake is located in Baturagung Mountain. Therefore, the analysis of this area was related to Baturagung Mountain.

The purpose of this study was to assess tectonic activities of Baturagung Mountain. The activities of Baturagung Mountain would be analyzed using quantitative analysis (morphometric).

2. Data and Method

Data used in this study are Shuttle Radar Topography Mission digital elevation model (SRTM DEM) data with spatial resolution of 30 m which is available freely from the USGS website [4], geological map (Yogyakarta and Surakarta – Giritontro sheets) data from geological research and development centre [5,6], and topographical map data from geospatial information agency [7]. All of the data were used to calculated geomorphic indices in morphometric analysis.

Morphometry is defined as quantitative measurement of landscape shape. At the simplest level, landforms can be characterized in terms of their size, elevation (maximum, minimum or average), and slope. Quantitative measurements allow geomorphologists to objectively compare different landforms and to calculate less straightforward parameters that may be useful for identifying a particular characteristic of an area such as level of tectonic activity [1].

This study applied morphometric analysis in Baturagung Mountain to evaluate relative tectonic activity rates. Considering the diversity of the morphotectonic features [1], we analyzed five geomorphic indices: drainage basin asymmetry (AF), hypsometric curve and integral (Hc and Hi), stream length gradient index (SL), basin shape index (Bs), and mountain – front sinuosity (Smf). All of parameter values computed in a single index as relative tectonic activity index (RTAI) [8] to characterize relative tectonic activity. Morphometric analysis used spatial tools geographic information systems (GIS) arcGIS 10 software and Microsoft Excel. This kind of methodology has been used to analyze various active areas such as Sarvestan area, Iran [8], Cimandiri Fault, Indonesia [9] and Lembang Fault, Indonesia [10]. The results from morphometric analyses were also validated in field based on geomorphological observations.

3. Result and Discussion

3.1. Morphometric analyses

Morphometric analyses were used to identify relative rates of active tectonics in study area. Five geomorphic indices parameters were analysed, including drainage basin asymmetry (AF), hypsometric integral (Hi), stream length gradient index (SL), basin shape index (Bs), and mountain-front sinuosity (Smf) [1,8]. Result of geomorphic indices values were classified into three classes tectonic activity: Class 1 (more active), Class 2 (active) and Class 3 (less active) (Table 1) [8]. Most of these indices are obtained for river basins. Two major rivers flow in study area (Dengkeng River and Oyo River) were subdivided into 65 subbasins.

Table 1. Relative tectonic activity index (RTAI) based on morphometric parameters [8].

No.	Morphometric Parameters	Relative tectonic activity index		
		Class 1	Class 2	Class 3
1	SL	$SL \geq 500$	$300 \leq SL < 500$	$SL < 300$
2	Hi	$Hi \geq 0.5$	$0.4 \leq Hi < 0.5$	$Hi < 0.4$
3	Bs	$Bs \geq 4$	$3 \leq Bs < 4$	$Bs \leq 3$
4	AF	$AF \geq 65$ or $AF < 35$	$35 \leq AF < 43$ or $57 \leq AF < 65$	$43 \leq AF < 57$
5	Smf	$Smf < 1.1$	$1.1 \leq Smf < 1.5$	$Smf \geq 1.5$

3.1.1. Drainage basin asymmetry (AF)

The drainage basin asymmetry (AF) can be used to evaluate tectonic tilting at the scale of a drainage basin [1]. AF defined as (1):

$$AF = 100\ (Ar/At) \quad (1)$$

Ar is the area of a part of a watershed on the right of the master stream and At is the total area of the watershed. AF equal about 50 indicate tectonic relatively stable, AF greater or less than 50 may suggest tilt [1]. In the study area AF varies from 23.97 (subbasin 60) to 86.13 (subbasin 9). AF values were grouped into three classes: 1 (AF≥65 or AF<35), 2 (35≤AF<43 or 57≤AF<65), and 3 (43≤AF<57) [8] (Figure 1; Table 2). Class 1 and 2 of AF dominantly located at faults zone (Figure 1).

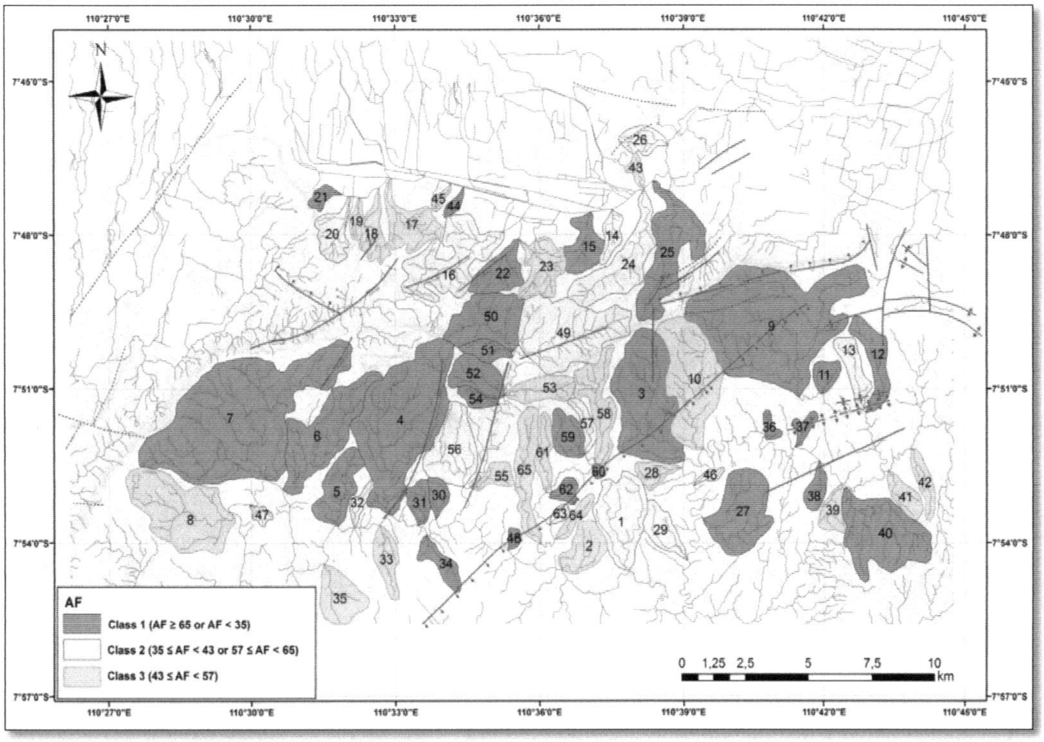

Figure 1. Map of AF analyses. Colour red, yellow, and green area indicate class 1, class 2, and class 3.

3.1.2. Hypsometric curve and integral (Hc and Hi)
The hypsometric integral (Hi) describes the relative distributions of elevations in a given area of a landscape particularly a drainage basin. The index is defined as the relative area below the hypsometric curve and thus expresses the volume of a basin that has not been eroded. A simple equation to approximately calculate the index is (2):

$$Hi = (\text{average.elev.} - \text{min.elev.}) / (\text{max.elev} - \text{min.elev}) \qquad (2)$$

Hi can be used as an indicator of a landscape stage in the cycle of erosion and divided into 3 stages: young, mature, and old. In the study area Hi varies from 0.16 (subbasin 16) to 0.8 (subbasin 35). Hi values were grouped into three classes with respect to the convexity or concavity of the hypsometric curve (Hc): Class 1 with convex hypsometric curves (Hi≥0.5), indicates young stage; Class 2 with concave-convex hypsometric curves (0.4≤Hi<0.5), indicate mature stage; and Class 3 with concave hypsometric curves (Hi<0.4), indicate old stage [1] (Figure 2; Figure 3; Table 2). Calculation of Hi indicated area of study influenced by erosion, tectonic and rock resistance.

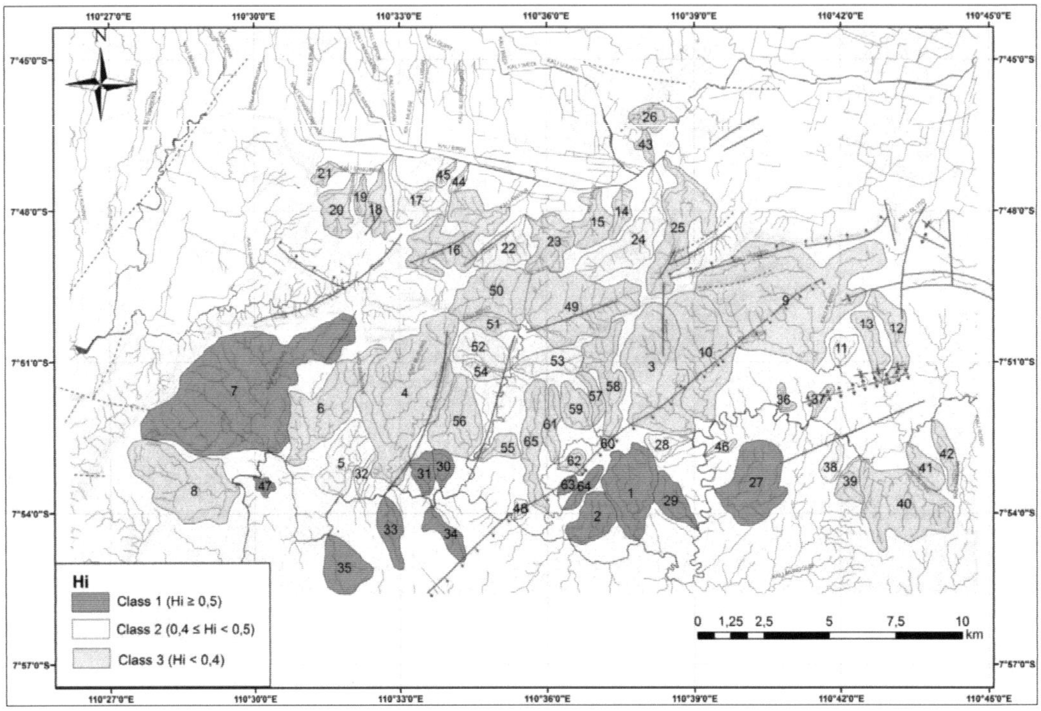

Figure 2. Map of Hi results. Colour red, yellow, and green area indicate young, mature, and old stages.

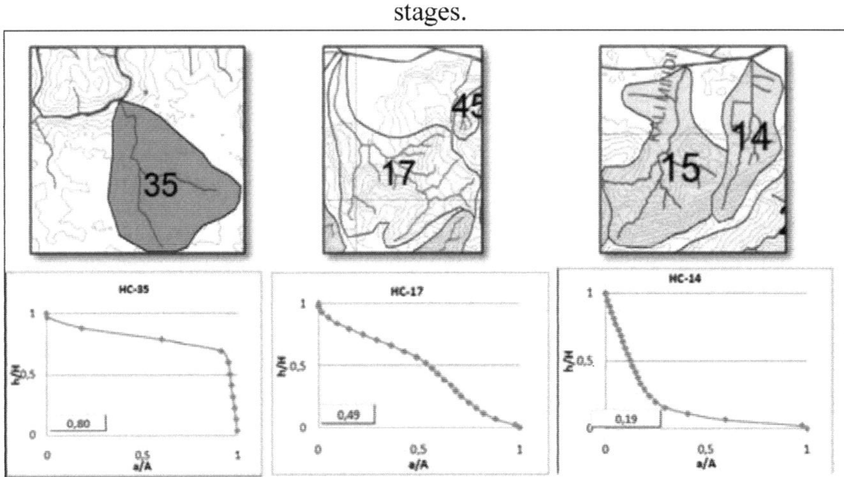

Figure 3. Hypsometric curves of three subbasins.

3.1.3. Stream length gradient index (SL)

SL correlates to stream power. SL index is sensitively to change in channel slope by tectonic activity, rock resistance, and topography. SL index defined as (3):

$$SL = (\Delta H/\Delta L)\ L \tag{3}$$

$\Delta H/\Delta L$ is the channel slope and L is the total channel length from the watershed divide to midpoint of the reach. The SL index can be used to evaluate relative tectonic activity. Although an area on soft rocks with high SL values indicates recent tectonic activity, anomalously low values of SL may also represent such activity when rivers and stream flow through strike – slip fault. High SL value indicates that valley profile steeply, deep incision, and possible as a fault zone and represent tectonic activity [1]. Lithology in study area is related to rock resistance which showed in geological map (Figure 4).

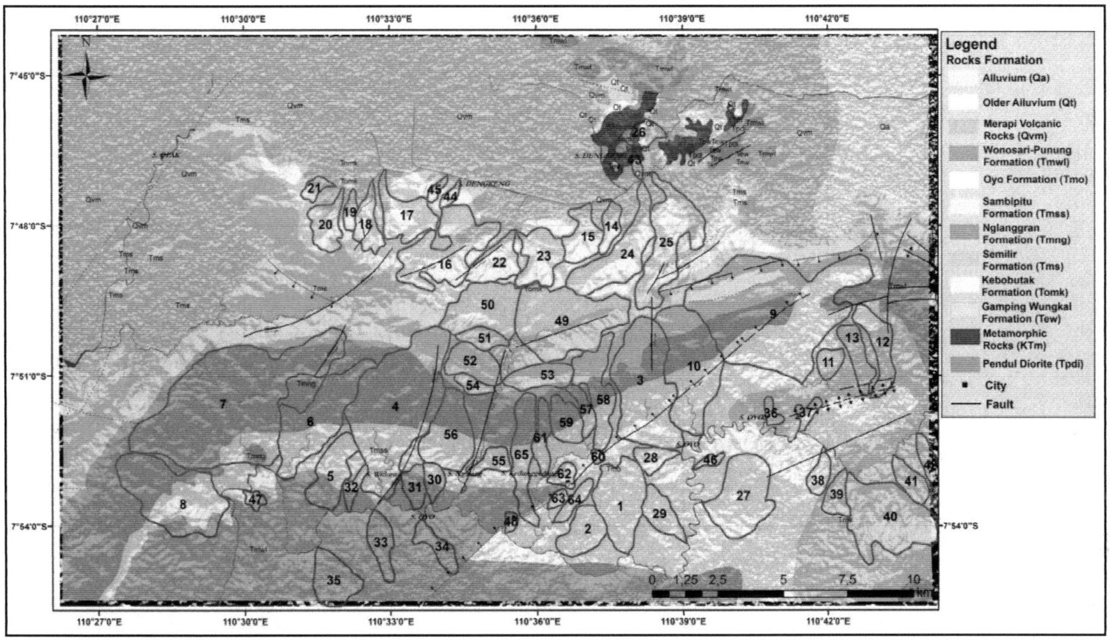

Figure 4. Geological map of study area.

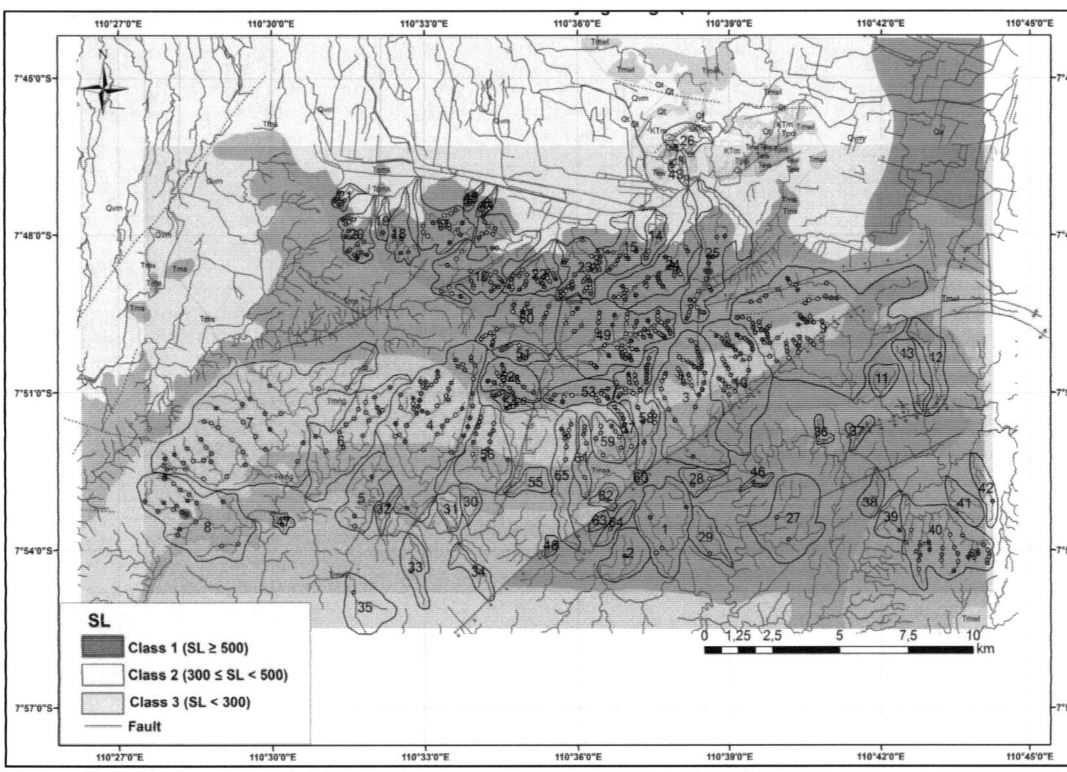

Figure 5. SL index along the drainage network. Contour area red, yellow and green area indicate class 1, class 2, and class 3.

SL is calculated along rivers using a digital elevation model (extracted from a digitized 1:25000 topographic map) and GIS (Figure 5 and 6) to compute its value for each subbasin. The value ranges from 16.70 to 995.85. The values were classified into three categories: Class 1 (SL≥500), Class 2 (300≤SL<500) and Class 3 (SL<300) [1]. The result of the classification is shown in Table 2. Calculation of SL at subbasins show high additional value with steep topography that flowed by the stream. The high changing of SL value occurred because of structural zone not because of lithological changing.

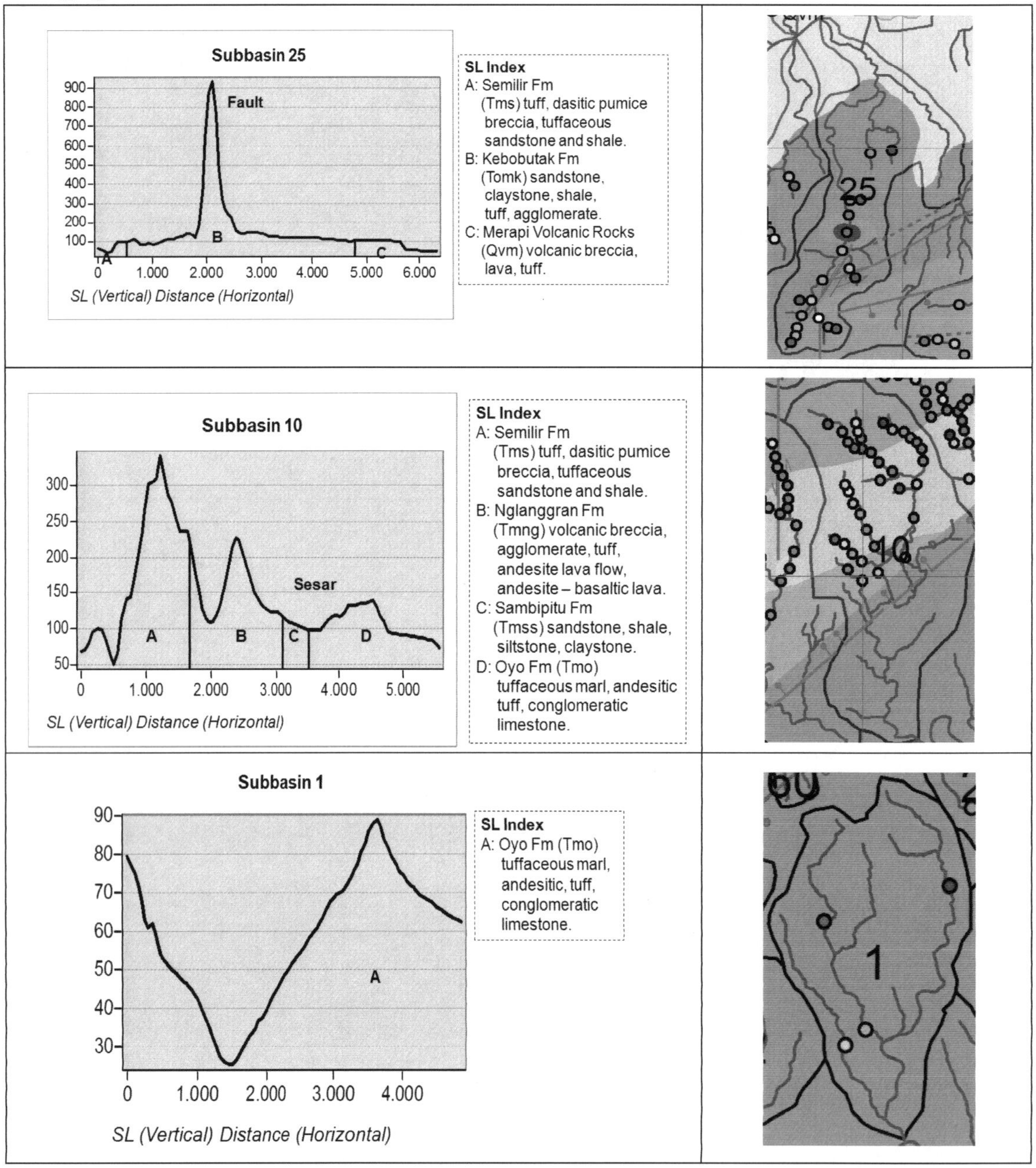

Figure 6. Longitudinal river profiles and measured SL values for three subbasins in the study area.

3.1.4. Basin shape index (Bs)

The horizontal projection of a basin may be described by the basin shape index or the elongation ratio, Bs defined as (4):

$$Bs = B_l / B_w \quad (4)$$

B_l is the length of a basin measured from the highest point and B_w is the widest point. Relatively young drainage basins in tectonically active areas tend to be elongated in shape, normal to the topographic slope of mountain [1].

Bs was computed using DEM and classified into three classes: 1 (Bs≥4), 2 (3≤Bs<4) and 3 (Bs≤3) [4]. Bs ranges from 0.44 (subbasin 36) to 6.31 (subbasin 65). The result of the classification is shown in (Figure 7; Table 2). Generally, based on Bs analysis, study area dominated by class 3, presence of class 1 and class 2 were influenced by tectonic activity at subbasin.

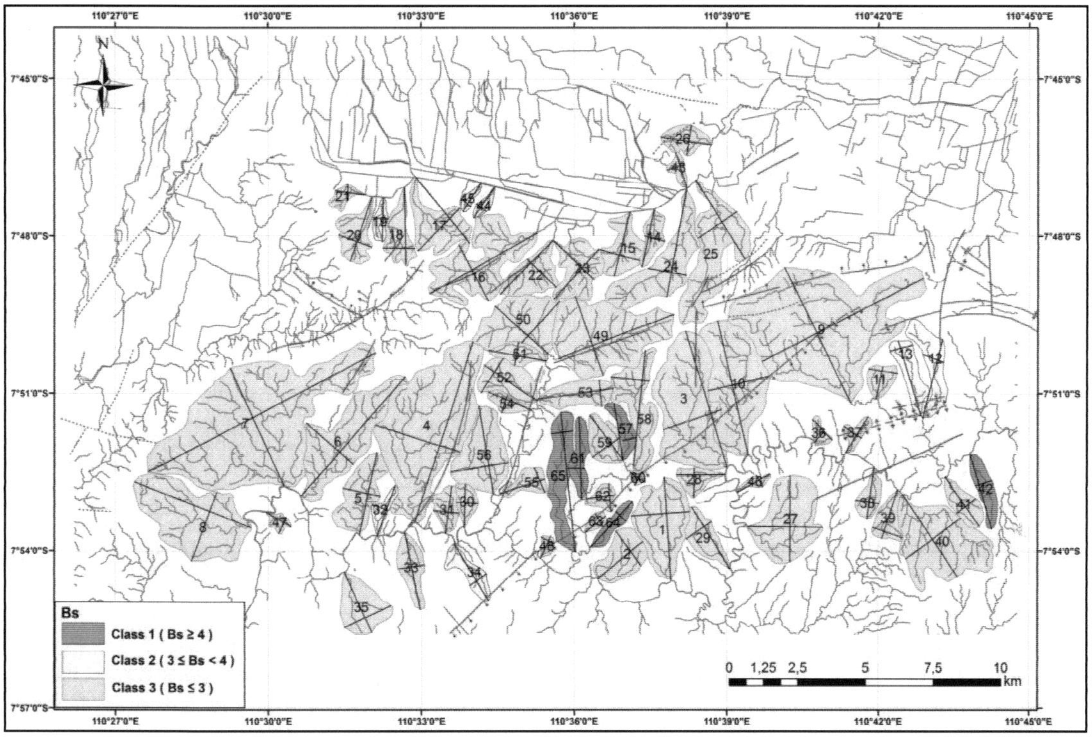

Figure 7. Basin shape index. Colour red, yellow, and green area indicate class 1, class 2, and class 3.

3.1.5. Mountain – front sinuosity (Smf)

Mountain-front sinuosity is defined as a topographic transition zone between mountains and plain [11]. Smf is define as (5):

$$Smf = Lmf/Ls \quad (5)$$

Lmf is the length of the mountain front along the foot of the mountain, at the pronounced break in slope; and Ls is the straight line length of the mountain-front. Smf index reflects the balance between erosional forces that tend to cut embayment into a mountain-front and tectonic forces that tend to produce straight mountain front coincident with an active range bounding range [4]. The value of Smf can be detected by tectonic activity. The low value of Smf is associated with active tectonics and uplift. High value of Smf associated with reduces of uplift and erosional processes that will carve a more irregular mountain-front [8].

The values of Smf was calculated for 17 mountain front (Figure 8; Table 2) using Smf and Ls values measured from SRTM image, and divided into three classes: 1 (Smf<1.1), 2 (1,1≤Smf<1.5), and 3 (Smf≥1.5). Smf values ranges from 0.99 to 4.12. Generally, class 1 and class 2 of Smf occurred with structural zone in study area.

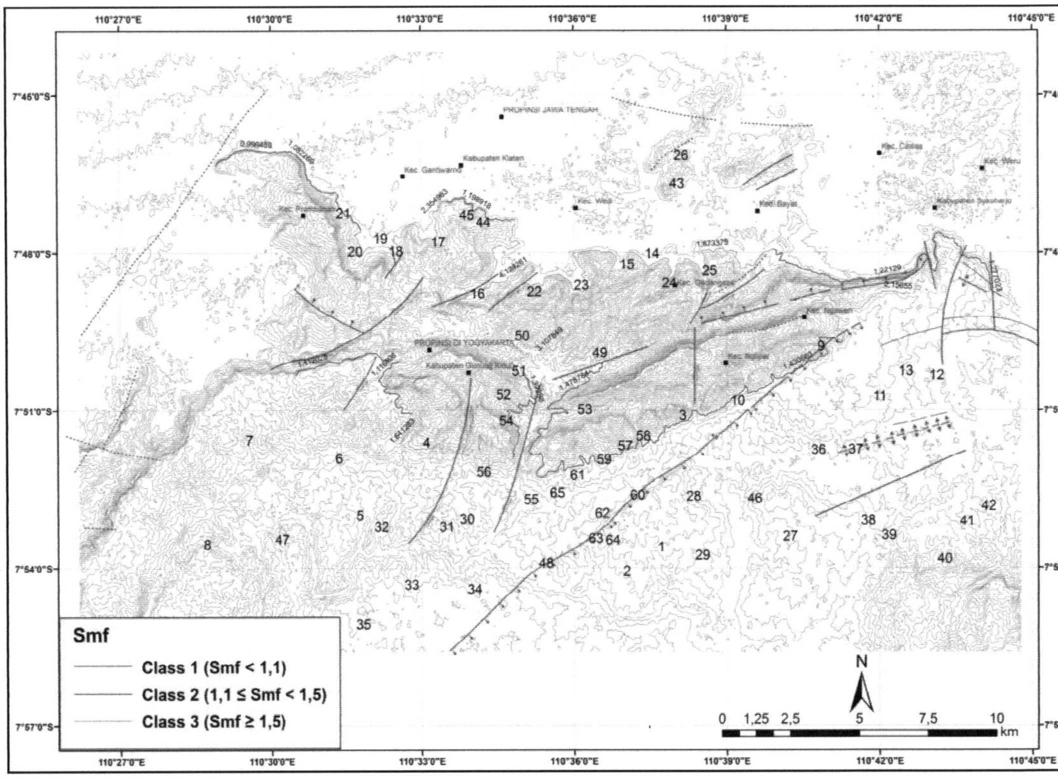

Figure 8. Smf index. Colour red, blue, and green indicate class 1, class 2, and class 3.

3.2. Spatial distribution of index values

Five morphometric parameters and lineament have been calculated to analyze the activity of Baturagung Mountain. The result found that AF is far from value 50 indicates the presence of tectonic tilting, hypsometric curve shows that this area belongs to young-mature topography, SL more than 300 indicating of active tectonic, Bs ≥ 4 indicating elongated shape of basin related to active tectonic, and Smf value is less than 1.5 indicating straight of mountain-front sinuosity as a lineament fault zone.

This study tried to evaluate tectonic activity in Baturagung area by using some of geomorphological parameters. The average of the five parameter measured geomorphic indices (RTAI) was used to evaluate the distribution of relative tectonic activity in the study area [8]. The values of the index were divided into four classes to define the degree of active tectonics: 1 – very high (1.0≤RTAI<1.5); 2 – high (1.5≤RTAI<2.0); 3 – moderate (2.0≤RTAI<2.5); and 4 – low (2.5≤RTAI) [8].

The distribution of the four classes is shown in Figure 9 and Table 2 shows the result of the classification for each subbasins. About 0.6% of the basin area (about 1.32 km^2) belongs to Class 2; 58.9% (122.1 km^2) to Class 3; and 40.4% (83.75 km^2) to Class 4.

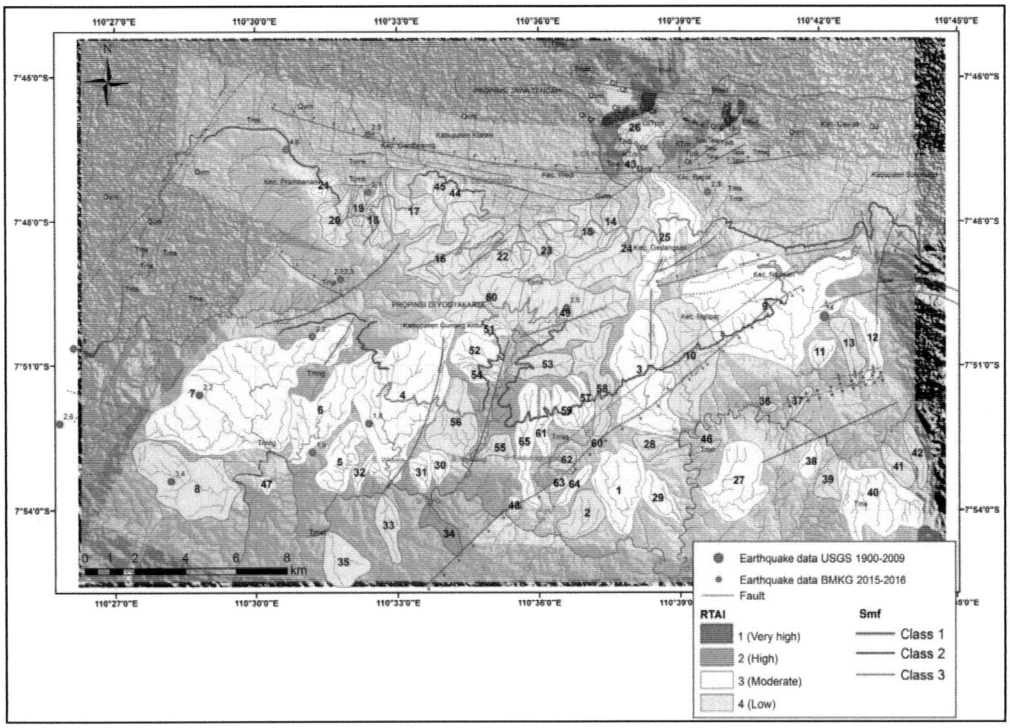

Figure 9. Relative tectonic activity index (RTAI) map. Colour pink, yellow, and green area indicate class 2, class 3, and class 4.

4. Conclusions

The result found that RTAI in the study area are divided into three classes: Class 2 (high 0.6% of the watershed area (1.32 km^2)); Class 3 (moderate 58.9% (122.1 km^2)); and Class 4 (low 40.4% (83.75 km^2)). All of morphometric analysis generally indicates this area is more influenced by tectonics than erosion. The results are consistent with geomorphological observations (Figure 10). Class 3 is with more steep hills than Class 4.

Figure 10. Geomorphological observations; Class 4 (A) and Class 3 (B).

Table 2. Values and classes of RTAI.

Subbasin no.	At (km^2)	Morphometric Parameters					Sum	RTAI	RTAI Class	
		Hi	AF	SL	Bs	Smf				
1	5.0	1	2	3	3		9	2.3	3	Oyo Fm
2	2.4	1	3	3	3		10	2.5	4	Oyo Fm
3	9.0	3	1	1	3	2	10	2.0	3	Oyo Fm, Sambipitu Fm, Ngalanggran, Semilir Fm
4	14.1	3	1	2	3	3	12	2.4	3	Sambipitu Fm, Ngalanggran, Semilir Fm
5	3.0	2	1	3	3		9	2.3	3	Wonosari-Punung Fm, Sambipitu Fm
6	6.6	3	1	2	3	2	11	2.2	3	Sambipitu Fm, Ngalanggran Fm
7	22.1	2	1	2	3		8	2.0	3	Sambipitu Fm, Ngalanggran Fm
8	8.3	3	3	1	3		10	2.5	4	Sambipitu Fm, Ngalanggran Fm, Wonosari-Punung Fm
9	20.3	3	1	1	3	2	10	2.0	3	Oyo Fm, Ngalanggran, Semilir Fm
10	78	3	3	2	3	2	13	2.6	4	Oyo Fm, Sambipitu Fm, Ngalanggran, Semilir Fm
11	1.0	2	1	3	3		9	2.3	3	Oyo Fm, Semilir Fm
12	2.4	3	1	3	2		9	2.3	3	Oyo Fm, Semilir Fm
13	1.7	3	2	3	2		10	2.5	4	Semilir Fm
14	1.0	3	2	3	3	3	14	2.8	4	Kebobutak Fm, Merapi Volcanic Rocks
15	2.2	3	1	3	3	3	13	2.6	4	Kebobutak Fm, Merapi Volcanic Rocks
16	6.3	3	2	3	3	3	14	2.8	4	Kebobutak Fm, Merapi Volcanic Rocks
17	2.9	2	3	3	3	3	14	2.8	4	Kebobutak Fm, Merapi Volcanic Rocks
18	1.7	3	3	3	3	3	15	3.0	4	Kebobutak Fm, Merapi Volcanic Rocks
19	0.6	3	3	3	2	3	14	2.8	4	Kebobutak Fm, Merapi Volcanic Rocks
20	1.9	3	2	3	3	3	14	2.8	4	Kebobutak Fm, Merapi Volcanic Rocks
21	0.6	3	1	3	3	2	12	2.4	3	Kebobutak Fm, Merapi Volcanic Rocks
22	2.3	3	1	3	3	3	13	2.6	4	Kebobutak Fm, Merapi Volcanic Rocks
23	2.8	3	3	3	3	3	15	3.0	4	Kebobutak Fm, Merapi Volcanic Rocks
24	4.8	2	2	3	3	3	13	2.6	4	Kebobutak Fm, Merapi Volcanic Rocks
25	5.9	3	1	1	3	3	11	2.2	3	Kebobutak Fm, Merapi Volcanic Rocks, Semilir Fm
26	1.3	3	2	3	3		11	2.8	4	Metamorphic rocks, Diorite, Merapi Volcanic Rocks, Older Alluvium
27	5.5	1	1	3	3		8	2.0	3	Oyo Fm
28	1.1	2	3	3	3		11	2.8	4	Oyo Fm
29	1.8	1	2	3	3		9	2.3	3	Oyo Fm
30	0.8	1	1	3	3		8	2.0	3	Wonosari-Punung Fm, Sambipitu Fm
31	1.1	1	1	3	3		8	2.0	3	Wonosari-Punung Fm, Sambipitu Fm
32	0.7	2	2	3	2		9	2.3	3	Wonosari-Punung Fm, Sambipitu Fm
33	1.7	1	3	3	3		10	2.5	4	Wonosari-Punung Fm
34	1.3	1	1	3	2		7	1.8	2	Wonosari-Punung Fm
35	2.7	1	3	3	3		10	2.5	4	Wonosari-Punung Fm
36	0.4	3	1	3	3		10	2.5	4	Semilir Fm
37	0.6	3	1	3	3		10	2.5	4	Semilir Fm, Oyo Fm
38	1.0	2	1	3	3		9	2.3	3	Oyo Fm
39	1.2	3	3	3	3		12	3.0	4	Semilir Fm
40	7.3	3	1	2	3		9	2.3	3	Semilir Fm
41	1.4	3	3	3	3		12	3.0	4	Semilir Fm
42	1.3	3	3	3	1		10	2.5	4	Semilir Fm
43	0.5	3	3	3	3		12	3.0	4	Metamorphic rocks, Merapi Volcanic Rocks
44	0.4	2	1	3	2	2	10	2.0	3	Kebobutak Fm, Merapi Volcanic Rocks
45	0.3	2	3	3	2	2	12	2.4	3	Kebobutak Fm, Merapi Volcanic Rocks
46	0.4	2	2	3	3		10	2.5	4	Oyo Fm
47	0.4	1	2	3	3		9	2.3	3	Wonosari-Punung Fm, Sambipitu Fm
48	0.3	2	1	3	3		9	2.3	3	Wonosari-Punung Fm, Oyo Fm
49	8.5	3	2	3	3	3	14	2.8	4	Kebobutak Fm, Semilir Fm
50	4.5	3	1	3	3	3	13	2.6	4	Kebobutak Fm, Semilir Fm
51	1.1	3	1	3	3	2	12	2.4	3	Semilir Fm
52	2.3	2	1	3	3	2	11	2.2	3	Semilir Fm
53	2.0	2	3	3	3	2	13	2.6	4	Semilir Fm
54	0.7	2	1	3	3	2	11	2.2	3	Semilir Fm
55	1.0	3	3	3	3		12	3.0	4	Sambipitu Fm
56	4.3	3	2	3	3	3	14	2.8	4	Sambipitu Fm, Ngalanggran Fm
57	0.9	3	2	3	1	2	11	2.2	3	Sambipitu Fm, Ngalanggran Fm
58	3.5	3	3	2	3	2	13	2.6	4	Sambipitu Fm, Ngalanggran Fm
59	1.7	3	1	3	3	2	12	2.4	3	Sambipitu Fm, Ngalanggran Fm
60	0.2	2	1	3	3		9	2.3	3	Sambipitu Fm, Oyo Fm
61	1.3	3	3	3	1	2	12	2.4	3	Sambipitu Fm, Ngalanggran Fm
62	0.7	3	1	3	3		10	2.5	4	Sambipitu Fm, Oyo Fm, Wonosari-Punung Fm
63	0.4	1	2	3	3		9	2.3	3	Oyo Fm, Wonosari-Punung Fm
64	0.8	1	3	3	1		8	2.0	3	Oyo Fm
65	3.1	3	3	3	1	2	12	2.4	3	Nglanggran Fm, Sambipitu Fm, Oyo Fm, Wonosari-Punung Fm

IOP Conf. Series: Earth and Environmental Science **71** (2017) 012006 doi:10.1088/1755-1315/71/1/012006

5. References

[1] Keller EA and Pinter N 2002 *Active Tectonics: Earthquakes, Uplift and Landscapes* (New Jersey: Prentice Hall) p 338

[2]Supartoyo, Surono and Putranto ET 2014 *Katalog Gempabumi Merusak di Indonesia Tahun 1612 –* 2014 (Bandung: Pusat Vulkanologi dan Mitigasi Bencana Geologi, Badan Geologi, Departemen Energi dan Sumber Daya Mineral) p 133

[3]Supartoyo, Hidayati S and Subandriyo 2016 Opak fault inferred from morphotectonic analysis (preliminary result) *Proceedings of International Conference (Jogja Earthquake in Reflection 2016)* pp 28-35

[4]United State Geological Survey (USGS): Shuttle Radar Topography Mission (SRTM) 30 M, http://earthexplorer.usgs.gov/ (Retrieved June 20, 2016)

[5]Rahardjo W, Sukandarrumidi and Rosidi HMD 1995 *Geological Map of Indonesia Quadrangle: Yogyakarta, Jawa, Scale 1: 100000* (Bandung: Geological Research and Development Centre)

[6]Surono, Toha B, and Sudarno I 1992 *Geological Map of Indonesia Quadrangle: Surakarta – Giritontro, Jawa Scale 1: 100.000* (Bandung: Geological Research and Development Centre)

[7]Badan Informasi Geospasial (BIG): Peta Rupa Bumi Indonesia (RBI), Scale 1:25000, http://tanahair.indonesia.go.id/ (Retrieved June 20, 2016)

[8]Dehbozorgi M, Poukermani M, Arian M, Matkan AA, Motamedi A and Hosseiniasi A 2010 Quantitative analysis of relative tectonic activity in The Sarvestan Area, Central Zagros, Iran *Geomorphology 03284* pp 1 – 13

[9]Supartoyo 2008 Tektonik aktif Sesar Cimandiri, Kabupaten Sukabumi, Provinsi Jawa Barat *Thesis* Institut Teknologi Bandung (unpublished)

[10]Hidayat E 2009 Analisis morfotektonik Sesar Lembang *Thesis* Institut Teknologi Bandung (unpublished)

[11]Bull WB 2007 *Tectonic Geomorphology of Mountain: A New Approach for Paleoseismology* (UK: Blackwell Publishing) p 316

Impact of the 1815 Tambora Eruption to global climate change

Achmad Djumarma Wirakusumah[1] and Heryadi Rachmat[2]

[1] Institute of Energy and Mineral, Ministry of Energy and Mineral Resource, Jalan Gajahmada 38, Cepu, Blora, Central java, Indonesia
[2] Geological Museum, Geological Agency, Ministry of Energy and Mineral Resource, Jalan Diponegoro No.57, Bandung, Jawa Barat, Indonesia
E-mail: ade.wirakusumah@gmail.com

Abstract : Tambora volcano is located at Sumbawa island, Indonesia. Geological study shows a successive of geomorphological development of Tambora Volcano. During 190 to 86 K-Years BP, shield-like or effusive volcano were formed; During 86 to 4 K-Years BP, a strato or explosive-volcano was formed; However, during 80 to 4 K-Years BP flank eruptions occurred intermittently and cinders were formed; In April 1815, a paroxysmal destructive eruption occurred which were followed by caldera forming; Since 1815, lava domes and solphataric fields were formed. The 1815 Tambora eruption emitted 60 to 80 megatons of SO_2 to the stratosphere (44 km high). The SO_2 spread the tropics, circled the world and it was oxidized to form H_2SO_4 so called sulphate aerosols protecting the sunlight to reach the earth surface causing global change effects. The Year of 1816 as the year without summer in Europe, the depressed situation in Europe, the epidemic disease of Benggal were three of examples of the impacts of the 1815 Tambora paroxysmal eruption. Therefore, characteristics of Tambora activity before paroxysmal should be learned for mitigation purposes.
Keywords : paroxysmal eruption; caldera forming; global climate change.

1. Introduction

Tambora Volcano, Sumbawa Island, West Nusa Tenggara, Indonesia is geographically located at the middle of the 7000 km long Indonesian volcanic belt (Figure 1). The volcanic belt lies starting from Aceh of the North Sumatera continue to volcanoes along the west side of Sumatera towards the south, continues to volcanoes along Java island from west to east, continues to volcanoes on Bali and Nusa Tenggara islands, volcanoes on the East Nusa Tenggara islands, continues to volcanoes on Banda Sea, and continues to volcanoes on Maluku islands and North Sulawesi. A number of 127 active volcanoes are located in Indonesia, that means about 13 % of active volcanoes all over the world are located in Indonesia. The volcanic belt of Indonesia is a part of the Pacific Ring of Fire (Figure 2) including New Zealand and North America and South America [1].

According to the data base of active volcanoes in Indonesia [2], there are 3 Indonesian active volcanoes showing catastrophic eruption producing caldera (7 km in diameter) during the human life history. They consist of Tambora in 1815, Rinjani in 1257, and Krakatau in 1883. Tambora is a strato volcano with a caldera morphology that was caused by a production of a catastrophic eruption in 1815 [2]. According to the human life history, the 1815 Tambora Eruption is the biggest eruption in the world, resulting totally more than 200,000 people dead at around the volcano and many parts of the world, and damages of many kinds of casualties locally, regionally, and globally [1] and [2]. No

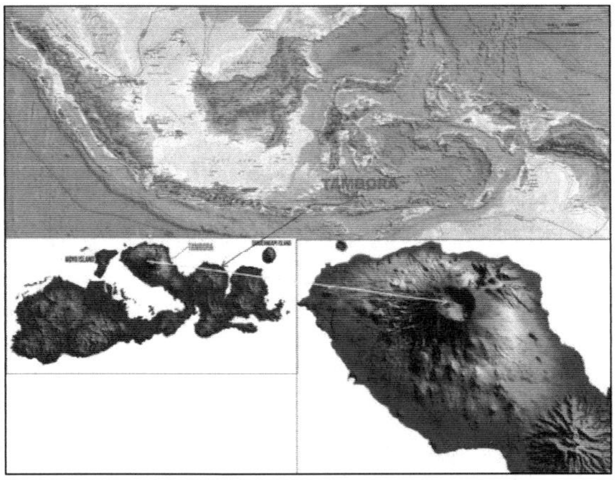

Figure 1. Location map of Tambora Volcano among the Indonesian active volcanoes.

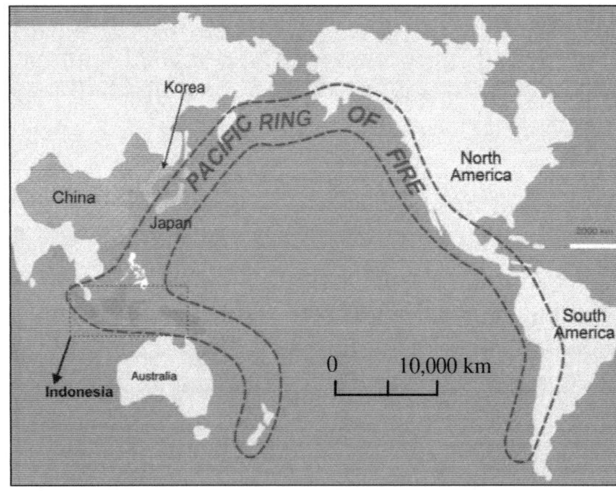

Figure 2. Pacific Ring of Fire, including all volcanoes in Indonesia [1].

adequate volcanic mitigation facility was applied at Tambora volcano in 1815, and no responsible institution took care of active volcanoes in Indonesia at that time. Consequently, no volcanic information concerning the 1815 Tambora Eruption could be distributed to all people all over the world, nor within in Indonesia at that time.

Learning from the 1815 Tambora Volcano eruption event is very important to study as a part of the volcanic mitigation study in Indonesia and all over the world. That is why, this study will be focused into geological condition especially before, during, and after the Tambora 1815 eruption occurred, how strong the 1815 eruption was, what is the effect of the catastrophic eruption to the world, what kind of method and management that can be applied for volcanic mitigation purposes at Tambora active volcano.

2. Data and Method.

No one knew that Tambora would erupt catastrophically in the world even in Indonesia before the eruption occurred. Only some people heard that a very big blasting sound at far away from Sumbawa (at Bengkulu, Sumatera island, 1775 km away, also Ternate, Maluku islands, 1400 km away) and some looked directly at the giant eruption of Tambora volcano in April 1815 at areas close to the volcano. A huge number of evens were not understandable in the world for several years after 1815, although finally a lot of people in the world knew after more than 6 decades that the cause of those events were actually the results of the 1815 eruptions of Tambora volcano.

By learning a similar activity to the 1815 Tambora that can be happened in the future at Tambora volcano or any volcano in Indonesia even all over the world, some methods in this study were carried out as follows.

The method was firstly preceded by a detailed geological study of Tambora volcano through a geologic map study of the volcano especially using volcanic-stratigraphy method. This study can be followed by making its Volcanic Hazard Map. Secondly, a collecting of technically by finding authentic information of deposits of Tambora eruption products all over the world was conducted. Thirdly, collecting several non-technically information by finding some prooven histories through any kind of mass media from all over the world that could be related with the 1815 Tambora eruption were also conducted. Study and realization of proposed plans concerning the Tambora volcano Mitigation were also carried out.

3. Result and Discussion.

3.1. Geology of Tambora Volcano.

Tectonically, Tambora volcano includes volcanoes on Sumatera, Java, Bali, West Nusa Tenggara, and East Nusa Tenggara were closely related with the collision between the Indian-Australian Ocean Plate from the South and the European Continental Plate from the North. The collision part is tilting towards the north. A partial melting generates a molten rock of magma produced volcanic rocks, so called Tambora volcano. Tambora is located at a very complicated tectonic position generates the magma moves towards the surface.

A classification of Volcanic Stratigraphy Unit level based on volcanic eruption source, geography, and tectonic setting was introduced [3]. The classification consist of Volcanic Bregada, Volcanic Khuluk, Volcanic Gumuk, and Volcanic Manggala. Volcanic Khuluk is a basic unit in grouping a volcanic stratigraphy unit for volcano, i.e Volcanic Khuluk Tambora. Volcanic Gumuk is a part of the Volcanic Khuluk which was formed as eruption products on the Volcanic Khuluk both for central or flank eruptions, i.e Volcanic Gumuk Apitoi, Volcanic Gumuk of Katupa etc. Volcanic Bregada is a volcanic stratigraphic unit covering volcanic eruption products originating from two or more Volcanic Khuluk or related to a caldera forming. The volcanic Bregada always has a Volcanic Khuluk or more.

According to the Geological Map of Tambora Volcano complex (Figure 3) [4], which is combined with the implemented Volcanic Stratigraphy Unit Level [3] at Tambora, the Geological History of Tambora volcano forming can be explained as follows. A simple and general geological history sequantly based on the Volcanic Stratigraphy Unit Level of Tambora is showed in Figure 4.

A great volcanic complex of Tambora which is located on Sumbawa island at 50 to 100 km north-west ward from Dompu, was preceded by forming a large volcano complex on a large area namely "Volcanic Bregada of Sanggar". The Volcanic Bregada was formed at least 690 KY BP consist of firstly, "Volcanic Khuluk of Labumbum" (690 – 410 KY BP) at the south-east part of the Volcanic Bregada, lithologically consisting of lava flow and pyroclastic rocks (andesitic in composition); secondly, "Volcanic Khuluk of Kawinda Toi" (410 – 190 KY BP) at the north-east part, lithologically consisting of lava flow and pyroclastic rocks with basaltic in composition; and thirdly, "Volcanic Khuluk of Tambora" (190 KY BP – now) at the center of the Bregada, lithologically consisting of lava flow and pyroclastic rocks (andesitic to andesitic basalt in composition). During the activities

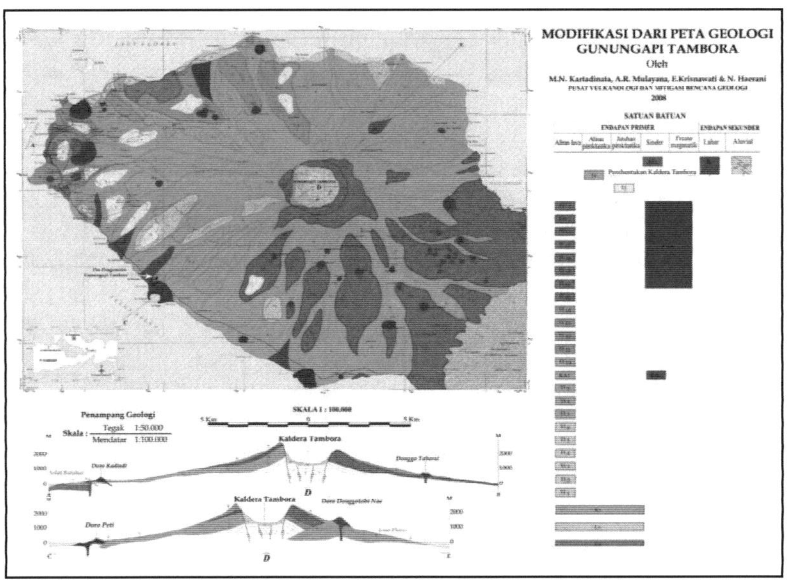

Figure 3. Geological map of Tambora volcano [4].

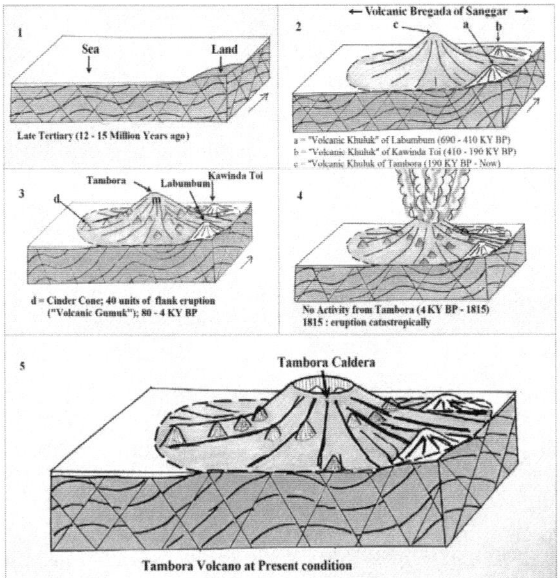

Figure 4. A Simple and General Geological History of Tambora Volcano Complex based on Volcanic Stratigraphy Level Unit, sketched by Wirakusumah.

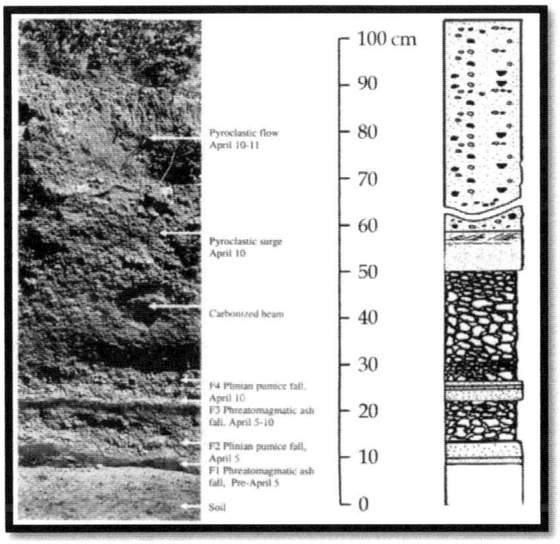

Figure 5. Stratigraphy coloum of the eruption products during the 1815 Tambora activity [1].

through Tambora vent to form strato volcano, eruptions were also intermittently occurred between central eruption (Tambora crater) and the flank eruption starting from 80 KY BP, producing 40 cinder cones at all directions of Tambora flank especially at elevation between 0 and 1600 m asl. Lithologically, they consist of lava flow, pyroclastic crocks, phreatomagmatic, and phreatic eruption products. On the body of Kawinda Toi, two small flank eruptions were formed in 80 KY BP, that means they were formed during the Volcanic Khuluk of Tambora forming. According to the Volcanic stratigraphy [3], these flank eruptions are called "Volcanic Gumuk" e.i "Volcanic Gumuk of Katupa" as the child of the Volcanic Khuluk Tambora which is located on the Kawinda Toi body. The youngest Volcanic Gumuk of Tambora is the Volcanic Gumuk of Api Toi as central eruption which was produced after the 1815 Tambora eruption (between 1847 and 1914). This volcanic complex lies on a Late Tertiary sedimentary rocks consisting of mainly limestones (about 12 to 15 Million years ago) as the basement.

According to the data above, there were two times of the main central eruption sources shifting within the Volcanic Bregada of Sanggar. The eruption source were firstly at around Labumbum Volcano body (690 KY BP), secondly the eruption source moved to around Kawinda Toi Volcano body (410 KY BP), and finally the eruption source moved to Tambora Volcano (since 190 KY BP). The formed 40 cinder cones on around the Tambora flank especially produced freatomagmatic materials indicates that when magma were came up towards the Tambora flank surface through some diatremas, the magma or the magma heat contacted water (ground water originated from sea water or land ground water) producing freatomagmatic eruptions and some of them are produced freatic eruptions. Another explanation, that when lava flows were contacting ground water, then phreatomagmatic or phreatic eruptions occurred.

Oldest period of Tambora lithology (pre-caldera forming) were dominated by andesite basalt to basaltic-andesite lava with 49 to 53 % SiO_2 content. These lava flows formed between 190 KY BP and 9 KY BP covering older rocks. This condition effect to form Tambora morphology of a shield volcano. However, starting 86,000 years some pyroclastic rocks were found producing a strato volcano characteristic. This indicates that there were shifting from lava dominantly (effusive eruptive characteristics) to an intercalation between effusive and explosive eruptions. This effect that Tambora shows a break slope to be a steeper slope at elevation of 1,800m asl, event such kind condition was continued reaching up to an extrapolation elevation of 4,200 m asl [4]. No eruption activity during a big gap of age (between 4000 years BP and the year of 1815) was recorded, that probably means during about 2000 year the Tambora had been building energy which was finally in 1815 the Tambora produced a catastrophic eruption producing a great caldera (diameter = 7 km). A stratigraphic coloum of the 1815 Tambora eruption product is shown in Figure 5 [1].

3.2. Magnitude and Ditribution of the Eruption.

Tambora erupted catastropically in 1815. The detail chronology of the eruption is shown in the Table 1.

Table 1. The detail chronology of the 1815 Tambora eruption based on the eye witnesses [1].

DATE	TIME	OBJECT	COMMENTS
(1812)	A few days	Black ash eruption column from the summit accompanied by rumbling.	It indicated the Pre-initial phase.
April 5	Sun set	34 m high black eruption column. The eruption sound were heard for several times from Jakarta, Makasar, and Ternate.	The main eruption produced loud sounds due to the partial destruction of lava plug.
April 6-9	Afternoon	Ash fell on Surabaya (600 km to the west) for 4 days.	The wind blew from east to Wrest-Northwest, to Java.
April 10	19:00 pm	Fire column rose from 3 places, followed by glowing ash looked like fire.	Stage of collapse of the Tambora summit part.
	20:00 pm	The three ash column was calesced, and were followed by a total fire covering all summit until flank area. Glowing rocks fell at 20 km away from the summit.	This stage is the strongest stage producing pyroclastic flow.
	21:00-22:00 pm	Volcanic ash & strong wind from all directions. The explosion sound was heard from Jakarta.	High velocity Turbulence wind indicate thepyroclastic flow (V = 80 km/hour)
	22:00 pm	Tsunami (4 m high) at Sanggar coast.	Pyroclastic flow reached the coast and generated tsunami (4 m) at Sanggar and at East Java (2 m) & Maluku islands (2m).
	24:00 pm	2 m di Jatim dan Maluku.	
April 11	07:00 am	Erption continued. The summit was claudy.	Paroxismal terminated. Eruption decreased.
	13:00 pm	Darkness for few days.	Ash covered sun ray.
	22:00pm	Ash fell on Java for 3 days.	The wind blew to the west and northwest.
	24:00 am	Eruption decreased significantly.	Magma began to retreat.
April 12-17	All the day	Ash fell on Java. Eruption terminated on17/4	Magma retreated& Eruption terminated.

The 1815 Tambora eruption produced a number of > 100 km^3 of ejected material [2] and Volcanic Explosivity Index (VEI) of 7 from the highest scale of 8. Comparing to some other big volcanic eruption of many volcanos in the world based on the ejected eruption material, the 1815 Tambora eruption is classified into very large eruption that is illustrated in Figure 6 [5].

The 1815 Tambora eruption sound was heard from Bengkulu (> 1700 km to the west), Makasar (400 km to the north), Ternate island (> 1400 km to the north east). The ejected material of the 1815 Tambora eruption was probably reached 44 km high vertically [1] that means reaching the stratosphere layer. The volcanic ash was pushed by the wind blowing to the west-northwest as illustrated by Figure 7 [6]. The plume was probably circling over the earth globe for years. This indicates that the volcanic ash originated from the 1815 Tambora eruption was deposited all over the world.

The 1815 Tambora eruption emitted 60 – 80 megatons of SO_2 (sulfur dioxide) to the stratosphere, which means 3-4 times as much as during the 1991 Pinatubo of the Philippines the biggest eruption of the 20th century [7].

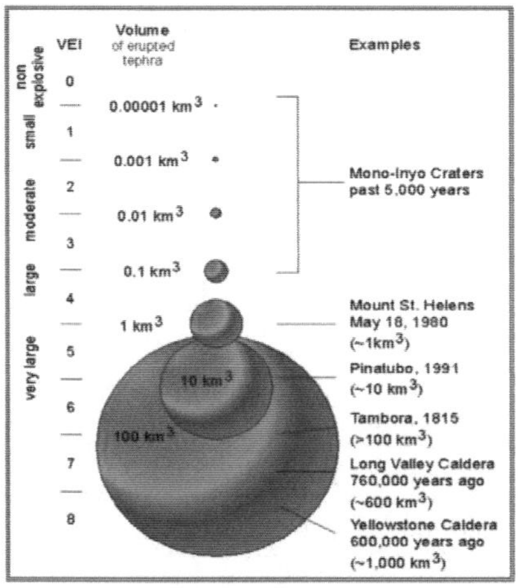

Figure 6. A Volcanic Explosivity Index has been proposed in 1982 [5]. (The source of figure: www.volcanoes.usgs.gov).

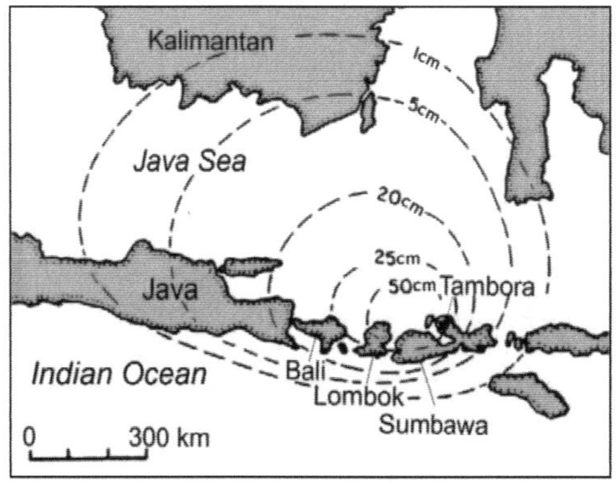

Figure 7. Map of volcanic ash movement originated from the 1815 Tambora eruption [6].

In the stratosphere, the SO_2 spread across aerosol the tropics, circled the globe and was oxidized within weeks to form H_2SO_4, a substance that condenses to tiny droplets called sulphate aerosol. These aerosols dimmed the sunlight and primary caused to the global climatic effects. The case of Tambora can be explained that the ejected ash and gases reaching the stratosphere, which after 2 weeks the suphur gases circled the tropics in Stratosphere. After 8 weeks they were rapidly oxidized to Sulphate Aerosols, and after 1 year the Aerosols slowly spread poleward [7]. That is why the aerosol originated from 1815 Tambora eruption was found in the ice core in Greenland. This means that the product of the 1815 Tambora eruption spread to all over the world.

3.3. The Impact of the 1815 Tambora Eruption.

3.3.1. The Impact of Pyroclastic Flow

Three kingdoms on Sanggar Peninsula close to Tambora volcano so called Kingdoms of Pekat, Tambora, and Sanggar were destroyed because of the 1815 Tambora eruption through pyroclastic flow deposit and tsunami. A total number of fatalities in these location of 10,000 people died directly was reported [1].

The other impact of the pyroclastic flow was the vanishing of houses, villages, cities, as well as kingdoms at Sanggar peninsula, consequently the population and its culture and language were also perished. A lot of folklore and caesura concerning all those stories were collected. One of great evidences is an ancient manuscript telling about the 1815 Tambora eruption written in Melayu laguange in Arabic character which is now is kept by Dr. Hj. Maryam S, a philologist in Sumbawa.

Some excavation of locations suspected as buried artifacts were carried out revealed those things including human remnant were really vanished and buried. That is why it is called as "The Pompeii of the East" (Figure 8).

3.3.2. The Impact of Pyroclastic Fall.

In other place on Sumbawa island such as at Dompu Kingdom, at Bima Kingdom, and at Sumbawa Kingdom, some 38,000 people died indirectly through starvation and disease. Starvation and disease were appeared after the pyroclastic fall deposit originated from the eruption destroyed all plantation including agriculture belong the people or government. Consequently, food and many important things were difficult to provide. This resulted the Sumbawa's inhabitants reduced for about 50 % (Figure 8).

Starvation and disease conditions also occurred in Lombok and Bali islands resulting a number of 44,000 people died in Lombok and that of 25,000 in Bali. Consequently, The total fatalities (directly and indirectly) in Indonesia was 117,000 people [1].

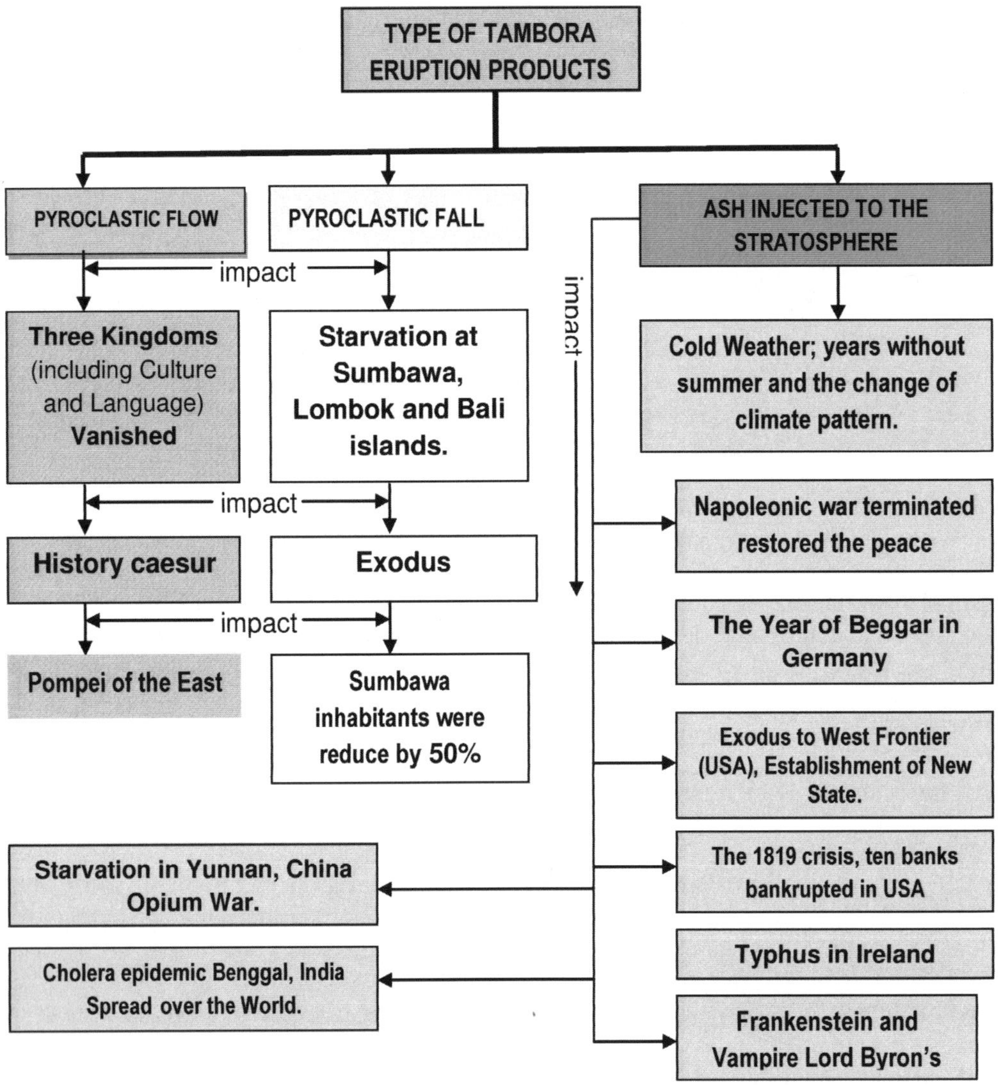

Figure 8. Schematic global impact of the 1815 Tambora eruption, modified from [1]

3.3.3. The Impacts of the Injected Ash to Stratosphere.
A number of thousands of people were died indirectly in Ireland (Europe) because of Typhus and food shortages. In Yunnan, China, a number of tens of thousands were died indirectly because of starvation and chilling summer. Another place, in Bengal, India, a number of ten thousands of people died indirectly caused by Cholera due to abrupt change of weather (Figure 8). The total fatalities in the world was > 200,000 people [1].

As it has been explained that a year after the 1815 Tambora eruption, the aerosol which was originated from Tambora eruption covered all of the globe. The aerosol became an obstacle for the sun ray to reach the ground surface freely. Consequently, the sky never showed a sunny day, people feel the weather on the earth very cold any where and every time especially at sub-tropic areas.

The global temperature in the world dropped 1° to 2.5° C and in some place reached 10° C. That is why in 1816 is called as "The Year without a summer" in Europe as well as in USA although a year after, the summer did not appear yet at both places [1]. This condition finally changed the climate pattern in the world.

The changed climate pattern resulted a lot of odd phenomenon but also a wisdom or philosophy such as rainy, cold and dry resulting the crops did not grow, the harvest failed, and food (including animals) shortages everywhere included in European countries, Asian countries, and USA. In Germany flood occurred along Rheine River caused food shortages in many cities the river's banks. In Britain and Ireland weather so cold, snow fell down. The starvation spread over the South Ireland. In 1816, typhus epidemic broke out in Ireland. The American sky was disturbed by the eruption ash of Tambora resulting the climate changed significantly in 1816. A lot of harvest failed and many of the farmed animals dead. Some of the arriving immigrant had to turn back to Europe because of food shortages condition occurred in the USA.

3.4. Volcanic Hazard Mitigation

Volcanic Hazard mitigation before, during and after the eruption should be done for reducing the volcanic risk purposes.

Volcanic Hazard mitigation activities consist of:

- Identification of volcanic activity
- Volcanic monitoring
- Establishing a statement of activity level of the volcano
- Volcanic Hazard Mapping
- Geological Mapping of the volcano
- Socialization concerning the development of volcanic activity
- Technical Reporting and based on Law Number 24 concerning Disaster Mitigation in Indonesia.
- Coordinating with the National Agency of the Natural Disaster Mitigation of Indonesia.

Volcanic Hazard Mitigation is technically preceded by volcanological study through methods of geology, geophysics, geodesy, geochemistry. To reduce a risk of volcanic hazard such as Tambora

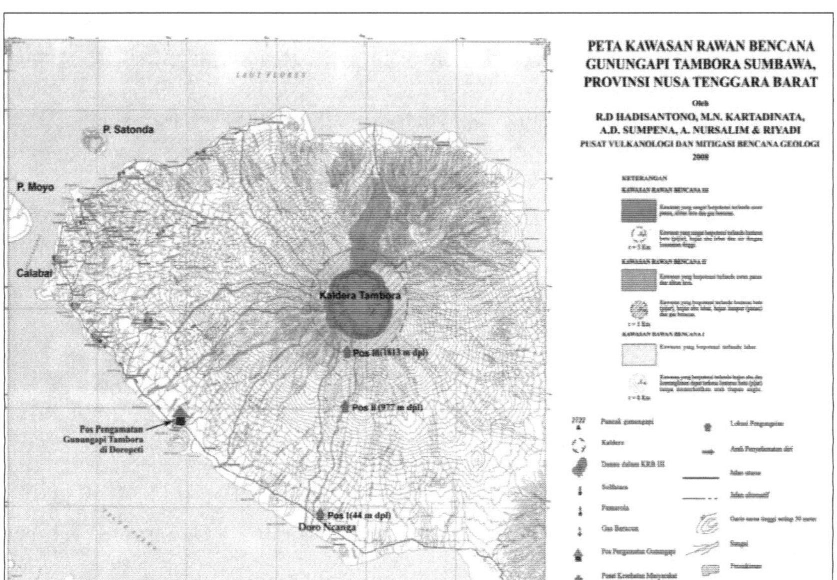

Figure 9. Volcanic Hazard Map of Tambora Volcano [8].

volcano, a geological study is required as Tambora shows short and long periods of repose time.

The long period of repose time can be studied based on data base along the human life history, but it is probably also required a carbon dating method through rock samples to understand the age of the eruption products. This should be related with the magma composition chemically through studying on every eruption product. This will help to understand the characteristic and volcanic forming history. A catastrophic eruption like the 1815 Tambora eruption is usually preceded by having a long repose time and changing of magma composition chemically.

The short period of repose time can be studied by geology, geochemistry and geophysics including geodesy. The integrated data of them will help in making decision of the volcanic activity as this information will be taken as a reference to act the following mitigation activity.

Volcanic hazard mitigation for Tambora volcano need to make 2 versions both for long and short periods. The Volcanic Hazard Map of Tambora Volcano (short period) is shown in Figure 9 [8].

4. Conclusion.

The 1815 Tambora catastrophic eruption with VEI of 7 resulted a great hole 7 km diameter and 1 km depth so called Tambora Caldera. The ash plume reached stratosphere, 44 km high covering all over the world for few years.

The 1815 Tambora eruption impacted a lot of fenomenon in the world especially the changed climate pattern to the globe such as the 1816 is called as The Year Without a Summer etc. The eruption was finally resulting a number of > 200,000 people dead all over the world.

Tambora volcano is a good example volcano to study in Indonesia even in the world as the volcano has 2 kinds of eruption characteristics which are related to a long period repose time and a short repose period, and having changing in magma chemical composition. Until now, the 1815 Tambora is classified as the most catastrophic eruption in the world within the human life history. Research and collecting information of Tambora volcano globally should be continued in order to make the data base more complete. Information of the volcano should be distributed to the involved institutions for volcanic mitigation purposes. This methods can be applied to study other volcanos in Indonesia even in the world.

References.

[1] Sudradjat A and Rachmat H 2015 Greetings from Tambora A Potpourri of the Stories on the Deadliest Volcanic Eruption, Geological Museum, Center for Geological Survey, Geological Agency, Bandung, Indonesia, 173 p.

[2] Geological Agency, Ministry of Energy and Mineral Resources of Indonesia 2011 "Data Dasar Gunung Api Indonesia", Bandung, 2nd Edition, p : 47-59.

[3] Wirakusumah A D, Bacharudin R, Sujanto, Bronto S, 1997 Satuan Stratigrafi Gunungapi Indonesia, Direktorat Vulkanologi, Direktorat Jenderal Geologi dan Sumber Daya Mineral, Departemen Pertambangan dan Energi, Bandung, 7 p.

[4] Kartadinata M N, Mulyana A R, Kriswati E and Haerani N 2008 Geological Map of Tambora Volcano, Sumbawa, West Nusa Tenggara, Indonesia, 1 : 100,000, Center for Volcanology and Geological Hazard Mitigation.

[5] Newhall C G and Self S 1982 The Volcanic Explosiovity Index (VEI): an estimate of explosive, magnitude for historical volcanism. *J. Geophys Res* **87**: 1231-1238.

[6] Sigurdsson H, Carey S, and Erfan R D 1986, The Deposits from the 1815 Tambora Eruption: Preliminary.

[7] Bronnimann S and Kramer D 2016 *Tambora and the "Year Without a Summer" of 1816 A Perspective on Earth and Human Systems Science*, Institute of Geography, University of Bern, Switzerland , 48 p.

[8] Hadisantono R D, Kartadinata M N, Sumpena A D, Nursalim A and Riyadi 2008 Vocanic Hazard Map of Tambora Volcano West Nusa Tenggara.

Petrological studies of volcanic ash from Sakurajima volcano in 2013, Southern Kyushu, Japan

Idham Andri Kurniawan[1,3], Masayuki Sakakibara[1,2], and Emmy Suparka[3]

[1] Department of Earth Science, Graduate School of Science and Engineering, Ehime University. 2-5 Bunkyo-cho, Matsuyama, 790-8577, Japan.
[2] Faculty of Collaborative Regional Innovation, Ehime University
[3] Department of Geology Engineering, Faculty of Earth Science and Technology, Bandung of Institute Technology. Jl Ganesha 10, 40132, Bandung, Indonesia
Email: idham@sci.ehime-u.ac.jp

Abstract. Many petrological studies of volcanic rocks have attempted to better predict future volcanic activity. In this study, we examined volcanic ash samples from Showa Crater in Sakurajima, Japan, which were erupted from January to October 2013. This study was used two types of juvenile material; black vesicular volcanic rock (BVVR) and black non-vesicular volcanic rock (BNVVR) to determine the variable of magmatic processes in the magma conduit during the 2013 eruption. We divided the duration into three periods such as January to April, April to July, and July to October based on the variable tendency of SiO_2 contents and volume abundance of the interstitial glasses in BVVR and BNVVR. The estimated melt compositions demonstrate that the variability of SiO_2 contents and volume abundance of microcrystal in the groundmass (BVVR and BNVVR) over time were caused by micro-crystallization processes. The estimated of water contents and vesicle volume shows the influence of degassing process in the rock forming. Also, the correlation of petrological features with the number of eruptions and eruptive volumes, exhibit the variable of magma supply into conduit in each period. Consequently, careful monitoring of the petrological features of erupted materials may provide useful constraints on future eruptive activity and thereby assist in the mitigation of volcanic hazards.
Keywords: Juvenile material; volcanic rock; interstitial glass; microcrystallization; conduit magma

1. Introduction

Petrological studies of volcanic ash have been extensively used to investigate recent volcanic eruptions and have been linked to geophysical observations of volcanoes. The volcanic glass and microlite compositions of juvenile material [1,2], crystal zoning [3], ash texture [4–6], and magmatic properties such as temperature, pressure, and volatile content [7] are all petrological monitoring tools. The integration of these studies with those involving timescales and geophysical observations is important for understanding magmatic processes.

Sakurajima is located in southern Kyushu, Japan, and is one of the most active volcanoes in the world. The types of eruption and nature of the crater of Sakurajima have changed through time. Previous studies have reported daily and inter-event variations in the eruptive mechanisms of small eruptions, including vulcanian, phreatic, and strombolian events, based on petrological studies of volcanic ash [2,4,6,8].

The objective of this study was to link petrological studies of volcanic ash and geophysical observations of the Sakurajima eruption in 2013, in order to reconstruct and understand the magma conduit system. This study presents compositional data for interstitial glass and microcrystal, calculations of groundmass melt compositions from 28 January to 29 October, 2013, and estimates temperature of magma, water content, and decompression- and magma ascent rates for each eruption. These data are used to develop a picture of cyclical variations of magma condition in a conduit, and to link with the number of eruptions and volume of erupted ejecta. Finally, these studies propose a simple model to explain the variable magmatic processes in the magma conduit of Sakurajima during the 2013 eruption.

2. Data and Methods

2.1. Data

The analyzed materials are all volcanic ash (Figure 1) from the Sakurajima eruption in 28 January – 29 October 2013. Seventeen samples representative of each month during the eruption were studied.

Figure 1. Photomicrograph of volcanic ash of 28 January 2013 from Showa crater.

2.2. Methods

2.2.1. Ash analysis

Volcanic ash samples from frequent Vulcanian explosions were periodically collected at distances between 2.8 and 10 km from the Showa crater (Figure 2). The volcanic ashes were sieved into different size fractions from 2.0 to 0.125 mm, and the 1.0–0.25 mm size fraction was used to clarify their components and to perform geochemical analysis. The grain components were examined using a stereoscopic microscope by counting at least 300 grains per sample. Several juvenile materials were handpicked, mounted in resin, polished, and carbon coated for scanning electron microscopy (SEM) and energy dispersive spectrometry (EDS) analysis.

2.2.2. Scanning electron microscopy and energy dispersive spectrometry analysis

The petrography of the samples was examined by SEM using a JEOL JSM-6510LV scanning electron microscope at Ehime University, Japan. Quantitative crystal textural analysis was conducted on backscattered SEM images, with variable magnification depending on the microlite size and density. The groundmass microcrysts were outlined manually and crystal measurements were made using imageJ (version 1.48) on at least 100 crystals in each sample. Mean crystal area, crystal volume fraction, and crystal dimensions (short and long axes) were measured with a crystal size distribution (CSD) software package. The crystal dimensions were calculated using a best-fit ellipsoidal method [9,10]. Microlite number density was calculated by dividing the number of whole crystals in the analyzed area by the area of the studied groundmass [7,9].

The major element compositions of the glasses and microcrystals were determined using a X-max 50 mm^2 Oxford electron microprobe. The acceleration voltage and sample current were 15 kV and 0.8 nA, respectively. The glass compositions were normalized to 100% and calculated using the INCA quantitative analysis program. A range of glasses (K-glass, obsidian, and Na-glass) and silicate (enstatite, clinopyroxene, and K-feldspar) standards were used for calibration of the data.120° E

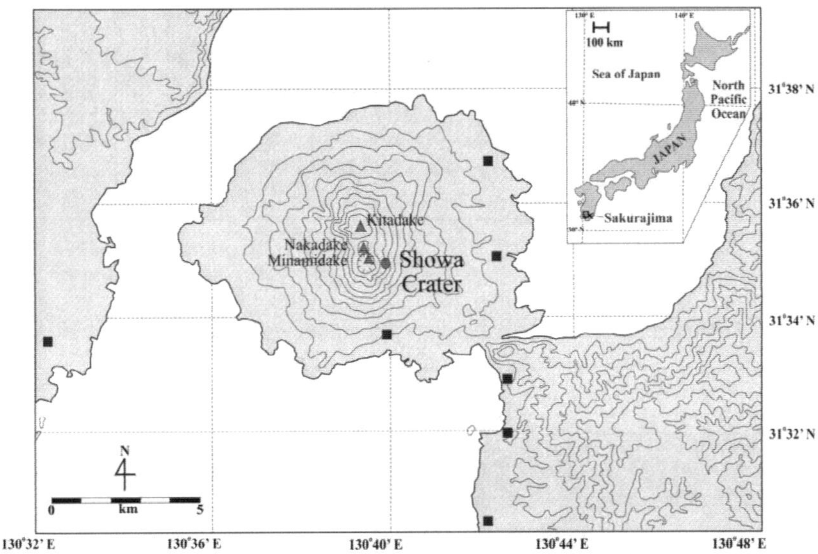

Figure 2. Sampling location map (solid rectangle)

3. Results

3.1. Volcanic ash component

Our previous studies have divided the volcanic ash into eight characteristic materials, which are black vesicular volcanic rock (BVVR), black non-vesicular volcanic rock (BNVVR), light brown volcanic rock (LBVR), white vesicular volcanic rock (WVVR), silicified rock (SR), oxidized rock (OR), altered volcanic rock (AVR), and isolated phenocrysts (IP) [11,12]. The BVVR, BNVVR, LBVR, WVVR, and IP are juvenile materials, and the SR, OR, and AVR are additional materials [11,12]. The juvenile material is component without a secondary process such as a hydrothermal or weathering process [11]. In this study, only BVVR and BNVVR were analyzed, given that both of these components are the same color and the dominant components (59.5%–96% BNVVR and 4%–40.5% BVVR; Figure 3).

The BVVR has an angular shape and is vesicular with an intersertal texture (Figure 4). The modal mineralogy of the groundmass comprises plagioclase (11.4%–38.3%), pyroxene (1.78%–15.8%), Fe–Ti oxides (0%–2.37%), and interstitial glass (56.5%–79.9%). The modal abundance of vesicles is 1.78%–15.8%. Microlites in this material are nominally acicular in shape (short/long axis ratio of 0.30), and the average microlite number density is 38,773 mm^{-2}.

The BNVVR material, contains plagioclase and pyroxene phenocrysts that are angular in shape, and has an intersertal texture. The groundmass components are interstitial glass (47.7%–63.9%) and microcrysts of plagioclase (22.6%–43.2%), pyroxene (5.34%–18.3%), and Fe–Ti oxides (0.13%–4.69%) (Figure 4). The microlites are nominally acicular in shape (short/long axis ratio of 0.26) and the average microlite number density is 41,654 mm^{-2}. The result of each grain per eruption could be seen in Appendix 1.

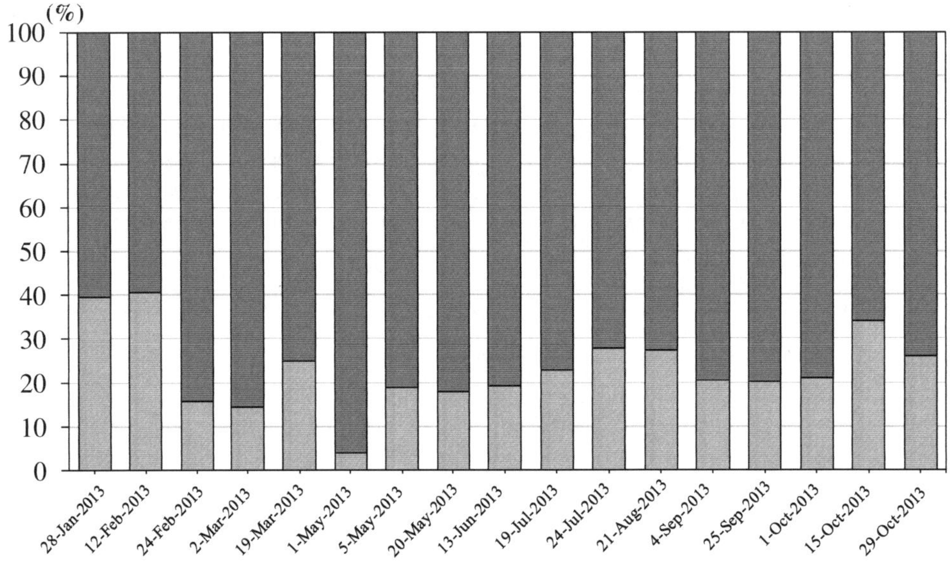

Figure 3. The volume abundance of black vesicular volcanic rock (light gray) and black non-vesicular volcanic rock (dark gray).

a b

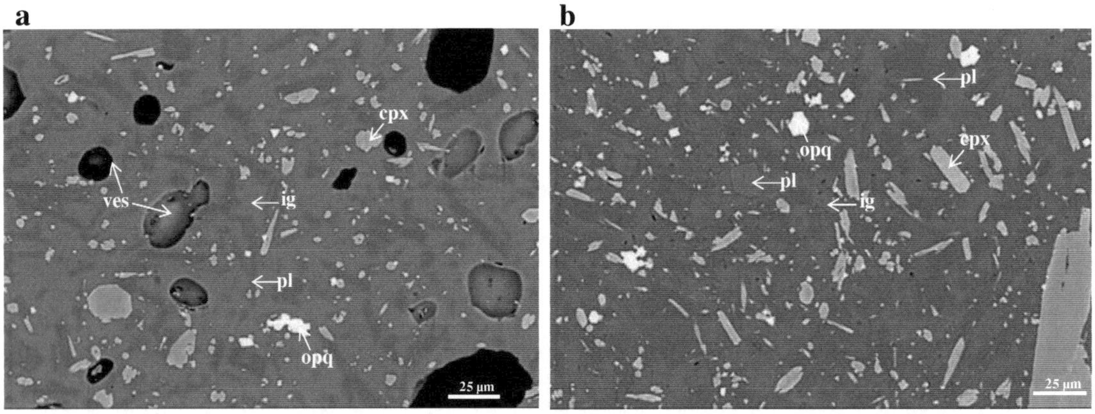

Figure 4. Backscattered electrons images of groundmass of BVVR (a) and BNVVR (b) (ig: interstitial glass, pl: plagioclase, cpx: clinopyroxene, opx: orthopyroxene, opq: opaque mineral, and ves: vesicular).

The volume abundance of interstitial glass of BVVR and BNVVR, and vesicular of BVVR demonstrates the variable tendency over time and could be divided into three periods in 2013 as such January – April, April to July, and July to October (Figure 5). The volume abundance of interstitial glass of BVVR is more glass than BNVVR. The variable tendencies of volume abundance of interstitial glass of BVVR and BNVVR, and vesicular of BVVR vesicles are enrich in Period 1 and 2. Then, Period 3 has different tendency with depleted of interstitial glass and vesicles.

3.2. Geochemical features

The interstitial glass compositions of the BVVR and BNVVR components differ markedly. The SiO_2 contents of interstitial glass in the BVVR are more mafic than in the BNVVR. The SiO_2 contents of both of them show variable tendency over the time. Further, the border between the tendency is consistent with the volume abundance of interstitial glass and vesicles in the groundmass and comprise into three periods in 2013 (Figure 6a). In the first period (January–April 2013), the BNVVR has SiO_2 contents of 71.1 to 76.8 wt.%, which are generally higher than the BVVR (68.3–72.1 wt.%) and tend to decrease with time. In the second period (April–July 2013) the BNVVR has SiO_2 contents of 69.5 to 78.1 wt.%, which are generally higher than the BVVR (67.6–74.7 wt.%) and tend to decrease with time. In the third period (July–October 2013) the BNVVR has SiO_2 contents of 69.6 to 76.6 wt.%, which are generally higher than the BVVR (68.5–73.8 wt.%) and increase with time.

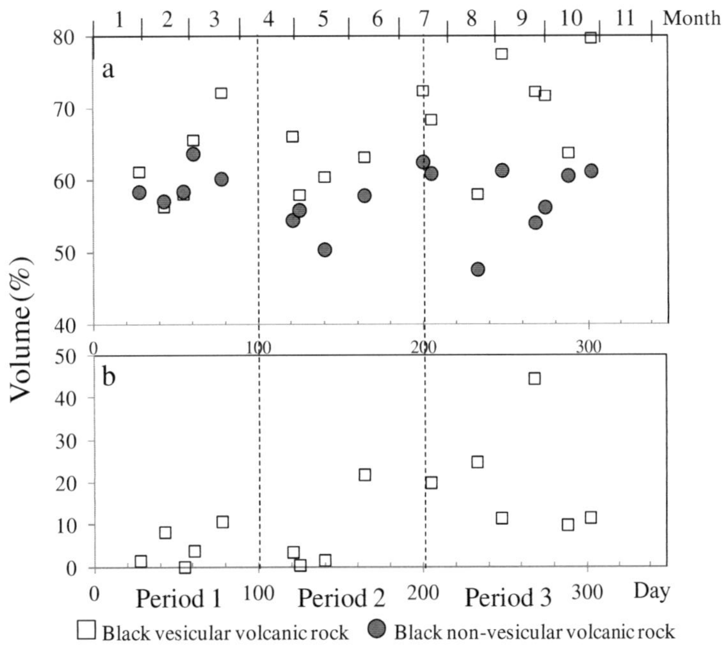

Figure 5. The modal abundance of (a) interstitial glass and (b) vesicles (vol.%).

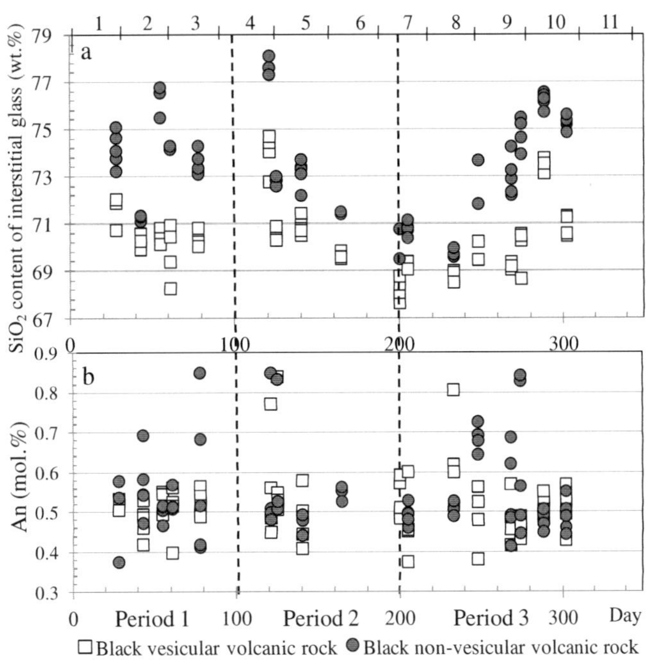

Figure 6. (a) SiO_2 content of interstitial glass and (b) An content of plagioclase (mol.%) of BVVR and BNVVR.

The pyroxene microcryst can be classified into orthopyroxene (enstatite) and clinopyroxene (pigeonite and augite). The Mg# values (100 × Mg/(Mg + Fe + Mn)) of the analyzed pyroxene microcrysts in the BVVR and BNVVR are relatively similar in all three periods (Appendix 2). No systematic chemical zoning was observed in the microcrystal cores. However, the microcrystal rims mostly have higher Mg contents than the cores (i.e., the light color in the BSE image in Figure. 4). The pyroxene Mg# values of the BVVR are higher than those of the BNVVR in all three periods, but do not change significantly over time. In the first, second, and third periods, pyroxene Mg# values for BVVR and BNVVR are 40.0–62.3 and 37.1–60.7, 44.0–60.3 and 37.1–47.3, and 39.2–64.3 and 37.1–62.3, respectively.

Plagioclase microcrystal from the two types of juvenile material exhibit similar range of An (100 × Ca/(Ca + Na + K)) contents (Figure 5) with a few additional high An content in several eruption. In the first, second, and third periods, the plagioclase An contents for BVVR and BNVVR are 0.40–0.57 and 0.38–0.85, 0.41–0.84 and 0.44–0.85, and 0.38–0.81 and 0.42–0.84, respectively.

4. Discussion

Mass-balance calculations using microcrystal phases and interstitial glass compositions were used to estimate the melt composition of each rock type. The groundmass compositions for each rock fragment were assumed to be the final melt composition (new melt), and the composition of the microcrystal phases was assumed to have been removed from the original melt composition [11,12]. Point and/or area counting analysis using the ImageJ 1.48 software package was employed to calculate the total volume of microcrystals. The modal compositions (volumes) of microcrystal and interstitial glass were converted into a weight fraction basis using assumed densities for the minerals and interstitial glass. The densities of the minerals used in these calculations were as follows: bytownite = 2.58 g/cm^3; labradorite = 2.84 g/cm^3; andesine = 3.11 g/cm^3; enstatite = 3.2 g/cm^3; pigeonite = 3.38 g/cm^3; augite = 3.51 g/cm^3; ilmenite = 4.75 g/cm^3; and interstitial glass = 2.35 g/cm^3 [2,11–13].

The SiO_2 contents of the estimated melt compositions are relatively similar for each eruption (Figure 7). The range in SiO_2 contents between BVVR and BNVVR in each eruption is 2–3 wt.%, and the range of SiO_2 in all periods is 60.5–65.6 wt.%. From these results, we suggest that the magma was the same in the three periods and that the differences between the BVVR and BNVVR resulted from micro-crystallization processes in the groundmass.

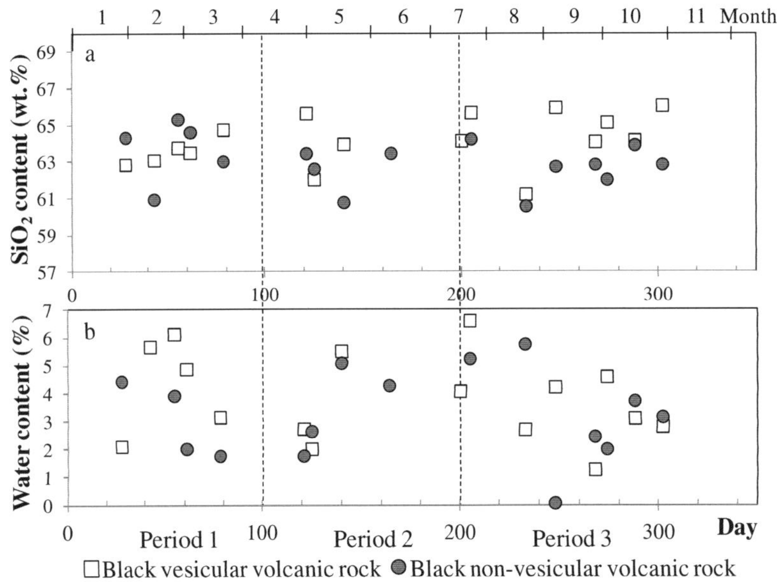

Figure 7. The result of estimation of melt composition (a) and water content (wt.%) (b).

Magmatic temperatures and approximate water contents were estimated from the experimentally calibrated plagioclase–melt geothermometer and hygrometer of [14]. In these calculations, several hypothetically value was determined, like the initial water content was 2.6 wt.% [15] and An content was used the majority An content as the equilibrium with the melt.

Based on these results, the different magma conditions for each period are shown. Periods one and two have relatively similar conditions. Period one is characterized by: (1) generally decreasing SiO_2 contents of interstitial glass in the BVVR; (2) melt SiO_2 contents for the BVVR and BNVVR of 62.9–64.8 and 61.0–65.4 wt.%, respectively, with SiO_2 contents increasing with time; (3) plagioclase temperatures of 1061–1118°C, and (4) generally decreasing water contents (1.75–6.12 wt.%). Period two is characterized by: (1) generally decreasing SiO_2 contents of interstitial glass in the BVVR; (2) melt SiO_2 contents for the BVVR and BNVVR of 62.0–65.7 and 60.8–63.5 wt.%, respectively, with melt SiO_2 contents in both rock types increasing over time; (3) plagioclase temperatures of 1086–1131°C; and (4) increasing water contents (1.75–5.51 wt.%). Period three is different to the other periods and shows: (1) generally increasing SiO_2 contents of interstitial glass in the BVVR and BNVVR; (2) melt SiO_2 contents for BVVR and BNVVR of 61.2–66.1 and 60.6–64.3 wt.%, respectively, with SiO_2 contents increasing with time; (3) plagioclase temperatures of 1050–1147°C, and (4) decreasing water contents (0.06–6.60 wt.%).

The estimates of melt temperatures and water contents in periods 1 and 2 are similar, but are different for period 3. This suggests that the change from period 2 to 3 was influenced by water enrichment. The enrichment in water and change in vesicle volume show that each period had different degassing processes, particularly during the change from period 2 to 3.

Therefore, we consider that the variations between BVVR and BNVVR over time are due to micro-crystallization processes in the groundmass, and the geochemical variations in the interstitial glass were caused by different degassing processes. This inference is consistent with a previous study [12] that showed that the different types of juvenile material likely resulted from cooling- and decompression-induced crystallization in the conduit.

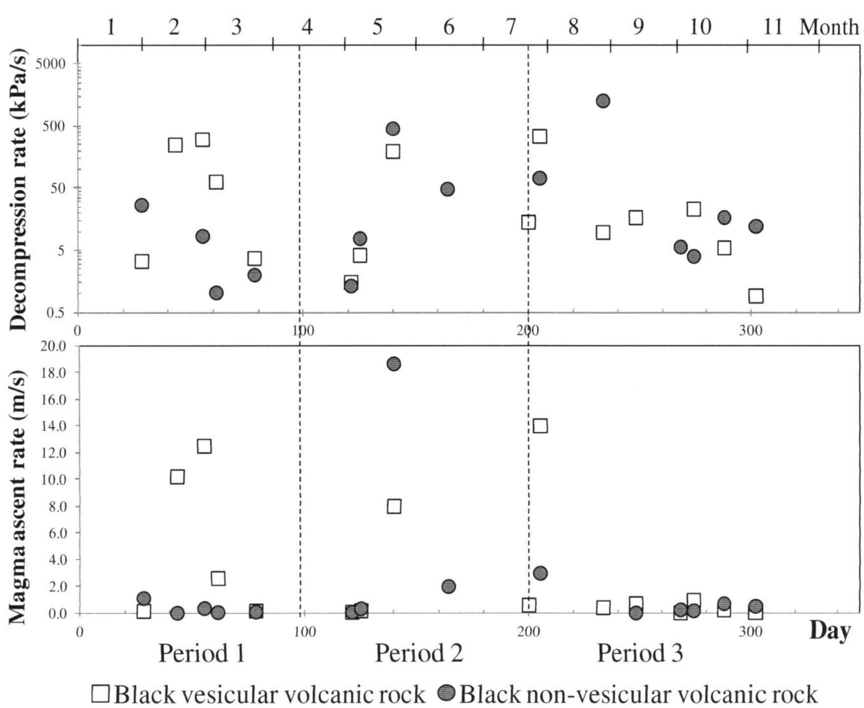

Figure 8. The result of estimation of decompression rate (kPa/s) and magma ascent (m/s).

We have also calculated the decompression and magma ascent rates required to explain the decompression processes that induced the variations in rock and geochemical features during the 2013 eruption. These calculations used the water exsolution rate meter model of [16], and used the anorthite content of the plagioclase microcrystal as an indicator of the water content at the nucleation. The results show that the decompression and magma ascent rate varied during each period (Figure 8). The decompression rate continuously changed over the time. In period 1, the decompression tended to decrease, whereas in period 2 it increased. The increase in the decompression rate was related to water enrichment, which decreased again in period 3. The magma ascent rates in periods 1 and 2 are similar, but differ in period 3. Period 3 was characterized by a slower magma ascent rate compared with the other periods, and it resulted in the SiO_2 content of interstitial glass changing differently (i.e., increasing) compared with the other periods (i.e., decreasing).

We then compared our petrological results with the eruption number and volume (Figure 9). The comparative data were obtained from the Kagoshima Meteorological Agency [17]. The total number of eruptions and volume of erupted material correlate with the three periods defined from the SiO_2 contents of interstitial glass. The timing of the boundaries between the three periods coincide with changes in the total number of eruptions and volume of erupted material.

In periods 1 and 2, when the SiO_2 contents decreased with time, the total number of eruptions and volume of erupted material have a normal distribution with a different total number of eruptions and eruptive volumes. Period 1 appears to have had a larger eruptive volume and was more active than period 2. Period 3 had a greater number of eruptions and a larger eruptive volume than the previous period (Figure 9). This indicates that the number of eruptions and eruptive volume were related to the total volume of magma supply in the conduit.

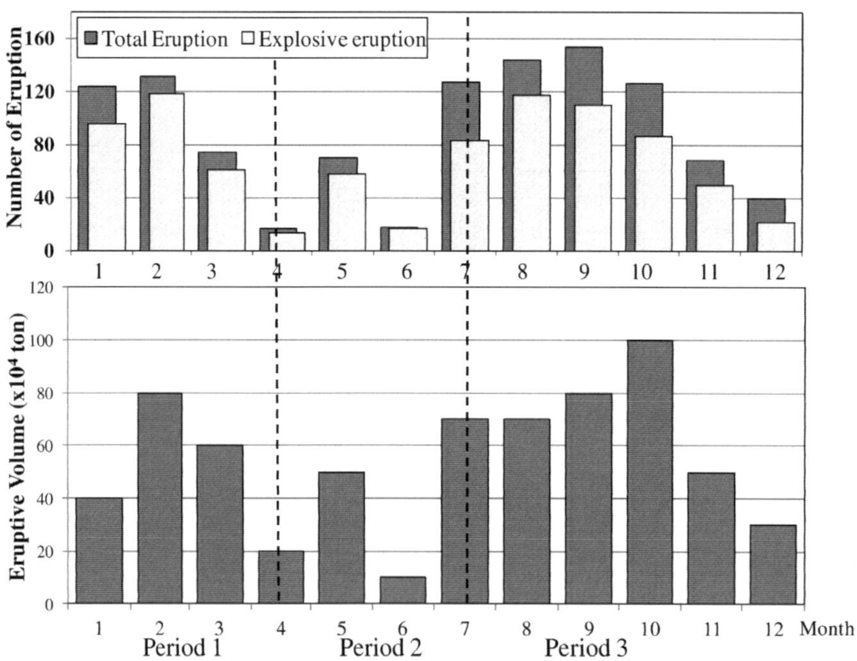

Figure 9. The number of eruption and erupted volume material [17].

In generally, we propose a model that explains the magma conditions in the conduit during 2013 (Figure 10). Three periods can be identified from both petrological studies and volcanic activity. In periods 1 and 2, SiO_2 contents decreased over time, reflecting microcrystal crystallization in the magma conduit. The difference between periods 1 and 2 is related to the total volume of magma supply, with this being greater in period 1 than period 2. Period 3 involved water enrichment and an enhanced magma supply. The water enrichment and increased vesicle volume were related to degassing processes. The water enrichment towards the end of period 2 resulted in an increase in the number of eruptions and eruptive volume of period 3 compared with periods 1 and 2.

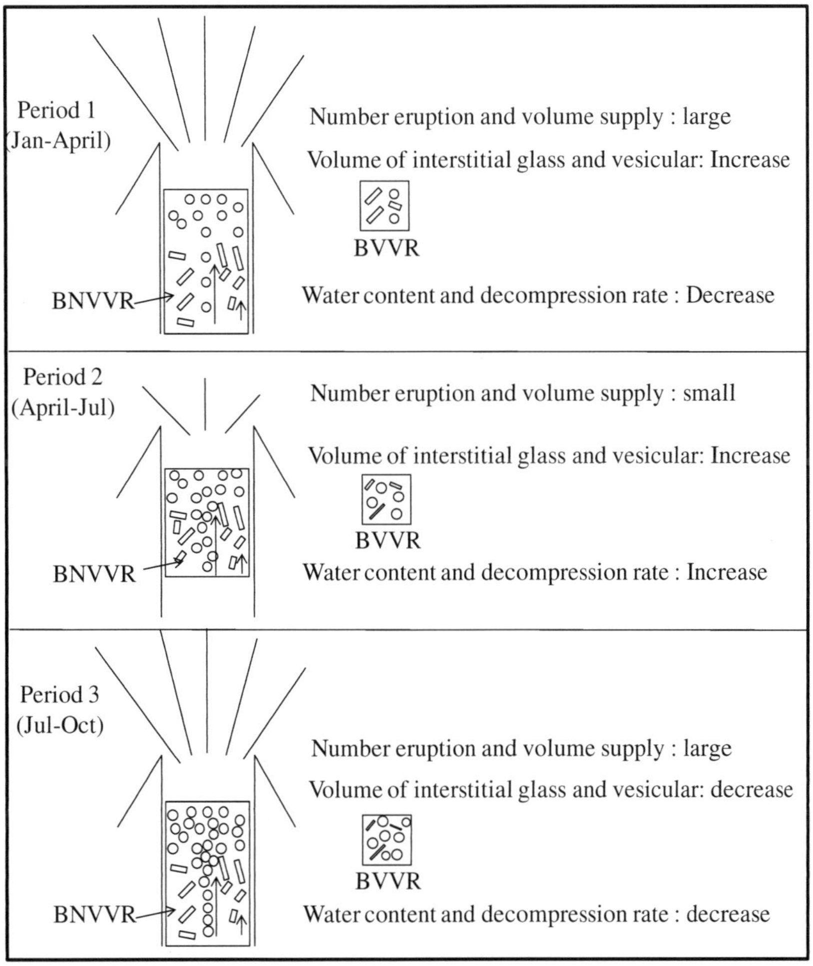

Figure 10. The model of magma conditions in the conduit during 2013.

5. Conclusions

This study has shown led to the following findings. (1) The SiO_2 contents of interstitial glass of the BVVR and BNVVR varied in three periods between January and October 2013; i.e., period 1 from January to April, period 2 from April to July, and period 3 from July to October. (2) The estimated melt compositions demonstrate that the variations of petrological feature in the volcanic rocks (BVVR and BNVVR) over time were caused by micro-crystallization processes in the groundmass. (3) The magmatic temperatures, water contents show different trends in each period. In particular, from period 2 to 3 the water content and vesicle volume increased. It is suggested that this reflects different degassing processes in each period. (4) The decompression and magma ascent rates influenced the rock-forming processes in the magma conduit; depleted of decompression and magma ascent rate will result in the enrichment of SiO_2 content of interstitial glass (i.e period 3). (5) Petrological features correlate with the number of eruptions and eruptive volumes, show the magma supply into conduit in each period is variable; period 1 and 3 are large, but period 2 is small magma supply. As such, it is apparent that careful monitoring of the petrological features of erupted materials can provide useful information for assessing future eruptive activity and can thereby be used to mitigate volcanic hazards.

Acknowledgments
The Authors best thanks to R. Uenosono and Kagoshima Meterological Agency for sharing the collection of sample. And we do not forget to declare no conflict of interest in this paper.

References

[1] Sakakibara M, Uenosono R and Takakura S 2013 Temporal cycles in glass composition within volcanic ash from Showa Crater, Sakurajima volcano, southern Kyushu, Japan *Abstract IAVCEI 2013, Kagoshima, Japan*
[2] Matsumoto A, Nakagawa M, Aiyasaka M and Iguchi M 2013 Temporal Variations of the Petrological Features of the Juvenile Materials during 2006 to 2010 from Showa Crater, Sakurajima Volcano, Kyushu, Japan *Bull. Volcanol. Soc. Japan* **58** 191–212
[3] Druitt T H, Costa F, Deloule E, Dungan M and Scaillet B 2012 Decadal to monthly timescales of magma transfer and reservoir growth at a caldera volcano *Nature* **482** 77–80
[4] Miwa T, Toramaru A and Iguchi M 2009 Correlations of volcanic ash texture with explosion earthquakes from vulcanian eruptions at Sakurajima volcano, Japan *J. Volcanol. Geotherm. Res.* **184** 473–86
[5] Miwa T and Geshi N 2012 Decompression rate of magma at fragmentation: Inference from broken crystals in pumice of vulcanian eruption *J. Volcanol. Geotherm. Res.* **227–228** 76–84
[6] Miwa T, Geshi N and Shinohara H 2013 Temporal variation in volcanic ash texture during a vulcanian eruption at the Sakurajima Volcano, Japan *J. Volcanol. Geotherm. Res.* **260** 80–9
[7] Blundy J and Cashman K 2008 Petrologic Reconstruction of Magmatic System Variables and Processes *Rev. Mineral. Geochemistry* **69** 179–239
[8] Miyagi I, Shinohara H and Itoh J 2013 Variations of Color and Leachate Contents of Volcanic Ashes from Sakurajima Volcano , Japan *Bull. Volcanol. Soc. Japan* **58** 213–26
[9] Higgins M D 2000 Measurement of crystal size distributions *Am. Mineral.* **85** 1105–16
[10] Hammer J E, Cashman K V., Hoblitt R P and Newman S 1999 Degassing and microlite crystallization during pre-climactic events of the 1991 eruption of Mt. Pinatubo, Philippines *Bull. Volcanol.* **60** 355–80
[11] Kurniawan I A, Sakakibara M and Suparka E 2015 Petrological and mineralogical studies on A . D . 2013 tephra from Sakurajima volcano , Japan : Contributions to reconstruction of magma conduits in an active volcano and its petrological monitoring *Proceeding Int. Conf. Transdiscipl. Res. Environ. Probl. Southeast. Asia* 33–42
[12] Kurniawan I, Sakakibara M and Suparka E 2016 Petrological Monitoring of the AD 2011–2012 Volcanic Ash from Sakurajima Volcano, Southern Kyushu, Japan *Geosciences* **6** 12
[13] Suzuki Y, Yasuda A, Hokanishi N, Kaneko T, Nakada S and Fujii T 2013 Syneruptive deep magma transfer and shallow magma remobilization during the 2011 eruption of Shinmoe-dake, Japan-Constraints from melt inclusions and phase equilibria experiments *J. Volcanol. Geotherm. Res.* **257** 184–204
[14] Putirka K D 2008 Thermometers and Barometers for Volcanic Systems *Rev. Mineral. Geochemistry* **69** 61–120
[15] Okumura S 2011 The H2O content of andesitic magmas from three volcanoes in Japan, inferred from the infrared analysis of clinopyroxene *Eur. J. Mineral.* **23** 771–8
[16] Toramaru A, Noguchi S, Oyoshihara S and Tsune A 2008 MND(microlite number density) water exsolution rate meter *J. Volcanol. Geotherm. Res.* **175** 156–67
[17] Japan Meteorological Agency 気象庁 | 火山活動解説資料 http://www.data.jma.go.jp/svd/vois/data/tokyo/STOCK/monthly_v-act_doc/monthly_vact_vol.php?id=506 (accessed Jan 10, 2016).

Appendix

Appendix 1. Modal composition of BVVR and BNVVR

No	Time (day)	Date	Type rock	SiO_2 content of interstitial glass (wt.%)	Volume percent (%)				
					Plagioclase	Pyroxene	Fe-Ti mineral	Interstitial glass	Vesicular/ bubble
1	28	28-Jan-2013	BVVR	71.7	23.7	13.7	1.22	61.4	1.52
2	43	12-Feb-2013	BVVR	70.3	35.1	7.47	0.97	56.5	8.28
3	55	24-Feb-2013	BVVR	70.4	38.3	3.30	0.12	58.2	0.07
4	61	2-Mar-2013	BVVR	69.8	27.1	6.03	1.16	65.8	3.87
5	78	19-Mar-2013	BVVR	70.5	19.2	7.57	0.92	72.3	10.7
6	121	01-May-2013	BVVR	74.0	24.3	8.62	0.81	66.2	3.54
7	125	05-May-2013	BVVR	70.5	32.0	9.26	0.60	58.2	0.44
8	140	20-May-2013	BVVR	70.9	32.7	5.61	1.00	60.6	1.61
9	164	13-Jun-2013	BVVR	69.7	34.7	1.78	0.16	63.4	21.8
10	200	19-Jul-2013	BVVR	68.2	22.9	4.50	0.04	72.6	0.01
11	205	24-Jul-2013	BVVR	69.2	28.9	2.55	0.00	68.5	19.9
12	233	21-Aug-2013	BVVR	68.8	34.1	7.54	0.16	58.2	24.8
13	248	04-Sep-2013	BVVR	69.9	14.4	7.92	0.02	77.7	11.4
14	268	25-Sep-2013	BVVR	69.3	11.4	15.8	0.34	72.4	44.4
15	274	01-Oct-2013	BVVR	70.1	22.2	4.87	1.10	71.9	0.01
16	288	15-Oct-2013	BVVR	73.5	26.2	7.55	2.37	63.9	9.83
17	302	29-Oct-2013	BVVR	70.8	15.2	3.65	1.24	79.9	11.5
18	28	28-Jan-2013	BNVVR	74.17	33.4	6.30	1.79	58.6	0
19	43	12-Feb-2013	BNVVR	71.28	22.6	18.3	1.80	57.3	0
20	55	24-Feb-2013	BNVVR	76.27	29.6	10.6	1.09	58.6	0
21	61	2-Mar-2013	BNVVR	74.23	24.5	10.1	1.54	63.9	0
22	78	19-Mar-2013	BNVVR	73.55	23.6	14.1	1.86	60.4	0
23	121	01-May-2013	BNVVR	77.59	33.1	10.2	2.07	54.6	0
24	125	05-May-2013	BNVVR	72.82	36.4	6.55	1.05	56.0	0
25	140	20-May-2013	BNVVR	73.16	39.1	6.61	3.77	50.5	0
26	164	13-Jun-2013	BNVVR	71.47	32.6	9.26	0.13	58.0	0
27	200	19-Jul-2013	BNVVR	70.14	24.1	12.8	0.47	62.7	0
28	205	24-Jul-2013	BNVVR	70.78	32.6	5.54	0.81	61.1	0
29	233	21-Aug-2013	BNVVR	69.76	43.2	7.40	1.62	47.7	0
30	248	04-Sep-2013	BNVVR	72.76	28.2	9.46	0.89	61.5	0
31	268	25-Sep-2013	BNVVR	73.00	31.0	14.1	0.78	54.2	0
32	274	01-Oct-2013	BNVVR	74.95	35.7	5.34	2.56	56.4	0
33	288	15-Oct-2013	BNVVR	76.23	29.1	5.47	4.69	60.8	0
34	302	29-Oct-2013	BNVVR	75.27	27.0	7.77	3.88	61.4	0

Appendix 2. The Mg# value (mol.%) of microcrystal of pyroxene

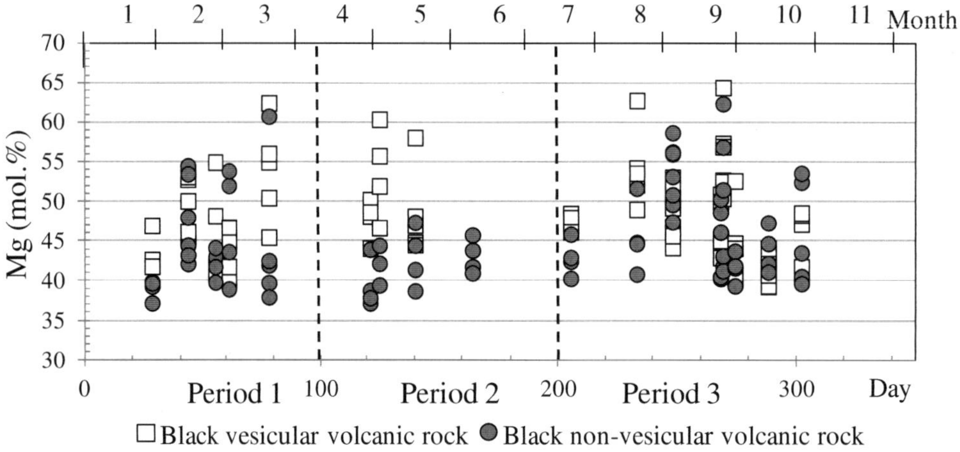

Sub Surface Geoelectrical Imaging for Potential Geohazard in Infrastructure Construction in Sidoarjo, East Java

Prihadi Sumintadireja[1,2] and Diky Irawan [2]

[1] Applied Geology Research Division, Geology Study Program, Institute of Technology Bandung, Jl. Ganesha 10, Bandung 40132, Indonesia
[2] Laboratory Application and Modeling of Geosciences (LVG), Geology Study Program, Institute of Technology Bandung, Jl. Ganesha 10, Bandung 40132, Indonesia.
Email: prihadi@gc.itb.ac.id

Abstract. Mud volcano remnants are identified in Surabaya and adjacent areas. The people in East Java based on historical report are custom and able to adjust with the natural phenomena within their areas. Sidoarjo mud volcano phenomena which coincident with drilling activity in 29 May 2006 is making people and government anxious for development a new infrastructure such as high rise building, toll road etc. An understanding of a geological hazard which can be single, sequential or combined events in their origin is the main key importance in subsurface imaging. Geological hazard can be identified by geophysical, geological, geotechnical method. The prompt selection of geophysical method to reveal subsurface condition is very important factor instead of survey design and field data acquisition. Revealing subsurface condition is very important information for site investigation consists of geological, geophysical and geotechnical data, whereas data analysis will help civil engineer design and calculate the construction safety.
Keywords: geoelectrical, mud volcano, geological hazard

1. Introduction

Geophysics is an application of physics sciences which measured the rock/earth properties in order to image the subsurface condition. Subsurface depth penetration result is depending on geophysical method selection and survey design. Regional scale subsurface condition can be used potential field method or passive geophysical method such as: gravity and geomagnetic. On the other hand for shallow and high resolution results are using active geophysical method which need generated source, such as: geoelectrical, georadar, control source audio magnetoteluric (CSAMT), seismic refraction/reflection etc.

Geoelectrical method is widely used in subsurface imaging due to simple and fast result for site investigation in infrastructure civil construction and development [1]. Commonly, well known as electrical direct current (DC) method which measured earth/rock electrical property, electrical direct current is injected to the earth, whereas the potential differences due to physical property measured within the earth. The measured geoelectrical potential differences value is depend on electrode array configuration and earth/rock medium as resistivity. The earth/rock resistivity value is highly correlated with porosity, fluid resistivity in pore, mineral content, water saturation, etc. [2]. Therefore, the geoelectrical method is one of the best geophysical methods for prediction geological subsurface imaging based on resistivity medium distribution.

The mud or clay with high porosity will represent as low resistivity features and clearly identified from the modeling. Subsurface mud volcano feature and distribution can be shown as map with geoelectrical method and interpreted as geological section. In this paper we elaborated the application of geoelectrical method for high rise building construction with deep foundation and vulnerable risk area in toll road relocation construction adjacent to recent mud volcano area in East Java province.

2. Mud Volcano distribution identification

There are many mud volcano found in East Java special Surabaya and adjacent area. At least a 14 has identified (Figure 1). There are many references elaborating and modeling the occurrences of mud volcano such as [3-8].

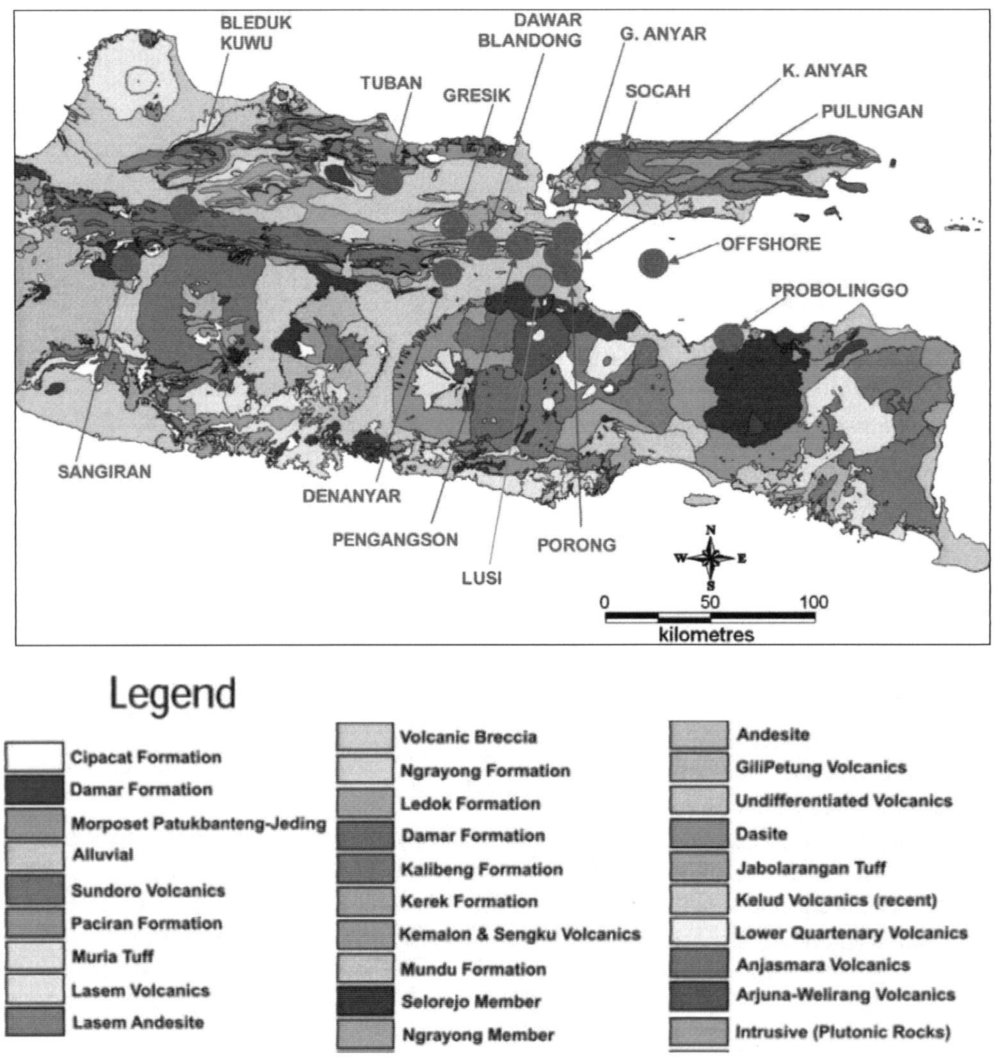

Figure 1. Geological map and mud volcano distribution in East Java [9].

The structural configuration of the East Java Basin includes two main structural lows of NE-SW and E-W orientation depocenters which are bounded by Bawean Arc to the northwest and North Madura Platform to the northeast directions. In the offshore part of the basin, the basement configuration is formed by northeast-southwest structural grain comprises a series of well define basement ridges with intervening grabens containing Tertiary sediments up to 18,000 feet in thickness. Two distinctive trends are recognized: the NE-SW orientation, parallel to the coastlines of East Kalimantan and West Sulawesi which form the flanks of the basin which continue towards SW direction into the East Java Basin where they define major sub-basins such as Central Basin and Muria Trough.

The origin of rifting in NE Java basin is interpreted to be relating to the onset of the NW subduction of oceanic crust along the SE margin of Kalimantan during the Early Cretaceous. The subduction is thought to have shifted with time in the southward in Eocene. The present location of the subduction trench runs parallel to the east-west striking wrench zone, which was reactivated in the Neogene period.

3. Case study 1: high rise building

Planning for high rise construction building is necessary to obtain special sub surface information better than ordinary building. Specific information dealing with soil/formation must be elaborate all of geology, geophysics and geotechnical data as hard data in order to mitigate risk and safety of the construction site. The anxiety people adjacent to center of mud volcano remnant must be consider as a stake holder who need clear explanation for both risk and safety of the construction related within their populated areas. Recent phenomena for mud volcano in May 2006 is located at Sidoarjo area East Java, make people worried if due to man-made construction will trigger mud volcano activity on their backyard.

Figure 2. Geoelectrical Measurement Line.

IOP Conf. Series: Earth and Environmental Science 71 (2017) 012009 doi:10.1088/1755-1315/71/1/012009

Six geoelectrical lines @ +235 m with 5 meter interval for 48 electrodes are designed as shown in Figure 2. The subsurface imaging is intended to convince people surrounding construction area about geohazard mud volcano potential due to high rise building construction plan.

Based on 2D geoelectrical model and well data, the survey area consists of 3 anomaly sub surface classification: landfill, very soft silty clay associated with mud volcano eruption and soft silty sand clay.

Resistivity 2D geoelectrical data for lines 5 and 6 (Figure 3) showing distribution of very soft silty clay (red-yellow) which thickening toward north part in western investigated area. Therefore, subsurface data indicated abundance clay deposit material of very soft silty clay which related with center of mud volcano extrusion in past. In the eastern part, upper layer interpreted as land fill (blue) as recent consolidated layer. All of 2D resistivity data showing thinning distribution of silty clay away from eruption center as shown in Figure 3a, whereas mud volcano distribution is shown in Figure 3b.

The geoelectrical model result is reflected lateral distribution of mud. Chronological sequence model for very soft silty clay is started from fault weak zone as mud pathways of mud volcano which the product found in site plan of the construction building (Figure 4).

Interpretation results based on resistivity subsurface imaging indicated that mud volcano adjacent to construction site is directly influence within 50 meter radius from center of mud eruption. On the other hand, the construction site is ±300 m away from the center is acceptable as long as civil engineering team select best foundation type and technique during building construction. Both geoelectrical and drilling data showing a good data match such as: characteristic of mud volcano deposits as very soft silty clay with high plasticity, still containing natural water with low SPT value (standard penetration test). Thickness of this layer is about 15 meter. Base of engineering bed rock is ranging between 65 and 145 meter below the surface.

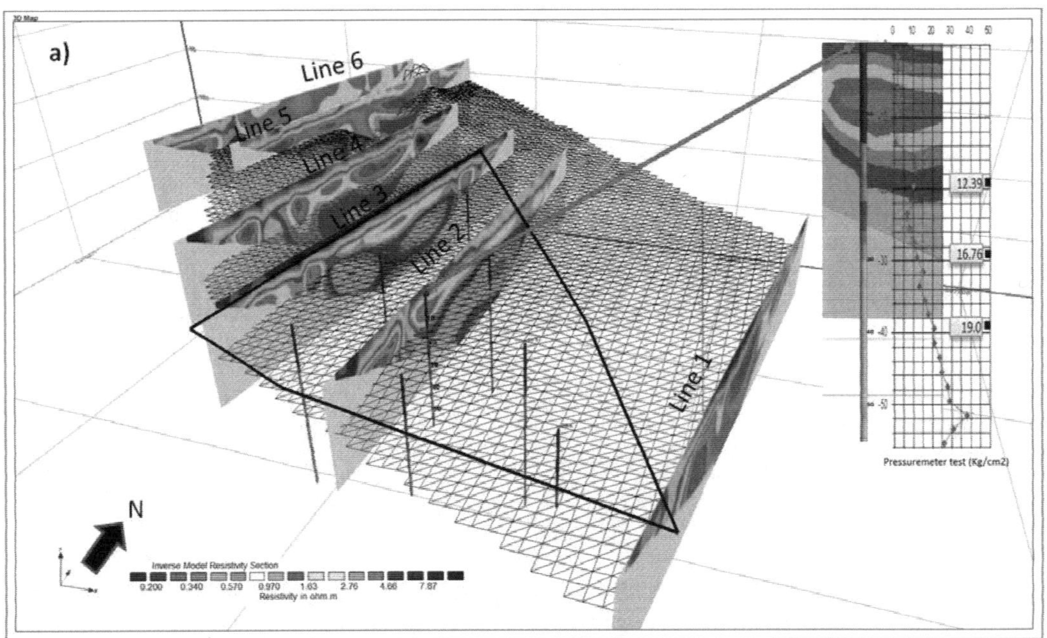

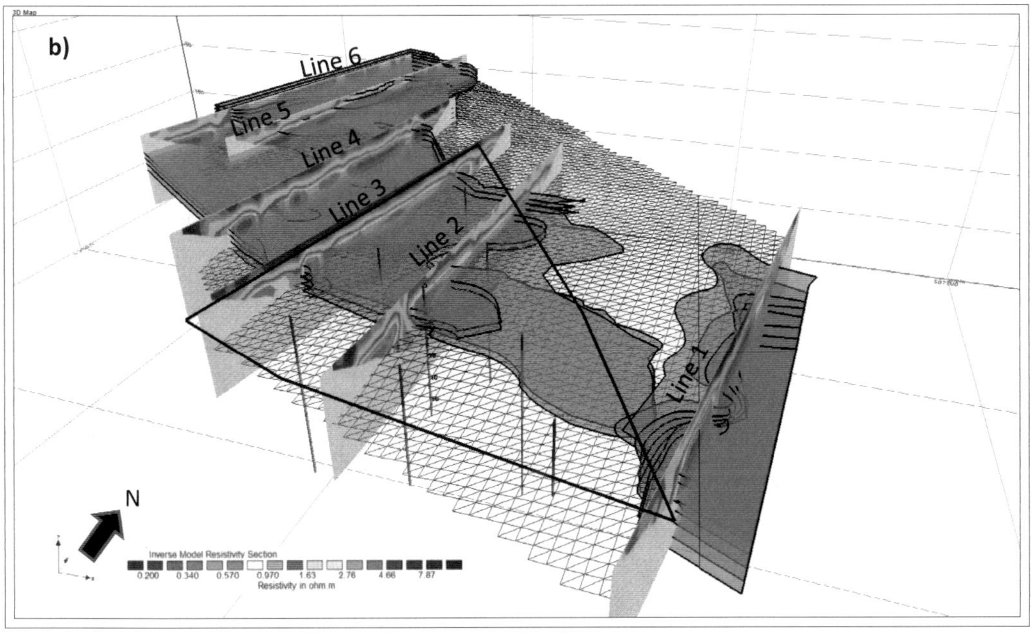

Figure 3. 2D Geoelectrical Model, well data and subsurface interpretation of lower boundary mud volcano product (a); Interpretation of mud volcano product distribution (b).

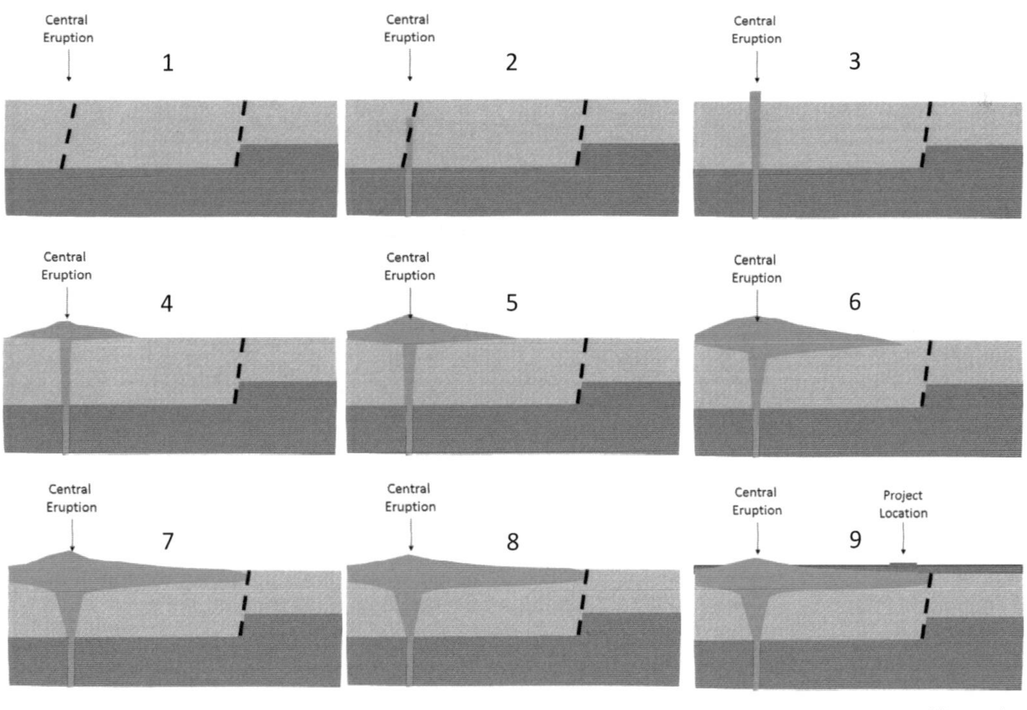

Figure 4. Chronological model very soft silty clay deposit erupted from weak zone of fault which is as mud pathways.

4. Case Study 2 Weak zone identification at toll road plan design

Mud extrusion at Porong, Sidoarjo, and East Java occurred 29 May 2006, about 150 m southwest of Banjarpanji 1 (BJP-1) Lapindo Brantas exploration well. Mud extrusion is still continuing until today (Figure 5).

Figure 5. Mud Extrusion Sidoarjo 28 Augusts 2009.

Based on analysis results carried out by [9 and 10], there is a possibility that the new toll road route planned within low deformation vulnerability zone. Therefore, deformation on the previous occurrence due to mud overburden surrounding can be found in the future. The presences of fault make a vulnerable potential for new source of mud extrusion since there is an indication weak zone as an active fault.

Geoelectrical using Wenner configuration of DC resistivity data acquisition carried out with 96 electrodes with 5 meter interval about 475 m length [11]. The objective is subsurface imaging about 30 meters depth with total 8 km line (Figure 6) The low resistivity features (red) in Figure 7 is consistently appears until + 27 meter depth in the northern part investigation area. Interpretation for the anomaly is a discontinuity of weak zone as fault within the adjacent area. The geoelectrical measurement result is confirming with geological survey agency study result [11-13]. Monitoring and special civil engineering construction around these areas must be considered the finding of the fault as identified previously.

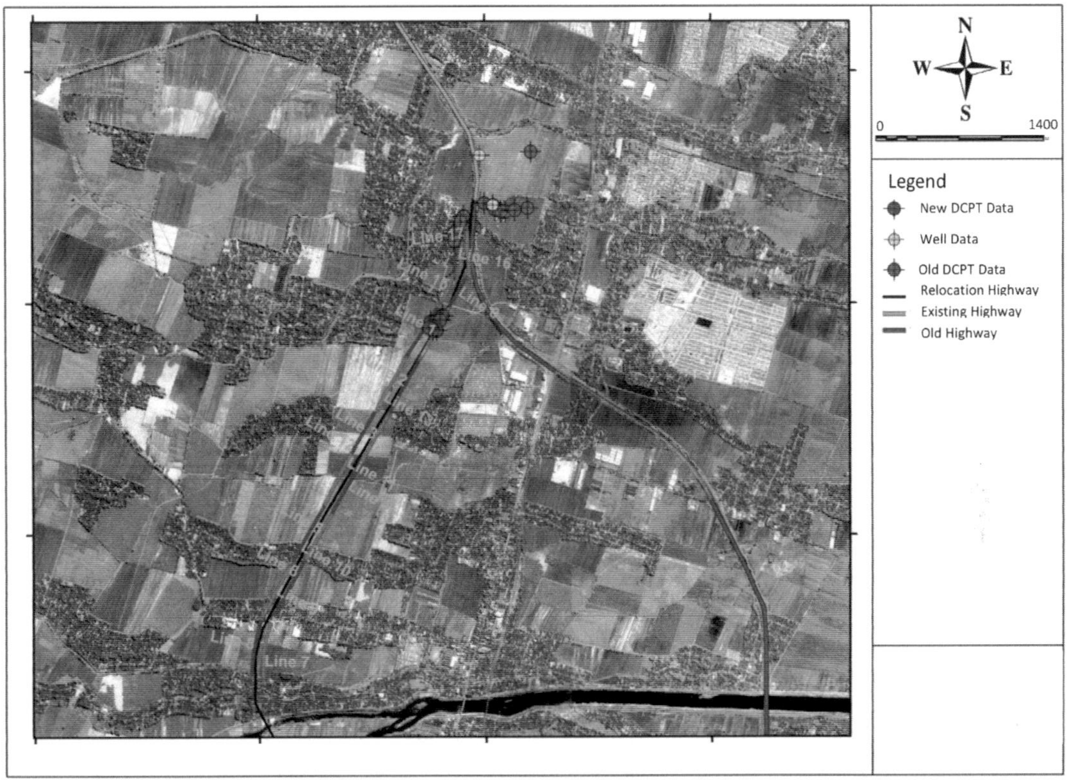

Figure 6. Geoelectrical line measurement [11].

Features of cross section compilation of 2D geoelectrical modeling as lateral resistivity is shown in Figure 7.

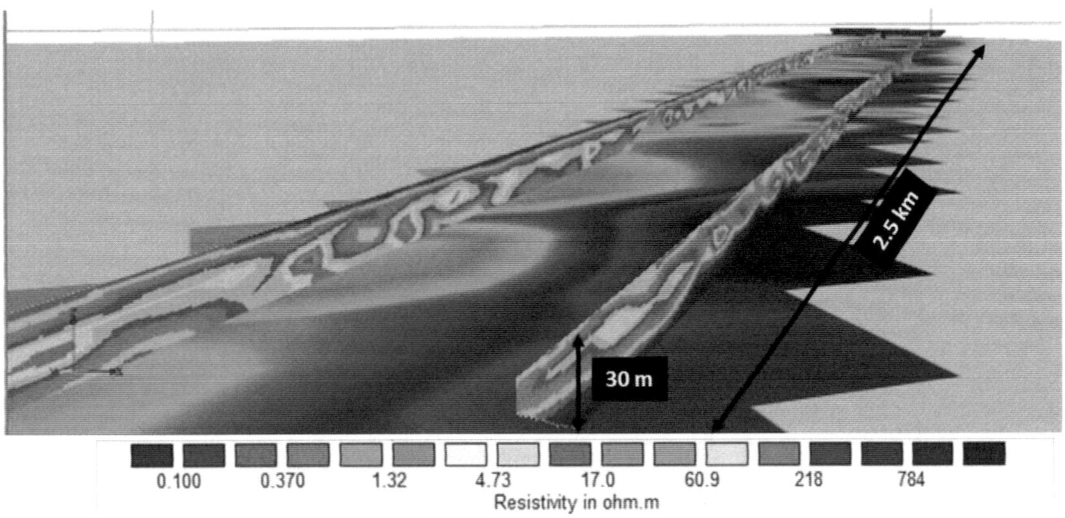

Figure 7. 3D Image of Lateral Resistivity Distribution.

5. Concluding Remarks

Based on two case study of planning high rise building and toll road which applying geoelectrical measurement and combine with well data is a powerful method subsurface imaging for both lateral and vertical of rock resistivity property distribution. The geoelectrical measurement is relatively give a very good result for shallow subsurface imaging, especially if combine with geotechnical data. It is easily to understand for common people to image distribution of rocks beneath their areas. Therefore, the contribution of the study results will help the stake holder who did the activity surrounding the area to explain their development/construction activity.

6. References

[1] Dahlin T and Zhou B 2004 A numerical comparison of 2D resistivity imaging with 10 electrode arrays *Geophysical Prospecting* **52** 379-398
[2] Reynolds JM 1997 An introduction to applied and environmental geophysics *John Wiley & Sons Ltd*
[3] van Bemmelen RW 1949 General Geology of Indonesia and Adjacent Archipelagos. In: The Geology of Indonesia vol IA *Martinis Nijhoff The Hague* 732 pp
[4] De Genevraye P and Samuel L 1972 Geology of the Kendeng zone (Central and East Java) *In: Proceedings of the Indonesian Petroleum Association 1st Annual Convention* pp 17–30
[5] Hall R 2002 Cenozoic geological and plate tectonic evolution of SE Asia and the SW Pacific: computer based reconstructions, model and animations *Journal of Asian Earth Sciences* **20** 353–432
[6] Sribudiyani, Prasetya I, Muchsin N, Sapiie B, Ryacudu R, Asikin S, Kunto T, Harsolumakso AH, Astono P and Yulianto I 2003 The collision of the East Java microplate and its implication for hydrocarbon occurrences in the East Java Basin *29th Annual Convention Proceedings of the ndonesian Petroleum Association*
[7] Smyth H, Hall R, Hamilton J and Kinny P 2003 East Java Cenozoic Basins, volcano and S–N cross section line ancient basement *Proceedings of the 13th Annual Convention & Exhibition*. Indonesian Petroleum Association
[8] Prasetyadi C, Suparka ER, Harsolumakso AH and Sapiie B 2006 The occurrence of a newly found Eocene tectonic Melange in Karangsambung Area, Central Java *Proceedings of the IAGI 35th Annual Meeting and Exhibition*
[9] Istadi B, Gatot P, Sumintadireja P and Alam S 2009 Modeling study of growth and potential geohazard for LUSI mud volcano International *Journal Marine and Petroleum Geology* pp 1-16
[10] Sumintadireja P, Indro PB and Istadi B Darmoyo, 2007 Geology and geophysics study in revealing subsurface condition of Banjarpanji mud extrusion, Sidoharjo, East Java, Indonesia *IUGG* **XXIV** Perugia Italy
[11] Lapi ITB 2012 Laporan Jasa Konsultasi Pekerjaan Kajian Geologi & Geofisika Trase Relokasi Jalan Tol Porong-Gempol PT Jasa Marga Report (unpublished)
[12] Lapi ITB 2006 Intergrated Geological, Geophysical and Geochemical Study of Mud Extrusion and Its Controlling Factors in Banjarpanji Area Brantas Inc Report (unpublished).
[13] Lapi ITB 2006 Geohazard Surveillance of Sidoarjo Mud Extrusion Area, East Java Lapindo Brantas Inc Report (unpublished)

Analyses of surface deformation with SBAR InSAR method and its relationship with aquifer occurrence in Surabaya City, East Java, Indonesia

Mushoddaq Mochammad and Asep Saepuloh

Study Program of Geology Engineering, Faculty of Earth Sciences and Technology, Bandung Institute of Technology (ITB), Indonesia
Jl. Ganesha No. 10, Bandung, West Java, Indonesia. 40132
Corresponding author e-mail address: Mshodaq@gmail.com

Abstract. Rapid development of Surabaya City requires space for industry and settlement. The city development is followed by increasing of water demand for various purposes. Large amount of groundwater extraction may cause ground subsidence. According to groundwater extraction, we tried to delineate the possibility of aquifer potential based on ground surface deformation using Interferometric Synthetic Aperture Radar (InSAR) and Electrical Resistivity methods. The InSAR is a method to estimate surface deformation in millimeter scale based on different phase of SAR data between acquisition times. In this paper, surface deformation combined with ground resistivity was used to analyse the potential of aquifer related to subsidence at Surabaya City, East Java, Indonesia. The Small Based Subset (SBAS) method was used to reduce phase delayed in the interferogram due to atmospheric condition. Based on the SBAS InSAR the deformation was detected at Surabaya City in January 2007 to February 2008. The largest and lowest deformation about 40 mm/year and about 0-10 mm/year were located in North to Southeastern and Western part of the city, respectively. The geodetic Global Positioning System (GPS) campaign in Surabaya confirmed that the detected deformation correlated with the ground subsidence with correlation coefficient about 0.96. To obtain the subsurface structures at subsidence zones, the Electrical Resistivity survey with Schlumberger configuration was performed in Surabaya city. According to the SBAS InSAR and subsurface resistivity, the subsidence around Pabeancantikan, Kenjeran, Simokerto and Tenggilismejoyo sub-districts agreed to high resistivity about 7-18 Ωm. The Sandstone of Kabuh and Pucangan formations in this area served as ground water aquifer which used by more than 200 wells. Ground water extraction probably triggered the subsidence at ground surface. The high resistivity at subsidence zones originated from high porosity of rocks served as aquifer with high groundwater extraction.
Keywords: Subsidence, SBAS, Aquifer, Surabaya, GPS, Schlumberger

1. Introduction

Land subsidence is a common phenomenon in big cities of Indonesia, such as Bandung [1], Jakarta [2], and Semarang [3]. Each city has different subsidence characteristics and their causes. The excessive exploitation of ground water is one of the causes of the land subsidence [4].

The Interferometric Syntetic Aperture Radar (InSAR) is a method to determine the surface deformation based on the difference in phase signal [5]. This method is effective to detect subsided or uplifted zones [6]. To characterize the land subsidence caused by groundwater exploitation; we also conducted the Electrical Resistivity survey with Schlumberger configuration. The Electrical Resistivity method could be used to interpret the location of ground water aquifers in the depth based on resistivity parameter of the rock formations. Characterizing land subsidence in Surabaya city has been being studied due to the amount of groundwater extraction. The connection between groundwater extractions with subsidence occurrence were analysed to interpret the cause possibilities of land subsidence in Survaya city including groundwater extraction.

Based on Shaded map from SRTM 1 arc second, the Surabaya City is located at low topography between 0-30 m above mean sea level (amsl). Surabaya city have height of ground ranging from 0-10 m that is occupying about 25,919 hectares (80,72%) spreads in the east, north, south, and centre. In other areas of the city have ground height of 10-20 m (12.53%) and above 20 m above sea level (6.76%) which is generally present in the west and south of the city of Surabaya. The location of study area is depicted in Figure (1).

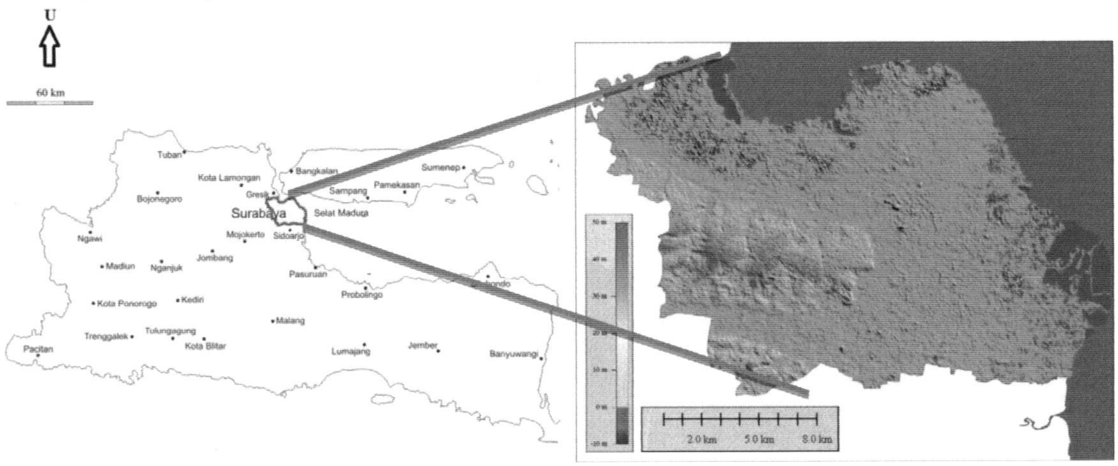

Figure 1. Study area at Surabaya city, East Java showed by the red line square (Shaded map from srtm 1arc second)

The geomorphology of the study area is dominated by paneplain of alluvial about 80% from total area. Therefore, the outcrop is very limited. Subsoil has a grain size of clay to sand. Differences in soil color and gradation of soil grains found locally in the eastern part. Light brown soil with grain size of clay to sand and gravel, found in small amounts. Based the geological map (Figure 2), this soil is interpreted as weathering rock of Pucangan formations. The geological structures could not be found clearly at the field. Therefore, we used geological maps (Figure 2) to know structure of rock, in the location trending anticline is the east - west.

Based on Geological map of Surabaya and Sepulu [7] Surabaya has 4 lithologies with the youngest lithology are are Alluvium unit plains that has unconformity contact with all the older (Figure 2). Detail stratigraphy could be explained as follows:

1. Aluvium Unit

Alluvium unit composed of clay to gravel sized loose material Alluvium unit composed of clay-sized material loosened up gravel, consisting of clay, sandy clay, sand gravel, loose, soft, water passing medium - high. This sediment is the result of the deposition of sediments by the river flow. Alluvial Deposition is spread in most areas of Surabaya starting from the north, south, east, and in the area around the coast. The maximum depth is 30 m and occupies a land area reaching more than 60% of the research area

2. Sandy mudstone unit

This unit consists of sandstone, tuffaceous sandstone and conglomerate. Commonly, colored gray - brown, fine grained – rough, angled - rounded, disaggregated bad - medium, fragile, passing the water is medium - high, containing of conglomerate insertions This unit has weathered up generally in the form silty clay containing a bit of sand, brown, soft and plastic when it expands, friable when dry, less than 1 m thick. Actually occupy several places in the southern part, approximately 17% of the area of mapping according to the distribution Pucangan Formation and Kabuh Formation

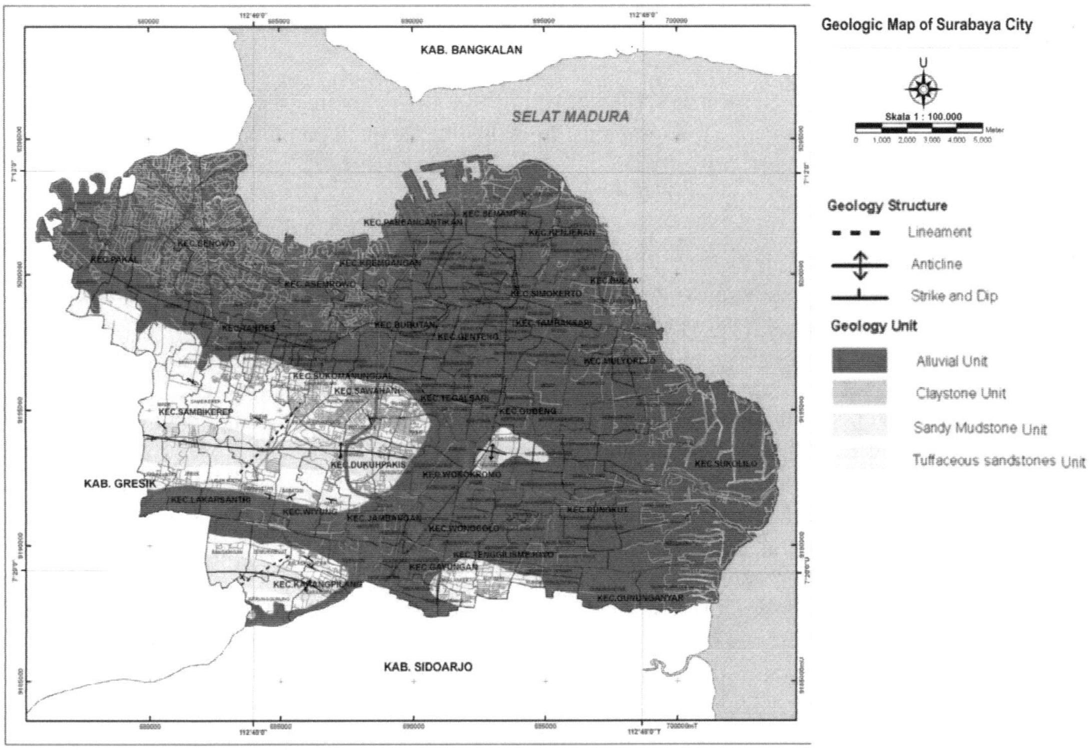

Figure 2. Geologic Map of the city of Surabaya scale of 1: 100.000 (modified from [7]).

2. Data and Method

2.1. Data

The data used in this study consists of two main data. The first data is Synthetic Aperture Radar (SAR) from Advanced Land Observation Satellite (ALOS) Phased Array type L-band Synthetic Aparture Radar (PALSAR) level 1.0. The second data is Electrical Resistivity from 20 measurement points. There are nine ALOS PALSAR data used and acquired on 3rd January 2007 to 21st February 2008. These data are HH polarimetric mode. We also used GPS campaign measurements from 20 stations in October to December 2010. The ALOS PALSAR data used in this study were listed in Tabel 1.

The correlation and interpretation of the ground surface deformation and electrical resistivity were used to evaluate and characterize the aquifer potential zones.

Table 1. Detail ALOS PALSAR data used in this study

No	Name of data	Date of record	Direction
1	ALPSRP050183770-L1.0	03-Jan-07	Descending
2	ALPSRP056893770-L1.0	18-Feb-07	Descending
3	ALPSRP063603770-L1.0	05-Apr-07	Descending
4	ALPSRP070313770-L1.0	21-Mei-07	Descending
5	ALPSRP083733770-L1.0	21-Agust-07	Descending
6	ALPSRP090443770-L1.0	06-Okt-07	Descending
7	ALPSRP097153770-L1.0	21-Nop-07	Descending

IOP Conf. Series: Earth and Environmental Science **71** (2017) 012010 doi:10.1088/1755-1315/71/1/012010

8	ALPSRP103863770-L1.0	06-Jan-08	Descending
9	ALPSRP110573770-L1.0	21-Feb-08	Descending

2.2. Method

The Small Baseline Subset (SBAS) InSAR algorithm was first demonstrated by [8] as a means to mitigate atmospheric artifacts and topographic errors in time-sequential interferograms and thus to obtain time-series deformation information. The algorithm uses only interferograms with small baselines that overlap in time in order to reduce spatial decorrelation. The SBAS technique relies on a proper combination of unwrapped interferograms characterized by a small normal baseline between the acquisition orbits and a short time interval (temporal baseline) between the acquisition epochs. Following [9], the SBAS processing work flows depicted in Figure 3 could be explained as follows:

1) Evaluation of the orbital parameters associated to each SAR acquisition (subsequently used within the estimation of the spatial baseline value) and reference Master image selection.
2) Generation of a set of single look SAR images (SLC) from the available raw data files (focusing).
3) DEM conversion into reference Master image SAR coordinates and computation of sensor to target distance files (range files) for each acquisition.
4) Optimal interferometric data pair distribution selection.
5) Co-registration of each SLC data pair with respect to the selected reference Master image
6) Differential interferograms and corresponding spatial coherence maps generation
7) Noise-filtering of the generated Differential InSAR (D-InSAR) fringes
8) Generation of the mean deformation velocity map and the corresponding time series, for each coherent pixel of the investigated area, via the inversion of the computed sequence of D-InSAR interferograms. At the same time, an estimate of the possible residual topographic components and the atmospheric contributions is also accomplished
9) Geocoding of the obtained results and projection onto a universal cartographic grid.

Resistivity method is one of geophysical exploration method that utilizes resistivity properties of medium to learn the situation subsurface. In principle, the current flowed / injected into the earth through a pair of the current electrodes and then the response received is the potential difference on pair of electrodes potential then the response received is the potential difference a pair of electrodes potential. From this response, can then studied the electrical properties of the media under the surface, thus enabling the geological interpretation to create a subsurface geological models based on the electrical properties of the medium (rock). In this study, the geological model used to determine the groundwater aquifer. Here are the steps performed in the resistivity method.

1) Measurement in the study area, the measurement is performed at 20 points using Schlumberger configuration. This configuration is already well known in determine the position groundwater (aquifers).
2) 1D modeling preparation with inverse modeling methods Ridge Regression (RR) which can be expressed by equation from [10] as follows:

$$\delta g = (J^T J)^{-1} J\, \delta f \qquad (1)$$

   Where δg is the difference between the initial parameters and allegations parameter, δf is the difference in pseudo resistivity model and pseudo allegations resistivity models, J is the Jacobian matrix.
3) From the 20 points 1D models correlated to become a 2D cross sections to determine the distribution of potential rock aquifers in the area of research.
4) From that 20 points, was made contour of thickness of potential aquifer areas on entire research area.

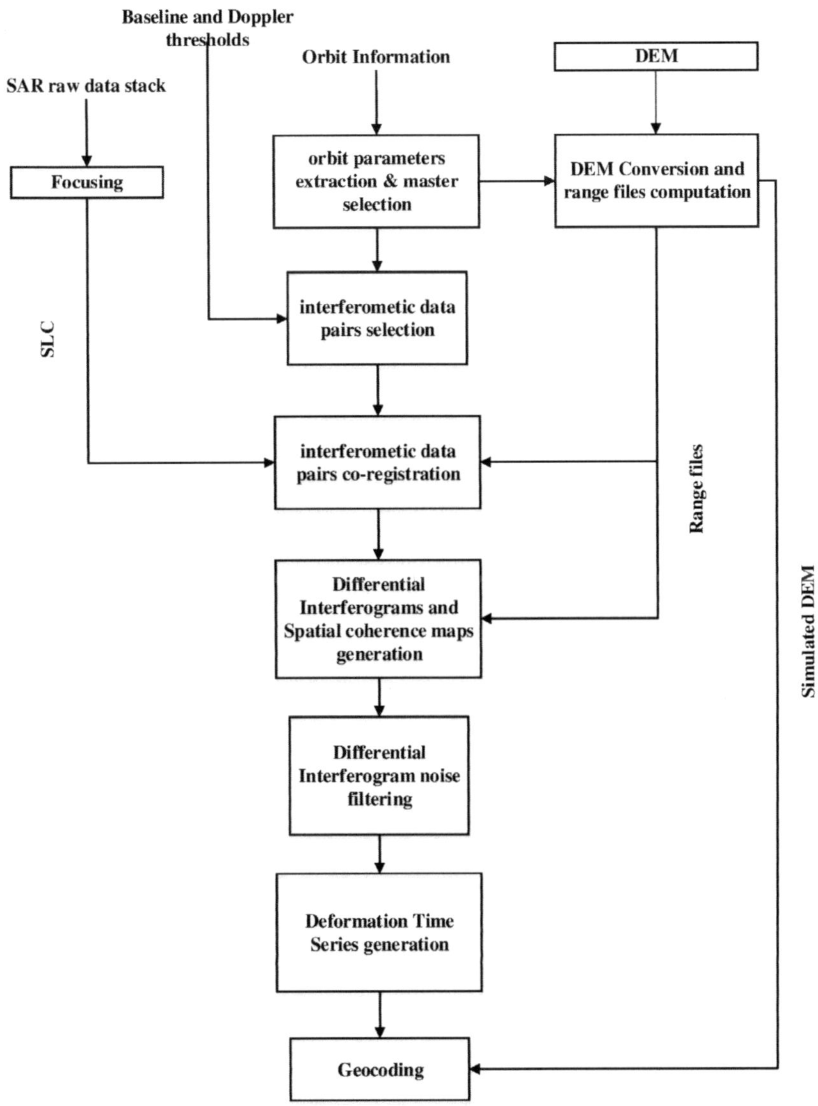

Figure 3. SBAS processing work flows [9].

3. Result and Discussion

3.1. SBAS InSAR

- Orbit parameters extraction and Master selection

By exploiting the orbital information, it is possible to plot each normal baseline of SAR data pair, as shownSAR image in the Baseline plan, as shown by Figure 4

By benefiting of this graphical representation, it is possible to properly select the reference master data with the shortest normal baseline. In order to help the subsequent co-registration step, it is useful to select as a reference the SAR image with a baricentrical position in the acquisition distribution plot. It implies that the normal of data pair are the smallest.

- Data Pairs Selection and co-registration

Data pair selection was obtained by imposing a threshold on the minima temporal and normal baseline. Therefore, only smallest perpendicular baseline interferograms would be generated. Each

interferogram corresponds to an arc connecting two green point, i.e., two SAR acquisitions. Blue arcs are relevant to interferograms belonging to the Triangulated Irregular Network.
For a generic point in the master image, the SAR coordinates of the same point in the slave image is given by:

$$x'_s = x'_m + \delta\, x'\,(x', r') \quad (2)$$

$$r'_s = r'_m + \delta\, r'\,(x', r') \quad (3)$$

wherein (x'_m, r'_m) and (x'_s, r'_s) are the coordinates of the point in the master and slave geometry, and $(\delta\, x', \delta\, r')$ is the difference coordinates in the SAR to be applied to the master image coordinates. The registration process can solve the problem of computing the geometric image transformation and the problem of re-sampling the second image with respect to the master one, in such a way that each ground point is located at the same position in the two different images.

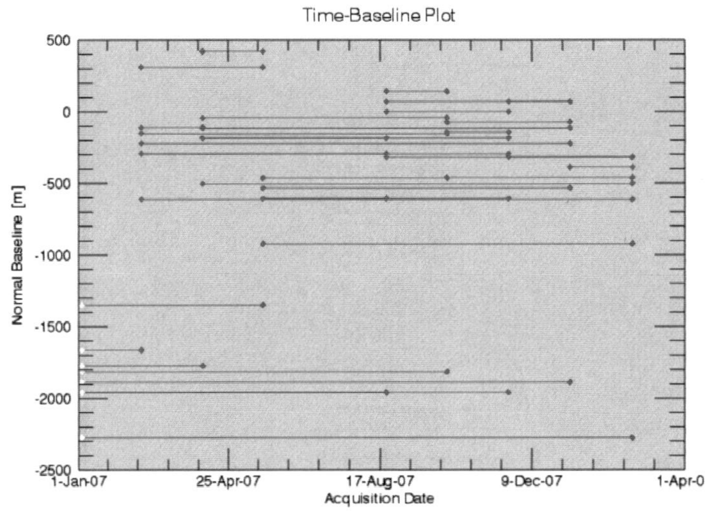

Figure 4. The normal baseline for data pair distribution from nine scenes of ALOS PALSAR.

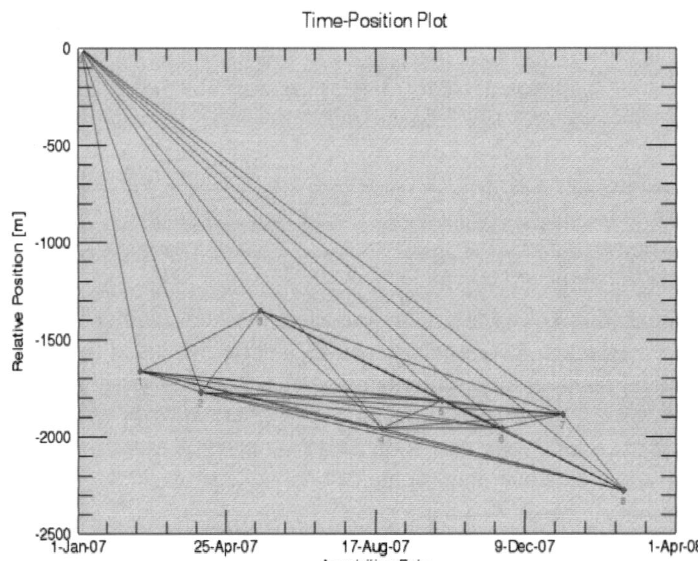

Figure 5. The ALOS PALSAR selected data pair distribution in the normal baseline plane.

- Differential interferogram generation

The generation of the differential interferograms is carried out by subtracting the estimated topographic phase contribution from the interferometric phase term the differential interferometric phase $\Delta\varphi$ is related to the deformation occurred between the two passes of the SAR sensor by the following relation:

$$\Delta r = \frac{4\pi}{\lambda}\,\Delta\varphi \tag{4}$$

where Δr represents the slant range sensor-to-target difference (after removing the synthetic fringes) between the two passes and λ is the sensor wavelength (Figure 6).

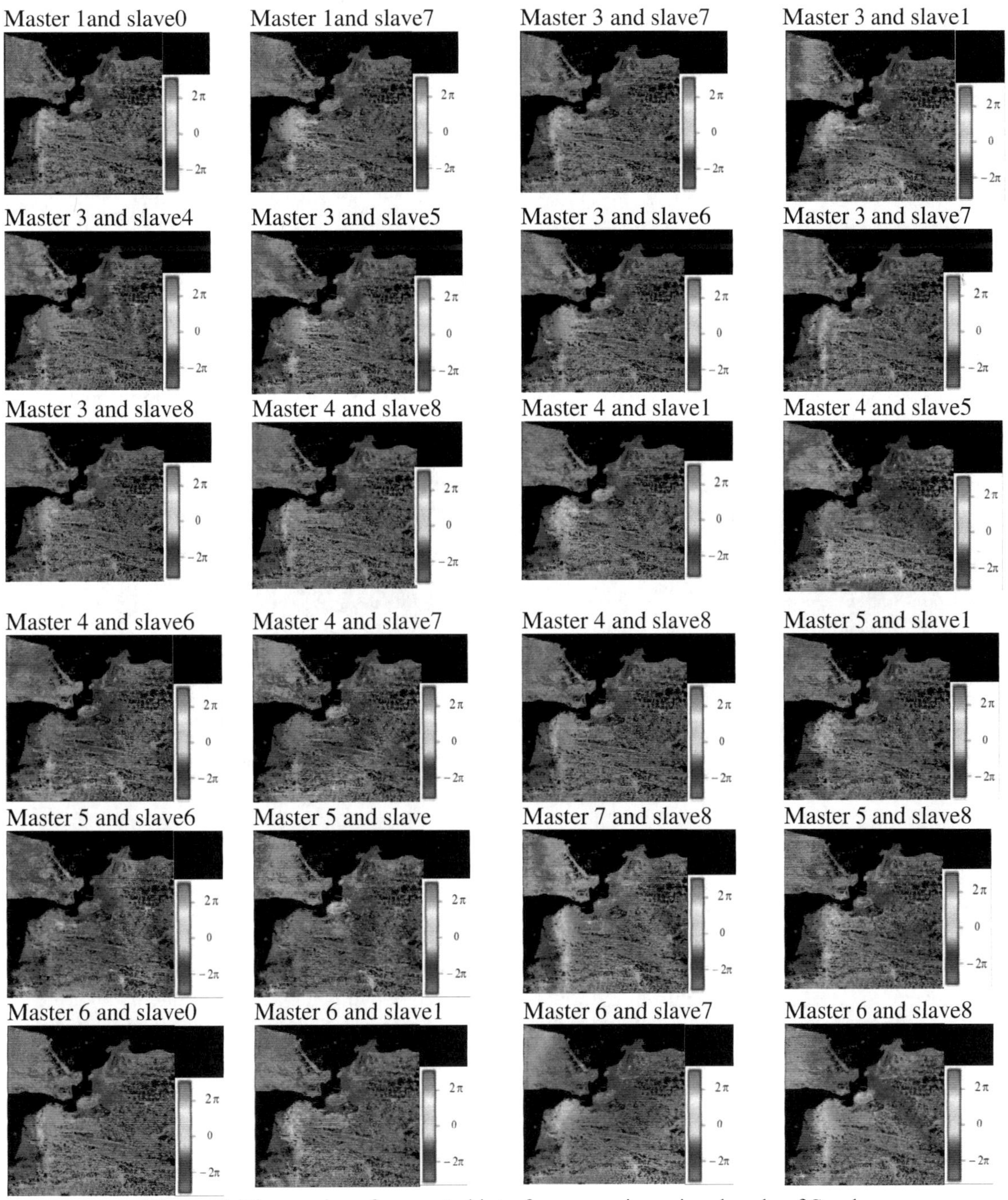

Figure 6. Time series of generated interferograms in regional scale of Surabaya.

- Deformation time series generation

Deformation time series generation is the SBAS INSAR processing can be summarized as follows:

1) In order to extract from each generated differential interferogram an information relevant to the actual deformation, the computed differential phases, representing the modulo-2π restriction (wrapped) of the original interferometric phase signal.
2) The unwrapped interferograms, previously filtered from the residual topography artifacts, are used within the time series generation. Indeed, the selected DInSAR pairs are characterized by a small spatial and baseline as well as by a small frequency shift between the Doppler centroids [11]. As a consequence of these constraints, the SAR images involved in the interferogram generation could be grouped in several independent small baseline subset, thus an appropriate "link" of such subsets is required.
3) Finally, the atmospheric phase signal is evaluated and removed from the computed deformation time series uses the Singular Value Decomposition (SVD) approach based on a minimum-norm criterion of the deformation rate to derive time-series deformation estimates.

In this context, we just remark that by exploiting the previously unwrapped filtered interferograms $\{\psi_j\}_{j=1}^{M}$ it is possible to obtain the phase values $\{\varphi_i\}_{i=1}^{N+1}$ at each SAR epoch, i. e., the temporal evolution of the deformation for each investigated pixel.

Since interferometric products are differential measurements, before to start, it is necessary to select one of the identified points as a spatial reference for the whole interferogram set. It would be characterized by high spatial and temporal coherence and its deformation behavior should be a priori known (typically, it is located in a non-deforming zone). Indeed, note that any error affecting this reference point will also affect the overall results; in this context, particularly critical can be the impact of the atmospheric artifacts that are spatially correlated thus to be confused with deformation signals.

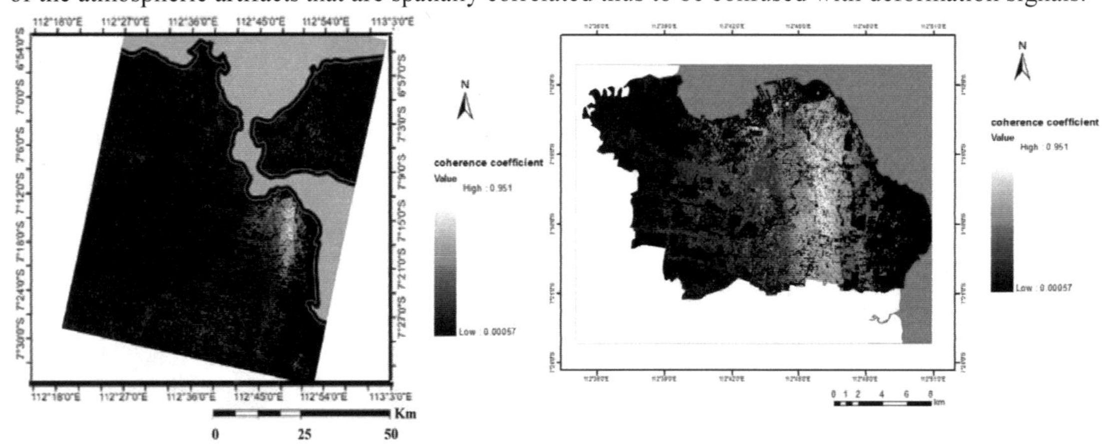

(a) (b)

Figure 7. The coherence value of regional area (a), the coherence value of research area (b).

Figure 7a shows the coherence value of 24 data pair. The high coherence presented by bright portions indicated high correlation of the phase signal in data pair. Figure 7b is a subset coherence image for Surabaya city. It shows that the coherence of the study area is high. Therefore, the data pairs are plausible to generate high quality of deformation. Pictured above is the value of coherence of the study area. A high coherence presented by bright portion indicated that the InSAR deformation could be calculated properly. Figure (7a) is the appearance of the value of coherence in regional areas. This figure shows that the study area is located at high coherency. Figure (7b) is detail image the coherence value, the image is an area of research site.

- Phase Unwrapping

The Phase Unwrapping (PhU) step, i.e., the problem to retrieve,starting from the computed modulo-2π restricted differential phases, the full phase variation signal, the latter associated to the occurred deformation, is probably the most critical task in SAR interferometry. However, in a multi-temporal DInSAR scenario, where large interferogram data set are involved, exploitation of temporal relations among the different interferometric phases can help Phase Unwrapping procedures.

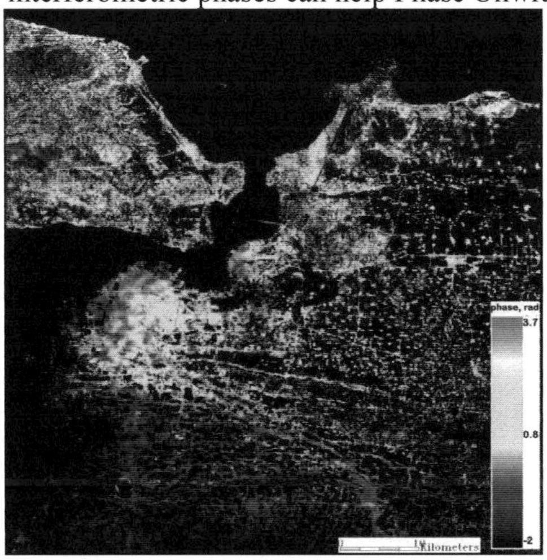

Figure 8. Unwraphed phase master 1 and slave 7.

- Atmospheric filtering and orbital ramps estimation

Detecting possible atmospheric artifacts is the final step of the SBAS processing. This processing is based on the observation that the atmospheric phase signal is highly correlated in space but poorly in time [12]; [13]. Consequently, the undesired atmospheric phase signal is estimated from the computed time series by a lowpass filtering step in the two-dimensional spatial domain and a temporal highpass filtering operation. The filtering is implemented with spatial extent of approximately 1 km in azimuth and range, consistent with the spatial correlation length of the atmospheric phase signal [14]. the atmospheric artifacts are removed and the generation of the final deformation time series is computed.

- Geocoding and SBAS Invertion

All the SBAS deformation products are obtained onto the SAR reference geometry but they can be represented with respect to an easier output reference grid.

The SBAS invertion steps are performed to obtain the LOS displacement of each time period. The LOS displacement can be converted into vertical and horizontal components. In this case, the GPS measurements indicated that the displacement in east–west direction is very small. In consideration of vertical component is a one dimensional vector, the conversion of the LOS displacement to vertical displacement can be expressed as $d_{disp} = d_{LOS}/\cos\alpha$ where d_{disp} is the vertical displacement; d_{LOS} is the LOS displacement; α is the incident angle, which can be obtained from the SAR data. The geocoded vertical displacement mapped in WGS84 coordinate system is shown in Figure 9 in which the colour points are the spread of groundwater wells in the Surabaya city

3.2. Aquifer potential zone

The 20 points resistivity data were obtained from field measurement at study area. In interpreting subsurface geological conditions, principally the value of resistivity reflects physical condition of layers or lithology. The interpretation of the geological conditions initially was obtained by 1D models. According to the model, we interpreted the depth and thickness of lithological layers (Table 2). The value of the same thickness in the area of research then a contour map created. After the process of making the aquifer contour, unconfined aquifer presented by thickness contour was depicted by figure 10a and confined aquifer by Figure 10b. The blue portion of the contour is an aquifer with high thickness and green does not have the potential to be aquifer aquifers due to green color does not have a sufficient thickness to keep the water in the pores. the unconfined aquifer seen to have four zones between Northern Surabaya, covering Bulak districts and Asemrowo districts. The Southern Surabaya has potential aquifers in Gununganyar and Karangpilang districs. The confined aquifer seen to have two zones. Major aquifers found in Gubeng-Bubutan districs and minor aquifer foind in Gununganyar districs.

Table 2. The thickness of the aquifer were obtained from measurements of geoelectric

Drill point	Depth	Thickness	Lithology
1	4.6m – 16.5m	10.97m	sandy mudstone
2	7.37m – 18.8m	11.13m	sandy mudstone
	41.7m – 56.7m	15m	sandy mudstone
3	9.6m – 15.8m	6.2m	sandy mudstone
4	8.45m – 13.3m	4.85m	sandy mudstone
	22.7m – 41.4m	18.8m	sandy mudstone
5	4.59m – 11.3m	6.71m	tuffaceous sandstones
6	3.99m – 10.6m	6.61m	tuffaceous sandstones
	51.1m – 66.1m	15m	sandy mudstone
7	6.05m – 18.5m	12.45	tuffaceous sandstones
8	4.38m – 7.48m	3.1m	sandy mudstone
	17.3m – 30.2m	12.9m	tuffaceous sandstones
	65.2m – 80.2m	15m	sandy mudstone
9	3.99m – 10.7m	6.71m	tuffaceous sandstones
	23.3m – 38.3m	15m	sandy mudstone
10	8.68m – 10.4m	1.72m	sandy mudstone
	14.2m – 32.2m	18m	sandy mudstone
11	6.13m – 10.7m	4.57m	tuffaceous sandstones
12	6.88m – 9.91m	3.03m	sandy mudstone
	76.5m – 90.5m	5.28m	sandy mudstone
13	9.43m – 16.8m	7. 37m	conglomerate
	39.1m – 54.1m	15m	sandy mudstone
14	12.4m – 19.2m	6.8m	conglomerate
	36.7m – 51.7m	15m	sandy mudstone
15	10m – 16.8m	6.8m	sandy mudstone
16	17.2m – 30.8m	13.6m	sandy mudstone
17	15.2m – 25.8m	10.6m	sandy mudstone
18	6.56m – 11m	4.44m	sandy mudstone
19	6.94m – 16.5m	9.56m	sandy mudstone
	35.3m – 37.1m	5.97m	sandy mudstone
20	49.8m – 64.8m	15m	sandy mudstone

3.3. Land subsidence and the relationship with the aquifer

Once known of potential unconfined aquifer and confined aquifer in Surabaya city, and known to the subsidence that occurred in the city of Surabaya, further analysis is the correlation between the two of them. Both could have happened any correlation when excessive groundwater extraction, meaning that the fluid (water) previously filling the aquifer zone is reduced. With this reduction, the layer above it will be down for filling the empty space below, of course, the influence of the load and the compactness of rock above it will take effect.

It can be seen in Figure 9a that the subsidence that occurred in the city of Surabaya, most occurred areas are Wonokromo, Gubeng, Pabeancantikan to Bubutan districts. Even the Subsidence that occurred in the city of Surabaya can reach 40mm per year.

The results of InSAR showed some highest in the city of Surabaya, the biggest subsidence zone is indicated by the red color. This area is most common in eastern and surrounding areas. Data from GPS measurements in 3 months (Figure 9b), the east part on the city of Surabaya is also a significant subsidence. Although the data were taken at different times, but the pattern of subsidence in the area of Surabaya can viewed in same.

To determine the cause of the subsidence (surface deformation), should be viewed on a variety of parameters such as geological factors, the loading rate up to mining activities. But in this study, focused on the parameters of existence the aquifer. The results of measurements, obtained contour unconfined aquifer and confined aquifer. unconfined aquifer shown in Figure 10a and confined aquifer shown in Figure 10b. Color contour on both these drawings are the thickness of the rock that has potential as aquifers.

Contur map in Figure 10b is The thickness value of the confined aquifer, in this picture there are two aquifer, which is located in downtown Surabaya. Seeing the results of the processing of InSAR and GPS measurements, confined aquifer is the most match layout of the land subsidence occurred because in confined aquifer, land subsidence caused by the extraction of groundwater, pores or aquifers that should be fills with water will be empty and no water which it replaces [15]. The existence of empty space that can not be filled again, things like this happen when drilling for water too deep, thus bypassing the impermeable zone and aquifer that is called a confined aquifer [16]. So that surface water can not fill this area. Another cause of course due to the extraction of ground water that is too excessive.

In addition to the aquifer contour maps and map decline in the city of Surabaya, there is also a map of drilling locations in the city of Surabaya (Figure 11), this map is a representation of groundwater carried out in the city of Surabaya. The groundwater extraction even up to more than 4000 m^3 per month. When looking at this map, we will know where the most extracted groundwater. So with the spread of drilled wells (Figure 11) seen that the distribution of wells drilled numerous in the confined aquifer (the most) in the area Gubeng to district Bubutan and the free aquifers are several wells drilled in the aquifer zone districts of Bulak.

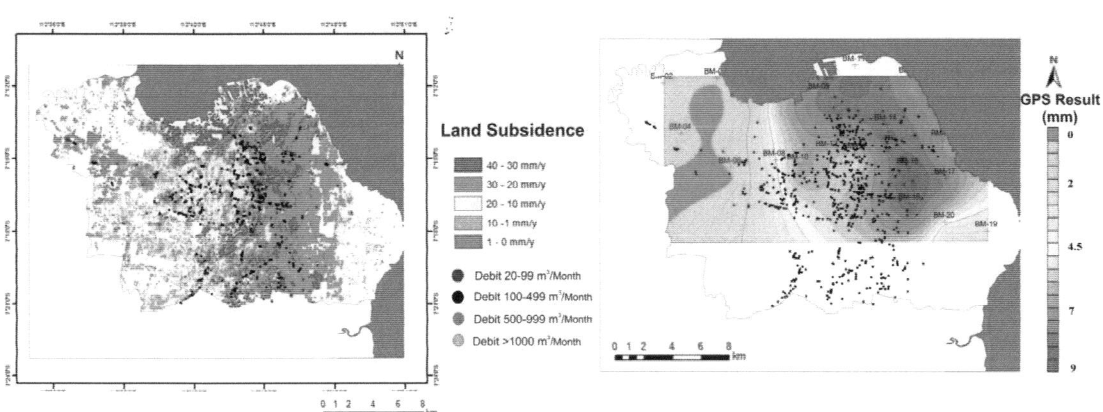

(a) (b)

Figure 9. (a) Subsidence maps obtained from SBAS InSAR processing, (b) Subsidence maps obtained from GPS measurement. Map shown in Surabaya area with a color scale that indicates the amount of subsidence that occurred.

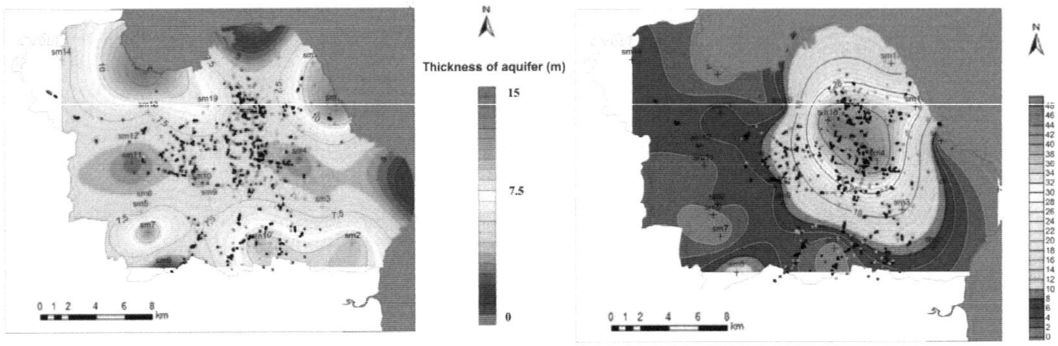

(a) (b)

Figure 10. **(a)** Thickness contour of unconfined **(b)** confined aquifer generated from each point geoelectric correlation process. Red portion indicates the thickness of the aquifer is higher than blue portion.

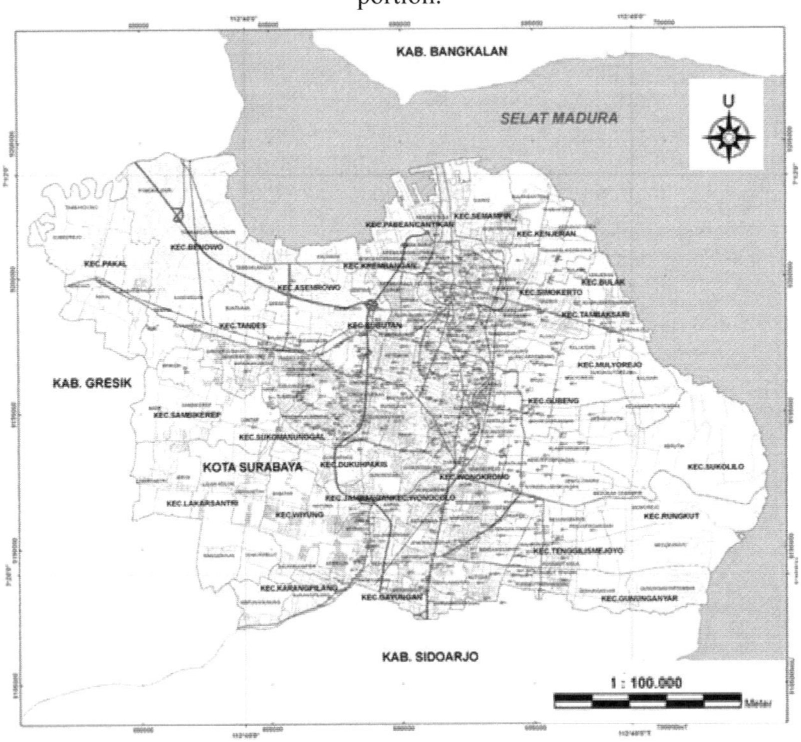

Figure 11. The distribution map of groundwater wells in Surabaya presented by green color.

4. Conclusions

SBAS Insar method has proven effective to estimate the deformation in the city of Surabaya. According to the SBAS InSAR, the general subsidence was occurred in the city of Surabaya. The highest subsided zones about 30-40 mm per year are presented by red portion and located at Wonokromo district, Gubeng district, Bubutan district, and Pabeancantikan district. On the unconfined aquifer zones, there are four zones of potential aquifers at Asemrowo districts, Bulak districts, Karangpilang district, and Gununganyar district. The four potential zones of the aquifer were located at the same position with the land subsidence in Surabaya, Gununganyar district, and Bulak district. On the confined aquifer zones, there are two zones of potential aquifers at Gununganyar, and large aquifers area at Gubeng and Bubutan districts. This large aquifer is the nearest one to the land subsidence in the city of Surabaya.

References

[1] Abidin, H.Z., H. Andreas, M. Gamal, A. D. Wirakusumah, D. Darmawan, T. Deguchi, Y. Maruyama. 2008. *Land Subsidence Characteristic of the Bandung Basin, Indonesia, as Estimated From GPS and InSAR*. Journal of Applied Geodesy 2, 167-177.

[2] Abidin. H.Z, R. Djaja, D. Darmawan, S. Hadi, A. Akbar, H. Rajiyowiryono, C. Subarya. 2000. *Land Subsidence of Jakarta (Indonesia) and its Geodetic Monitoring System*. Natural Hazards 23: 365–387.

[3] Kadir W.G.A., Santoso D., dan Sarkowi M. 2004. *Time lapse Vertical Gradient Microgravity Measurement for Subsurface Mass Change and Vertical Ground Movement (subsidence) Identification, Case Study: Semarang Alluvial Plain, Central Java, Indonesia*. Proceeding of the 7th SEGJ International Symposium.Sendai-Japan.

[4] Hutasoit, L.M., M.H. Pindratno. 2004. *land Subsidence at DKI Jakarta. (P3TPSLK) BPPT and Hanns Seidel Foundation (HSF)*.

[5] Saepuloh A., Urai M., Aisyah N., Sunarta, Widiwijayanti C., Subandriyo, and Jousset P., 2013. *Interpretation of ground surface changes prior to the 2010 large eruption of Merapi volcano using ALOS/PALSAR, ASTER TIR and gas emission data*, Journal of Volcanology and Geothermal Research, 261, 130-143.

[6] Amighpey, M. and Arabi, S. 2016. *Studying land subsidence in Yazd province, Iran, by integration of InSAR and levelling measurements Remote Sensing Application*, vol 4, 1–8.

[7] Sukardi. 1992: *Geological map of Surabaya dan Sapulu* (16308 - and 1609-1), scale 1:100.000. Departemen pertambangan dan Energi, Direktorat jenderal Geologi dan Surberdaya Mineral, Pusat Peneliti dan Pengembangan Geologi. Bandung

[8] Berardino, P., Fornaro, G., Lanari, R., and Sansosti, E. 2002. *A new Algorithm for Surface Deformation Monitoring based on Small Baseline Differential SAR Interferograms*. IEEE Trans. Geosci. Remote Sens., 40, 11, pp. 2375-2383.

[9] Casu, F. 2009. *The Small Baseline Subset technique: performance assessment and new developments for surface deformation analysis of very extended areas* , University of Cagliari, Italia.

[10] Faizin Nur and mustopa E.J. 2015. Pemodelan dan Inversi 1-D Metode Geolistrik Konfigurasi Schlumberger dengan Menggunakan Matlab. SNIPS 2015. Bandung. Indonesia.

[11] Franceschetti, G., Lanari; R. 1999. *Synthetic Aperture Radar Processing*, CRC PRESS, Boca-Raton .

[12] Ferretti, A., Prati, C., Rocca, F. 2001. *Permanent scatterers in SAR interferometry*. IEEE Trans. Geosci Remote Sens., 39, 8-20.

[13] Goldstein, R. M. 1995. *Atmospheric limitations to repeat-track radar interferometry*. Geoph. Res. Lett., 22, 2517–2520.

[14] Hanssen, R. 2001. *Radar interferometry – data interpretation and error analysis*. Kluwer Academic Press, Dordrecht, The Netherlands.

[15] Setyawana, A.,Fukuda, Y., Nishijima, J. and Takahito Kazama, T. 2015. *Detecting Land Subsidence Using Gravity Method in Jakarta and Bandung Area, Indonesia*. Procedia Environmental Sciences, 23. 17 – 26.

[16] Kodoatie, R.J. 1996: *Pengantar Hidrologi*. Andi Offset, Yogyakarta.

Groundwater and solute transport modeling at Hyporheic zone of upper part Citarum River

Irwan Iskandar[1,2], Hendy Farazi[1], Rahmat Fadhilah[2], Cipto Purnandi[2], Sudarto Notosiswoyo[1,2]

[1]Earth Resources Exploration Research Group Faculty of Mining and Petroleum Engineering
[2]Groundwater Engineering Master Program, Faculty of Earth Science and Technology
Email: irwan@mining.itb.ac.id

Abstract. Groundwater and surface water interaction is an interesting topic to be studied related to the water resources and environmental studies. The study of interaction between groundwater and river water at the Upper Part Citarum River aims to know the contribution of groundwater to the river or reversely and also solute transport of dissolved ions between them. Analysis of drill logs, vertical electrical sounding at the selected sections, measurement of dissolved ions, and groundwater modeling were applied to determine the flow and solute transport phenomena at the hyporheic zone. It showed the hyporheic zone dominated by silt and clay with hydraulic conductivity range from 10^{-4}~10^{-8} m/s. The groundwater flowing into the river with very low gradient and it shows that the Citarum River is a gaining stream. The groundwater modeling shows direct seepage of groundwater into the Citarum River is only 186 l/s, very small compared to the total discharge of the river. Total dissolved ions of the groundwater ranged from 200 to 480 ppm while the river water range from 200 to 2,000 ppm. Based on solute transport modeling it indicates dissolved ions dispersion of the Citarum River into groundwater may occur in some areas such as Bojongsoang-Dayeuh Kolot and Nanjung. This situation would increase the dissolved ions in groundwater in the region due to the contribution of the Citarum River. The results of the research can be a reference for further studies related to the mechanism of transport of the pollutants in the groundwater around the Citarum River.

1. Background

Citarum River flows from Cisanti Lake, Wayang Mountain to Java Sea. This river flows through 12 cities and many people rely on Citarum water for their daily life. Citarum River is known as one of the most polluted river in the world [1]. It was caused by urban and industrial waste along the riverbanks. Other than taking their water supply directly from the river, people also taking the water from dug-well along the river.

Interactions between Citarum River and the shallow groundwater (hyporheic zone) is very important to understand. Release and sink of dissolved metals or ions between the two environment i.e. surface water of Citarum River and groundwater in the hyporheic zone should be known to understand environmental problems. One of the method that can be applied to monitors transfer ions between river and groundwater is by observing total dissolved solid (TDS) at Citarum River and the hyporheic zone. In this study, the method was applied and combined with vertical electrical sounding at hyporheic zone of Citarum River and hydrogeology numerical simulation.

2. Research Area

In this study, research area is limited only to upper part of Citarum River, from Cisanti Lake to Saguling Reservoir. Regional geological map of research area shows that research area consists of 4 main formations: Kosambi Formation (lacustrine deposits, mostly is clay), Cibeureum Formation (volcanic deposits), Tertiary Volcanic and Intrusive Rocks (**Figure 1**). Typical hydrostratigraphy along the hyporheic zone is shown in drilling log at **Figure 1**.

One of drilling log from Gedebage area were analyzed and shows alternate layer between sandstone and claystone.

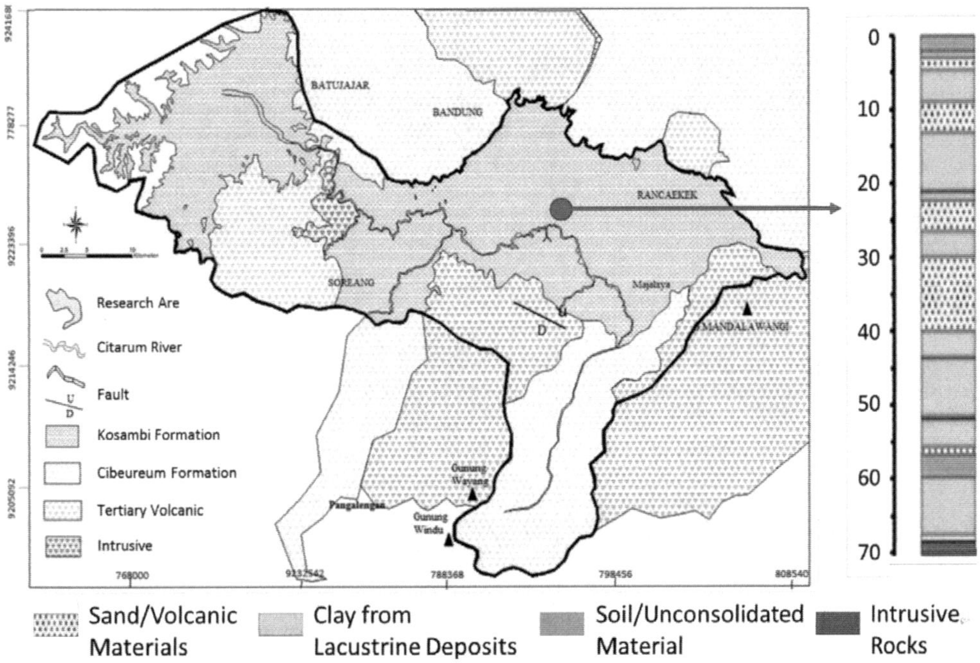

Figure 1 Geological Map of Research Area (modified from [2]).

3. Field Works

Field works were conducted in year 2014-2015. The field works are including: vertical electrical sounding (VES) and also mapping of groundwater level and TDS value along the hyporheic zone consist of TDS value measurements and resistivity method. VES was conducted at 3 locations: Cisanti Lake, Citarum River, and Saguling Reservoir; the results were used to get sub-surface condition at hyporheic zone at (1) Cisanti Lake, (2) Bojongsoang, and (3) near Saguling Reservoir (**Figure 2**).

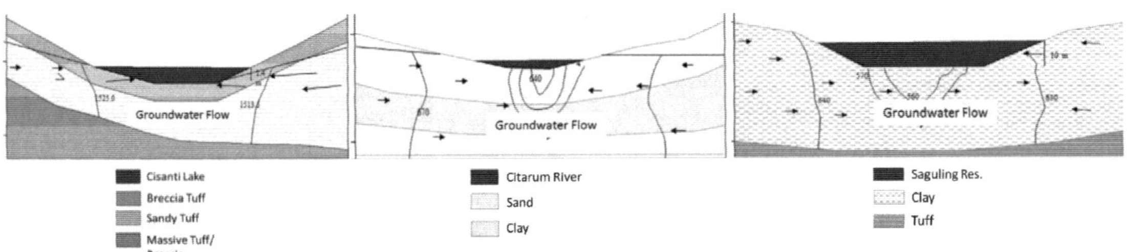

Figure 2 Lithology based on VES Results, Left: Cisanti Lake; Middle: Bojongsoang; Right: Saguling Reservoir. The arrows show groundwater flow direction based on groundwater level measurements at the 3 locations.

TDS values and groundwater level were measured from the dug-wells along the river (**Figure 3**). Total dissolved ions of the groundwater ranged from 200 to 480 ppm while the river water range from 200 to 2,000 ppm. Based of spatial distribution of TDS values, it can be distinguished that southern part of Citarum River has higher TDS values than the northern part. High value of TDS (>1200 ppm) from the surface water i.e. Citarum River are located at Nanjung and Bojongsoang area (**Figure 3**). These areas are known as industrial zone and high population density.

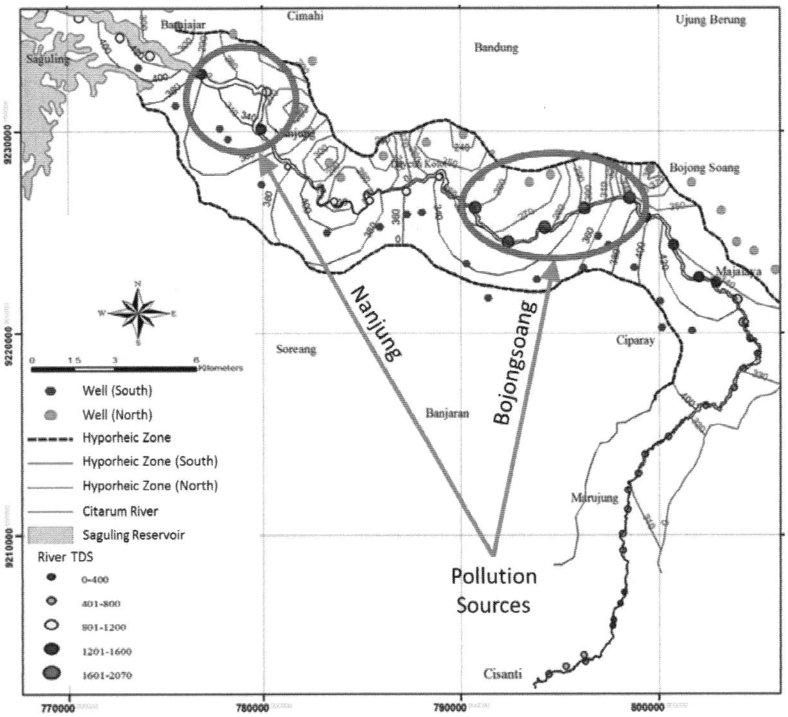

Figure 3 TDS Value Distribution

4. Groundwater Flow and Contaminant Transport Simulation

Hydrogeological model was built to simulate the interactions between Citarum River and groundwater at hyporheic zone. The simulation aims to predict dissolved ion transport between surface and groundwater, by advection and dispersion equations. Rock samples from the hyporheic zone were tested to get hydraulic conductivity values. The results show that the hydraulic conductivity at hyporheic zone ranged from 10^{-4} to 10^{-8} m/s. The parameters for simulation are summarized in **Table 1** and the 3D physical model is shown at **Figure 4**.

Table 1. Parameters Used in Hydrogeological Model

Parameter	Value
Hydraulic Conductivity	Aquifer (low permeability): K = $10^{-6} \sim 10^{-4}$ m/s; in hyporheic zone, silty sand
	Aquitard (lower permeability) : K = 8 x 10^{-8} m/s; in hyporheic zone in deeper part
Storativity	Aquifer: Ss = 7.8 x 10^{-4}; Sy = 0.21; Eff. Porosity = 0.2; Tot. Porosity = 0.56
	Aquitard: Ss = 1 x 10^{-6}; Sy=0.08; Eff. Porosity = 0.11; Tot. Porosity = 0.24

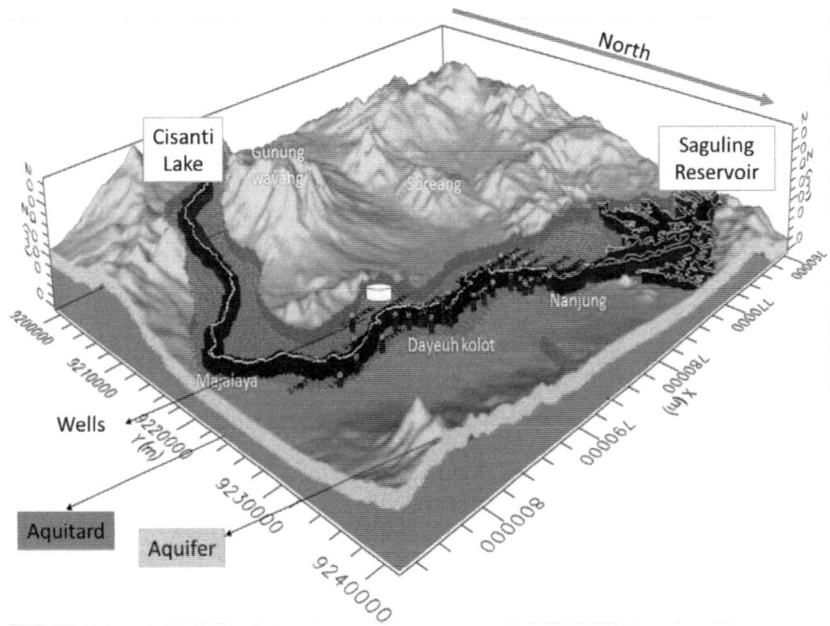

Figure 4. 3D Physical Model of Citarum River

5. Simulation Results

The groundwater simulation model shows that groundwater from hyporheic zone flows into the river with very low gradient, and it shows that the Citarum River generally is a gaining stream. Direct seepage along hyporheic zone from groundwater to Citarum River is only 186 l/s. It is very small compared to the total discharge of the river, it is concluded that the water at Citarum River mostly is from the run-off.

Based on solute transport modeling, it indicates that dissolved ions from the river can be dispersed to groundwater. The phenomena of solute transport at Citarum hyporheic zone is caused by dispersion mechanism due to high concentration gradient between surface water and groundwater. The dispersion mechanism causing TDS value at shallow groundwater at hyporheic zone can be elevated from 300 ppm to 500 ppm caused by pollution at Citarum River. This condition would be occurred mainly at Bojongsoang-Dayeuh Kolot and Nanjung area (**Figure 5**).

6. Conclusions

The study of groundwater and solute transport modeling at hyporheic zone of upper part of Citarum River can be concluded as follows:

- There is interaction between Citarum River and shallow groundwater; flows direction from groundwater to Citarum River, with very small hydraulic gradient (about 186 l/s).
- In a very small gradient between groundwater and river water, dispersion of contaminant from Citarum River to shallow groundwater may still be occurred. In this case pollution in river water also causing pollution in hyporheic zone.
- Based on simulation, TDS in hyporheic zone can be elevated from 300 ppm to 500 ppm caused by pollution in river water.

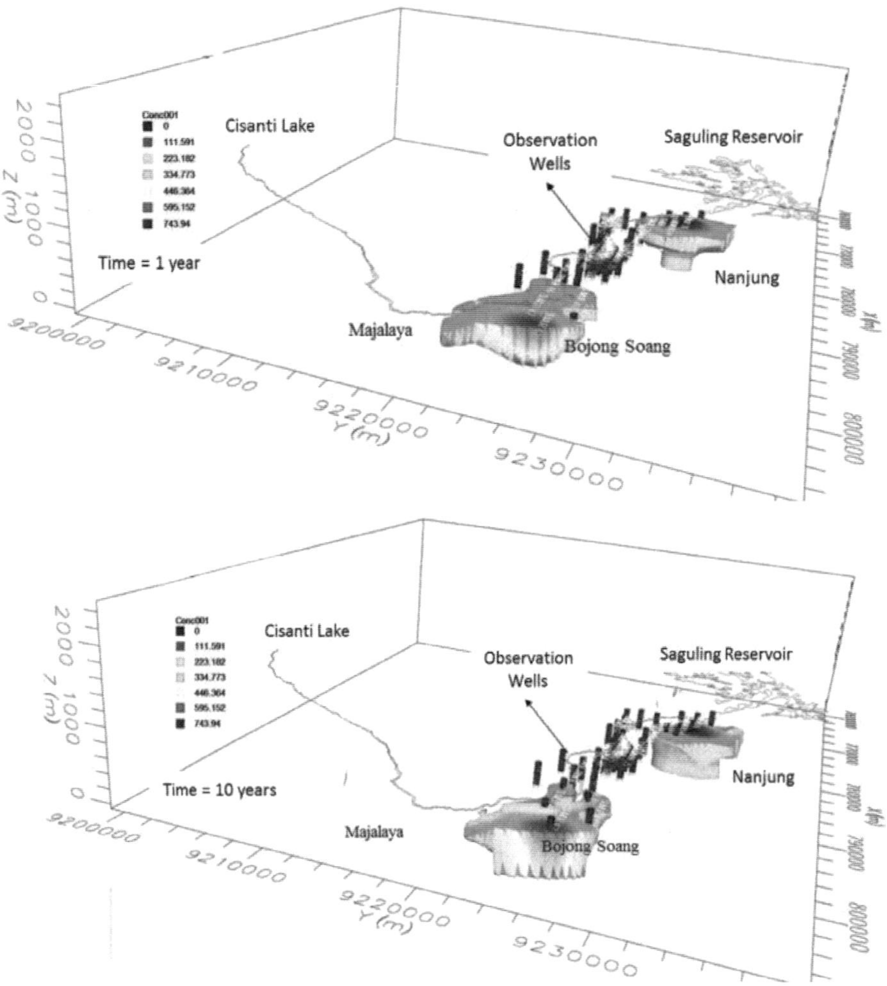

Figure 5. Hydrogeological modeling results of Citarum River (top: after 1 year, bottom: after 10 years)

7. References

[1] Richard Shears (5 June 2007). "Is this the world's most polluted river?". Mail Online. Retrieved 24 May 2010

[2] Hutasoit, L. M. and Ramdhan, A. M., 2006. Recharge area and the origin of brackish water in East Bandung: Result of exploration well. *Proceedings of 9th International Symposium on Mineral Exploration*. Bandung, September 2006

A review on the geoenvironmental and geoecological integrated technology for environmental remediation in Vietnam: approaches, contributions, challenges and perspectives

Mai Trong Nhuan, Nguyen Thi Hoang Ha, Ta Thi Hoai and Tran Dang Quy

VNU University of Science, 334 Nguyen Trai, Hanoi, Vietnam
Email: nhuanmt@vnu.edu.vn

Abstract. Geoenvironmental and geoecological integrated technology (GGIT) is a cost-effective and environment-friendly technology that encompasses the applications of earth science principles and functions of geological environment and ecosystems to assimilate and minimize the spread of pollutants, to enhance the sorption capacity and environmental remediation. On the basis of the integrated approaches such as system, anthropogenic activities – ecosystem – environment interaction, effectiveness and feasibility, GGIT has provided significant applications in Vietnam such as waste containment and remediation and environmental protection. The results of a pilot scale using iron mine drainage sludge and common reed (*Phragmites australis*) for wastewater treatment in a Pb-Zn mine in northern Vietnam indicated the effective and potential application of GGIT. However, GGIT has many challenges in limited funding conditions, constraints in the initial development of GGIT, incomplete transfer to users, and quantitative assessment of pollutant cleanup by natural environments and ecosystems. Environmental pollution quote, impacts to exposed organisms, increasing demands for application of low-cost technologies, the availability of potential sorbents, indigenous plants, and ecosystems for environmental remediation, and collaboration will promote development, contribution, and implementation of GGIT applications in Vietnam.
Keywords: geoecology, geoenvironment, pollutant, remediation, Vietnam.

1. Introduction

The geogenic, industrial, agricultural, pharmaceutical, and domestic effluent sources have generated considerable amounts of waste materials to the surrounding environment [1]. An increasing ecological and global public health concerns associated with contaminants have highlighted the need for environmental remediation. The requirements of cost-effective and environment-friendly methods have created new demands for geoenvironmental and geoecological integrated technology (GGIT). Geoenvironmental technology refers to the application of the principles of earth sciences and geological environment functions for minimizing dispersion and assimilating pollutants and for designing remediation measures [2]. Geoecological technology implies the ecosystem functions for assimilating and remediating pollutants in geological environment [3]. GGIT is defined as the technology applying earth science principles and functions of geological environment and ecosystems aiming to minimize the dispersion, to assimilate pollutants, and to enhance the resilience capacity of geological environment by using the geological formations and ecosystems.

Geoenvironmental technology has been applied as for remediation of contaminated environments by using various geological formations (e.g., iron oxides, laterite, clay minerals, and calcite) [4, 5, 6,

7]. Typically, this technology has been demonstrated in the practical sites including ferrihydrite adsorption for Se cleanup of Kennecott Utah Copper mine; permeable reactive barriers for As, Se, Mo removal from Monticello Mill tailings site; successive alkalinity producing system for Al, Cu, Fe, Mn, Zn remediation in Summitville mine, and anoxic limestone drains for Pb, Zn, Cu removal from Valzinco mine [8, 9, 10, 11]. Constructed wetland systems with different designs (horizontal and vertical flow, free water surface, and integrated constructed wetland) as well as plant species themselves have been recognized as geoecological technology and widely used for environmental cleanup [12, 13]. The integration of geoenvironmental and geoecological technologies for higher treatment effectiveness has been implemented [14, 15]. Practical results of GGIT applications are often found to be cost-effective and environment-friendly.

With the low nation's GDP per capita point, Vietnam has ranked as a developing country where industrialization is increasingly being embraced as the primary mean to prosperity. According to the Vietnam national report at the United Nations Conference on Sustainable Development (Rio+20), during the 20 years' implementation of national sustainable development, Vietnam has obtained remarkable achievements in socioeconomic development. However, the country has also faced many inherent or emerged challenges, particularly in the fields of environment, including (1) serious pollution due to underdeveloped industrial technology and excessive use of chemical fertilizers, (2) environmental degradation as a consequence of the wars (bombs, landmines, agent orange, or dioxin), (3) wasteful and fossil-based production and consumption, (4) inefficiency energy in supply, building, planning and consumption, (5) depletion of natural resources, particularly mineral and water resources, and (6) biodiversity reduction [16]. Being one of the countries that possess favorability of potential contaminant sorbents and high biodiversity ecosystems, the perspectives of GGIT application may help Vietnam overcome many the country's challenges and shortcomings towards environmental protection and sustainable development.

This paper aims to introduce the overview of GGIT in Vietnam with special emphasis on environmental pollution quote, approaches, contributions, challenges, and perspectives for environmental remediation.

2. Geoenvironmental and geoecological integrated technology in Vietnam

2.1. Demands for GGIT application

The socio-economic development, together with negative impacts of climate change and natural disasters, has caused much pressure on the environment. It is reported by the Ministry of Natural Resources and Environment of the Socialist Republic of Vietnam that water quality in 3 river basins of Cau, Nhue - Day and Dong Nai has exceeded the allowable limits; dust pollution in urban areas keeps increasing, environmental pollution in industrial zones/clusters and craft villages is increasing; agricultural pollution due to inappropriate use of fertilizers and pesticides has not improved; oil pollution in coastal waters keeps deteriorating; solid waste has not been collected and treated appropriately, especially hazardous wastes [17]. Biodiversity is degrading rapidly, species are facing risks of reducing in quantity and a higher level of threats. In addition, the degradation of ecosystems and the decrease in number of species and genetic resources have not reduced. Environmental security is being threatened, especially water resources, transboundary pollution and losses of minerals [17].

Water quality in upstream of rivers is generally good [18]. However, in cities and towns, industrial and domestic wastewater without treatment is discharged directly into most rivers, lakes, and canals. Consequently, water quality declined sharply, many parameters such as BOD_5, COD, coliform, total N, total P were higher than allowable limits regulated by the Ministry of Natural Resources and Environment [18].

Surface water in some rural areas in the Red River delta and the Mekong River delta has been polluted by organic matters and microbial from discharge of domestic and livestock wastewater [18]. Soil environment in agricultural areas is polluted and degraded due to the lack of scientific cultivation, use of fertilizers and pesticides [18]. Domestic solid wastes in rural areas continue to increase but the management has not been received adequate attention [18].

Many industrial areas have been severely degraded. Approximately 70% of the more than 1 million m^3 of wastewater per day from industrial zones is discharged directly into the receiving water without treatment that causes environmental pollution of surface water [19]. Approximately 57% of operating industrial zones does not have treatment system [19]. The amount of solid wastes in industrial areas increased. Waste disposal or treatment in the industrial areas was inadequate, especially management, transportation and registration for waste sources.

Environment of the craft villages was also degraded. Water pollution was seriously contaminated in various villages for food processing, paper recycling (organic pollution), textile and dyeing (wastewater contains many chemicals), and metal recycling (heavy metals contamination) [20]. Solid wastes were generally discharged directly into the environment [20].

Vietnam possesses a wide range of mineral resources. The country has some of the world's biggest resources of phosphate (apatite), bauxites, rare earths, and large, commercially viable deposits of oil, coal, gold, gemstones, copper, zinc, tin, chromite, manganese, titanium (mineral sands), graphite and other minerals [22, 23]. Vietnam produces approximately 3.5 million tons of crude oil/year and its mineral production including coal (about 10.7 million tons/year), cement (1.6 million tons), phosphate (300,000 tons), chromite (3.5 million tons), iron ores, gold, cassiterite (3,000 tons), graphite, kaolin and many other minerals produced to serve domestic demand [22]. Mining activities also release pollutants that pose adverse effects on exposed environment and ecosystems. Typically, most soil and water samples in some of the largest mines in Vietnam such as Pb-Zn Cho Don, Cu Sinh Quyen, Sn Ky Lam and Sb Mau Due have been contaminated with heavy metals such as As, Pb, Zn, Cu, and Cd (Tables 1, 2).

Table 1. Average concentrations of heavy metals in soil samples in mine sites

Parameter	Unit	Pb-Zn Cho Don	Cu Sinh Quyen	Sn Ky Lam	Sb Mau Due	Regulated limits
As	mg/kg	**542**	**27.5**	**479**	**60.6**	12
Zn	mg/kg	**970**	149	184	**242**	200
Pb	mg/kg	**2040**	31.6	**71.8**	55.8	70
Cd	mg/kg	**2.35**	0.3	0.3	0.9	2
Cu	mg/kg	**126**	**327**	**235**		50
Ni	mg/kg	39.0	-	-	-	-

Source: [21]

Table 2. Average concentration of heavy metals in water samples in mine sites

Parameter	Unit	Pb-Zn Cho Don	Cu Sinh Quyen	Sn Ky Lam	Sb Mau Due	Regulated limits
pH	-	6-8.5	**3.49-4**	**4.02**-8.08	**3.78-5**	6-8.5
As	ppm	**0.01**	0.004	**0.01**	0.0002	0.01
Zn	ppm	0.06	**0.64**	0.03	**3.75**	0.5
Pb	ppm	**0.03**	0.002	0.003	0.01	0.02
Cd	ppm	0.0004	**0.006**	0.0002	**0.04**	0.005
Cu	ppm	0.002	**0.14**	0.02	**0.13**	0.1
Ni	ppm	0.009	**0.48**	0.01	**1.24**	0.1

Source:[21]

IOP Conf. Series: Earth and Environmental Science **71** (2017) 012012 doi:10.1088/1755-1315/71/1/012012

Environmental pollution causes synchronous impacts on the community health [17]. It also brings environmental conflicts in the community due to poor management policies and mechanisms [17]. Environmental pollution and biodiversity degradation directly influence the country's sustainable development and national security, causing significant economic losses and adverse impacts on natural ecosystems [17] which highlight need for effective application of remediation technologies.

2.2. Approaches

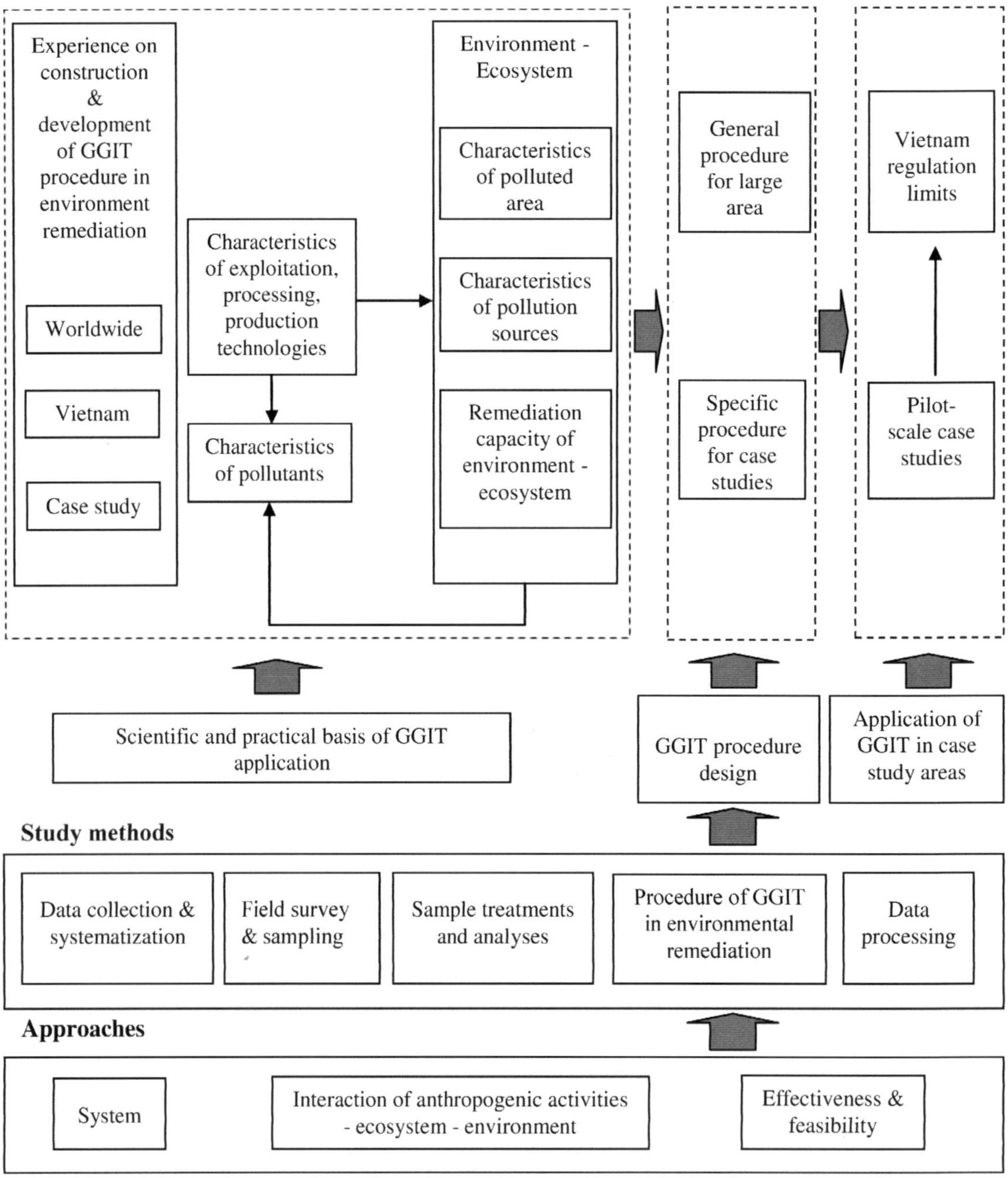

Figure 1. Framework for solving environmental problems based on GGIT

2.2.1. System approach. System approach is the recognition of the earth through their system structure, hierarchy and motivation; thereby, it is a comprehensive and dynamic approach [24]. This is the most dialectical approach for dealing with environmental problems.

Anthropogenic activities such as industrial, agricultural, pharmaceutical, domestic effluents have discharged large amounts of wastes into the surrounding environments. Therefore, they have established a system of the industry – ecosystem – environment. In case without effective treatment system, the pollutants from these activities will be impossible to disperse widely and quickly into other ecosystems and surroundings. The generation and movement of pollutants depend on the interaction of three components. GGIT is the solution to prevent and cleanup contaminants as well as to minimize their spread among components. In addition, it is a sustainable and integrated technical solution for environmental remediation towards sustainable development. It is important to determine the most vulnerable zones that are seriously affected from anthropogenic activities for GGIT applications.

2.2.2. Interactive approach to anthropogenic activities – ecosystem - environment. The balance of natural ecosystems as well as geological environment has existed which may be disrupted by anthropogenic activities. Each anthropogenic activity can release specific type of pollutants that will react with the containing environment and ecosystem. Natural environment and ecosystem can adsorb or absorb, assimilate, metabolize and decompose pollutants or minimize the dispersion or leaching of these pollutants. When the pollution loads exceed resilience and recovery of environment and ecosystem, it is necessary to apply additional geological formations and constructed ecosystems for capacity enhancement such as planting species themselves or in constructed wetlands (phytoremediation), using limestone or calcium sulfate to neutralize environment, using clay rich and non-fracture layers for landfill, and using natural sorbents for sorption of pollutants. The application of natural formations and ecosystems makes GGIT a cost-effective and environment friendly technology that can remediate or minimize the spread of pollutants in the ecosystem - environment systems.

2.2.3. Effective and feasible approach. Procedure of GGIT ensures the effectiveness in the prevention and treatment of environmental pollution: the ability to prevent and cleanup pollutants as well as minimize expense and duration of remediation. In addition, it also ensures high feasibility: accepted by treatment systems and institutions; successfully deployed and applied in the study sites and replicated in similarly polluted areas; consistent with the characteristics and culture of contaminated areas.

2.3. Contributions

Since 1986, Vietnam has been implementing "Doi moi" policy. As results, Vietnam has sustained an impressive growth in access to infrastructure services, particularly since the early 1990s. The country's GDP growth during 2000-2008 averaged at 7.85%. Both the size of the economy and the productive capacity of all sectors increased. GDP per capita in 2016 was over 2164 USD (current price), as 6 times increase from the 2000 level. Population has increased from approximately 61 to 95 million people (2016). Consequently, a volume of water and solid wastes has dramatically increased. Along with the development of economics and infrastructure, the awareness on environmental protection has been strengthened and integrated in all development activities as implementation of Strategic Environment Assessment as stipulated by the Law on Environmental Protection. Accordingly, all national projects have to report the environmental impact assessment. In this context, GGIT has had new application that requires a need for waste containment and remediation, and environmental protection.

GGIT has contributed to find suitable areas for sanitary landfill of solid wastes. Many landfill sites have been implemented including Ha Noi (Nam Son, Kieu Ky, Xuan Son, Nui Thong, Cau Dien), Ho Chi Minh (Phuoc Hiep, Da Phuoc), Hai Phong (Trang Cat, Dinh vu, Do Son), Da Nang (Khanh Son), Hue (Thuy Phuong), Thai Nguyen (Da Mai, Song Cong), Nam Dinh (Loc Hoa), Khanh Hoa (Ru Ri, Luong Hoa), Tien Giang (Tan Lap, Long Binh, Binh Phu, Vinh Binh, Tan Thuan)…[25]. The following characteristics should be considered for landfill site selection: (1) Any material with high hydraulic conductivity, high permeability and with a high water table is probably an unfavorable site; (2) Flat areas are favorable sites, provided an adequate layer of material with low hydraulic conductivity, low permeability, such as clay and silt, is present above any aquifer.

The use of indigenous plant species for phytoremediation of contaminated soils and water has been studied and applied, mainly in pilot scales. Screening of indigenous plant species in metal mine sites showed potential species for phytoremediation such as *Ageratum houstonianum* Mill., *Brassica juncea, Cynodon dactylon, Cyperus rotundus, Eleusine indica, Equisetum ramosissimum, Houttuynia cordata* Thunb., *Pennisetum purpureum, Potamogeton oxyphyllus* Miq., *Pteris vittata* L., and *Vetiveria zizanioides* [26, 27].

Constructed wetland ecosystems have also used for environmental remediation. *Phragmites australis, Saccharum spontaneum* L., and *V. zizanioides* are common species used in the constructed wetlands in Vietnam [26, 28, 29]. This technology has been applied in wastewater treatment from hospital, breeding areas, industrial zones, and craft villages.

Natural sorbents such as laterite, bentonite, zeolite, goethite, and clay minerals have been applied for sorption of contaminants. The application of As natural sorbents at household scale has been used popularly as the solution of seriously arsenic problem in the Red River delta. The sorption capacity of Pb, As, Cd, Zn, and Mn by natural laterite from Vinh Phuc Province (Northern Vietnam) was 1550, 756, 397, 281, and 143 mg/kg, respectively [30]. High removal efficiency of heavy metals by laterite was also reported [31]. The modification of laterite, basalt, and clay sorbents by addition of sodium silicate solutions showed high sorption capacity of heavy metals such as Cd, Cu, Pb, Zn, Cr and As [8]. The highest sorption capacity of As (III), As (V), Cu, Pb, Zn, Cd, and Cr (VI) by modified basalt from Phuoc Long (Central Highland, Vietnam) was 1200, 4090, 16950, 73790, 15720, 30370, and 6830 mg/kg, respectively; those by modified laterite from Hanoi was 1330, 1750, 7610, 41400, 46960, 11090, and 8880 mg/kg [8].

The preliminary application of GGIT has conducted at pilot scale for remediation of wastewater from a lead-zinc mine in northern Vietnam. In this pilot site, drainage sludge form an iron processing mine was used as sorbents of multiple heavy metals and a common species was used at constructed wetland system (*P. australis*). The results demonstrated that the removal efficiency of Zn, Pb, Mn, and As by iron mine drainage sludge - surface constructed wetland system was 78.9, 73.5, 91.2, and 80.5%, respectively; that by iron mine drainage sludge – subsurface flow constructed wetland system was 81.7, 81.1, 94.1, and 83.1% which reflected that subsurface flow constructed wetland showed higher removal efficiency than the surface system.

While much advancement has been made in GGIT, it is mainly applied for landfill site selection and remediation.

2.4. Challenges

In Vietnam, researches and technologies applied in the environmental remediation are commonly organic matters and suspended solids in water. Particularly, remediation technologies of heavy metal contaminated environment have performed at small scales. The enforcement of Environmental Law and investment or funding conditions is possibly the main obstacles for the applications of environmental remediation in general and GGIT in particular. In addition, quantitative assessment of assimilating and remediating pollutants by natural environments and ecosystems is also one of big challenges for GGIT.

There are constraints in the initial development of GGIT and in particular uncertainties exist for long-term performance, reducing confidence level of GGIT application. GGIT transfer to users is incomplete. Standardization of materials and test procedures and development of guidelines for construction and operation are required for various GGIT applications. Regulatory agencies as well as industry have significant depositories of data on GGIT facilities that have not been made available in a format that can be used by stakeholders or fully analyzed for obtaining representative conclusions.

The major challenge for academia is to retain and expand GGIT in both research and education to meet new demands and requirements of environmental protection and sustainable development, in particular under reduced funding conditions. GGIT research, educational initiatives and its effective application can be undertaken through multi-component multidisciplinary platforms with participation of universities as well as industry and professional organizations, policy makers, and users and other stakeholders. Better integration and collaborations between these stakeholders would contribute to the advancement of the field but will need very great efforts from all of them.

2.5. Perspectives

Increasing demand in GGIT due to cost effectiveness: remediation cost is one of the most important criteria for choosing treatment technologies. The use of natural materials and plant species indicates the low cost by GGIT application.

Fixation of pollutants: Co-disposal and co-utilization of hazardous waste materials are the mainstream development to avoid the harmful effects of the pollutants to be disposed.

Increased application of vegetation and diverse ecosystems: The adequate selection of plants and ecosystems eliminates or reduces the use of more costly engineering artifacts in pollution remediation. Screenings of indigenous plant species indicate the high potential of using these plants for environmental remediation such as *A. houstonianum* Mill., *B. juncea*, *C. dactylon*, *C. rotundus*, *E. indica*, *E. ramosissimum*, *H. cordata* Thunb., *P. australis*, *P. oxyphyllus* Miq., *P. purpureum*, *P. vittata* L., *S. spontaneum* L. , and *V. zizanioides* [26, 27, 28].

Increased application of natural minerals: A number of hundreds of iron deposits/occurrences and other minerals (e.g., bentonite, vermiculite, laterite) [22, 23, 32] is widely distributed in Vietnam which reflects the great potential for using these materials as sorbents of contaminants.

Collaboration between GGIT stakeholders (e.g., scientists, engineers, policy makers, managers, and community) will promote development and contribution as well as implementation of new GGIT application.

Enforcement of law and implementation of policies, strategies concerning environment and sustainable development creates institutional fundamentals for GGIT development and new application.

Renovation of GGIT education for development of human and engineering attitude, vision, creativity capacity and skills based on strong fundamental natural, liberal arts and interdisciplinary knowledge and approaches will provide high qualification human resources for GGIT.

3. Conclusions

Geoenvironmental and geoecological integrated technology is rather new in Vietnam, but it has many contributions to solve environment problems. The reduced funding conditions, initial development, incomplete transfer to users, and quantitative assessment are the main challenges for GGIT. The increasing demand for application of low-cost and environment friendly technologies, the availability of potential sorbents, indigenous plants, and ecosystems for environmental remediation, and collaboration will promote perspectives for GGIT applications in Vietnam.

4. Acknowledgments

The authors are grateful for the support of the National Project coded KHCN-TB.02C/13-18 of the National Science and Technology Program for Sustainable Development of Vietnam' Northwest Region (KHCN-TB/13-18).

5. References

[1]. Tchounwou P B, Yedjou C G, Patlolla A K and Sutton D J 2012 Heavy metals toxicity and the environment *EXS* **101** 133-164

[2]. Bhandari A, Reddi L N, Erickson L E, Hutchinson S L and Steward D R 2004 Research-integrated curriculum in geoenvironmental engineering *Proc. of the 2004 American Society for Engineering Education Annual Conference and Exposition*

[3]. Porter C M and Nairn R W 2008 Ecosystem functions within a mine drainage passive treatment

system *Ecol. Eng.* **32** 337-346

[4]. Maiti A, Dasgupta S, Basu J K and De S 2007 Adsorption of arsenite using natural laterite as adsorbent *Separation and Purification Technology* **55** 350-359

[5]. Han R, Zou L, Zhao X, Xu Y, Xu F, Li and Wang Y 2009 Characterization and properties of iron oxide-coated zeolite as an adsorbent for removal of copper (II) from solution in fixed bed column *Chemical Engineering Journal* **149**(1) 123-131

[6]. Bhattacharyya K G and Gupta S S 2008 Adsorption of a few heavy metals on natural and modified kaolinite and montmorillonite: a review *Advances in Colloid and Interface Science* **140**(2) 114-131

[7]. Liu H, Chen T and Frost L R 2014 An overview of the role of goethite surfaces in the environment *Chemosphere* **103** 1-11

[8]. Minh N T 2010 Modification of sorbents using natural basalt, laterite, and clay for treatment of wastewater contaminated with heavy metals and arsenic *National Project Report (KC.02.25/06-10)* In Vietnamese

[9]. USEPA 2008 Field Application of a Permeable Reactive Barrier for Treatment of Arsenic in Ground Water Office of Research and Development *EPA 600/R-08/093* http://www.epa.gov/nrmrl/pubs/600r08093.html

[10]. USEPA 1998 Permeable Reactive Barrier Technologies for Contaminant Remediation http://clu-in.org/download/rtdf/prb/reactbar.pdf

[11]. Watzlaf G, Schroeder K T and Candace L K 2000b Long-Term Performance of Anoxic Limestone Drains *U.S Department of Energy, National Energy Technology Laboratory* www.imwa.info/bibliographie/19_2_098-110.pdf

[12]. Wang Q, Cui Y and Dong Y 2002 Phytoremediation of polluted waters: potentials and prospects of wetland plants *Biotechnol. Acta* **22**(1-2) 199-208

[13]. Vymazal J and Brezinova T 2016 Accumulation of heavy metals in aboveground biomass of *Phragmites australis* in horizontal flow constructed wetlands for wastewater treatment: a review *Chem. Eng. J.* **290** 232-242

[14]. Gusek J J, Wildeman T R, Miller A and Fricke J 1998 The challenges of designing, permitting and building a 1,200-GPM passive bioreactor for metal mine drainage, West Fork Mine, MO *Proc. of the 15th Annual Meeting*, ASSMR, St. Louis, MO 203–212

[15]. United States Environmental Protection Agency 2014 Reference guide to treatment technologies for mining influenced water Available at https://clu in.org/download/issues/mining/Reference_Guide_to_Treatment_Technologies_for_MIW.pdf

[16]. Nhuan M T 2013 Geoenvironmental engineering in Vietnam: contributions, challenges, opportunities, and perspectives for sustainable development and climate change response *Proc. Int. Conf. Hanoi Geoengineering* 2013 Vietnam

[17]. MONRE (Ministry of Natural Resources and Environment of the Socialist Republic of Vietnam) 2010 National State of Environment 2010: An overview of Vietnam Environment

[18]. MONRE 2012 National State of Environment 2012: Vietnam Surface Water Environment

[19]. MONRE 2009 National State of Environment 2009: Vietnam Industrial Area Environment

[20]. MONRE 2008 National State of Environment 2008: Vietnam Craft Village Environment

[21]. Xuan P T 2010 Impacts of mine tailings on environment and human health for proposing mitigation solutions *National Project Report* (KC.08.27/06-10) In Vietnamese

[22]. Khoi N N 2014 Mineral resources potential of Vietnam and current state of mining activity. *Applied Environmental Research* **36**(1) 37-46

[23]. Nhan N V 2004 Ore Deposits *Vietnam National University Publisher, Hanoi* In Vietnamese

[24]. Hoe N D and Hieu V V 2007 Systematic approach in environmental research and development 23-35

[25]. MONRE 2011 National State of Environment 2011: Vietnam Solid Waste

[26]. Anh B T K, Kim D D, Kien N T, Minh N N, Trung N Q and Chuyen N H 2014 Phytoremediation of heavy metal polluted soil and water in Vietnam *Journal of Vietnamese*

Environment **6**(1) 47-51

[27]. Ha N T H, Sakakibara M, Sano S and Nhuan M T 2011 Uptake of metals and metalloids by plants growing in a lead - zinc mine area, Northern Vietnam *J. Hazard. Mater.* **186** 1384-1391

[28]. Nga T T H and Ha N T H 2016 Simultaneous removal of some heavy metals and arsenic from aqueous solutions by *Phragmites australis Vietnam J. Sci. Technol.* **54**(2A) 259-264

[29]. Ha N T H, Anh B T K, Ha T T T 2016 The potential of plant species growing around Cho Don lead - zinc mine for phytoremediation of arsenic *VNU J. Sci.* **32** (2S) 1-8

[30]. Thao N H P, Ha N T H, Thuy P T, Khai N M and Nga T T H 2016 Study the environmental pollution treatment ability of laterite *VNU J. Sci.* **32**(1S) 321-326

[31]. Con T H, Thao N P 2004 Activatation of modified clay and laterite as sorbent of arsenic in domestic water *VNU J. Sci.* **1** 212-217

[32]. Department of Geology and Minerals of Vietnam 2014 Geology and Resources in Vietnam *Science and Technology Publisher*

Majalaya Flood Early Warning System: A Community Based Approach

I Dewa Gede A Junnaedhi[1], Edi Riawan[1], Rusmawan Suwarman[1], Tri Wahyu Hadi[1], Atika Lubis[1], Nurjanna Joko Trilaksono[1], Rahmawati Rahayu[1], PrawiraYudha Kombara[1], Riki Waskito[2], Hendra Ekalaya Oktora[2], Rahmat Supriatna[2], Aan Anugrah[2], Abdul Haq Mudzakkir[2], and Wawar Setiawan[2]

[1]Research Group of Atmospheric Science, Faculty of Earth Sciences and Technology, Institut Teknologi Bandung, West Java, Indonesia
[2]Garda Caah, Community Volunteers, Jl. Raya Laswi, RT 01/10 B, Majalaya, Kab. Bandung, West Java, Indonesia
E-mail: dewa108@meteo.itb.ac.id

Abstract. Majalaya, a small city to the south-east of Bandung, was hit by flood almost every year. From January to June 2016, up to 5 severe floods and 4 moderate floods have hit this city. Although it usually not last for long, but the flood stream could be very rapid, thus have a high potential to bring damage to the city. Starting from 2012, ITB through Weather and Climate Prediction Laboratory (WCPL) has support Garda Caah (flood watcher society in Majalaya) with weather prediction system. In the late 2015, ITB also enhancing Garda Caah observation system by installing several Automatic Weather Station (AWS) and Automatic Water Level Recorder (AWLR) throughout Majalaya upstream area. The instruments itself was supported by a re-insurance company MAIPARK and some was built in house by WCPL. The collaboration between ITB, Garda Caah, and Majalaya citizens has been proved to be mutually beneficial. Garda Caah could get more accurate and faster observation and enhanced knowledge, thus could provide a better flood warning for Majalaya citizens. On the other hand, ITB could get data from observation network, with more efficient way to maintain observation instruments as it done by Garda Caah and other Majalaya citizens.
Keywords: Majalaya, flood, Early Warning System, community based, weather monitoring, weather prediction

1. Introduction

Majalaya is a small city located to the south east of Bandung (Figure 1). Majalaya also the name of the district, which is part of Bandung Regency, West Java. The city was famous for their textile industries which propel its economy since 1960ies. Beside textile industry, farming and other agriculture activity also plays important role in Majalaya's economy [1-2].

Majalaya also among the cities that crossed by Citarum River, the longest and largest river in West Java. The Citarum River is spanning from Gunung Wayang in the Kertasari area, to Muara Gembong in Bekasi area, passing through three reservoirs, namely Saguling, Cirata and Jatiluhur [3]. The city itself was part of the upper Citarum watershed, extending from Situ Cisanti in the Kertasari area to the first reservoir, Waduk Saguling. Majalaya is situated in the first 30 km of this upper Citarum watershed with the main river path is passing Kertasari, Pacet and then Majalaya (see Figure 1). In this part of Citarum, many tributaries also exist and contribute to Citarum River just before Majalaya. In case of heavy rain, these situations often cause the river to have excessive discharge, thus flooding the city. This condition was getting worse by excessive ground water extraction in Majalaya which might causing land subsidence [4] and landuse change in the upstream area which reduce soil ability to hold rainfall [5].

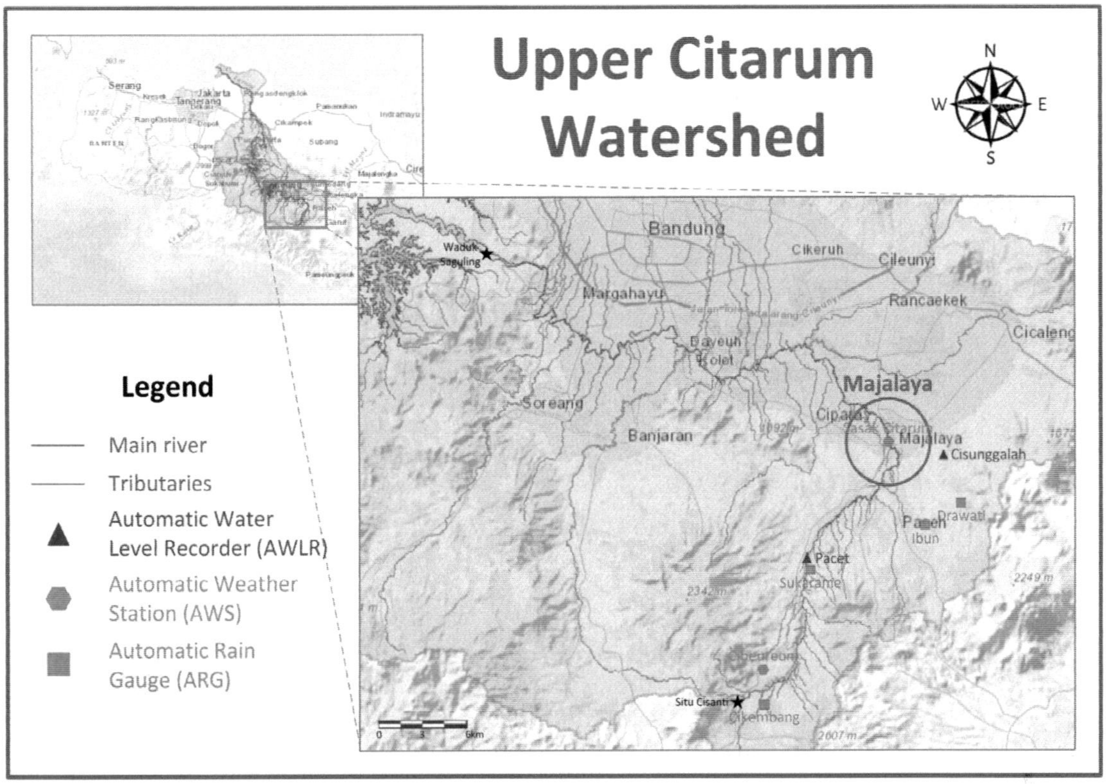

Figure 1. Map of Upper Citarum Watershed with Majalaya area marked with red circle. Purple triangle, brown hexagon, and green square mark the position of AWLR, AWS and ARG, respectively. Black star indicate the edges of Upper Citarum. Basemap, and river and tributaries database was acquired from ArcGIS^{TM} webmap viewer.

Floods occur almost every year in Majalaya. Based on record by a small group of Majalayan volunteer, it was happening annually since 2008 (Figure 2). Flood also occurs several times in one year. The minimum number of flood occurrence was 6 events (on 2012), and the maximum number of flood occurrence was 20 events (on 2008). However, from January until mid-June 2016, there have been 17 flood occurrences in Majalaya, and this number night still increase until the end of the year as the wet season is yet to come. These numbers indicate that citizen of Majalaya were constantly threatened by floods from Citarum River.

Floods in Majalaya usually happen in relatively short time. The flood water also rise rapidly with quite fast stream, thus it could be categorized as flash flood. When flood happen, some of Majalaya's main roads were inaccessible, paralyzing the transportation. Most of textile factories, especially those located near the riverbanks also halt their operation. Government and private offices, banks, shops, and other activities were also interrupted by the flood (see Figure 3). Although casualties were minimum, but many citizens lost their goods and belonging, especially when the floods happen without warning.

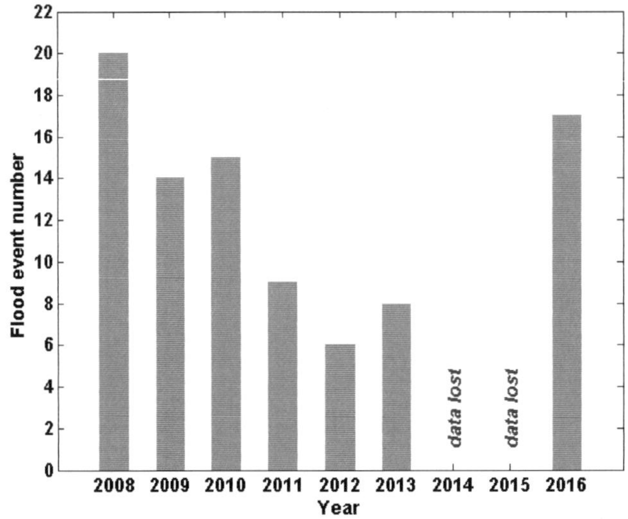

Figure 2. Annual number of flood occurrence in Majalaya as recoded by Garda Caah. Data on 2014 and 2015 were lost during a flood event in 2015.

To minimize the floods impact, since 2008, several Majalaya citizens took the initiatives to form a flood watcher group, named Garda Caah. Their main activities were manually monitoring rainfall and water level in the upstream area, and issue a warning when the condition was in favor for floods to happen. Their warnings were quite effective to reduce the impact of floods by giving Majalaya's citizen enough time to prepare for the incoming flood. However, because they do the monitoring manually, sometimes floods happen without any warning issued. Nevertheless, with very limited tools and knowledge, they managed to create a network of observers, a communication system, and flood warning network by cooperating with other groups and societies.

Although Garda Caah's flood early warning system was quite succeeding, their limited time to do monitoring has limiting the availability of the warning system. An improvement to the system, especially on the monitoring system, is necessary to make a better flood early warning system (FEWS), thus increasing Majalaya's citizen preparedness against floods.

Figure 3. Photographs of Majalaya flood on 28-29 March 2016 (a) and on 23 May 2016 (b). The photographs were taken from local newspaper websites, Pikiran Rakyat [6] and Galamedia [7], respectively.

This research is aimed to support, improve and enhance the flood early warning system by implementing hydro-meteorological knowledge into the existing system operated by Garda Caah. In this paper, the development of the Majalaya flood early warning system, abbreviated M-FEWS, and the implementation are presented. The system itself is still under development and the skill of the system is yet to be assessed, thus this paper will also described some further research and development that could possibly improve the early warning system.

2. Majalaya Flood Early Warning System

An early warning system (EWS) was defined by UNISDR as "the set of capacities needed to generate and disseminate timely and meaningful warning information to enable individuals, communities and organizations threatened by a hazard to prepare and to act appropriately and in sufficient time to reduce the possibility of harm or loss" [8]. By this definition, an early warning system, including flood early warning system (FEWS), must be developed with aim to raise preparedness of the people that affected by the disaster far before it happen. As noted by United Nations in research by Cools et al. [9], an EWS commonly consist of four major components: 1) risk knowledge, 2) monitoring, forecasting and warning, 3) communication of an early warning, and 4) response capability. Many studies and implementation of FEWS usually focus on second and third components [9-15]. In the existing M-FEWS however, the response capability (fourth component) has been long become the part of its implementation. The communication of the early warning (third component) also has been establish very well on previous system by chaining the information to local government, private sectors and other groups and societies. Thus, to strengthen the system, this research will focus on the second component, with a little enhancement to the third components.

To improve the M-FEWS which previously built by Garda Caah, Institut Teknologi Bandung (ITB) through Weather and Climate Prediction Laboratory (WCPL) proposed to add weather prediction and automatic hydro-meteorological observations into the system. The weather prediction sub-system could give an outlook of the weather condition in the future, especially rainfall, thus increasing people alertness and enabling an earlier preparedness for the flood [12],15,16]. The automatic hydro-meteorological observation sub-system will increase M-FEWS availability and reduce the human observer misinterpretation. These two sub-systems will play important role for the enhancement of M-FEWS. Together with the warning dissemination sub-system that already exist, these three will become the key components of the new M-FEWS.

2.1. Weather Prediction

The weather prediction sub-system was developed in-house by WCPL since 2006, but implemented for M-FEWS since 2012. The weather prediction was done by downscaling the Global Forecast System (GFS) model output. GFS is a global weather forecast model run by the US National Center for Environmental Prediction (NCEP) under National Oceanic and Atmospheric Administration (NOAA). GFS was run four times daily, named cycle, on 00, 06, 12 and 18 UTC. The global forecast is available for 16 days lead time, starting from each cycle time, and available to download from NOAA server. The model spatial resolution could reach 28 km, however for this purpose the 50 km resolution was chosen to save space on server storage and maintain download speed.

The downscaling of GFS output was done using mesoscale numerical weather prediction (NWP) models, i.e. PSU/NCAR Mesoscale Model version 5 (MM5) [17] and Weather Research and Forecasting (WRF) [18] model. These models were run twice a day, using GFS cycle 00 and 12 as input, in WCPL's 64 cores computing server. The model simulation was run for 72 hours forecast lead time, increasing its input model resolution to 27 and 9 km spatial resolution, covering Indonesia and Java respectively. Although the models could be run for longer simulation, many research suggest that our current ability to produce good prediction, especially for rain, is still bound to the first 48-72 hour, unless ensemble prediction method was used [10],12,13]. The spatial resolution of the prediction was also limited to 9 km due to limited computational resource. For longer but less accurate outlook, up to 7 days rain probabilities were also calculated using GFS output for whole Indonesia in 50 km resolution.

The downscaled weather prediction results, along with the rainfall probabilities, were disseminated through WCPL Weather website, with URL address http://weather.meteo.itb.ac.id/prediksi.php. Most of the predictions were presented in a maps (**Figure 1**Figure 4a, b, and c), which require some

IOP Conf. Series: Earth and Environmental Science 71 (2017) 012013 doi:10.1088/1755-1315/71/1/012013

knowledge to fully understand it. In order to give an easier-to-understand forecast, a clickable, specific-location forecast was provided, including several locations over Majalaya and Upper Citarum area (Figure 4d).

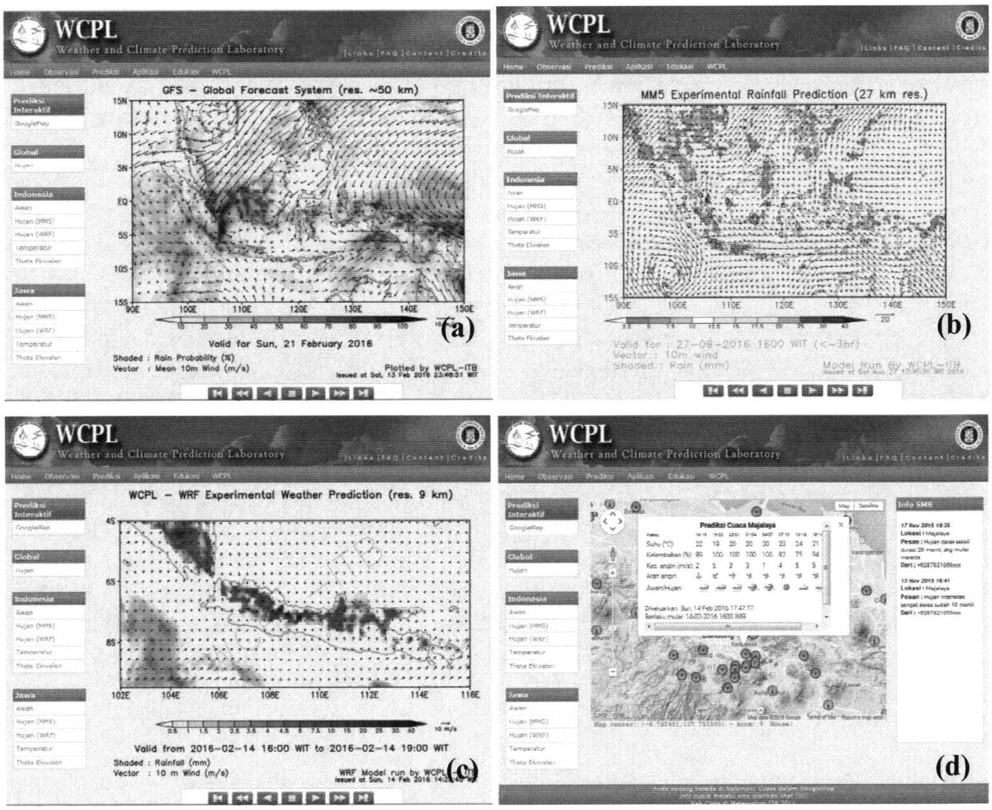

Figure 4. Screenshots of weather prediction results which disseminated on WCPL Weather website. (a) GFS rain probabilities map. (b) MM5 27 km rainfall prediction. (c) WRF 9 km rainfall prediction. (d) Clickable weather forecast on specific location, built on top of GoogleMaps^{TM}.

2.2. Hydro-meteorological Observations

The hydro-meteorological observation sub-system was constructed using automatic instruments that measure weather and hydrological parameters. The addition of this observation system was started since December 2015. Three types of instruments were used in this observation network, namely Automatic Weather Station (AWS), Automatic Rain Gauge (ARG) and Automatic Water Level Recorder (AWLR). The instruments were placed in certain location on the Upper Citarum area, especially in the potential upstream area (see Figure 1 for the location and the type of instrument installed). Kertasari and Pacet were selected as main observation site because the main discharge to Majalaya was come from these locations. While Ibun, Drawati and Cisunggalah were also selected because it was thought that discharge from these locations also give significant contribution to Citarum section that passing Majalaya.

All of the observation instruments were self-assembled using open-source hardware and software platform. The choice of using self-assembled instruments rather than industrial built-up instruments was to reduce the instrument cost per unit. By doing so, more instruments could be used in the observation network. The AWS was built using Arduino, and equipped with quite complete weather sensors such as temperature, humidity, pressure, wind and rain sensor. Arduino is a credit card size microcontroller board that was developed in open-source paradigm, which also accompanied with open-source software to program it. On the other hand, the ARG is a stripped version of AWS where only rain sensor that remained. To simplify programming, the ARG was built using BeagleBone, an open-source single-board computer that could run Linux operating system. The AWLR also use the same platform as ARG, since it only use one sensor, i.e. ultrasonic based water level sensor. The bottom-right part of Figure 6 shows the example of the instruments that were used in the observation sub-system.

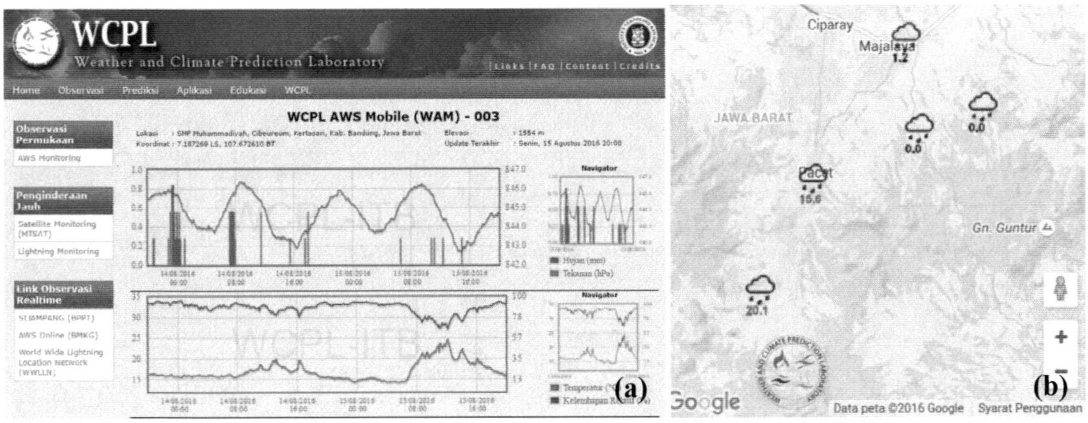

Figure 5. (a) Screenshot of the real-time weather monitoring using AWS on Cibeureum, Kertasari. The page could be accesed on address http://weather.meteo.itb.ac.id/observasi.php. (b) Screenshot of simultaneous, real-time rainfall observations presented on GoogleMap^{TM}. Red number mark the location with heavy rainfall. The page could be accessed on the url address http://weather.meteo.itb.ac.id/wcplfsys.

All instruments were equipped with telemetry module to send the data into the data server that located in ITB Bandung. The telemetry module use Short Message Service (SMS) to send the data. The data itself was collected every 5 minutes, however to reduce telemetry cost, the data was sent every 10 minutes. SMS was used for the telemetry because stability and availability. In some places, especially in the upstream area of Citarum, cellular network signal often very poor, thus cannot maintain internet based connectivity. On contrary, SMS service usually quite stable, even on poor network condition, thus become the only option to transmit the data.

Data from observation instruments was received by a SMS Gateway, and archived on WCPL data server. The data also automatically plotted and displayed on WCPL Weather web for real-time monitoring (see Figure 5a). For simultaneous view of the the rainfall data also plotted on top of GoogleMap^{TM}. This page serves as the main rainfall monitoring display. It will show a red color when heavy rain detected (Figure 5b).

The archived data could be used for further research by WCPL team or by other scientific community. This data will become valuable source of information to gain the risk knowledge, the first component of EWS, which have shown extensively by Borga et al. [14]. It could also be used to better understand the nature of the floods through development of rainfall-runoff model [19]. The data could also feed to the weather prediction model through data assimilation to improve forecast results [20].

2.3. Information Flow and Warning Dissemination

On previous FEWS, developed by Garda Caah, the main communication platform was the open amateur radio channel using handy talkie (HT). It was low cost and quite effective. However, because it was using open channel, it was quite hard to exchange disclosed information, especially flood possibility, in order to not create panic on society. To overcome this, the information exchange on M-FEWS was done through several methods. SMS and Internet Messenger (IM) service was used to exchange disclosed and undisclosed information, while web (http) service and amateur radio was used only for undisclosed information.

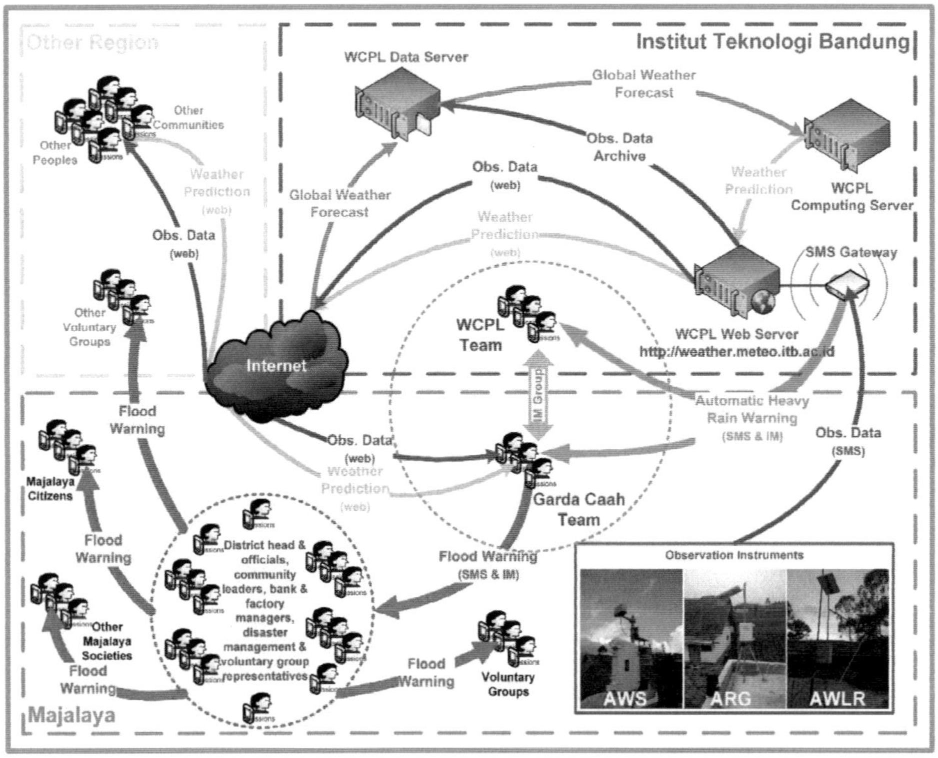

Figure 6. The schematic diagram of information flow and warning dissemination on Majalaya Flood Early Warning System. Internet web (http), Short Message Service (SMS) and Internet Messenger (IM) was used as the main communication platform.

Figure 6 show the schematic diagram of information flow and warning dissemination on M-FEWS. Every day, the weather prediction sub-system is downloading the global forecast and running the weather prediction model to produce rainfall forecast. The result was disseminated through web service so it could be accessed by all stakeholders and also by other people and communities as well. The hydro-meteorological observation sub-system also supplies the hydro-meteorological information every day, with 10 minutes time resolution. This information is also automatically disseminated through web service for real time monitoring. All these information could be used by Garda Caah and other stakeholders to monitor the hydro-meteorological condition over Majalaya and other Citarum Watershed area.

The flood early warning itself is managed by a closed group consist of Garda Caah and WCPL members, which named Ring 1 (blue dotted circle on Figure 6). This group communicates and discuss through IM forum. The final decision to issue warning was on Garda Caah team, as the team already has the capability to assess the flood favor condition with their local knowledge. The group existence also served as a forum for strengthens the risk knowledge of Garda Caah team, especially on weather and climate expertise. The local knowledge and capability of local stakeholders, complemented with experts qualitative opinion, were playing important role in flood warning system [15,19,21].

Flood early warning with different level of status could be issued by Garda Caah based on weather prediction and or hydro-meteorological observation. Flood warning is send to Ring 2 (red dotted circle on Figure 6), which consist of government officers, including the Head of Majalaya District, community leaders, bank and factory managers, and also the disaster management agency and voluntary groups. The warning then forwarded by the Ring 2 to other Majalaya citizens and societies. The level of status itself is subject of further research, thus would not be discussed on this paper. By the time of this paper was written, a set of standard operational procedures (SOP) on issuing flood warning, including the level of status, was being made by collaborative effort between Garda Caah, ITB and Badan Pengelola Lingkungan Hidup (BPLHD) Jawa Barat (The Environment Management Agency of West Java Province).

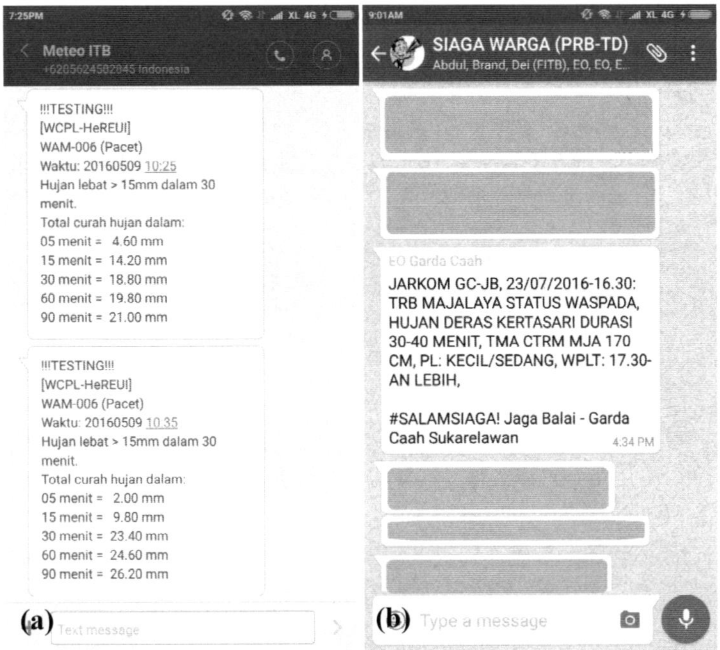

Figure 7. Screenshot of an automatic heavy rain warning that was send using SMS (a). Screenshot of flood warning that issued through IM (b).

Most of the time, flood warning was start from observation data, especially rainfall observation. Upon received, the rainfall data was checked against certain threshold. If it exceed the threshold, an automatic heavy rain warning will be send to Ring 1 through SMS and IM (the heavy rain warning example was shown on Figure 7a). This is the first stage of warning. If the warning continue, and have confirmed and assessed by Ring 1, then a flood warning will be issued. This is the real flood early warning. The examples of flood early warning that issued by Garda Caah is shown on Figure 7b. This two stage warning is necessary to prevent false alarm warning because of hardware error or failure.

Currently, the rainfall threshold that being used to trigger heavy rain warning is 15 mm per 30 minutes. This value was chosen semi-arbitrary, based on heavy shower criteria by American Meteorological Society and UK MetOffice. This value also been used by Cools et al. [15] as rainfall threshold on Egypt FEWS, although for longer duration (i.e. 6 hour), as the rain rate on the study area is much smaller. This threshold value is subject to further research, because the relation between rainfall intensity-duration and flood on Majalaya is yet to known.

3. Community Involvements and the Collaboration Results

From beginning, the Majalaya Flood Early Warning System was initiated by the Majalaya community itself, i.e. Garda Caah and several other communities. As mentioned above, the initial system was operated entirely by Garda Caah. The local government of Majalaya and the provincial government of West Java have recognized that the M-FEWS was truly a community initiative and support it in various ways. The Disaster Management Agency of Bandung Regency also has a tight relation to community, especially for disaster response activities.

The enhancements of Majalaya Flood Early Warning System by ITB, which started since 2012, did not reduce the community involvement to the system. In fact, it was expanding the "community" term because academic society is also part of the community itself. During the enhancement process, the role of the communities, not only those who live in Majalaya, but also those who lived in the upstream area, cannot be taken lightly. In the process of instruments installation, ITB team was not only helped by the Garda Caah member, but also by local people who live near the installation site (Figure 8a).

Figure 8. Photograph of AWS installation in Drawati by Garda Caah member and local people (a). Photograph of capacity building workshop on flood hazard for various societies around Majalaya and Bandung (b).

The replacement of manual observation system with technological based observation, did not remove the local observer from the FEWS. Instead, it adds their duty to look after the instruments and become the first person to contact in case of hardware failure. This necessity requiring more active involvement of the person, especially on upgrading their knowledge over the instrument itself. As noted by Baudoin et al. [21] and Gautam and Phaiju [19], local people involvement in monitoring, observation, and instrument maintenance is one key factor that sustain the existence of the Early Warning System.

In the core of M-FEWS, community still hold the biggest portion. The real flood warning still managed by Garda Caah and their partner communities. The prediction and observation sub-system enhancement was also results of collaborative effort between Garda Caah, and ITB, with support from local government and private sector (see Acknowledgement). The enhancements process of M-FEWS not only consists of system upgrade, but also a capacity building for Majalaya communities as well as other communities around Bandung (Figure 8b).

The collaborative effort to build the M-FEWS was beneficial to all stakeholders. The Majalaya communities have a quite reliable Flood Early Warning System, while ITB and or other scientific community could have dataset for research purpose. Currently, the M-FEWS is still in very early stage and not fully functioning yet. By the time of this paper written, new instruments is still being installed. The data acquisition and dissemination system is still in active development. The automatic warning system still need an improvement, and the effectiveness of the early warning system was not yet evaluated properly. Nevertheless, the existence of the flood warning system already has impact on Majalaya communities.

4. Possible Improvements for the M-FEWS

One important and basic subject to improve M-FEWS is to understanding the nature of the flood itself. This will need various dataset that related to flood, for example rain, water level and stream flow, discharge debit, sedimentation rate, river path and cross-section, land use, soil type, elevation, slope, etc. To provide these datasets, an observational campaign or research need to be conducted by corresponding institution or scientific community. Technological advancement might ease these observations, but community participatory will be very advantageous. The effort to understand the nature of the flood will also require trans-disciplinary and cross-institution collaborative research. By conducting these researches, problems like where to install the instruments, how to define rainfall threshold value, and how to estimate flood discharge arrival could be addressed more easily.

Another important research that could be done to improve the M-FEWS is to simulate the river discharge using hydrologic and or hydraulic model. As suggested by Plate [11], hydrological conversion of rainfall into runoff and discharge for large river basin cannot be ignored. Thus predicting flood using hydrological and hydraulic simulation is necessary for a better flood early warning system. A complete knowledge of the watershed that mentioned earlier will also be needed in this research. Such complex modeling will also require various skill and knowledge from different discipline, so collaboration is suggested.

Improvement on the hydro-meteorological observation is still a necessity, as the current stage of the sub-system is far from reliable. A better design on the observation system, especially power efficiency and the telemetry system is an open area to be explored. Designing cost-effective instruments will also open the possibility to expand the observation network. Rainfall anomaly detection system also need to be enhanced, as the simple single-threshold method might not suitable for triggering flood warning due to different characteristic of the observation sites. As shown by Krzhizhanovskaya et al.[22], the use of Artificial Intelligence in the rainfall (and possibly water level) anomaly detection could bring the FEWS into the next level.

The weather prediction sub-system of the M-FEWS also a broad subject of improvements. As previously mentioned, data assimilation of the existing observation into the model may improve the rainfall forecast result. Assimilation of a remotely sensed data, such as radar, might also have good impact on the forecast [14]. Various researchers, i.e. Gouweleeuw et al. [10], Pappenberger et al. [12] and Thielen et al. [13], also suggest the use of the now trending ensemble forecast to provide better rainfall forecast to be used in flood early warning system. However, as demonstrated by those researchers, a data assimilation and ensemble prediction system need a lot of resources which may not be have by single institution, thus require collaboration with other institution.

Various scientific and technical problems need to be solved and fixed before the M-FEWS could be said complete. In the future, M-FEWS development could use as a framework for various research and development. It could also use as platform to raise people awareness to their environment. It could also use to strengthen cooperation between authorities and various institutions. Beyond that, it could use to assist adaptation against environmental change.

5. Summary

Majalaya Flood Early Warning System (M-FEWS) has been initiated by Garda Caah since 2008. Since 2012, ITB through WCPL, has enhance the system by adding the weather prediction system developed in-house. In the late 2015, an automatic hydro-meteorological observation system and an automatic heavy rain warning system were also added to the system.

Currently, the M-FEWS is still in very early stage and need many fix and improvement. Various possible research and development could be done to improve the early warning system. Many of those researches will require collaborative works from trans-disciplinary researcher and various institutions. The M-FEWS is a result of a collaborative effort between community, academic institution, governments and private sector. It was developed with the spirit of collaboration and cooperation, thus any collaborative works to improve it are welcomed.

6. Acknowledgements
This research was partially funded by ITB through Pengabdian Masyarakat (PM) program for the year 2015 and 2016. Several observation instruments and capacity building program were funded by PT. Asuransi MAIPARK Indonesia through their Corporate Social Responsibility program. The authors would also express a gratitude to local government of Majalaya, all members of Garda Caah, Jaga Balay, other Majalaya societies, and many other people in Majalaya for their help and support in this research.

7. References
[1] Antlöv H and Svensson T 1991 In the Shadow of Agriculture: Non-Farm Activities in the Javanese Economy, Past and Present (Amsterdam: Royal Tropical Institute) pp 113–26.
[2] Hardjono J 1993 Indonesia Assessment 1993 - Labour: Sharing in the Benefits of Growth? (Canberra: Australian National University) pp 273–89
[3] Apip, Takara K, Yamashiki Y and Ibrahim A.B, 2009 Study on Early Warning System for Shallow Lanslides in the Upper Citarum River catchment, Indonesia. Annuals of Disas. Prev. Res. Inst. **52** 9–18
[4] Gumilar I, Abidin H Z, Hutasoit L M, Hakim D M, Sidiq T P and Andreas H 2015 Land Subsidence in Bandung Basin and its Possible Caused Factors Proc. 3rd Int. Symp. on Earthquake and Disaster Mitigation (Bandung: Elsevier) pp 47–62
[5] Watung R L, Tala'ohu and Dariah A 2005 Land use change and flood mitigation in Citarum and Kaligarang Watersheds, Prosiding Seminar Nasional Multifungsi Pertanian dan Ketahanan Pangan (Bogor: Pusat Penelitian dan Pengembangan Tanah dan Agroklimat, Badan Penelitian dan Pengembangan Pertanian, Departemen Pertanian)
[6] Sukirman, E.(2016, March 28) Banjir Kepung Majalaya dan Paseh Pikiran Rakyat Retrieved from : http://www.pikiran-rakyat.com/bandung-raya/2016/03/28/365196/banjir-kepung-majalaya-dan-paseh
[7] Kosasih, E., (2016, May 23) Sungai Citarum Kembali Meluap, Majalaya Darurat Banjir. Galamedia Retrieved from : http://m.galamedianews.com/bandung-raya/91823/sungai-citarum-kembali-meluap-majalaya-darurat-banjir.html
[8] United Nations Office for Disaster Risk Reduction (UNISDR) 2009 UNISDR terminology on disaster risk reduction (Switzerland:United Nations)
[9] Cools J, Innocenti D and O'Brien S 2016 Lessons from flood early warning systems Environmental Science & Policy **58** 117–122
[10] Gouweleeuw B T, Thielen J, Franchello G, De Roo A P J and Buizza R 2005 Flood forecasting using medium-range probabilistic weather prediction. Hydrol. Earth Syst. Sci. **9** 365–380
[11] Plate E J 2007 Early warning and flood forecasting for large rivers with the lower Mekong as example J. of Hydro-environment Res. **1** 80–94
[12] Pappenberger F, Bartholmes J, Thielen J, Cloke H L, Buizza R and de Roo A 2008 New dimensions in early flood warning across the globe using grand-ensemble weather predictions Geophys. Res. Lett. **35** L10404
[13] Thielen J, Bartholmes J, Ramos M H and de Roo A 2009 The European Flood Alert System – Part 1: Concept and development Hydrol. Earth Syst. Sci. **13** 125–140
[14] Borga M, Anagnostou E N, Blöschl G and Creutin J-D 2011 Flash flood forecasting, warning and risk management: the HYDRATE project. Environmental Science & Policy **14** 834–844
[15] Cools J, Vanderkimpen P, El Afandi G, Abdelkhalek A, Fockedey S, El Sammany M, Abdallah G, El Bihery M, Bauwens W and Huygens M 2012 An early warning system for flash floods in hyper-arid Egypt Nat. Hazards Earth Syst. Sci **12** 443–457
[16] Alfieri L, Salamon P, Pappenberger F, Wetterhall F and Thielen J 2012 Operational early warning systems for water-related hazards in Europe. Environmental Science & Policy **21** 35–49
[17] Grell G A, Dudhia J, Stauffer D R, others 1994 NCAR Technical Note (NCAR/TN-398+STR):

A Description of the Fifth-Generation Penn State/NCAR Mesoscale Model (MM5) (Boulder, Colorado: National Corporation for Atmospheric Research)
[18] Skamarock, William, Klemp, Joseph, Dudhia, Jimy, Gill, David, Barker, Dale, Wang, Wei, Huang, Xiang-yu, Duda and Michael 2008 NCAR Technical Note (NCAR/TN–475+STR): A Description of the Advanced Research WRF Version 3 (Boulder, Colorado: National Corporation for Atmospheric Research)
[19] Gautam D K and Phaiju A G 2013 Community Based Approach to Flood Early Warning in West Rapti River Basin of Nepal. IDRiM Journal **3** 155–169
[20] Kalnay E 2003 Atmospheric Modeling, Data Assimilation, and Predictability (New York: : Cambridge University Press)
[21] Baudoin M-A, Henly-Shepard S, Fernando N, Sitati A, 2014 Early warning systems and livelihood resilience: Exploring opportunities for community participation UNU-EHS Working Paper Series No.11 (Bonn : United Nations University Institute of Environment and Human Security (UNU-EHS)).
[22] Krzhizhanovskaya V V, Shirshov G S, Melnikova N B, Belleman R G, Rusadi F I, Broekhuijsen B J, Gouldby B P, Lhomme J, Balis B, Bubak M, Pyayt A L, Mokhov I I, Ozhigin A V, Lang B and Meijer R J 2011 Flood early warning system: design, implementation and computational modules Proc. Int. Conf. on Computational Science 2011 (Singapore : Elsevier) pp 106–115

IOP Conf. Series: Earth and Environmental Science **71** (2017) 012014 doi:10.1088/1755-1315/71/1/012014

Sediment trapping analysis of flood control reservoirs in Upstream Ciliwung River using SWAT Model

Mirwan Rofiq Ginanjar[1] and Santosa Sandy Putra[2]

[1] PUSAIR, Ministry of Public Works & Housing, Jl. Ir. H. Djuanda 193, Bandung 40135, Indonesia

[2] Balai Sabo, Ministry of Public Works & Housing, Yogyakarta 55282, Indonesia

E-mail: mirwanrofiq@gmail.com

Abstract. The plans of Sukamahi dam and Ciawi dam construction for Jakarta flood risk reduction purpose had been proposed as feasible solutions to be implemented. However, the risk of the dam outlets clogging, caused by the sediment, is important to be anticipated. The prediction of the max sediment concentration in the reservoir is crucial for the dam operation planning. It is important to avoid the flood outlet tunnel clogging. This paper present a hydrologic sediment budget model of The Upstream Ciliwung River Basin, with flood control dam existence scenarios. The model was constructed within SWAT (Soil and Water Assessment Tools) plugin and run inside the QGIS framework. The free hydrological data from CFSR, soil data from FAO, and topographical data from CGIAR-CSI were implemented as the model input. The model resulted the sediment concentration dynamics of the Sukamahi and Ciawi reservoirs, on some suspended sediment parameter ranges. The sediment trapping efficiency was also computed by different possible dam capacity alternatives. The research findings will give a scientific decision making base for the river authority, in term of flood control dam planning, especially in The Upstream Ciliwung River Basin.

Keywords: sediment, flood, swat, ciawi, dam, ciliwung

1. Introduction

Watershed damage occurred almost in all watersheds in Indonesia as well as in the Upstream Ciliwung [1]. Watershed Damage, such as land conversion from native forests into plantations and the construction number of residential, hotels, and villas cause a reduction trend of the water infiltration areas and caused water flows directly into the river. This situation is causing flood in the downstream area, which is The Special Region of Jakarta, The Nation Capital. Floods occurred almost in every year, which causing the loss of property, interrupt daily economic activities, and the lack of comfort for the people that live in the area [2]. One of the effort to reduce the flooding is to build dams in the upstream Ciliwung. The aim is to use the reservoir storage as much as possible, so that the water coming into the river does not flow directly into the downstream area [3–5]. The flood water volume will be hold for a certain period in the reservoir. This strategy can at least minimize the flooding that occurred in Jakarta. Ciawi Dam is proposed to be located in the Ciawi village at Ciliwung River, while Sukamahi dam will be located in the Sukamahi village at Sukabirus River. Sukamahi and Ciawi Dam are located at Megamendung District, Bogor, West Java Province, Indonesia. Ciawi Dam proposed location is at 106°52’20” East Longitude, 06°39’28” South Latitude and the Sukamahi Dam proposed to be located in 106°52’20” East Longitude, 06°40’12” South Latitude. The location map of the proposed reservoirs are depicted in Figure 1.

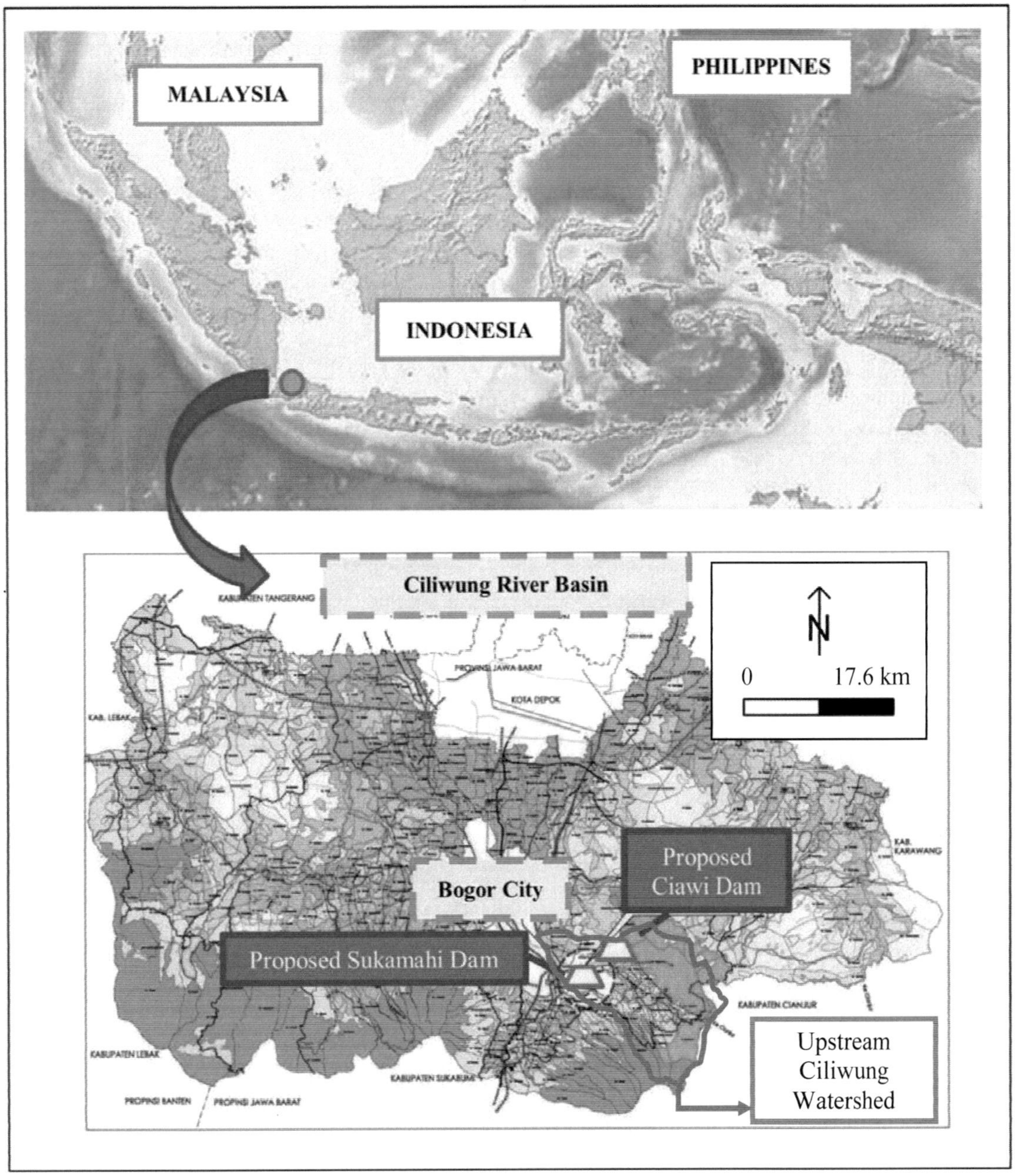

Figure 1. The maps of Ciliwung River Basin as a case study location [6].

The proposed planned dams to be built in Upper Ciliwung Watershed are Ciawi Dam and Sukamahi Dam. Besides as flood control dams, the dams are expected to provide other benefits such as water supplies to the surrounding area, in to some certain limitations [7]. Land use changes on upstream Ciliwung watershed cause erosion and lead to sedimentation in the river. A percentage of the projected sediment will be settled in the Sukamahi and Ciawi dam. Sedimentation that occurs and settles in reservoirs will reduce the life of service ability of the reservoirs. In this paper, analysis and prediction of Sukamahi and Ciawi Dam sedimentation will be considered. The research results hopefully can define the proper Ciawi and Sukamahi dam operational use in the near future.

2. Data and Method

The model that is used to simulate the hydrological condition of the Upstream Ciliwung River basin is the Soil and Water Assessment Tool (SWAT). The software is free used and can be easily operated using QSWAT in GIS environment. The tools can be downloaded from the SWAT TAMU webpage (http://swat.tamu.edu/software/qswat/). The QSWAT module helps the user in formatting the input file to match with the SWAT required format. The user just needs to determine the outlet of the watershed and the reservoir location. Later on, QSWAT will divide the watershed into several Hydrological Response Units (HRU). Each HRU represents a certain hydrological condition.

In order to build up the Ciliwung hydrological model, the required data are the land cover map, the soil map, the digital elevation model, and the hydro climatological data. The land cover, soil and topographic of the river basin are used to generate the watershed HRU, as figured out in Figure 2. The MODIS LC map is used as the land cover data source for this project [8]. The FAO Digital Soil Map of The World data are occupied as a soil data source [9]. The CGIAR CSI SRTM DEM is adopted for the topographical base in this project [10]. As the measured hydro climatological data, the National Centres for Environmental Prediction Climate Forecast System Reanalysis (CFSR) data are used for the SWAT simulation [11].

The detail design of the Ciawi and Sukamahi dam had been done by the river authority in 2015. The detail design report provides the information of the proposed reservoir locations and the reservoir parameters [12]. In order to reveal the proposed reservoir performances, in terms of its sediment budget, several set of parameter scenarios were set up, as listed in Table 1. Each of the parameter has descriptions that are noted in the Appendix section. The parameter can be inputted to the model by modifying the default parameter values range in the SWAT reference database beforehand.

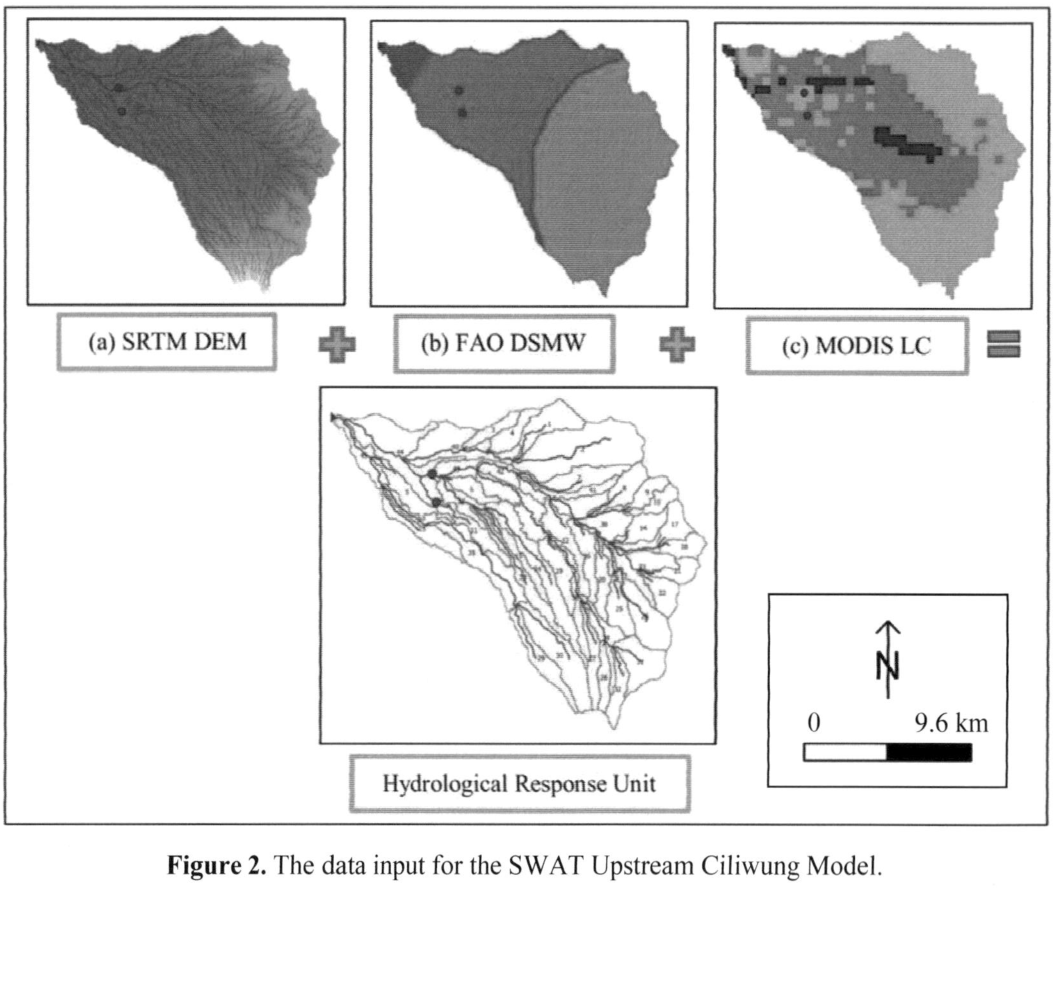

Figure 2. The data input for the SWAT Upstream Ciliwung Model.

Table 1. The parameter value arrangements for each reservoir modelling scenarios.

	Common Parameter			Ciawi Dam Parameter						Sukamahi Dam Parameter					
Scenario	RSED	RNSED	RD50	RESA	REVOL	RPSA	RPVOL	RVOL	R_RR	RESA	REVOL	RPSA	RPVOL	RVOL	R_RR
	(mg/l)	(mg/l)	(um)	(ha)	(1E4 m^3)	(ha)	(1E4 m^3)	(1E4 m^3)	(m^3/s)	(ha)	(1E4 m^3)	(ha)	(1E4 m^3)	(1E4 m^3)	(m^3/s)
Basic	4000	4000	10	38.56	503	25.707	385	385	10.48	9.85	132	6.567	42	42	0.284
NoStorage	4000	4000	10	38.56	503	3.856	5.03	5.03	10.48	9.85	132	0.985	1.32	1.32	0.284
CoarseD50	4000	4000	63	38.56	503	25.707	385	385	10.48	9.85	132	6.567	42	42	0.284
FineD50	4000	4000	2	38.56	503	25.707	385	385	10.48	9.85	132	6.567	42	42	0.284
Mildly Turbid	1200	1200	10	38.56	503	25.707	385	385	10.48	9.85	132	6.567	42	42	0.284
Clear	300	300	10	38.56	503	25.707	385	385	10.48	9.85	132	6.567	42	42	0.284
NoStorage1	4000	4000	10	77.12	1006	2.571	1.006	1.006	15.72	14.76	198	0.657	0.198	0.198	0.332
NoStorage2	4000	4000	10	115.7	1509	2.571	1.006	1.006	20.96	19.70	264	0.657	0.198	0.198	0.379

The main consideration that is faced by the river authority is to ensure the reservoir standard operations. The main option is to utilize the proposed reservoir as a flood reduction dam only. However, there is also another view based on the needs of the Ciliwung River Community, which is to utilize the exceed water for domestic purposes [13]. The two leading options have different implication to the reservoir operation. The first idea desires the reservoir to contain as less water as possible. The following option insists the reservoir operator to control the reservoir water storage. The first scenario is the basic scenario, where most of the reservoir parameter values are adopted from the river authority design report. Some parameters, such as the equilibrium sediment concentration was leaved by default in the basic scenario due to the no data availability. The basic scenario still accommodates some portions of the reservoir volume as a storage column. The second scenario is the No Storage Scenario, where the reservoir is set to have no space for storing water during the non-flood event. The extreme definition requires the reservoir to have 0 (zero) m3 of storage. However, in order to avoid the SWAT model crash due to zero reservoir storage, a small minimum storage is set in the parameter table. The No Storage 1 and No Storage 2 scenarios adopt the same setting with the No Storage scenario. However, the reservoir capacity and the reservoir release parameters were set to be proportionally higher compared to the one in the No Storage scenario. The rest of the scenarios explore the effect of sediment turbidity and sediment grain size to the sediment budget. The water sediment density criteria were adopted from the WHO standard [14]. The clear water is assumed to have 300 mg/l sediment concentration and 1200 mg/l for the mildly turbid water. The sediment median grain size is categorised as coarse D50 (63 μm) and fine D50 (2 μm) [15]. The simulation was conducted from 1 Jan 1979 until 31 July 2014 based on the hydro climatological data existence. It was assumed that the reservoir was constructed since in the beginning of the simulation. This approach was used to see the dam's performance based on long term measured data (± 30 years).

3. Result and Discussion

The upper Ciliwung Watershed SWAT model results sediment budget and sediment concentration within the designed reservoirs. The model results present detail profiles of the sediment budget fluctuation in different reservoir release conditions. The model also provides the correlation of sediment concentration with the reservoir water storage conditions. The result analyses neglect the first five years of the simulation results. The purpose is to remove the initial condition effects from the results. It was confirmed that the sediment dynamics in the reservoir were highly dependent on the reservoir storage capacity, reservoir release rate, and the suspended sediment concentration of the water. In this case study, the sediment profile is distinct, especially between the condition when the reservoirs were operated in its flood control mode and water storage mode.

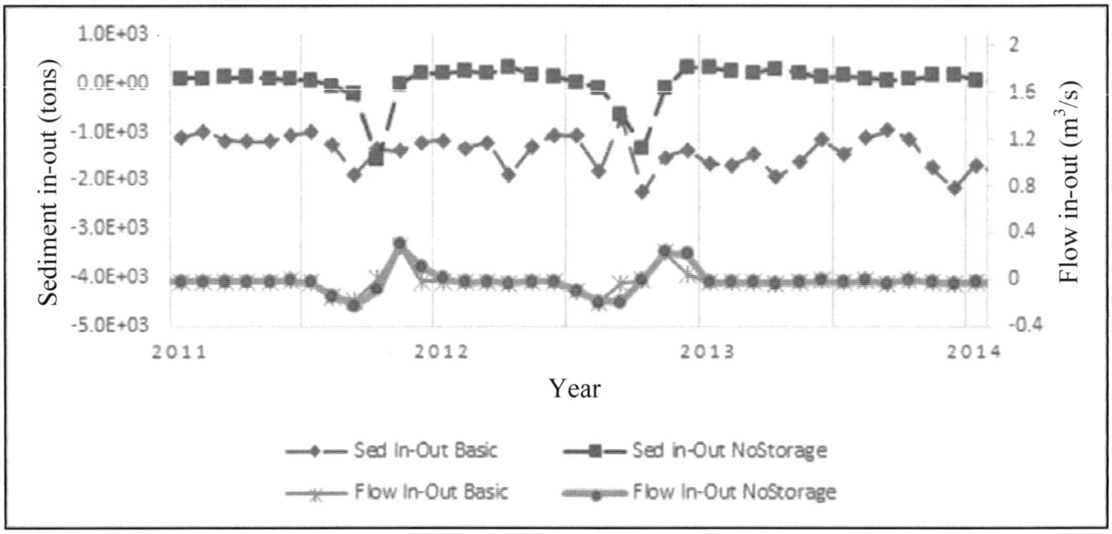

Figure 3. Flow – sediment budget time series profile of designed Sukamahi Reservoir.

In Figure 3, there is a correlation between the sediment in-out and the flow in-out of the reservoir. When there is a sharp decrease on the flow in-out value, there would be a sharp increase on the sediment in-out value too. If the reservoir is filled up, there would be a sedimentation in the reservoir body. The different situation happens for the basic storage mode, where there is no diverse link among the sediment in-out and the flow in-out variables. The situation was caused by the reservoir storage damping effect. The reservoir input flow is not always linear with the reservoir release flow. It means the reservoir operation affects the sedimentation in the reservoir that has storage volume. In this case, the basic scenario assumes that the average reservoir annual release is 10.48 m^3/s (Ciawi Dam).

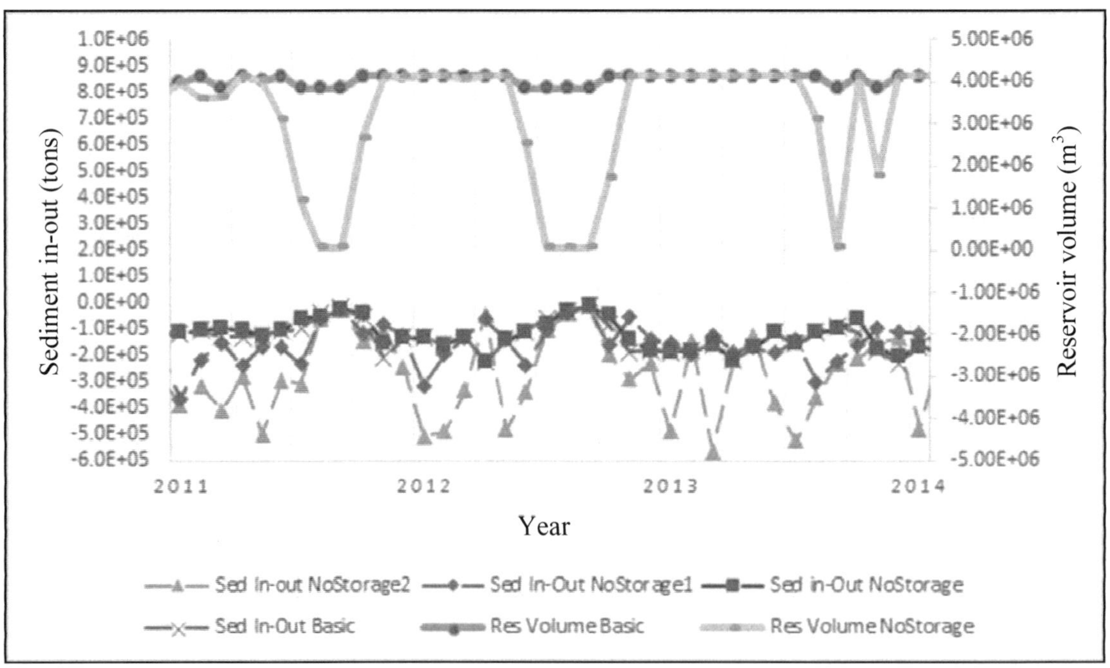

Figure 4. Flow – sediment budget time series profile of designed Ciawi Reservoir.

The result still shows the risk of reservoir sedimentation in the scenario of no storage reservoir. Therefore, the result of no storage reservoir scenarios were explored and displayed in Figure 4. In the no storage scenario, the reservoir can hold the flood for a certain period [16]. Afterwards, the captured flood volume will be released based on the reservoir release parameter. The wider the flood tunnel opening will correlate with higher sediment release from the reservoir. It can be shown by comparing the sediment budget of the no storage 1 scenario and no storage 2 scenario. It means that there would be a negative margin between the sediment entering the reservoir and the sediment released from it. The flood trapping efficiency of a flood control reservoir can be assessed based on the different of the flow entering the reservoir and the flow that is releasing from the spillway. In this scenario, the flood was captured until the reservoir filled up to the spillway elevation. In the time the water reach the reservoir primary spillway, the primary reservoir release is starting. If the flood is exceeding the primary storage, the reservoir will continuously filled up until the emergency spillway. The food that is exceeding this level will be directly flowed down to the downstream.

In the NoStorage2 scenario, the Ciawi Reservoir capacity was tripled. It can be shown from Figure 4 that the Ciawi reservoir was projected to capture the extreme flood up to ± 4.073 million cubic meter in the 2012-2013 period. The Figure 4 shows that there is a correlation between the reservoir filling periods with the sediment in-out pattern. The reservoir dewatering will be followed with the decrease in the reservoir sediment budget variable. However there is no significant different of the reservoir sediment budget state between the storage and no storage scenario. It means that the reservoir scenario will produce almost similar sediment transport to the downstream. The suspended sediment trapping efficiency of the reservoir is less than 2%, which is found as less risky. The reservoir release is more sensitive to the sediment trapping efficiency rather than the reservoir storage capacity. The average sediment concentration in the Sukamahi reservoir depends on the reservoir volume. If the reservoir storage decrease, then the sediment concentration will be increased, as shown in Figure 5. The average reservoir sediment concentration parameter has strong impact to the sediment concentration within the period of simulation. It is especially on suspended D50 < 10 μm condition or on clear to mildly turbid condition. If there is a stable reservoir volume, than the modelled sediment concentration will recover to the initial inputted average value. There is also another finding that in 10 μm < D50 < 63 μm condition, which is mostly silt suspended material, the Sukamahi reservoir will have a stable sediment concentration range of about 1.000 to 2.000 ppm. The recognised implication is that fine sediment management in the upstream of the reservoir catchment area will highly reduce the Sukamahi average reservoir sediment turbidity.

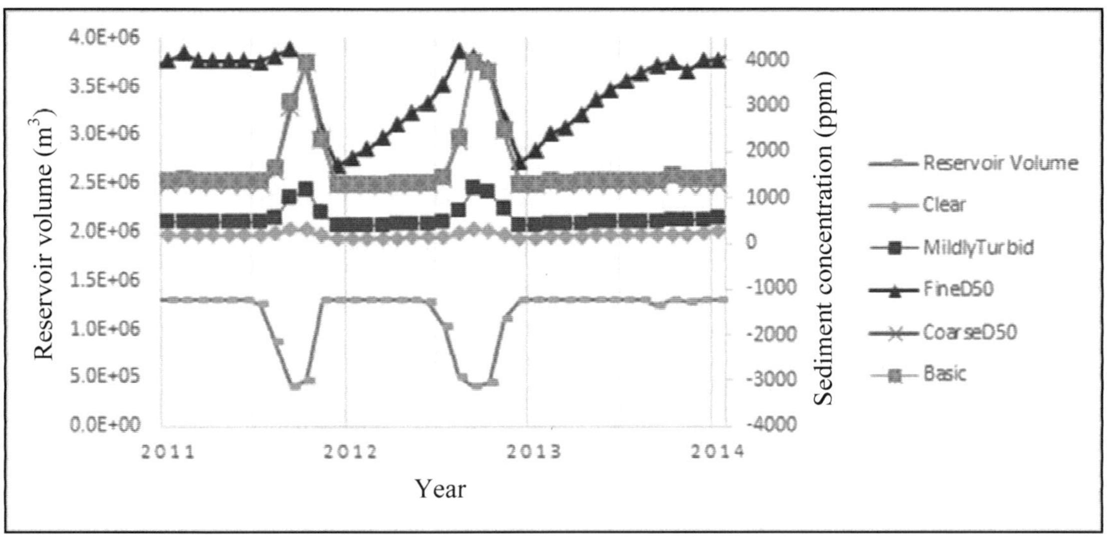

Figure 5. Designed Sukamahi reservoir sediment concentration correlation with the reservoir volume.

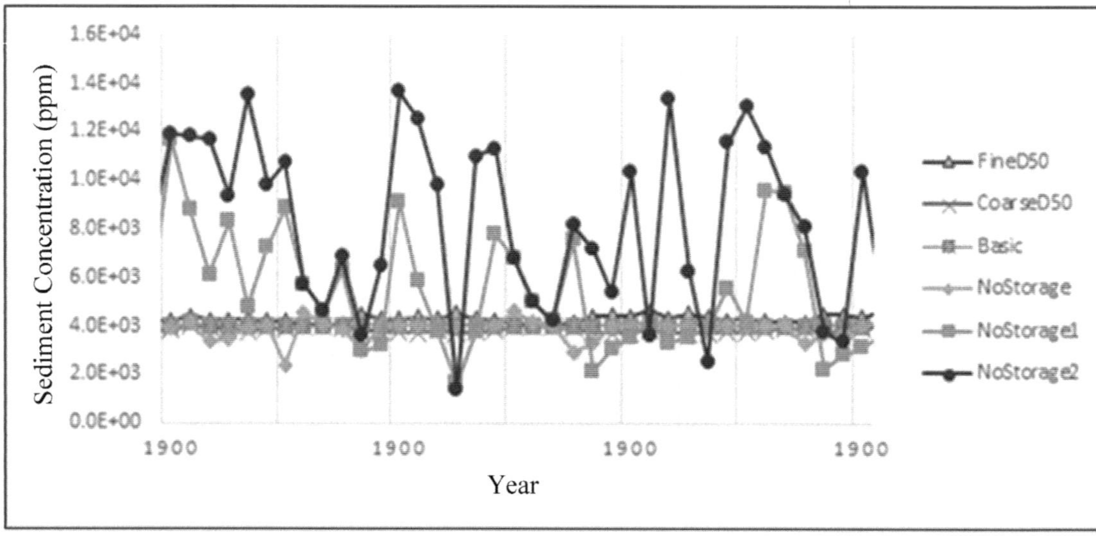

Figure 6. Designed Ciawi Reservoir sediment concentration correlation with the reservoir volume.

In the larger reservoir application, as in this case is the Ciawi reservoir, the average reservoir sediment concentration is highly governed by the average sediment input parameter value. Moreover, the variation of the suspended particle D50 parameter will also affect the sediment concentration in the reservoir in the range of ± 500 ppm, as depicted in Figure 6. The model tries to reduce the sediment concentration out of the reservoir, which also similarly shown from the negative sediment budget pattern in Figure 4 before.

The reservoir annual release rate was also proofed as a dominant cause of turbid reservoir condition. The higher reservoir release means there is a higher water circulation in the reservoir. It is caused by the high transition of water flowing in and releasing out of the reservoir. Therefore there is no time for the sediment to settle in the reservoir water body. The reservoir sediment concentration is almost nearly the same with the input water sediment concentration from the river, which located of the upstream of the reservoir. Finally, it can be summarized that there is a chance to keep some small storage in the designed Ciawi and Sukamahi reservoir. The reason is that the sedimentation risk can be overcome with higher reservoir release strategy and fine sediment management in the catchment.

4. Conclusion

The Upper Ciliwung SWAT (Soil and Water Assessment Tool) Hydrological model can result a thorough sediment budget profile with proposed Ciawi and Sukamahi reservoir scenario. The limitation of data availability in the sediment modelling effort can be answered by the use of free hydrological data from the online accountable data sources. It can be concluded that the sediment budget within the reservoir is mainly govern by the reservoir release parameter. Although the NoStorage reservoir is the common strategy to avoid the reservoir sedimentation, the reservoir design that is proposed by the river authority still allows the reservoir to store some amount of water for other domestic purposes. However, the acceptable storage must be decided by the optimum sediment in-out and flow in-out correlation as proposed in this paper. It is also recommended for a good fine sediment management in the up streamed river basin, before the sediment enters the reservoir. The fine sediment is a main source for high sediment concentration in the proposed reservoirs.

5. Acknowledgements

A great thanks was propounded to Research Centre for Water, Ministry of Public Works and Housings Indonesia, for the data and funding for this research. The authors also utter special thanks for to Mr.

Hatma Suryatmojo, Forestry - UGM, Mr. Faris, BBWS Ciliwung Cisadane, Mr. Andi Siswandi, Dinas Pengairan Bogor, Dr. Samuel J. Sutanto, and all colleagues from Balai Sabo and Balai HITA.

6. References

[1] Pawitan H and Haryani G S 2011 Water resources, sustainability and societal livelihoods in Indonesia *Ecohydrol. Hydrobiol.* **11** 231–43

[2] Budiyono Y, Aerts J, Brinkman J, Marfai M A and Ward P 2015 Flood risk assessment for delta mega-cities: a case study of Jakarta *Nat. Hazards* **75** 389–413

[3] Martdianto R and Kadri T 2013 Prioritas penentuan lokasi waduk pada DAS Ciliwung untuk banjir Jakarta *JATI UNDIP J. Tek. Ind.* **7**

[4] Sawiyo 2013 *Development of Suitability Criteria and Hydrological Conditions Impact Assessment of Channel Reservoir in Upper Ciliwung Watershed* (Institute Pertanian Bogor)

[5] Tarigan S D 2016 Modeling effectiveness of management practices for flood mitigation using GIS spatial analysis functions in Upper Cilliwung watershed *IOP Conf. Ser. Earth Environ. Sci.* **31** 012030

[6] Demis 2006 Map of Indonesia

[7] Padawangi R, Turpin E, Herlily, Prescott M F, Lee I and Shepherd A 2016 Mapping an alternative community river: The case of the Ciliwung *Sustain. Cities Soc.* **20** 147–57

[8] Savtchenko A, Ouzounov D, Ahmad S, Acker J, Leptoukh G, Koziana J and Nickless D 2004 Terra and Aqua MODIS products available from NASA GES DAAC *Adv. Sp. Res.* **34** 710–4

[9] The International Union of Soil Science (IUSS) 2007 Digital Soil Map of the World v.3.6 *FAO*

[10] Jarvis A, Reuter H I, Nelson A and Guevara E 2008 Hole-filled SRTM for the globe Version 4 *available from CGIAR-CSI SRTM 90m Database (http//srtm. csi. cgiar. org)*

[11] Saha S, Moorthi S, Wu X, Wang J, Nadiga S, Tripp P, Behringer D, Hou Y-T, Chuang H, Iredell M, Ek M, Meng J, Yang R, Mendez M P, van den Dool H, Zhang Q, Wang W, Chen M and Becker E 2014 The NCEP Climate Forecast System Version 2 *J. Clim.* **27** 2185–208

[12] Balai Besar Wilayah Sungai Ciliwung Cisadane 2015 *Laporan Pekerjaan Detail Desain Bendungan Ciawi dan Bendungan Sukamahi* (jakarta)

[13] Hartono D M, Novita E, Gusniani I and Oriza I I D 2010 The role of water supply and sanitation during floods: Case study of flood disaster in five regions of jakarta *Int. J. Technol.* **1** 29–37

[14] WHO 1996 Total dissolved solids in Drinking-water Background document for development of *Heal. criteria other Support. Inf.* **2** 8

[15] International Standard Organization 2002 *ISO 14688-1:2002* vol 14688-1 (International Standard Organization)

[16] Liu X, Guo S, Liu P, Chen L and Li X 2011 Deriving Optimal Refill Rules for Multi-Purpose Reservoir Operation *Water Resour. Manag.* **25** 431–48

7. Appendixes

RSED The initial average sediment concentration in the reservoir (mg/l).
RNSED The equilibrium of average sediment concentration in the reservoir (mg/l).
RD50 The median of the suspended particle in the reservoir (μm).
RESA The impounded surface area when the reservoir is filled till the emergency spillway (ha).
REVOL The required water volume to fill the reservoir up to the emergency spillway (10^4 m^3).
RPSA The impounded surface area when the reservoir is filled till the principal spillway (ha).
RPVOL The required water volume to fill the reservoir up to the principal spillway (10^4 m^3).
RVOL The initial reservoir volume (10^4 m^3).
R_RR The average daily principal spillway release discharge rate (m^3/s).

IOP Conf. Series: Earth and Environmental Science **71** (2017) 012015 doi:10.1088/1755-1315/71/1/012015

The Suitable Index of Flow and Density in the Mixed Traffic

Fadly Arirja Gani[1], Toshio Yoshii[2] and Shinya Kurauchi[2]

[1] Graduate School of Science and Engineering, Ehime University, Bunkyo-cho, Matsuyama 790-8577, Japan.

[2] Department of Civil and Environmental Engineering, Ehime University, Bunkyo-cho, Matsuyama 790-8577, Japan.

Corresponding Author: fadly_agani@yahoo.com

Abstract. Some traffic indices were developed to consider the existence of vehicle variance either in the homogeneous or mixed traffic. This study investigates the proper index of flow and density in the mixed traffic. There are eight pairs of indices of traffic flow and density that will be compared, in which it was varied by the concept and the measurement interval. The concept was distinguished according to the respond to the variance of vehicles in the traffic analysis, and the measurement interval was differentiated over time and time-space. The traffic was observed over a short section of a road which set as the detection zone. All pairs of indices will be applied and evaluated on the Fundamental Diagram to describe the actual traffic state in the density-flow plane. Then, the R-square of each density-flow curve will be compared. The result shows that the existence of vehicle variance in the share of traffic can be described more properly by considering the projected area of vehicles. The index of flow and density which considers the projected area of vehicles can explain the actual traffic state more appropriately by increasing the R-square of the model. In addition, the result does not provide a significant difference between the measurement over time and time-space in a short section of the road.
Keywords: Vehicle Variance; Mixed Traffic; Flow; Density; Occupancy; Area Occupancy

1. Introduction

According to the existence of vehicle variance, the condition of traffic is distinguished in two terms, which known as homogeneous and heterogeneous or mixed traffic. In each condition, traffic has a different characteristic. The terminology of homogeneous traffic refers to the condition which follows the lane rule, and the composed vehicle has a similar characteristic. This condition was addressed to the traffic in the developed countries. The traffic condition will be different in the developing countries, which associated by another terminology namely heterogeneous or mixed traffic. The composition of traffic in this condition is mixed, with a variety of vehicles, motorized and non-motorized, using the same right of way [1]. In addition, it was also classified as a non-lane based traffic due to the poor lane-discipline of many road users [2]. The term of lane discipline refers to the driver behavior in the share of traffic: changing and following the lane rule of a road.

In the reality, it shows that the traffic entity is composed by vehicles, which varies in terms of physical and behavior. Even in the homogeneous traffic, the presence of vehicle variance cannot be ignored. The term of physical in the vehicle variance refers to the different size or vehicle dimensions, and the behavior variance refers to the different of vehicle speed. There is a little different effect of the vehicle variance between the homogeneous traffic and the mixed traffic. The existence of vehicle variance will be responded differently in both conditions of traffic due to the characteristic. Therefore, the vehicle variance should be considered in the analysis either in the homogeneous or heterogeneous traffic.

Generally, there are two approaches that can be used to analyze the traffic such as microscopic and macroscopic perspective. However, it is not known yet, which one the approach is superior for analyzing the traffic in both homogeneous and heterogeneous or mixed condition. In this study, the index of macroscopic traffic flow variable will be investigated. In the macroscopic perspective, traffic state can be described by the relationship between two variables such as flow, density and speed. The proper relation can be obtained by the suitable index of each variable. In the analysis, the index of macroscopic traffic flow variables should be able to consider the different effect due to the vehicle variance.

In homogeneous condition with lane based traffic, the length of vehicles was considered to affect the traffic. So, the index of macroscopic traffic flow variable was developed to consider the effect of variance of vehicle length. It was different in the mixed condition, with non-lane based traffic. In addition to the length, the width of vehicles was also considered to influence the traffic. So, over this study, the projected area of vehicles will be considered to develop the index of macroscopic traffic flow variable in the mixed traffic. There are some improvements related to the development of the index of traffic flow and density. In the mixed condition, mostly the index of traffic flow and density was adopted from the homogeneous condition. So, the proposed index of traffic flow and density was developed based on the existing index in homogeneous condition. However, it was not clarified yet whether all the vehicle variance has the significant effect to the traffic analysis or vice versa. This point is important to be known that the variance of vehicles needs to be considered or can be ignored. This knowledge can be used to validate the accuracy of the proposed index on the mixed condition.

Specifically, this study focuses to investigate the proper index of macroscopic flow variable in the mixed traffic. It was needed to describe the flow rate and the density of traffic which able to obtain the effect of the vehicle variance. To establish the proper index of these variables, the specific criteria of vehicle variance should be identified. So, the index of flow rate and density will be varied, which considers and not considers the difference of traffic entity. Then, the effect can be shown by comparing all the indices to describe the traffic state on the fundamental diagram (density-flow plane). The analysis result will show whether the variance needs to be considered or not in the traffic flow and density. Finally, it will lead to the suitable index of flow and density in the mixed traffic.

2. Literature Review

The first mathematic model was developed to estimate the actual traffic state by considering all the vehicles are similar, which expressed in vehicle unit. In the reality, vehicles have a different characteristic. So, in the estimation of actual traffic state, the characteristic of vehicles was important to be considered. It may produce a better model and possible to explain the variation in the share of traffic. Since the first traffic flow model was introduced, several concepts have been developed at the macroscopic perspective. These developments are the effort to consider the characteristic of vehicles in the traffic analysis appropriately. The characteristic of vehicle can be considered as a group or individually. In the heterogeneous or mixed traffic, the concept of PCU was widely developed to obtain the effect of different characteristic of vehicles. The concept of PCU classified the entity of traffic into several categories which has a similar characteristic. In the PCU concept, several approaches are developed in the mixed traffic to derive an appropriate result in the traffic analysis [3–6]. By using numerical approaches, the others studies have been conducted to incorporate the different characteristics of vehicles in macroscopic models [7–9]. Both concepts show that the characteristic of vehicle should be considered in the traffic model. In the other concept, the characteristic of vehicles was considered individually. The concept of occupancy and area occupancy [10–12] are proposed to consider the different of vehicle in length and the projected area respectively. The concept was developed to provide the proper density variable. In addition, an extensive review of the heterogeneous traffic flow model has been conducted [1]. It was concluded that linear density measurements are inadequate in the mixed traffic and suggest adopting the areal density concept, which consider the variation of the projected area of vehicles. All the concept which provided from several literatures are related one another. By understanding the measurement interval of macroscopic variable, it can provide an illustration of the relation and the proper concept. Then, the appropriate index can be developed and summarized by showing the relation with each other.

2.1. The Measurement Interval of Macroscopic Variable

At the macroscopic perspective, three main variables are defined: density k, flow rate q and mean speed u. In this perspective, traffic state can be described by the relation between these variables. The three aggregate variables are always connected by the fundamental relation or identity relation in which it can be expressed by the Equation (1).

$$q = k \cdot u \tag{1}$$

This fundamental relation considers the traffic flow under stationary and homogeneous condition. So, by this equation, knowing two variables automatically leads to the third. The definitions of macroscopic variables can be explained in the measurement interval by using trajectory data in the time-space diagram, which illustrates in the **Figure 1**. In this illustration, macroscopic variables can be derived by three measurement intervals: measurement over time (local measurement), measurement over space (Instantaneous measurement), and measurement over time and space.

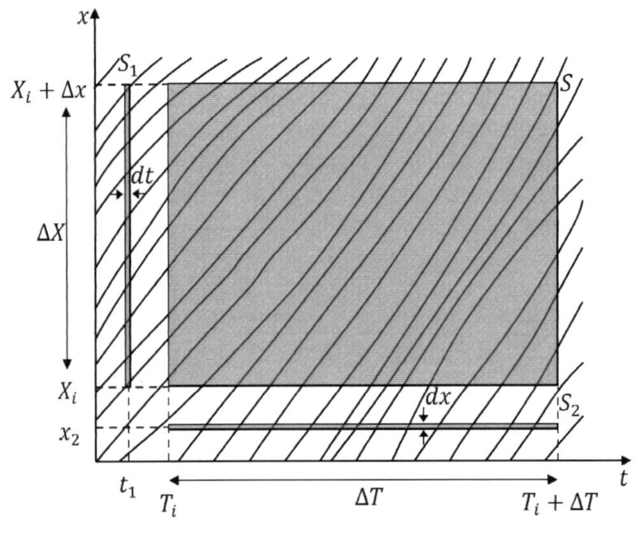

Figure 1. Measurement of macroscopic variables.

In the **Figure 1**, density was defined on the S_1 measurement interval. This rectangular measurement interval covers a road section of length ΔX during a slight time interval dt. Density k reflects the number of vehicles (m) per kilometer of the road, in which it was measured over space with ΔX length at any time instant (t_1). In this measurement interval, density k is formulated as the Equation (2):

$$k = \frac{m}{\Delta X} \tag{2}$$

The definition of density can be expanded into time-space S measurement intervals. In the rectangular measurement interval over time ΔT and space ΔX, the generalized density was defined as the Equation (3):

$$k = \frac{m \cdot dt}{\Delta X. dt} = \frac{Total\ time\ spent\ by\ all\ vehicles\ in\ S}{Area(S)} = \frac{\sum_i t_i}{\Delta X \cdot \Delta T} \tag{3}$$

Where, t_i – occupied time or time spent by vehicle i on the measurement interval.

In S_2 measurement interval, the definition of flow can be obtained. This rectangular measurement interval represents a slight road length dx during a time interval ΔT. Flow rate q represents the number of vehicles (n) that pass a certain cross-section per time unit. Flow was measured over time with time interval ΔT at any location(x_2). In this measurement interval, flow rate q is formulated as the Equation (4):

$$q = \frac{n}{\Delta T} \tag{4}$$

By expanding the meaning into time-space S measurement intervals, the generalized flow was obtained as the Equation (5):

$$q = \frac{n \cdot dx}{\Delta T \cdot dx} = \frac{Total\ distance\ covered\ by\ all\ vehicles\ in\ S}{Area(S)} = \frac{\sum_i d_i}{\Delta T \cdot \Delta X} \tag{5}$$

Where, d_i - the distance covered by vehicle i on the measurement interval.

Then, having known two previous variables, the third variable (speed, u) can be derived by using the identity relation. In this perspective, the speed u was defined as the quotient of the flow rate q and the density k. So, it was clear that the speed variable should comply with the identity relation.

2.2. The Index of Traffic Flow

According to the previous definition, the flow was measured and represented in the homogeneous vehicle unit index (vehicle unit/time unit) either measurement over time or time-space. In this flow variable, the traffic was considered under homogeneous condition. It means the entity was composed by the similar vehicles. However, the real traffic will never be in such a condition. Realizing the existence of vehicle variance in the traffic entity, a new index was developed to understand and deal with this condition. In Highway Capacity Manual 1965 [13], concept of passenger car equivalent (PCE) was first specifically discussed significantly as a measure index to represent the intensity of traffic. At the first, this concept classified the traffic entity into some types, which has a similar characteristic on speed. The term of "PCE" or also known as "PCU" was defined as the number of passenger cars displaced in the traffic flow by a truck or a bus under the prevailing roadway and traffic condition. To express the flow rate, all the vehicle in the traffic will be converted into a homogeneous equivalent number of passenger car by the Equation (6).

$$q = \frac{\sum_j (\alpha_j \cdot n_j)}{\Delta T} \tag{6}$$

Where, α_j - passenger car equivalent of a vehicle type j; n_j - number of vehicle category j that pass a cross section during time interval ΔT.

PCU method is widely used for modeling the heterogeneous traffic in developing countries. However, this method has a significant limitation. In Indonesian Highway Capacity Manual 1997 [14], the traffic entity was simplified into three types such as motorcycle, light vehicle and heavy vehicle. In this concept, this simplification is the advantage and the weakness. The analysis result by simplifying the traffic will deviate from the true condition. Number of vehicle category and equivalent value for each category will determine the accuracy of the method. So, it makes the method cannot apply in general due to the problem of setting PCE value in which it is possible to take the different values in each region/country. This method cannot cover the development of the kind of vehicle without improvement of PCE value, so that the improvement of PCE value is required to obtain a better result. By considering the limitation of PCU method, the other concept was adopted to capture the existence of the vehicle variance by treating the traffic entity as an individual vehicle. In the previous study [15], this concept measures the flow rate by adopting the dimension of vehicle or vehicle size, which obtained by the projected area of the vehicle in m^2/time unit /road. In this study,

IOP Conf. Series: Earth and Environmental Science **71** (2017) 012015 doi:10.1088/1755-1315/71/1/012015

adopting "vehicle size" as an index of flow variable will be measured per whole width of the road by following the expression as the Equation (7).

$$q = \frac{\sum_i a_i}{\Delta T}, a_i = w_i \cdot l_i \quad (7)$$

Where, a_i - projected area of vehicle i that passes a certain cross-section during time interval ΔT (m^2); w_i, l_i - width and length of the vehicle i.

Adopting "vehicle size" to describe the flow variable in traffic flow analysis can distinguish the effect each kind of vehicle based on the vehicle size and can be applied generally. This index can distinguish the effect of the vehicle variance in dimension that would be different. So, flow variable will describe a different value for each kind of vehicle based on the projected area by measuring dimension each vehicle on the road. The term of "vehicle size" refers to the projected area of each vehicle that obtained by multiplying the dimension (width and length) of the vehicle.

2.3. The Index of Traffic Density

Some index of density variable was developed to capture and deal with the existence of vehicle variance in share of traffic. Density is one of the macroscopic traffic flow variables, which explains the extent of usage of road space by vehicles. Density is useful in getting an idea about how crowded a certain section of the road is. Since density is essentially defined as a spatial measurement, it is one of the most difficult quantities to obtain. So, it can be derived simply from the identity relation by analyzing the flow and speed variable in advance. Basically, density can be defined as vehicles per length unit of the road. By adopting PCU concept, density of a certain section of the road also can be defined over time and space. It was defined as pcu per length unit of the road by the Equation (8), where the notation of the space interval ΔX was replaced by the length of the detection zone L.

$$k = \frac{\sum_j (\alpha_j \cdot \sum_{i \in j} t_i)}{L \cdot \Delta T} \quad (8)$$

Then, the new index to represent the density was developed in the measurement over time. The term of occupancy was introduced. It was developed to capture the length difference of the vehicle and make it easier to be obtained. The concept of occupancy is a function of speed and length of a vehicle. In this concept, occupied time was measured over a certain loop or a certain cross-section by using the additional device. The occupancy can be measured by using ultrasonic sensors, induction loops and image-processing techniques. Occupancy can be distinguished into two types. The first type, occupancy was defined as a non-dimensional variable that represents the percentage time of the road section (detection zone) which occupied by a vehicle over a given period. This occupancy was measured by using induction loops and can be formulated as shown in the Equation (9). The second type, occupancy was defined as a non-dimensional variable that represents the percentage time of the cross-section of the road which occupied by a vehicle over a given period. In this term of occupancy, the length of the detection zone (L) was ignored, and it was measured by ultrasonic sensors at certain cross-section and can be formulated in the Equation (10).

$$\rho_1 = \frac{\sum_i O_{1,i}}{\Delta T}, O_{1,i} \cong \frac{l_i + L}{v_i} \quad (9)$$

$$\rho_2 = \frac{\sum_i O_{2,i}}{\Delta T}, O_{2,i} \cong \frac{l_i}{v_i} \quad (10)$$

Where, $\rho_{1,2}$ - occupancy; $O_{1,i}$ - the time the vehicle i occupied the detection zone or road section; $O_{2,i}$ - the time the vehicle i occupied the detection line or cross-section of road; l_i - the length of the vehicle i; v_i - speed of vehicle i; L - length of the detection zone; ΔT - time interval.

The first type of method was not proper to be said as a measurement over time because it considers a space interval. In the equation, it seems to overestimate the occupied time in the detection zone by considering the length of the vehicle. So, its value will be changed by varying the length of the detection zone. However, in the traffic with different vehicle lengths, it is apparent that both occupancy methods are more meaningful than density (in homogenous traffic). Even though, the concept of occupancy is considered not proper for mixed or heterogeneous traffic as no-lane based traffic flow. The main reason is both density and occupancy developed for lane based traffic to represent the traffic density variable in homogeneous traffic.

In heterogeneous or mixed condition, the width of the vehicle was considered to affect the traffic. Instead of length difference, the variation of the projected area of a vehicle was more meaningful to be considered in this condition. As no-lane based traffic, the width of the vehicle can affect the performance of traffic. The index of traffic density variable was developed to capture the variance of the vehicle projected area (length and width). The term of "area occupancy" was proposed in the earlier study [11,12]. It was considered more suitable to represent the density variable in the mixed traffic condition. Area occupancy considers the horizontal projected area of the vehicle as the basis without any restriction on the length of detection zone and width of the road [16]. It was treating the detection zone which analyzed per length and width unit of the road. Area occupancy can be defined as a ratio of the detection zone area (in terms of time) that is occupied by vehicles. The concept of area occupancy as an index of traffic density variable in time-space measurement can be seen in the Equation (11) and (12).

$$\rho_a = \frac{\sum_{i=1}^{N} \frac{L - x_i}{v_i} \times w_i \times l_i}{T \times W \times L} \tag{11}$$

$$\rho_a = \frac{\sum_i (a_i \cdot t_i)}{A \cdot \Delta T} \tag{12}$$

Where, ρ_a- area occupancy on the observed road over time interval ΔT; A, W, L – area, width and length of the road section under consideration; x_i - denotes the distance between the front-end of the vehicle i and any of the two ends of the detection zone; $(L - x_i)$- denotes the distance traveled by the vehicle i over the road section under consideration; a_i, w_i, l_i - projected area, width and length of vehicle i; t_i - occupied time of vehicle i in the detection zone during the time interval ΔT; ΔT - time interval.

In the Equation (11), ${(L - x_i)}/{v_i}$ represents the time spent by each vehicle t_i over the road section under consideration. Since the difficulties to consider the variable of $(L - x_i)$ in the measurement, the area occupancy was measured by the temporal measurement [11] as seen in the Equation (13).

$$\rho_a = \frac{\sum_{i=1}^{N} \frac{l_i}{v_i} \times w_i}{T \times W} \tag{13}$$

In the Equation (12), the concept of area occupancy was formulated over time and space by the following definition [12]:

1. Area occupancy will be unchanged for any length of the detection zone. This result can be obtained by two conditions. The first condition: modeling the vehicle under stationary and homogeneous condition where the vehicles on this road share the same speed and traveling with constant speeds over the detection zone. The second, all the vehicles have traversed the entire detection zone during the observation period. So, at the beginning and at the end of the observation period there are no vehicles located in the detection zone.
2. Area occupancy considers the horizontal projected area of the vehicle as the basis without any restriction on the length of detection zone and width of the road (treating the whole of the width of the road as a single unit without consideration of traffic lanes).

3. Methodology

In this study, the proper index of flow and density will be investigated in the mixed traffic. The indices of traffic flow and density will be varied by the concept and the measurement interval. The concept was distinguished according to the respond to the variance of vehicles in the traffic analysis which measure over time and time-space. At the first, flow was measured over time, and density was measured over space. Measuring traffic over time is easier to be conducted rather than the measurement which considers a space interval, mainly over a long section of the road. So, traffic flow is relatively easier to be measured than density. By the difficulties, the density was proposed to be measured over time or time-space with a short space interval. The traffic flow and density will be analyzed on the observed traffic data in the mixed condition. The flow and density are measured over time and time-space in the detection zone which considers a short section of a road as seen in the **Figure 2**.

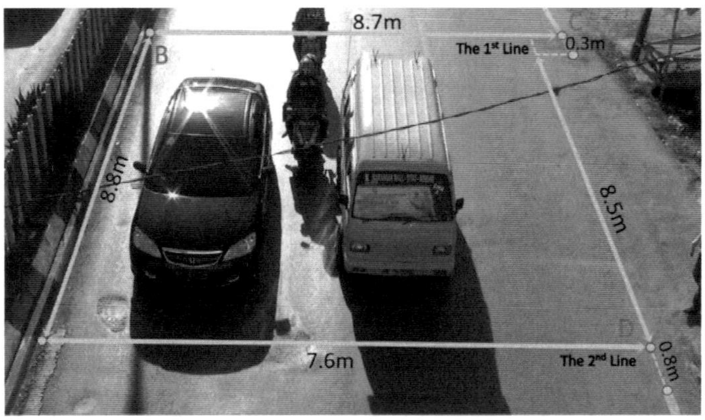

Figure 2. Detection zone of observation area.

The flow and density variables which measured over time are based on the second line of the detection zone. In order to define the flow and density, the front end of the vehicle is used as a reference point to establish the trajectory which will be analyzed within each time interval. Taking the measurement of density and flow over time and space requires defining the occupied time t_i and the distance covered by vehicle d_i. Both variables in the time-space interval can be described in **Figure 3**.

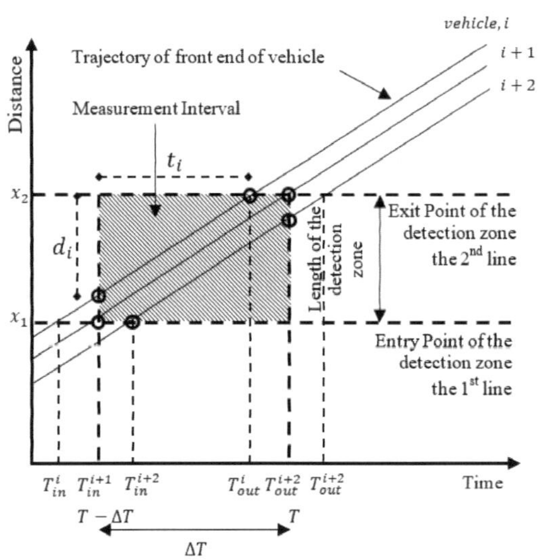

Figure 3. Occupied time over time and space.

According to the previous illustration in the **Figure 3**, the occupied time t_i and the distance covered by vehicle d_i relate each other. Between these variables, the occupied time t_i was easier to be defined than the distance d_i which also denoted by $(L - x_i)$ as seen in the Equation (11). The occupied time t_i will be used to obtain the density variable for all indices in the time-space measurement. It will be defined by knowing the period when the vehicle was at the first and the second line by the Equation (14):

$$t_i = t_{out} - t_{in}; t_{in} = max(t_{i,1}, T - \Delta T); t_{out} = min(t_{i,2}, T) \quad (14)$$

Where, t_i - occupied time of vehicle i in the detection zone during the time interval ΔT; $t_{i,1}, t_{i,2}$ - time when the trajectory line of vehicle i get the 1st and the 2nd line of detection zone.

Then, the variable of the distance covered by vehicle d_i can be defined. In time-space measurement, the distance covered by vehicle d_i was based on the occupied time t_i in the detection zone, which used to obtain the flow for all indices. By assuming the speed of the vehicle v_i is constant over the space interval, the distance covered by vehicle d_i can be formulated as the Equation (15), in which occupied time t_i refers to the Equation (14).

$$d_i = v_i \cdot t_i \quad (15)$$

The index of flow and density was developed to capture the existence of vehicle variance in share of traffic. In the development of this index, the important thing which should be noticed is showing whether this variance needs to be considered or can be ignored in the traffic analysis. The effect of vehicle variance on the traffic analysis at macroscopic perspective can be captured by treating the vehicle as an individual or simplifying them into several groups by using PCU concept in the index flow and density. By direct observation, the existence of vehicle variance was easier to be identified. The difference of the vehicle in the properties such as width and length is the clear evidence to show the existence of vehicle variance. It becomes a basis to develop the index of flow and density which able to show the respond of traffic. These indices will be used to identify the existence of vehicle variance and its effect analytically. Responding to the existence of vehicle variance can be obtained by comparing using the index of flow and density, which considers and not considers the variance of the vehicle in the traffic analysis.

In the basic relation, traffic flow and density are described in the Fundamental Diagram by the term of the density-flow plane. Generally, traffic flow variable can be explained by using different index of density. Flow was expressed by the index of vehicles per time unit, and density was expressed by the index of vehicles per length unit. In the mixed traffic, both flow and density were adjusted to being measured per whole road instead of per lane. In two-dimensional diagram, traffic flow and density are described on the y-axis and x-axis respectively. The relations of both variables are required to comply with the identity relation of Fundamental Diagram. In such case, the gradient of each point on the diagram has to show the speed of traffic. By keeping this rule, density also can be transformed into the other term which measured over time. It can be derived from the generalized density definitions by ignoring the length of the detection zone L as a space interval ΔX. By assuming the length of section was very small, the distance covered for each vehicle will be equal $d_i \approx L$. So, density over time can be expressed as the Equation (16).

$$k = \frac{\sum_i t_i}{\Delta X \cdot \Delta T} = \frac{\sum_i \left({}^{d_i}/_{v_i} \right)}{L \cdot \Delta T} = \frac{\sum_i \left({}^{L}/_{v_i} \right)}{L \cdot \Delta T} = \frac{\sum_i \left({}^{1}/_{v_i} \right)}{\Delta T} \quad (16)$$

Where, k - density; t_i - occupied time of vehicle i in the detection zone during the time interval ΔT; v_i - speed of vehicle i

The principal to derive the Equation (16) shows the relation between the Equation (11) and (13) which shown in the previous discussion. Basically, all developed indices which mention on the previous discussion can be used to describe the flow and density variable. However, the combination of traffic flow and density should be selected carefully to comply with the fundamental relation in Equation (1). At the first, both of flow and density were developed by adopting the passenger car unit (PCU) concept. It was also measured per whole road. On this concept, the relation between flow and density still complies with the fundamental relation. Traffic flow in this concept was expressed by the index of PCU per time unit, and density was expressed by the index of PCU per length unit. In this study, the PCU concept refers to the Indonesian Highway Capacity Manual 1997 [14] by the PCE value and number of vehicle category in the **Table 1**.

Table 1. Passenger car equivalent value.

No.	Type of Vehicle	Passenger Car Equivalent Value
1	Motorcycle	0.4
2	Light Vehicle	1
3	Heavy Vehicle	1.3

By adopting PCU, flow can be expanded into the measurement over time and space by determining a length of the detection zone L as the Equation (17).

$$q = \frac{\sum_j(\alpha_j \cdot \sum_{i\in j} d_i)}{L \cdot \Delta T} \tag{17}$$

Where, d_i – the distance covered by vehicle i in the detection zone during the time interval ΔT.

In addition, adopting PCU in density can be expressed on the measurement over time. It can be derived from the Equation (8) by eliminating the length of the detection zone L as the Equation (16), in which it can be formulated as the Equation (18).

$$k = \frac{\sum_j(\alpha_j \cdot \sum_{i\in j}(^1/_{v_i}))}{\Delta T} \tag{18}$$

On the further development, traffic density was measured over time by considering the variance of vehicle length, which known by the term of occupancy. In this study, the second occupancy was more appropriate to be used. It can be explained by expanding the meaning into time-space intervals. The definition can be derived by considering the length of the detection zone L as a space interval into the Equation (10) as the Equation (5), which can be expressed as seen in the Equation (19).

$$\rho = \frac{dx \cdot \sum_i\left(^{l_i}/_{v_i}\right)}{dx \cdot \Delta T} = \frac{\sum_i\left(l_i \cdot {^{d_i}/_{v_i}}\right)}{L \cdot \Delta T} = \frac{\sum_i(l_i \cdot t_i)}{L \cdot \Delta T} \tag{19}$$

Where, ρ - occupancy; l_i - the length of the vehicle i.

The definition of occupancy in time-space is similar with density by considering the difference of vehicle length. To adopt this occupancy either over time or time-space in the Fundamental Diagram, the flow also should be expressed by considering the difference of vehicle length as shown in the Equation (20) and (21).

$$q = \frac{\sum_i l_i}{\Delta T} \tag{20}$$

$$q = \frac{\sum_i(l_i \cdot d_i)}{L \cdot \Delta T} \tag{21}$$

In the mixed condition which no-lane based traffic, the index of density was developed by considering the projected area of the vehicle. This measure refers to the two terms namely area occupancy and areal density. Understanding this concept can be explained by the illustration in the **Figure 4**.

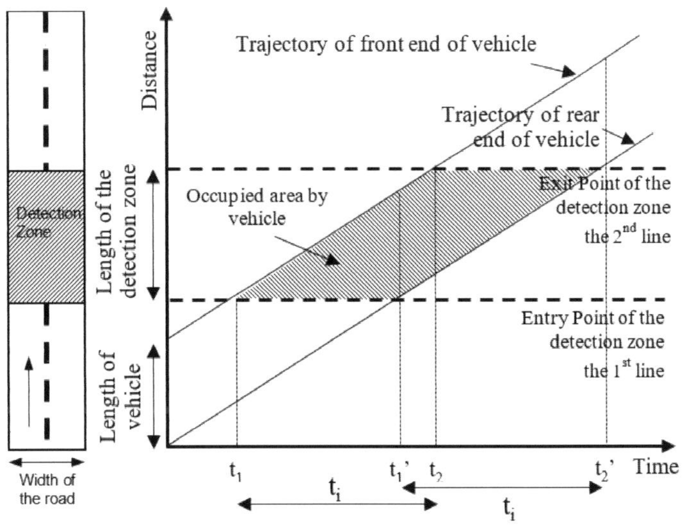

Figure 4. Concept of area occupancy in time-space diagram.

In the **Figure 4**, the occupied time t_i of each vehicle can be measured by two reference points of the trajectory: the front end and the rear end of the vehicle. So, this occupied time can be defined as the time difference to get the exit and the entry point or the time difference to leave the exit and the entry point. Area occupancy can be measured over time by local measurement, over space by instantaneous measurement, and measured over time and space. In the homogeneous and stationary condition, the concept of area occupancy is not restricted by the length of the detection zone, which mean the area occupancy will not differ by the change of the detection length. In the local or over time measurement, the length of detection zone L as a space interval ΔX will be ignored. It can be assumed that the length of section was very small. So, area occupancy over time as seen in the Equation (22) can be derived from Equation (12) by the same process to obtain the Equation (16) .

$$\rho_a = \frac{\sum_i (a_i \cdot t_i)}{A \cdot \Delta T} = \frac{\sum_i \left(a_i \cdot {}^{d_i}/_{v_i} \right)}{W \cdot L \cdot \Delta T} = \frac{\sum_i (a_i \cdot {}^{L}/_{v_i})}{W \cdot L \cdot \Delta T} = \frac{\sum_i ({}^{a_i}/_{v_i})}{W \cdot \Delta T} \tag{22}$$

Where, ρ_a- area occupancy on the observed road over time interval ΔT; W, L - width and length of detection zone; w_i, l_i - width and length of the vehicle i; v_i - speed of vehicle i; ΔT - time interval.

In the instantaneous measurement, area occupancy was measured over space at a time instant, in which time interval will be ignored. It also can be assumed that the time interval was very small. So, the occupied time for each vehicle will be equal $t_i \approx \Delta T$. In the measurement over space, area occupancy can be derived from Equation (12) as seen in the Equation (23).

$$\rho_a = \frac{\sum_i (a_i \cdot t_i)}{A \cdot \Delta T} = \frac{\sum_i a_i}{A} \tag{23}$$

By this equation, area occupancy over space can be defined as the ratio of the total projected area of vehicles to the area of roadway. This definition was similar with the other term of index traffic density variable namely areal density [10]. So basically, the term of area occupancy leads to the extended definition of areal density, which measured over time and space. Area occupancy over space

IOP Conf. Series: Earth and Environmental Science **71** (2017) 012015 doi:10.1088/1755-1315/71/1/012015

or time-space was easier to be measured on a short section of the road as the detection zone by knowing the time at the entry point and the exit point. However, it was difficult and needs more effort to measure area occupancy over the long section of the road.

To adopt this measure either over time or time-space in the Fundamental Diagram, the flow should be expressed by considering the projected area of the vehicle as shown in the Equation (7). This definition of flow can be extended into the time-space measurement by the Equation (24).

$$q = \frac{\sum_i (a_i \cdot d_i)}{L \cdot \Delta T} \quad (24)$$

By the equation, area occupancy or areal density are measured per width unit of the road and flow is measured per whole road for each direction. The combination of this index can be used by transforming one of them into the same treatment of the width of the road. In this study, area occupancy or areal density either over time or time-space will be transformed. By considering the width of the road W as one, Equation (12) and (22) can be rewritten as seen in the Equation (25) and (26).

$$\rho_a = \frac{\sum_i (a_i \cdot t_i)}{A \cdot \Delta T} = \frac{\sum_i (a_i \cdot t_i)}{W \cdot L \cdot \Delta T} = \frac{\sum_i (a_i \cdot t_i)}{L \cdot \Delta T} \quad (25)$$

$$\rho_a = \frac{\sum_i \left({a_i}/{v_i} \right)}{W \cdot \Delta T} = \frac{\sum_i \left({a_i}/{v_i} \right)}{\Delta T} \quad (26)$$

Where, ρ_a- area occupancy on the observed road over time interval ΔT; W, L, A - width, length and area of detection zone.

In this study, all the combination of traffic flow and density will be applied and evaluated on the Fundamental Diagram to describe the actual traffic state in the density-flow plane. The combinations that will be analyzed are summarized in the **Table 2**.

Table 2. The index of traffic flow and density.

No.	Concept and Measurement Interval	Variable	
		Flow	Density
1	Vehicle Unit - Over Time	$q_{1a} = \frac{n}{\Delta T}$	$k_{1a} = \frac{\sum_i ({1}/{v_i})}{\Delta T}$
2	Vehicle Unit - Over Time and Space	$q_{1b} = \frac{\sum_i d_i}{L \cdot \Delta T}$	$k_{1b} = \frac{\sum_i t_i}{L \cdot \Delta T}$
3	Passenger Car Unit - Over Time	$q_{2a} = \frac{\sum_j (\alpha_j \cdot n_j)}{\Delta T}$	$k_{2a} = \frac{\sum_j (\alpha_j \cdot \sum_{i \in j} ({1}/{v_i}))}{\Delta T}$
4	Passenger Car Unit - Over Time and Space	$q_{2b} = \frac{\sum_j (\alpha_j \cdot \sum_{i \in j} d_i)}{L \cdot \Delta T}$	$k_{2b} = \frac{\sum_j (\alpha_j \cdot \sum_{i \in j} t_i)}{L \cdot \Delta T}$
5	Vehicle Length Unit - Over Time	$q_{3a} = \frac{\sum_i l_i}{\Delta T}$	$k_{3a} = \frac{\sum_i \left({l_i}/{v_i} \right)}{\Delta T}$
6	Vehicle Length Unit - Over Time and Space	$q_{3b} = \frac{\sum_i (l_i \cdot d_i)}{L \cdot \Delta T}$	$k_{3b} = \frac{\sum_i (l_i \cdot t_i)}{L \cdot \Delta T}$

7	Vehicle Projected Area Unit - Over Time	$q_{4a} = \frac{\sum_i a_i}{\Delta T}$	$k_{3a} = \frac{\sum_i (a_i / v_i)}{\Delta T}$
8	Vehicle Projected Area Unit - Over Time and Space	$q_{4b} = \frac{\sum_i (a_i \cdot d_i)}{L \cdot \Delta T}$	$k_{4b} = \frac{\sum_i (a_i \cdot t_i)}{L \cdot \Delta T}$

Where, q - flow on the observed road over time interval ΔT; k - traffic density on the observed road over time interval ΔT; n, n_j - number of vehicle and number of vehicle category j that pass a cross section during time interval ΔT; α_j - passenger car equivalent of a vehicle type j; a_i, l_i - projected area and length of vehicle i; d_i - the distance covered by vehicle i in the detection zone during the time interval ΔT; t_i - the occupied time of vehicle i in the detection zone during the time interval ΔT; v_i - speed of vehicle i; L - length of detection zone.

In **Table 2**, the combination of the index of traffic flow and density can be distinguished into four different concept as follows:

- The first concept is the combination which not considers the existence of vehicle variance, so all vehicles are considered similar and it will be indexed by vehicle unit.
- The second concept is the combination which considers the existence of vehicle variance by simplifying the traffic entity into some categories. This combination adopts the concept of the passenger car unit.
- The third concept is the combination which considers the variance of vehicle length, so the vehicle will be indexed by vehicle length unit (m^1).
- The fourth concept is the combination which considers the variance of the projected area of the vehicle, so the vehicle will be indexed by vehicle area unit (m^2).

All the combination will be plotted in a two-dimensional diagram which describes the relation between flow and density. In addition, the combination of index of flow and density will be modeled by a polynomial of degree two equations to describe the actual traffic state. The goodness of fit on this model is measured based on the R-square value. These values will be a basis to evaluate the model, which represent each combination. The R-square value of each model will be compared to obtain the effect of vehicle variance in the estimation of actual traffic state. Then, the proper index of flow and density will be established based on the highest R-square of the model.

4. Observation Data

In the present study, the traffic in Makassar will be observed. It was selected due to the characteristic of the traffic in Makassar classified as the mixed traffic. The traffic was observed at the morning peak hours. The observed location was a main road, unaffected by major intersection, has a separator direction and having a good view for video recording. The observation was conducted on 20th September 2013 at 08.30-11.30 a.m. in sunny conditions. The traffic was observed by video recording at the pedestrian bridge of Urip Sumoharjo Street. Data for each vehicle will be collected during the observation period. These individual data consist of the projected area as representation of the vehicle dimension (width and length), the period when the vehicle at the first line $t_{i,1}$ and at the second line $t_{i,2}$. The video data will be extracted into image data for 10 frames per second. By using a computer programming which has developed, all these image data will be analyzed. The required time data will be tracked for each vehicle. In addition, the dimension for each vehicle can be obtained by collecting the image coordinates for the specific part of each vehicle. At this stage, the human resources as the operator still require for tracking each vehicle by clicking the pointer manually. Then, the coordinate of image will be converted to get the real coordinate by projective transformation. The method which called a numerical rectification will be used to transform the coordinate. This approach was also used in several studies to transform the screen coordinates to be a real coordinates [17–20].

5. Result and Analysis

In this study, all the raw data of each vehicle will be analyzed to derive the variable of traffic flow and density. The aggregate data of traffic flow and density will be analyzed for every minute. By this time interval, there are 148 observed data of density and flow, which have been collected during three-hour observation. The remaining data were not used due to the error at image processing stage. Briefly, the actual traffic state can be known by plotting the flow and density in the two-dimensional diagram. By using the first index in the **Table 2**, the relation between flow and density can be shown in the **Figure 5**. The first index of flow and density is expressed by vehicle unit, which measured over time.

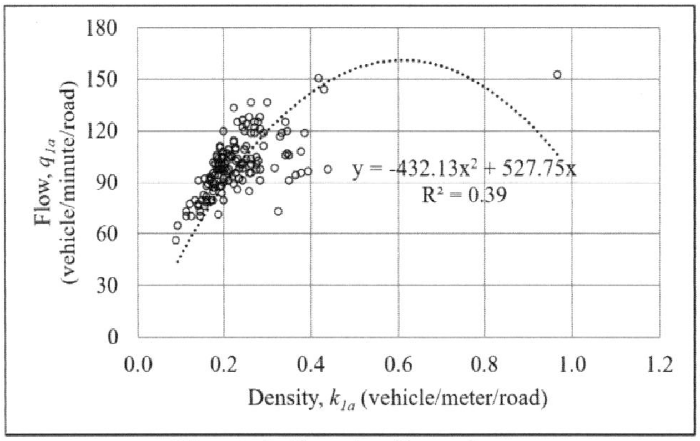

Figure 5. Density-flow plane.

In the **Figure 5**, the actual traffic state during the observation period shows that almost the observed traffic was on the uncongested phase. There is just a few sample on the congested phase. Therefore, the model should be applied only on the uncongested phase. The present study emphasizes to the comparison of several indices of flow and density to estimate the actual traffic state. The relation between flow and density was modeled on parabolic equation by using regression analysis method. By using different index of traffic flow and density, it will produce the different regression model. There are eight combinations of the index of traffic flow and density by variation on the concept and the measurement interval. All the indices will be used to model the actual traffic state. The analyzed regression model for each combination can be seen in the **Table 3**.

Table 3. The relationship between traffic flow and density.

No.	Combination of $q-k$	Measurement Interval	Coefficient		R	R-Square
			k	k^2		
1	$q_{1a}-k_{1a}$	Over Time	527.75	-432.13	0.62	0.39
2	$q_{1b}-k_{1b}$	Over Time and Space	527.77	-432.40	0.62	0.39
3	$q_{2a}-k_{2a}$	Over Time	541.89	-957.03	0.66	0.44
4	$q_{2b}-k_{2b}$	Over Time and Space	542.29	-960.25	0.66	0.44
5	$q_{3a}-k_{3a}$	Over Time	525.74	-228.02	0.72	0.52
6	$q_{3b}-k_{3b}$	Over Time and Space	526.12	-228.79	0.72	0.52
7	$q_{4a}-k_{4a}$	Over Time	517.49	-220.93	0.79	0.62
8	$q_{4b}-k_{4b}$	Over Time and Space	518.40	-222.55	0.79	0.63

Number of Observation: 148

In the **Table 3**, the comparison between the concepts which adopts different combination can be evaluated. This table shows the equation and the goodness of fit each model to estimate the actual traffic state. The purposes of this study can be obtained by comparing the goodness of fit all the models, which represent its own concept. The 1st and 2nd model describe the index of flow and density, which not consider the existence of vehicle variance. On this concept, all vehicles are considered equal. It will be set as a basis for comparison to the other concept. Both models as the representation of the first concept show the similar R-square: 0.39. The second concept which represented on the 3rd and the 4th model adopts the passenger car unit concept as an index of the traffic flow and density. This concept simplifies the entity of traffic into some categories based on the physical and the difference of speed. In this concept, the simplification makes the result deviates from the true condition. The accuracy of methods will depend on how well this simplification to represent the variance of the vehicle that exists in the share of traffic. Both models which consider this variance show a better R-square value: 0.44. It was increased as compared with the first concept. In this analytical method, it is quite enough to show that the variance of the vehicle in share of traffic exists. So, the entity of traffic should be differentiated either individually or as a group on any basis which appropriate. The proper index of flow and density should cover the existence of vehicle variance, which possible to explain the traffic behavior well.

By using the individual characteristic as a basis to cover the variation in the share of traffic, two concepts are introduced: occupancy and area occupancy. In the concept of occupancy, the length of vehicles becomes a basis to distinguish the variance of traffic entity as an individual vehicle. This concept was represented by the 5th and the 6th model. The models on this concept show R-square value: 0.52. It shows the better value as compared with the first and the second concept. This result describes that the variation of traffic behavior can be explained more properly by the difference of vehicle length. It also shows that the PCU method which adopted from IHCM, 1997 [14] was not enough to represent the variance of vehicles in the share of traffic. In the concept of area occupancy, the projected area of vehicles become a basis to explain the variance of the vehicle in share of traffic. The projected area is the function of width and length of the vehicle. This concept was represented by the 7th and the 8th model. The models on this concept show the improvement by the R-square value: 0.62 and 0.63 respectively. This result shows that the index of flow and density, which considers the projected area of vehicles can explain the actual traffic state more properly by obtaining the highest R-square values. So, instead of the vehicle length, projected area of the vehicle can be considered to explain the variation in the share of traffic flow by individual data.

In addition, the result in the **Table 3** also shows the comparison between the measurements over time and time-space, which differentiated for each concept. This comparison was used to verify the result on both density and flow, which measured over time. Taking the measurement over time is more convenient than the measurement which involves the space interval. By considering this reason, variable density and flow are preferred to be measured over time. The result shows a similar result between measurement over time and time-space on short section of the road. It can be obtained by comparing the R-square models which measure over time and time-space for each concept. The comparison for each concept shows the similar of R-square value. The similarity of the result can be explained by two reasons. The length trap which considers as a detection zone for the measurement over time and space is too short. The other reason is the speed variable of each vehicle, which used for the measurement over time. This speed variable was measured on the same longitudinal trap or detection zone for the measurement over time and space. Finally, by the similarity on the result, analyzing the actual traffic state can rely on the both variable density and flow, which measured over time.

6. Conclusion

This study investigates the existence of vehicle variance and the proper index of traffic flow and density to capture its effect in the mixed condition. It was needed to consider any specific criteria to distinguish the vehicle in the share of traffic either as a group or as an individual. As the result of this study, there were three different concept or criteria, which used to distinguish the vehicle: based on the group using PCU, the length of vehicle and projected area of the vehicle. By adopting these concepts as the index of flow and density, the model which obtained can estimate the actual traffic state well. It was analytically proof that the vehicle variance in the mixed traffic needs to be considered in the analysis. Among the three concept, the projected area of the vehicle, which used as a criterion to distinguish the vehicle can explain the variation of traffic behavior in the share of the road more properly. It also describes that the motions of the vehicle are affected by the dimension or the projected area of vehicles: length and width. Therefore, considering the projected area of the vehicle as the index of density and flow variable can produce a better model to estimate the actual traffic state on the Fundamental Diagram. Moreover, taking the measurement over time, and the measurement over time and space do not provide a significant difference on the traffic observation over a short section of a road.

7. Acknowledgements
The authors would like to acknowledge the financial support through the scholarship given by Directorate General of Higher Education of Indonesia.

8. References
[1] Khan S I and Maini P 1999 Modeling heterogeneous traffic flow *Transp. Res. Rec. J. Transp. Res. Board* **1678** 234–41

[2] Chandra S and Sikdar P 2000 Factors affecting PCU in mixed traffic situations on urban roads *Road Transp. Res.*

[3] Chandra S and Kumar U 2003 Effect of Lane Width on Capacity under Mixed Traffic Conditions in India *J. Transp. Eng.* **129** 155–60

[4] Nguyen Y C, Sano K, Tran T V and Doan T T 2010 Methodology for Motorcycle Equivalent Unit at Road-Segments in Urban Road *Transp. Res. Board 89th Annu. Meet. (No. 10-3026).* 1–20

[5] Mallikarjuna C and Rao K R 2006 Modelling of Passenger Car Equivalency Under Heterogeneous Traffic Conditions *22nd ARRB Conf. – Res. into Pract. Canberra Aust. 2006* 1–13

[6] Dhamaniya A and Chandra S 2013 Concept of Stream Equivalency Factor for Heterogeneous Traffic on Urban Arterial Roads *J. Transp. Eng. ASCE* 1117–23

[7] Chanut S and Buisson C 2003 Macroscopic Model and Its Numerical Solution for Two-Flow Mixed Traffic with Different Speeds and Lengths *Transp. Res. Rec. J. Transp. Res. Board* **1852** 209–19

[8] Wong G C K and Wong S C 2002 A multi-class traffic flow model–an extension of LWR model with heterogeneous drivers *Transp. Res. Part A Policy Pract.* **36(9)** 827–41

[9] Zhu Z, Chang G and Wu T 2003 Numerical Analysis of Freeway Traffic Flow Dynamics for Multiclass Drivers *Transp. Res. Rec.* **1852** 201–8

[10] Chari S R and Badarinath K M 1983 Study of mixed traffic stream parameters through time lapse photography *Highw. Res. Bull. (Indian Road Congr. Highw. Res. Board)* **20** 57–83

[11] Mallikarjuna C and Rao K R 2006 Area occupancy characteristics of heterogeneous traffic *Transportmetrica* **2(3)** 223–36

[12] Arasan V T and Dhivya G 2008 Measuring Heterogeneous Traffic Density *World Acedemy*

Sci. Eng. Technol. **46** 342–6

[13] Highway Research Board 1965 Special Report 87: *Highway Capacity Manual* (Washington D. C.: National Research Council)

[14] Directorate General Bina Marga 1997 Indonesian Highway Capacity Manual (IHCM) (Jakarta: Directorate General of Highway Ministry of Public Works)

[15] Gani F A, Yoshii T and Kurauchi S 2014 The Effect of Heterogeneity of Vehicle Size on the Fundamental Diagram in Mixed Traffic. (Case Study: Makassar Traffic on Urip Sumoharjo Street). *Proceeding of the International Conference of Transdiciplinary Research on Environmental Problem in Southeastern Asia* (Makassar, Indonesia: Penerbit ITB) pp 95–104

[16] Arasan V T and Dhivya G 2010 Methodology for Determination of Concentration of Hetrogeneous Traffic *J. Transp. Syst. Eng. Inf. Technol.* **10** 50–61

[17] Khan S I and Raksuntorn W 2001 Accuracy of numerical rectification of video images to analyze bicycle traffic scenes *Transp. Res. Rec. J. Transp. Res. Board* **177** 220–8

[18] Minh C C, Sano K and Matsumoto S 2005 Characteristics of Passing and Paired Riding Maneuvers of Motorcycle *J. East. Asia Soc. Transp. Stud.* **6** 186–97

[19] Minh C C, Sano K and Matsumoto S 2007 Maneuvers of Motorcycles at Red-lights of Signalized Intersections *11th World Conference on Transport Research*

[20] Matsuo K, Hirobata Y and Komatsu H 2013 An Analysis of the Effects of Simple Traffic Safety Measures at a Signalized Intersection: Before-and-after Study on Right-Turn Behavior *Proceedings of the Eastern Asia Society for Transportation Studies* vol 9

Economic features of the artisanal and small-scale gold mining industry in Bombana, Southeast Sulawesi, Indonesia

Basri[1,2], Masayuki Sakakibara[1, 3], Ratnawati[4]

[1]Graduate School of Science Engineering, Ehime University, 2-5 Bunkyo-cho, Matsuyama, Ehime Prefecture 790-8577, Japan
[2] Makassar School of Health Science (Sekolah Tinggi Ilmu Kesehatan Makassar), Jl. Maccini Raya No. 197, Makassar 90231, Indonesia
[3] Faculty of Collaborative Regional Innovation, Ehime University
[4] Health Department of Bombana Regency, Southeast Sulawesi, Indonesia
Corresponding Author: basrikesmas@gmail.com

Abstract. There is a consensus that the artisanal small-scale gold mining (ASGM) sector contributes to trapping individuals in a cycle of poverty and financial insecurity with a low standard of living. The purpose of this study was to analyze the economic circumstances of miners in the Bombana ASGM area in Indonesia using a mixed methodology approach that involved collecting, analyzing, and mixing both quantitative and qualitative data in a single study. Results show that the average and maximum monthly incomes of 201 miner households were Rp. 4,360,000 and Rp. 9,750,000, respectively. The average income in the ASGM area in Bombana was higher than that at similar sites on Java Island (~Rp. 2,900,000) and was much higher than the average monthly income of the inhabitants of Bombana (~Rp. 2,100,000). Because of the higher revenue generated in the ASGM sector, miners and their families depend long-term on the mining work, which makes it difficult to control the mining activity and the associated environmental degradation.
Keyword: economic features, artisanal and small-scale gold mining

1. Introduction

In developing countries, the informal artisanal small-scale gold mining (ASGM) sector has rapidly expanded over the past two decades, and more than 15 million people are dependent on the gold mining sector [1,2]. Gold production from ASGM accounts for between 20% and 30% of the total world output [1].

There is a consensus that, when engaged in ASGM, individuals remain trapped in a cycle of poverty [3,4], hampered by heavy metal exposure [5] financial insecurity and a low standard of living [6]. The ASGM sector's role in the economies of developing countries may have grown in response to the poverty caused by unemployment [3,7]. In some developing countries, this sector is designated as poverty alleviation and has been successful as a major driver of the economy where other industries have failed [8]. Social and economic benefits are obtained by simultaneously achieving maximum formalization of the artisanal and small-scale mining industry [7], promoting women's empowerment and gender equity [9], encouraging safer livelihoods, and improving the minimal infrastructure that exists [8].

In a regional context, the ASGM sector in Bombana, Indonesia, plays a pivotal role in the economy and society. Estimates suggest that the area contains sufficient resources for about 30 years of mining and that it can bring prosperity to the whole province [10]. This activity attracts people not only from the southeast province of Sulawesi, but from other parts of Indonesia who seek to try their luck as gold miners [11]. An estimated 800 miners and at least 6 mining companies were active at the end of 2015. Since decentralization of laws and regulations, the number of people working in this sector has doubled [12].

The higher revenue generated in Bombana means that miners and their families depend on the mining work for the long term, thereby making it difficult to control the gold exploration activity. This can produce negative environmental, and socio-economic effect. In the recent past, there have been ongoing problems of over-exploitation, poor working conditions, child labor, a lack of social security, and relationships with organized crime, and illegal mercury amalgamation process in Bombana. Previous studies have provided an overview of the negative economic, environmental, and social impacts of the mining activities, but have not compared the financial implications between the mining and non-mining communities. Equally, few studies have reviewed the economic structures that affect the income levels of the communities around mining sites. The aim of the present case study of the Bombana ASGM sites is to establish the economic features of the mining and non-mining communities. The specific objectives are to (1) identify the income levels in mining and non-mining communities, (2) determine disparities in environmental sanitation facilities, and (3) examine the various social and economic structures related to the ASGM sector [13].

2. Material and method

2.1. Description of the study area

Bombana Regency is situated on the southeast peninsula of Sulawesi, south of the equator (Figure. 1). The total population of 159,718 consists of 80,616 men and 79,102 women [14]. The Moronene and Bugis native tribes dominate, and most of other residents are from Java and Bali, having migrated from other islands through a transmigration program [11]. Likewise, in the mine areas, where various ethnic groups work together, and there is a diverse community; the ethnic Bugis and Makassar still dominate, followed by the Moronene and Tolaki indigenous tribes from Southeast Sulawesi.

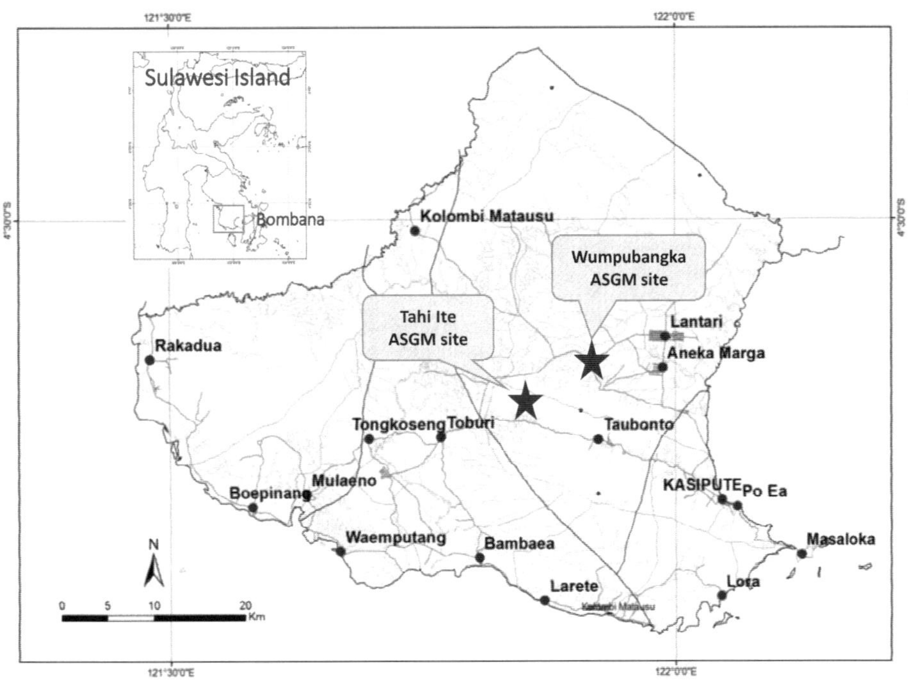

Figure 1. Location of study site on Bombana Regency

This area has witnessed a rapid expansion of gold mining activities in the past eight years. We sampled two sites in this area to determine their typical household economic situations. The first site, known as the Tahi ite ASGM site, is a reference site inhabited by around 210 families, among which are 800 active miners. The site is exploited by six mining companies. Land immediately bordering and surrounding this area comprises a forest, savannah grassland, and pastoral land. There are also geothermal hotspots nearby. The second, known as the Wumpubangka ASGM site, is inhabited by 100 households with 400 workers. This site is fully managed by three mining companies and is bordered by a residential area.

2.2. Data collection and analysis

A total of 25 households from the two sites were sampled by purposive sampling. To be included in the sample, household members had to have worked at the mine site for at least two years, and the head of the family had to be able to communicate well. A mixed methodology approach was employed, involving the collection, analysis, and mixing of both quantitative and qualitative data in a single study. Information about the economic characteristics of the miners was obtained during in-depth interviews with heads of households based on structured questionnaires.

3. Result and discussion

3.1. Demographic characteristics of the respondents

The 31-43-year-old group was the largest in the mining community, while the 44-56-year-old group was the largest in the non-mining community (Table 1). These age groups are included as the economically active [15]. There was no significant difference between the ages of the two groups (p-value < 0.05). The proportions of male and female survey respondents differed in both the mining and non-mining communities, but there was statistically no significant difference in gender within the communities (p > 0.05).

Table 1. Demographic characteristics of respondents interviewed in this survey.

Variable	Community status Mining community (n=25)	Non-mining community (n=18)	Total (n = 43)	χ^2 - value
Age category (years)				
<18	-	-	-	0.26*
18 - 30	4 (16)	5 (27.8)	9 (20.9)	
31 - 43	14 (56)	5 (27.8)	19 (44.2)	
44 - 56	7 (28)	7 (38.9)	14 (32.6)	
> 56	-	1 (5.6)	1 (2.3)	
Gender				
Male	17 (68)	16 (88.9)	33 (76.7)	0.1*
Female	8 (32)	2 (11.1)	10 (23.3)	
Education				
Elementary School	10 (40)	7 (38.9)	17 (39.5)	0.615*
Junior High School	8 (32)	3 (16.7)	11 (25.6)	
Senior High School	5 (20)	6 (33.3)	11 (25.6)	
University	2 (8)	2 (11.1)	4 (9.3)	

Source: Field survey (2016)
Figures in parentheses are percentages and those out of parentheses are frequency.
* = non-significant at p>0.05

The fact that there were quite a few female respondents in the mining area indicates that vulnerable groups are exposed to a high level of risk. The highest level of education achieved by the miners was generally either elementary or junior high school level. Some respondents had been attending school but then dropped out so that they could help their parents extract and transport the raw material from underground.

3.2. Economic status and domestic sanitation facilities

In developing countries, the availability of sanitation facilities is an indicator of a family's economic status [16]. A high economic level in the home generally means that there will be sanitation facilities. However, this studies have shown that even though the housing conditions of the two groups of the present study are very different, there was no significant difference in the availability of a water source or defecation facilities between the two sample groups. This result indicates that the economic level of the Bombana mining community did not fall below that of the non-mining communities.

Table 2. Disparities of environmental sanitation facilities.

Variable	Community status Mining community (n=25)	Non-mining community (n=18)	Total (n = 43)	χ^2 - value
Type of water source				
Tap water	1 (4)	4 (2.22)	5 (11.6)	0.154*
Wells/boreholes	11 (44)	5 (27.8)	16 (37.2)	
Water spring/river	13 (52)	9 (50)	22 (51.2)	
Defecation facilities				
Own facilities	14 (56)	14 (77.8)	28 (65.1)	0.313*
Public facilities	4 (16)	1 (5.6)	5 (11.6)	
Others	7 (28)	3 (16.7)	10 (23.3)	
House materials				
Concrete	0 (0)	8 (44.4)	8 (18.6)	0.000**
Wood/board/plywood	15 (60)	10 (55.6)	25 (58.1)	
Bamboo	10 (40)	0 (0)		

Source: Field survey (2016)
Figures in parentheses are percentages and those out of parentheses are frequency.
* = non-significant at $p>0.05$
** = significant at $p<0.05$

Nearly 100% of the shelter facilities in the mine area are temporary, and so are made of wood and bamboo. Moreover, workers move to other locations on a three-monthly cycle to pursue freshly excavated land for raw material. Migrants from neighboring areas do not have permanent settlements and tend to occupy low-quality houses. There was a significant difference between the housing facilities in the two locations ($p < 0.05$) (see Table 2). During field observations, we noticed that spring and river water were the main sources of water for the people in the two communities. The main water network does not extend to all homes yet, and so access to tap water is limited. This resembles the situation in the Ghana's upper east region, where as many as 70% of miners use surface water as their primary source of drinking water, and 30% use sachet water (i.e., water sold in small plastic bags) as their main drinking water source [3]. About 56% of households interviewed in the early mining community use the excavated area for recreation or as a defecation facility compared with 16% of public facility and 28% of other facilities. This proportion is lower than the percentage reported for the ASGM community in the Equador, where it is 100% [4]. Families that do not have toilet facilities use the forest, river, or beach for defecation. Human excreta contains scores of germs, some of which may cause ascariasis, hookworm infection, and trachoma [17–19]. Environmental sanitation involves raising community health by providing a clean environment and breaking the cycle of disease [19]. The criteria for this include a clean and safe water supply, clean air, efficient and safe waste disposal procedures, protection of food from chemical and biological contaminants, and suitable housing in safe and clean surroundings. On a global scale, it is estimated that 748 million women, men, and children lack access to a safe source of drinking water, and 1 billion people do not have access to safe water that is reliably and continuously delivered in sufficient quantities [20]. Lack of sanitation facilities and sources of safe drinking water in small-scale gold mining communities may contribute to the spread of infectious diseases. Also, once infected with diseases such as cholera, typhoid, and hepatitis A, the viruses can easily be transmitted to others in these communities.

3.3. Monthly income of respondents

Gold mining has a positive impact on the economic welfare of society in Bombana. As a result, there has been extensive economic development in recent years, attributable mainly to the amount of money generated by the ASGM sector [21]. With the government's permission, the gold mine's management company has helped boost the economy in the area around the mining site. This approach, with the creation of new jobs, increased income, and the ability to meet daily needs, encourages new economic potential [22].

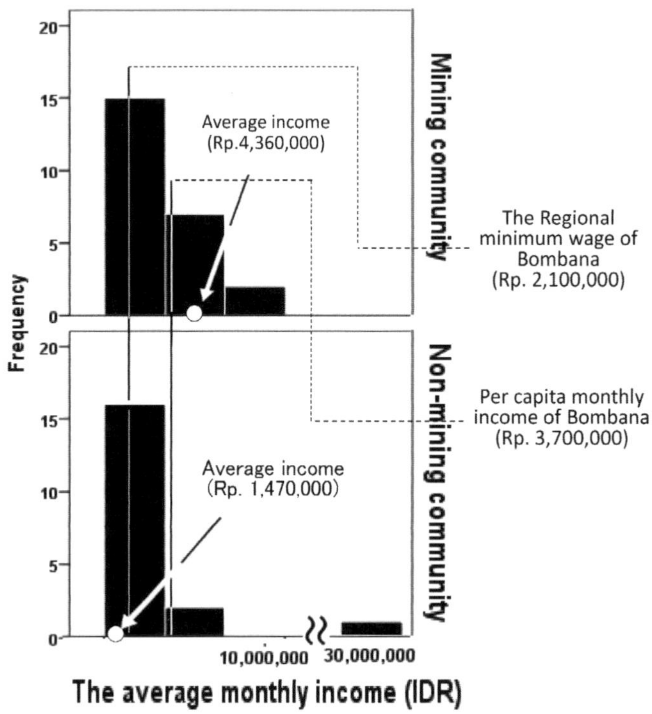

Figure 2. Comparison of regional minimum wage and per capita monthly income of Bombana with average monthly income of mining and non-mining community.

Table 3 shows significant differences among the monthly incomes of the general population of mine workers in Bombana ($p < 0.05$). The maximum income of the miners can be as much as Rp. 9,750,000 and the average income is Rp. 4,360,000, which is much greater than the maximum income of non-miners is Rp. 4,000,000. The miners' average income is much higher than the minimum wage of the Bombana people, which is around Rp. 2,100,000 per month, and is also higher than the per capita income of the population of Southeast Sulawesi, which is around Rp. 3,700,000 per month (Figure. 2). Meanwhile, non-miners who work as civil servants, fishers, farmers, and traders have income levels below both the minimum monthly income and the per capita income. The regional minimum wage and the per capita monthly income of the inhabitants of Bombana are Rp. 2,100,000, (equal to US $158.6) and Rp. 3,700,000 (equal to US $282.12), respectively. As comparison, the basic salary of mining worker determined by the Association of Artisanal Miners in Ecuador, is only US $79.20, which is equivalent to Rp. 1,029,000 per month. The ratio of the income levels between artisanal miners in Bombana and Equador is 4:1.

Table 3. Monthly income of respondent based on average, regional minimum wage, and per capita monthly income

Variable	Community status Mining community (n=25)	Non-mining community (n=18)	Total (n = 43)	χ^2 - value
Monthly average income (IDR)	4,360,000	1,470,000	3,150,000	0.041*
Regional minimum wage				
Above standard	11 (44)	16 (88.9)	27 (62.8)	0.004**
Below standard	14 (56)	2 (11.1)	16 (37.2)	
Monthly income per capita				
Above standard	16 (64)	16 (88.9)	32 (74.4)	
Below standard	9 (36)	2 (11.1)	11 (25.6)	

Source: Field survey (2016)
Figures in parentheses are percentages and those out of parentheses are frequency.
* = non-significant at p>0.05
** = significant at p<0.05

An increase in the gold price of nearly 400% between 2000 and 2012 has led to an increase in the incomes of all the families associated with gold mining [23], such as company owners, traditional landowners, traders, business people, military, and police officers. The economic boom has impacted on both local people and outsiders. Meanwhile, the monthly income of farmers has been decreasing because of decreased production from paddy fields and horticulture. About 26,000 ha of land in Bombana area previously used for aquaculture was taken over by mining at the end of 2013. Consequently, at least 40% of the productive land (paddy fields, orchards, and fields) has been converted to mining land [21]. The situation is exacerbated by damage to agricultural infrastructure, such as rivers and irrigation channels, so that there is less clean water, and the rice is not harvested [22].

3.4. Work structure and economic benefits

Since the implementation of the decentralization of policy in 2003 and the exploration of the gold mine site in 2008 in Bombana regency [12], the interest dynamics between the main groups in the region have changed. The central and local government, traditional landowners, trommel owners, informal miners, and companies fought each other for control of the mine site [10]. Some of these stakeholders contest each other while others collaborate. The common goal of the traditional landowners, trommel owners, and informal miners is to create work structures and to establish rules for the management of the mines. So far, the key players involved in this structure have gained direct and indirect economic benefits.

In this scheme, miners extract the gold from alluvial soil [22] by excavation, crushing, milling, amalgamation, and smelting (Figure. 3). Excavation and crushing are done manually, whereas during the milling step the material is processed using ball mills (trommels). After amalgamation and smelting, the gold ore must be sold to the trommel's owner to compensate for using the trammel. The trommel owners buy gold ore for Rp. 350,000 per gram and sell it to merchants for Rp. 400,000 per gram. In additional, the trade rates fluctuate and do not apply all the time, and is only valid in the area Bombana. In some cases, workers do not have the capital to start the process of extracting and transporting material, and so they borrow money and equipment from the land owner. When the gold ore extraction achieves maximum results, debt and equipment rent can be repaid with a surplus remaining. On the other hand, if the enterprise does not reach the breakeven point and is even negative, then the debt will accumulate for a longer period.

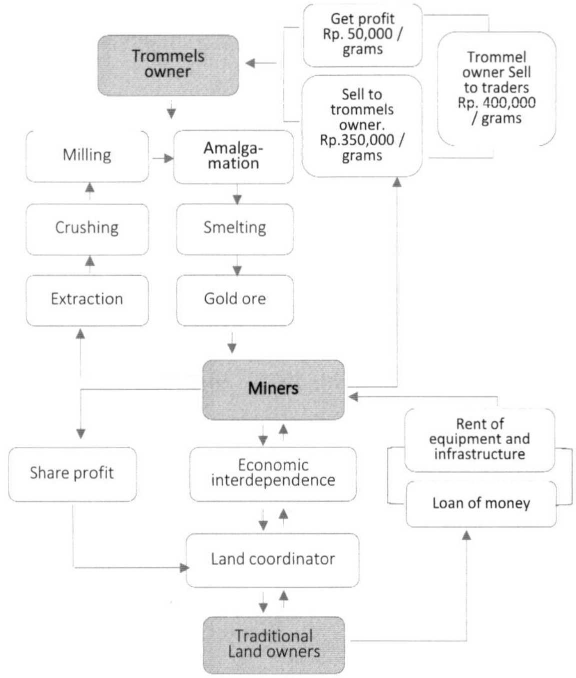

Figure 3. Scheme of work structure in mining communities.

There is a social change in the work structure, as the day workers gain experience and become new trommel and land owners. Several miners are saving capital and have started to buy the necessary equipment and infrastructure, which is mainly machinery and sets of trommels. Meanwhile, other miners have tried to buy new land from residents at competitive prices. Field coordinators have taken the opportunity to recruit new migrant workers and have provided them with equipment and money to undertake mining in a new location.

The property owner gives the miners authority to extract the gold under the supervision of the land coordinator, and will make an agreement to divide the profits obtained from the sale of gold. Although the structural work has defined rules, there are frequent internal conflicts. In workers' conflicts, a verdict is decided by the landowner with assistance from a local field coordinator. In some cases, conflicts that could potentially lead to riots are handled by the local police. Under this threat of police intervention, the capital owners and mine workers continue to work together, thus creating economic interdependence [10].

4. Conclusion

We established the economic situation in mining and non-mining communities in Bombana. The mine workers' incomes are higher than non-mining community. The exploitation of high-quality gold deposits in Indonesia has led to an economic boom in the Bombana Regency. Damage to land, agricultural infrastructure, and fisheries causes that production levels have decreased, and farmers' incomes have dropped dramatically. The higher revenue generated in the ASGM sector means that miners and their families depend on mining work over a long term, thereby making it difficult to control both the mining activity and the resultant environmental degradation. We suggest, therefore, that the government and other stakeholders should rise to the challenge of solving the problem at the local scale.

5. Acknowledgment

The authors grateful to the government of Bombana Regency, Indonesia for allowing us to conduct research and its support with sampling. One author (B) wishes to thank the Indonesia Government for providing Dikti Scholarship for graduate studies in Ehime University. This research was supported by (conducted as a part of) Feasibility Study "Social Acceptance of Regional Innovation for Reducing High-Impact Environmental Pollution", Research Institute for Humanity and Nature (RIHN). This work was supported by JSPS KAKENHI Grant Number 16H02706 to Prof. M. Sakakibara.

6. References

[1] Kristensen A K B, Thomsen J F and Mikkelsen S 2014 A review of mercury exposure among artisanal small-scale gold miners in developing countries *Int. Arch. Occup. Environ. Health* **87** 579–90

[2] Dooyema C A, Neri A, Lo Y-C, Durant J, Dargan P I, Swarthout T, Biya O, Gidado S O, Haladu S, Sani-Gwarzo N, Nguku P M, Akpan H, Idris S, Bashir A M and Brown M J 2012 Outbreak of fatal childhood lead poisoning related to artisanal gold mining in northwestern Nigeria, 2010. *Environ. Health Perspect.* **120** 601–7

[3] Basu N, Clarke E, Green A, Calys-Tagoe B, Chan L, Dzodzomenyo M, Fobil J, Long R, Neitzel R, Obiri S, Odei E, Ovadje L, Quansah R, Rajaee M and Wilson M 2015 Integrated Assessment of Artisanal and Small-Scale Gold Mining in Ghana—Part 1: Human Health Review *Int. J. Environ. Res. Public Health* **12** 5143–76

[4] Wilson M L, Renne E, Roncoli C, Agyei-Baffour P and Tenkorang E Y 2015 Integrated assessment of artisanal and small-scale gold mining in Ghana - Part 3: Social sciences and economics *Int. J. Environ. Res. Public Health* **12** 8133–56

[5] Basri, Sakakibara M and Sera K 2017 Current Mercury Exposure from Artisanal and Small-Scale Gold Mining in Bombana, Southeast Sulawesi, Indonesia—Future Significant Health Risks *Toxics* **5** 7

[6] Hilson G and Pardie S 2006 Mercury: An agent of poverty in Ghana's small-scale gold-mining sector *Resour. Policy* **31** 106–16

[7] Siegel S and Veiga M M 2009 Artisanal and small-scale mining as an extralegal economy: De Soto and the redefinition of "formalization" *Resour. Policy* **34** 51–6

[8] Hilson G 2002 Small-scale mining and its socio-economic impact in developing countries *Nat. Resour. Forum* **26** 3–13

[9] Heemskerk M 2003 Self-Employment and Poverty Alleviation: Women's Work in Artisanal Gold Mines *Hum. Organ.* **62** 62–73

[10] Amri U 2011 The dynamic of power relation over Bombana gold mining in southeast Sulawesi Indonesia *Kawistara* **1** 213–320

[11] Riase U 2014 The Impact of The Gold Mining onthe Social, Economic, and Cultural in the Bombana District Southeast Sulawesi Province *IJSTAS* **1** 53–65

[12] Amri U Power Contestation and Environmental Degradation: Evidence from Bombana Gold Mining Site

[13] UNEP 2012 Analysis of formalization approaches in the artisanal and small-scale gold mining sector based on experiences in Ecuador, Mongolia, Peru, Tanzania and Uganda 1–15

[14] Statistic of Bombana Regency 2015 *Bombana Regency in Figures 2015* vol 1(Statistic of Bombana Regency)

[15] International Labour Office 2011 *A Skilled Workforce for Strong, Sustainable and Balanced Growth: A G20 Training Strategy*

[16] Minh H Van and Hung N V 2011 Economic Aspects of Sanitation in Developing Countries *Evironmental Heal. Insights* **5** 63–70

[17] Esrey S a, Potash J B, Roberts L and Shiff C 1991 Reviews / Analyses Effects of improved water supply and sanitation on *Bull. World Health Organ.* **69** 609–21

[18] Carr R 2001 Excreta-related infections and the role of sanitation in the control of transmission *Water Qual. Stand. Heal.* 89–113

[19] Montgomery M a and Elimelech M 2007 Water and sanitation in developing countries: including health in the equation. *Environ. Sci. Technol.* **41** 17–24

[20] World Health Organization; UN-Water 2014 *Investing in water and sanitation: increasing access, reducing inequalities* vol 1

[21] Meisanti, Ali M S S, Jusoff K, Salman D and Rukmana D 2012 The impacts of gold mining on the farmer's community *Am. J. Sustain. Agric.* **6** 209–14

[22] Erman E 2015 *Informal Gold Mining and Miners: Work Characteristics, Property Rights, and Gold Trading Chains in Bombana District, Southeast Sulawesi, Indonesia*

[23] Seccatore J, Veiga M, Origliasso C, Marin T and De Tomi G 2014 An estimation of the artisanal small-scale production of gold in the world *Sci. Total Environ.* **496** 662–7

Ability of treated kapok (*Ceiba pentandra*) fiber for removal of clay particle from water turbidity

Nurfitri Abdul Gafur[1], Masayuki Sakakibara[1, 2], and Mohamad Jahja[3]

[1] Department of Earth Science, Graduate School of Science and Engineering, Ehime University, Japan
[2] Faculty of Collaborative Regional Innovation, Ehime University, Japan
[3] Department of Physics, Mathematics and Science Faculty, State University of Gorontalo, Indonesia
Email: nurfitri.gafur@ymail.com

Abstract. Kapok (*Ceiba pentandra*, family Bombacaceae) fiber (KF) is a by-product of traditional agriculture in tropical countries and is mainly used as fiberfill in fabric. The aim of this study was to explore the possibility of using KF to remove clay particles from turbid water. Firstly, KF was boiled at 100 °C in deionized water for 15 min to remove the surface oil. A suspension of montmorillonite powder mixed 1 L of deionized water, divided into volumes of 100, 200, 300, and 400 mL, was used as the turbid water source. The ability of KF to remove clay particles from the water was assessed by filtering the water through 60 g of treated KF in a ~397.9 cm^3 acrylic column. Results showed that the treated KF effectively removed the clay particles from the entire volume of turbid water in this experiment; the results also demonstrate that this KF fiber has value as a simple and inexpensive tool for water treatment, especially in developing countries.
Keywords: Kapok fiber; water turbidity

1. Introduction

In recent years, water management has become a topic of major concern for both scientists and the public worldwide because water, a primary source of life, is now contaminated in many developing countries, including those in Southeast Asia.

There is particular concern about the dumping of industrial waste into rivers as this may compromise the quality of water drawn for human consumption. Wastewater from industries such as metallurgy, leather tanning, metal finishing, steel fabrication, and mining is [1] a major concern because of its potential to produce serious environmental problems. Several approaches and materials have been tested for their capacity to remove heavy metals from industrial wastewater, including chemical precipitation, solvent extraction, adsorption, and resin ion exchange [2]. Surface water, however, is not only contaminated by heavy metals, but also contains various types of suspended materials that cause turbidity in the water [3].

The fibers of many natural plant varieties have been used to absorb heavy metal ions. Because of the high costs and lack of availability of adsorbents traditionally used for reducing the content of heavy metals in wastewater, alternative low-cost regenerated adsorbents such as sand, sepiolite, orange peel, banana peels, and even various kinds of fiber are now being studied [4]. The use of natural materials to reduce water turbidity is not a new idea, and, for example, chestnut, acorn, *Cactus latifaria, Prosopis juliflora,* and crumb rubber have served as natural coagulants in previous studies [5–7].

In 2012, it was found that heavy metal waste particles could be remediated using cotton fibers that had been chemically treated to transform them from hydrophobic to hydrophilic. When treated in this way, cotton fibers can provide effective adsorption of Cr (VI) [1] and heavy metal ions such as lead (Pb), copper (Cu), cadmium (Cd), and zinc (Zn) [8]. Despite their effectiveness, however, chemically treated cotton fibers may be an expensive material for remediating water pollution.

In this study, as an alternative to chemical treatment, a natural fiber obtained from *Ceiba pentandra* pods was first treated by boiling, and the boiled KF was then tested to determine (1) its suitability for use in a filtration process for cleaning turbid water and (2) whether it had potential as a low-cost material for treating turbid water destined for domestic use

2. Material and method

Valve-type kapok fibers (KF) were supplied by the Katsu Takashi Planning Co. Ltd. First, 30 g of KF were placed into pure water and boiled on a hotplate at 100 °C for 15 min. The characteristics of the KF and its composites were then investigated. Various impurities that caused the pure water to turn a light orange colour were discarded. Then, the boiled KF was rinsed thoroughly with high purity water.

The boiled KF was filled into a clear acrylic tube with a diameter of 7 cm to a height of about 20 cm. Once settled into the tube, the boiled KF was rinsed again with pure water until the outflow water from the tube was clear. The turbid water suspension was made by mixing 1 g of bentonite (montmorillonite k-10 sourced from Wako Pure Chemical Industries Ltd.) with 1 l of deionized water. The turbid water was shaken in an ultrasonic bath for 5 min. The turbid water was then divided into four different volumes (100, 200, 300, and 400 mL) for evaluating the filtration ability of the boiled KF. The synthetic water had a turbidity of 240 NTU.

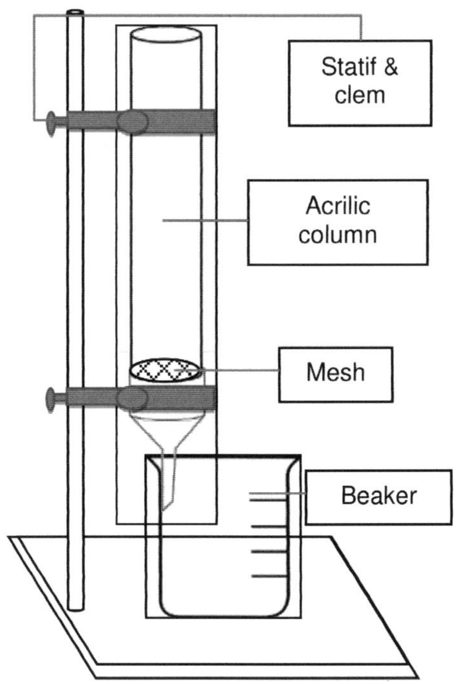

Figure 1. An illustration of filtering tube.

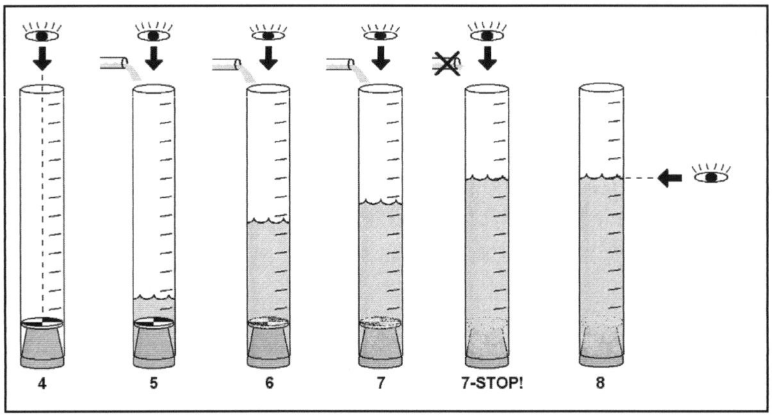

Figure 2. An illustration of processing water turbidity measurement [9].

First, the 100, 200, 300, and 400 mL of turbid water sample was added gradually to the center of the adsorbent using a syringe, respectively. Then measured the pH, turbidity, and electric conductivity of the turbid water before and after filtering to assess the effectiveness of the process. The turbidity of the water was measured with a turbidity tube and a Secchi disk, as shown in the figure below and following the water clarity assessment method. Standard Methods for Tap Water by APHA, 1998 were used for comparing the pH and turbidity of the filtered turbid water. The same procedure was followed for filtering the deionized water as a control.

3. Result and discussion

3.1. pH and electric conductivity effects on the water turbidity
The pH values of the control deionized water and the turbid water are presented in Figure 3. The results shows that pH values of deionized water and turbid water decreased slightly after filtering through the boiled KF. The pH values observed in the deionized water and turbid water were greater than 7 (Figure 3). To be suitable for human consumption, the pH of water should be between 6.5 and 8.5[10,11]. Comparison with the quality standards for tap water indicates that the turbid water was suitable for human activities other than drinking.

Table 1. Quality parameter of tap water

Parameter	Tap water
pH	7.4
Turbidity	0.2 NTU
Calcium (Ca2+)	88 mg/L
Magnesium (Mg2+)	9 mg
Sodium (Na+)	25 mg/L
Alkalinity	225 mg CaCO3/L
KMnO4 demand	6.7 mg KMnO4/L

As with the pH, the electrical conductivity of the turbid water also decreased after filtering, as shown in Figure 4. There was variation in turbidity among the different volumes of filtered turbid water (Table 3). Electrical conductivity is generally used as a measure of ionic processes in water or solutions that enable current transmission [12]. Conductivity is also dependent on water temperature, salinity, and total dissolved solids [13].

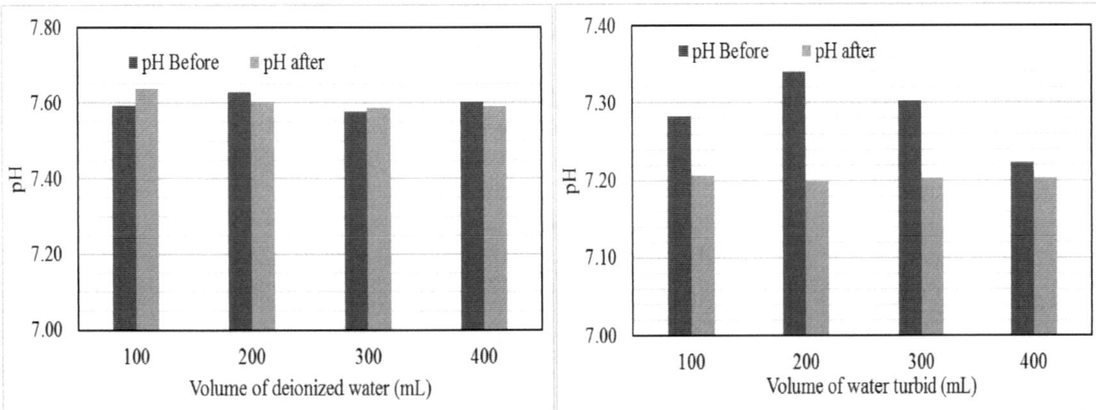

Figure 3. Graph of pH values of pure water and water turbid before and after filtration.

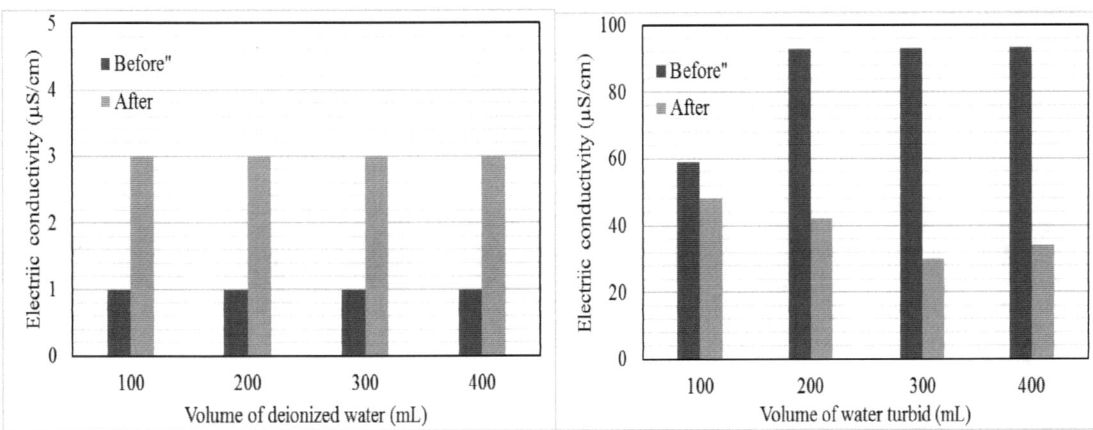

Figure 4. Graph of electric conductivity of pure water and water turbid before and after filtration.

The effectiveness of turbidity removal varied among the different volumes of turbid water. The results indicate that the boiled KF could remove particles from the turbid water. In the caption for Table 2 the term "clear" is not well-defined, because of the turbidity tube with a Secchi disk is also should be considered by the amount of the water volume to measure. The term "clear" shows that the water color is clear. The removal percentages and the figure of comparison initial and final turbidity are presented in Table 2 and Figure 3.

Table 2. Comparison initial and final turbidity after filtering using boiled KF.

Volume of water turbidity (mL)	Initial turbidity (NTU)	Final turbidity (mL)	Removal percentage (%)
100	240	Clear	Clear
200	240	40	16.67
300	240	100	41.67
400	240	200	0

3.2. Initial and final turbidity

Water turbidity removal using natural fiber is difficult directly to applicate in the inhabitant to get the clean water. Compared with the quality parameters of tap water by APHA, 1998 the final turbidity after filtering are not consumable for living things. The flow rate and the media sizes are also important thing to consider the ability of boiled KF for filtering. Comparing with other filtering using crumb rubber, the turbidity removal efficiencies were 92.7 % with the media size about 0.16 mm [7].

Some of the water filtering processess are using the natural plant for water turbidity removal by mixing with other material such as pilot plat mixing with crumb rubber, pilot plant mixing with hollow fiber microfiltration, etc. [7,14]. Water turbidity removal using only fiber is difficult to find, because it has many factors which must be considered such as the porosity, transmembrane area, transmembrane flow, density, permeability, etc. [5,7,14,15]. The percentage turbidity removal values which shown in Table 2 and 3. In this experiment, the best a 100 mL volume of water turbidity removal is using 60 g of boiled KF. Water turbidity level has no big effects the amount of pH in water. However, reducing level of water turbidity has several variation values on the electric conductivity. It is clearly shown in Figure 4. The reducing of electric conductivity related to the reducing of final turbidity. The result clearly indicates that the boiled KF has an ability to remove the water turbidity.

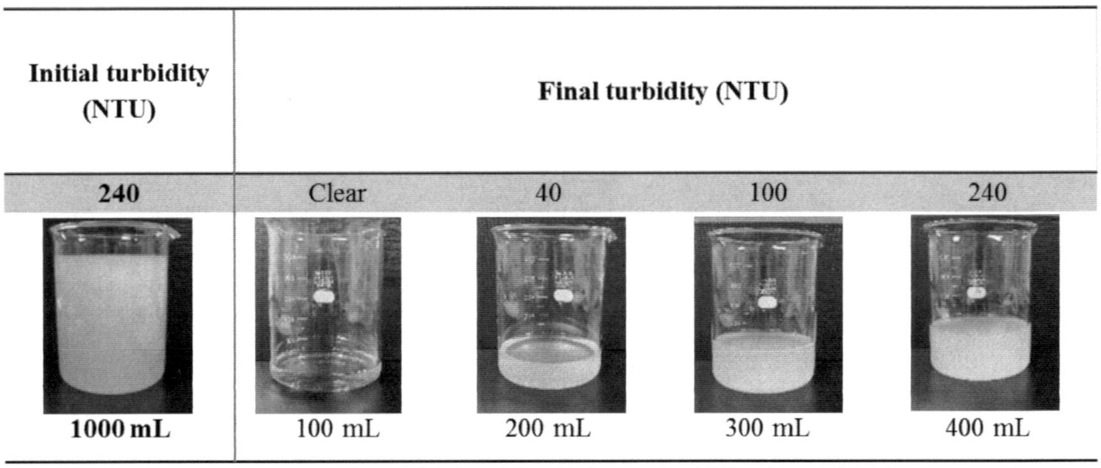

Figure 5. Comparison initial and final of water turbidity for each volume.

4. Conclusion

The experiment conducted to establish the ability of boiled KF for water turbidity removal. This experiment has three parameters to establish the ability of boiled KF such as pH, electric conductivity and turbidity. The initial and final pH of water turbidity is about 7.2. It shows that the final water turbidity levels have no big effects on the amount of pH. However, water turbidity levels have effects on the electric conductivity. Compare with the turbidity level of initial and final turbidity, the 60 g of boiled KF has an ability to remove a 100 mL of water turbidity, which has the best percentage of turbidity removal.

Acknowledgment
The authors would like to acknowledge the support of the following: JSPS KAKENHI Grant Number 16H02706 and Feasibility Study "Social Acceptance of Regional Innovation for Reducing High Impact Environmental Pollution", Research Institute for Humanity and Nature (RIHN).

References

[1] Zheng Y, Liu Y and Wang A 2012 Kapok fiber oriented polyaniline for removal of sulfonated dyes *Ind. Eng. Chem. Res.* **51** 10079–87
[2] Huynh H T and Tanaka M 2003 Removal of Bi, Cd, Co, Cu, Fe, Ni, Pb, and Zn from an Aqueous Nitrate Medium with Bis(2-ethylhexyl)phosphoric Acid Impregnated Kapok Fiber

Ind. Eng. Chem. Res. **42** 4050–4

[3] Shahriari, T., NabiBidhendi G . 2012 Starch efficiency in water turbidity removal *Asian J. Nat. Appl. Sci.* **1** 34–7

[4] Zhang J D, Liu Y W, Gao S M, Li C Z, Zhang F, Zen H M and Ye C S 2006 Pilot testing of outside-in UF pretreatment prior to RO for high turbidity seawater desalination *Desalination* **189** 269–77

[5] Šćiban M, Klašnja M, Antov M and Škrbić B 2009 Removal of water turbidity by natural coagulants obtained from chestnut and acorn *Bioresour. Technol.* **100** 6639–43

[6] Diaz A, Rincon N, Escorihuela A, Fernandez N, Chacin E and Forster C F 1999 A preliminary evaluation of turbidity removal by natural coagulants indigenous to Venezuela *Process Biochem.* **35** 391–5

[7] Sulaymon A H, Abdul-aha M Y and Alwared A I 2014 Removal Water Turbidity by Crumb Rubber Media **10** 23–31

[8] Chung B Y, Cho J Y, Lee M H, Wi S G, Kim J H, Kim J S, Kang P H and Nho Y C 2008 Adsorption of heavy metal ions onto chemically oxidized Ceiba pentandra (L.) Gaertn. (Kapok) fibers *J. Appl. Biol. Chem.* **51** 28–35

[9] Myre E and Shaw R 2006 The Turbidity Tube : Simple and Accurate Measurement of Turbidity in the Field 1–15

[10] Anon Guidelines for Drinking-water Quality Third Edition Incorporating The First And Second Addenda Volume 1 Recommendations Geneva 2008 WHO Library Cataloguing-in-Publication Data

[11] Oram, Brian Water Quality, Drinking water, Corrosion and Water pH. http://www.water-research.net/index.php/ph-in-the-environment (accessed on 2016-05-15)

[12] Meride Y and Ayenew B 2016 Drinking water quality assessment and its effects on residents health in Wondo genet campus , Ethiopia *Environ. Syst. Res.* 1–7

[13] Rose, Kevin; Kelly, Daniel; Kemker, Christine; Fitch, Katie; Card A 2016 Conductivity, Salinity & Total Dissolved Solids - Environmental Measurement Systems. http://www.fondriest.com/environmental-measurements/parameters/water-quality/conductivity-salinity-tds/ (accessed on 2016-05-15)

[14] Sorlini S, Gialdini F and Collivignarelli C 2013 Removal of cyanobacterial cells and Microcystin-LR from drinking water using a hollow fiber microfiltration pilot plant *Desalination* **309** 106–12

[15] Van Der Bruggen B, Vandecasteele C, Van Gestel T, Doyen W and Leysen R 2003 A review of pressure-driven membrane processes in wastewater treatment and drinking water production *Environ. Prog.* **22** 46–56

Phytoremediation of arsenic- and molybdenum-contaminated alkaline wastewater by *Eleocharis acicularis* in winter in Japan

Shusaku Yamazaki[1], Kenji Okazaki[1], Toshiyuki Kurahashi[1] and Masayuki Sakakibara[2]

[1] Civil Engineering Research Institute for Cold Region, Public Works Research Institute, National Research and Development Agency, 1-34, Hiragishi 1-3, Toyohira-ku, Sapporo, Hokkaido, 062-8602, Japan

[2] Graduate School of Science & Engineering, Ehime University, 0-13, Dogo-Himata, Matsuyama, Ehime, 790-8577, Japan

E-mail: yamazaki@ceri.go.jp

Abstract. Phytoremediation using aquatic plants is a sustainable, low-cost measure for remediating water contaminated by toxic heavy metals. In this study, we conducted a channel experiment using *Eleocharis acicularis* in heavy metal-contaminated mildly alkaline wastewater under unfavorable plant habitat conditions in winter in northeastern Japan. The wastewater from an embankment consisting of Neogene marine sediments had a temperature of 10-15 °C and a pH of about 9, and it contained ~0.02 mg/L of As and ~ 0.23 mg/L of Mo. About 16 kg (fresh weight) of *E. acicularis* was laid in a plastic channel measuring 30 cm in width by 20 m in length, and the channel was enclosed in a tunnel greenhouse. The experiment was conducted for the 3 months from November 2015 at an average flow rate of 0.3 L/min and an air temperature of -4 to 19 °C. No reductions in As or Mo concentrations in the outflow were detected. However, at 3 months, the *E. acicularis* showed accumulations of ~7 mg/kg As and ~18 mg/kg Mo as dry weight, indicating that this remediation method is workable in an unfavorable low-temperature, mildly alkaline environment.

1. Introduction

Toxic heavy metal elements, such as arsenic (As) and cadmium (Cd), in drinking water are a key issue of concern to human and animal health [1-2]. In light of this, a sustainable, low-cost remediation method is required for the treatment of contaminated wastewater that is drained from soil and rock excavated from building sites, tunnels and mines. Phytoremediation is method of remediating contaminated soil and water by using plants that are able to hyperaccumulate organic and inorganic contaminants, including heavy metal elements [3]. Phytoremediation is considered to be more sustainable than methods that use chemical media or physical removal and sealing of the excavation site [4]. Phytoremediation using aquatic plants has advantages in that some species can be grown in sized sheets in free-floating, submerged or emergent conditions, and can take up elements directly through their roots and leaves from flowing or pooled water [5]. The sized sheets are easy to install, replace and dispose of, making them convenient in terms of treatment site design and management. Some aquatic macrophytes can be grown in or on water without soil. This enables complex factors such as surface adsorption and desorption of heavy metal elements in a soil medium to be excluded, which is advantageous in simplifying the evaluation of element uptake efficiency from wastewater to aquatic macrophytes. However, a disadvantage is that the range of growth conditions favorable for aquatic plants in wastewater (e.g., chemistry, temperature and flow rate of wastewater) is more limited than in a soil environment and there is no buffering of pH and nutrients by soil material. Therefore, many case studies under various conditions of water and weather are needed before phytoremediation can be used for the various types of contaminated wastewater.

This paper focuses on the phytoremediation of wastewater contaminated with heavy metals by using the aquatic macrophyte *Eleocharis acicularis*. This species is known as a hyperaccumulator of elemental heavy metals such as As, cupper (Cu), zinc (Zn), lead (Pb), indium (In), silver (Ag) and Cd [6-10] and it is also good at accumulating antimony (Sb) [11]. *E. acicularis* grows as an autochthon worldwide, including in northern Japan, which has snowfall and subzero air temperatures. It can grow in submerged, emergent and soil conditions in nature, and in sheets it can grow in a water pool with freely flowing water. Many studies using aquatic plants have demonstrated its effectiveness at accumulating toxic heavy metal elements in the field and in the laboratory [5]. To our knowledge, no studies have addressed the phytoremediation of actual wastewater flow in a channel with aquatic macrophytes under winter weather conditions, where the low temperatures and shorter day length would be unfavorable for plant growth and element uptake. The main objectives of the present study are to understand a phytoremediation efficiency in a 90-day experiment conducted in the winter from November 2015 to February 2016 using *E. acicularis* in mildly alkaline wastewater contaminated with As and molybdenum (Mo) in a flow-controlled channel at a wastewater treatment site in Japan at about 38° North latitude.

2. Data and method

2.1 Site description and experimental design

In this study, an experiment channel was set up at a wastewater treatment site about 150 meters above sea level in Sendai City, northeastern Japan. At this site, an embankment mainly consisting of about 800,000 m^3 of Neogene (late Miocene to early Pliocene) marine sedimentary rocks steadily drains mildly alkaline wastewater contaminated with As at a flow rate of about 20 L/min. The wastewater from the embankment is first collected in a drainage tank before an As treatment plant.

2.2 Experimental design

The experiment channel, with a total length of 20 m, is set adjacent to the drainage tank. The experimental channel was constructed from two small channels that were 10 m in length (figure 1). The channels are made of 1 cm thick PVC resin plates with 30 cm in width and 10 cm in sidewall

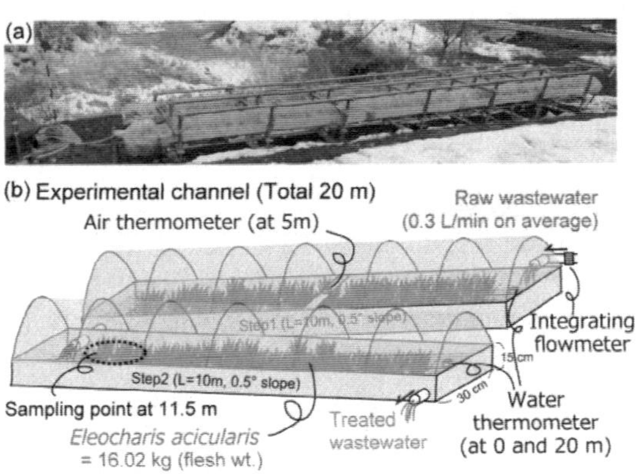

Figure 1. (a) The experimental channel set at a wastewater treatment site. (a) Photo. (b) Diagram.

Table 1. Cumulative volume of inflow, temperature and pH in raw and treated wastewater during the experiment.

Time	Cumulative	Raw (0m)		Treated (20m)	
(day)	flow vol. (L)	T(°C)	pH	T(°C)	pH
0.00	0	15.5	8.94	Unreached	
0.05	188	16.3	-	-	6.11
0.10	300	13.6	-	12.9	6.74
0.23	466	11.1	-	6.6	7.12
0.79	898	11.5	-	6	7.39
0.98	1061	15.2	9.02	-	7.46
1.06	1120	14.7	9.00	14.8	7.4
1.13	1148	13.4	9.08	14	7.36
42	29428	17.1	9.32	20.9	7.89
89	39371	12.5	-	14.8	-
90	40098	10.1	-	5.9	-

Table 2. The results of chemical analysis of the certified reference material and the list of analytical methods for *E. acicularis* and water samples (M1 and M2). Abbrebiations: NA, not analyzed; MS: ICP-MS, AES: ICP-AES, Mo: spectrophotometric determination using molybdeum blue.

Reference materials		Al	Mn	Fe	Zn	As	Se	Mo	Cd	Sb	Pb	Sr	Mg	Na	Ca	K	P
	Method	μg/L	μg/L	μg/L	μg/L	μg/L	μg/L	μg/L	μg/L	μg/L	μg/L	mg/L	mg/L	mg/L	mg/L	mg/L	mg/L
TMDA-64.2	Reference	290	295	306	310	162	155	290	266	128	288	0.641	-	-	-	-	-
	M1	300	297	303	306	159	156	287	274	133	306	0.654	5.14	8.94	21.4	0.99	0.023
	M2	302	NA	329	314	158	141	270	253	120	283	0.699	6.58	10.53	24.9	NA	NA
CRM-7202b	Reference	-	-	-	-	-	-	-	-	-	-	-	1.23	3.64	4.51	0.83	-
	M1	NA	NA	NA	NA	NA	NA	NA	NA	NA	NA	NA	1.03	3.42	4.39	0.99	3.35
	M2	NA	NA	NA	NA	NA	NA	NA	NA	NA	NA	0.04	1.19	3.63	4.48	0.88	NA
Analytical method		Al	Mn	Fe	Zn	As	Se	Mo	Cd	Sb	Pb	Sr	Mg	Na	Ca	K	P
E. acicularis.		MS	AES	AES	MS	MS	MS	MS	MS	MS	MS	MS	AAS	AAS	AAS	AAS	Mo
	M1	MS	MS	MS	MS	MS	MS	MS	MS	MS	MS	MS	AAS	AAS	AAS	AAS	Mo
	M2	MS	-	MS	MS	MS	MS	MS	MS	MS	MS	AES	AES	AES	AES	AES	-

height. The two channels were connected by hosepipe and placed with slope angle of 0.5° (figure 1b). To protect plants in the channel from low temperatures and snowfall in winter, the channel was covered with a tunnel greenhouse of 0.2-mm-thick clear plastic sheeting and the sidewalls and bottom were covered with 20-mm-thick urethane foam (figure 1). The rate of inflow of the raw wastewater could be roughly controlled by faucet handle at the tank and was measured by an integrating flowmeter (NW05-NTN of Aich Tokei Denki Co., Ltd.) at the inflow point.

2.3 Sample preparation

Sheets of *E. acicularis* with a plant height of 10-15 cm, a width of 30 cm and a length of 60 cm were weighed for fresh weight after being hung out for 1-2 hours to dry off the water that was added for shipping. The fresh weight of each sheet ranged from 261 to 906 g, and the total fresh weight in the channel was 16.02 kg. After weighing, the sheets were laid in the channel so as to leave 50 cm free of plants at the uppermost and lowermost parts of each channel, in order to keep clearance for water flow. The area density of the *E. acicularis* was 29.74 kg/m^2 as fresh weight.

2.4 Sampling and measurement

During the channel experiment, 7 inflow and 17 outflow samples were collected using a PP-bottle that was rinsed with ultrapure water after being soaked in 1M nitric acid for at least a week. At sampling, the water temperature and the pH of inflow and outflow were measured by portable tester (HI98121 Hanna Inc.). As pre-treatment, 0.6 ml of nitric acid (69% HNO_3 for ultratrace analysis from Wako Pure Chemical Industries, Ltd.) was added to a 50-ml sample after filtration using a syringe filter with a pore diameter of 0.45 μm in the laboratory.

Several pinches of *E. acicularis* were sampled from around 11.5 m from the upstream end of the channel 13 times during the experiment (table 3). The sample was dried at 50°C in a drying oven for at least 24 hours and then was measured for dry weight. The sample was powdered as whole plant using sample mixer (IKA A11 basic) with stainless steel blade. Acid digestion of *E. acicularis* was performed, according to the following steps. 0.1 g of dried powder was mixed with 5 ml ultrapure-grade HNO_3 and 2 ml H_2O_2 in a closed, 100-ml PTFE vessel, and then the mixture was digested with the assistance of a microwave (Titan MPS PerkinElmer Co., Ltd.), with the sequence of 5 minutes at 150°C, 15 minutes at 190°C and 10 minutes at 50°C. The analytical solution was prepared by dilution up to 50 ml with ultrapure water.

The chemical components in the wastewater and the *E. acicularis* samples were analyzed using one of four methods: Inductively-coupled plasma mass spectrometry (ICP-MS), Inductively-coupled plasma atomic emission spectrometry (ICP-AES), atomic adsorption spectroscopy (AAS) and adsorption spectroscopy using molybdenum blue. Table 2 lists the method that was used for each element and sample type. The methods for water samples are subdivided into M1 and M2. The

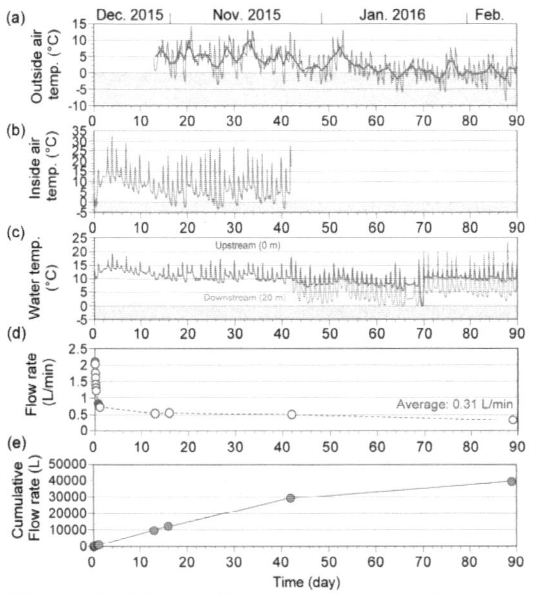

Figure 2. (a) Outside air temperture (blue: raw data; red: daily average), (b) inside air temperature, (c) water temperature, (d) flow rate and (e) cumulative flow volume during the experiment.

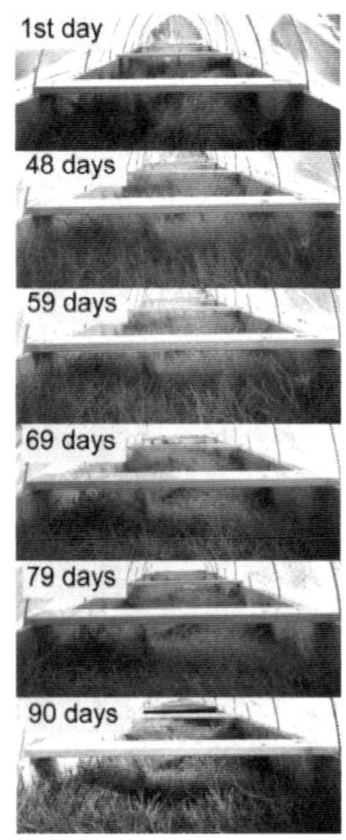

Figure 3. Photos taken by fixed-point camera downstream-ward at 16 m from the inflow point.

analytical accuracy of each element was evaluated by using certified reference materials of TMDA 64.2 (National Water Research Institute of Environment Canada) and CRM-7202b (National Institute for Environmental Studies, Japan). The elements Al, Mn, Fe, Zn, As, Se, Mo, Cd, Sb, Pb and Sr were measured within ±10% error, and the elements Sr, Mg, Na, Ca and K were measured within ±6%. The elements of Mg and K as determined by the M1 method show greater errors within ±20%.

3. Results

3.1 Experimental conditions and plant growth

The change in air temperature outside of the channel from day 12 to day 90 is graphed in Figure 2a. The outside temperature ranged from -8.7 to 14.1°C. On successive days between day 44 and day 51 and after day 56, the daily minimum air temperature was below zero at midnight and the daily mean was below 5°C. The air temperature inside the tunnel greenhouse in the upper channel ranged from -3.5 to 32.1°C from the start to day 43 (figure 2b). The tunnel greenhouse's contribution to the increase in inside air temperature can be estimated as +1.8 to +17.2°C from the daytime temperature difference between the inside and outside of the channel (figure 2b). The water temperature at the point of inflow during the 90 days was 11.6°C on average and ranged from 5.4-19.1°C. At the point of outflow, the range of water temperature was greater than at inflow. Between day 60 and day 69, the water temperature of outflow was almost 0°C. The total volume of flow into the experiment channel during the 90 days was 40,098 L. The flow rate decreased from about 2 L/min at day 1 to 0.5 L/min at day 43. The average flow rate during the experiment was 0.31 L/min (table 1 and figure 2d). Throughout the experiment period, the inflow showed no turbidity and the pH was roughly stable, at the mildly alkaline values of 9.0-9.3 (table 1). However, the treated wastewater showed pale green turbidity during the first several hours and turned transparent after that. The pH of the treated wastewater gradually changed from 6.11 at the start to 7.89 at day 43 (table 1). The recorded photos of *E. acicularis* growth at 16 m from the inflow point represent a lowering of water depth at day 61 and a further lowering at days 65-66, which is recorded as a decrease in flow rate between days 40 and 90

Table 3. The chemical compositions of the *Eleocharis acicularis*, the raw wastewater and the treated wastewater. Values with "<" were below the detection limit.

E. acicularis Time (day)	Al mg/kg	Mn mg/kg	Fe mg/kg	Zn mg/kg	As mg/kg	Se mg/kg	Mo mg/kg	Cd mg/kg	Sb mg/kg	Pb mg/kg	Sr mg/kg	Mg mg/kg	Na mg/kg	Ca mg/kg	K mg/kg	P mg/kg
0	1050	85.9	1727	149.5	0.51	0.38	4.82	11.09	5.47	3.00	29.5	1230	504	4357	17705	6323
0.30	792	95.1	1716	176.5	0.72	0.38	7.78	7.55	3.77	2.74	31.6	971	555	5614	15568	6142
0.73	305	99.7	803	164.9	0.53	0.25	6.61	7.31	5.50	1.86	27.1	1088	559	4816	23524	6880
1.13	1915	123.1	2486	206.3	1.34	0.26	9.80	7.55	6.59	4.10	31.1	1138	645	5602	16501	6183
4	1114	125.7	1463	141.0	1.28	0.22	12.46	5.92	0.62	1.72	28.7	1076	1021	5268	24247	6860
6	4543	170.2	3325	164.1	2.72	0.19	10.68	6.25	0.84	1.77	27.9	1224	1417	4903	25110	7132
8	373	173.0	435	139.6	0.98	0.13	9.74	4.25	0.31	0.60	23.4	1063	1588	4635	28545	7196
12	1129	250.0	824	132.8	1.65	0.10	11.66	4.26	0.93	0.69	25.4	1133	1679	4683	27189	6975
16	2947	240.6	2076	148.5	2.29	0.24	11.77	3.83	0.62	1.71	31.3	1109	1701	5887	21340	6259
32	154	708.4	461	117.9	2.64	0.18	17.37	3.25	1.27	0.73	32.8	991	2407	6275	22908	5539
42	177	511.8	283	109.7	2.71	0.14	16.54	2.74	0.34	0.59	29.5	970	2697	6040	22210	5419
63	1774	884.9	1811	109.6	9.51	0.22	26.62	1.80	0.73	1.33	35.5	942	2703	8058	14731	4350
89	5410	443.6	3877	105.0	8.79	0.33	17.84	1.18	0.67	1.96	36.8	1051	1187	8244	8134	4027

Raw wastewater Time (day)	Method	Al μg/L	Mn μg/L	Fe μg/L	Zn μg/L	As μg/L	Se μg/L	Mo μg/L	Cd μg/L	Sb μg/L	Pb μg/L	Sr mg/L	Mg mg/L	Na mg/L	Ca mg/L	K mg/L	P mg/L
0	M1	<2	<2	<2	<1	26.6	<0.2	253	0.06	0.413	<0.2	0.860	3.98	88.4	259	11.6	0.045
1.13	M1	<2	<2	<2	1.6	26.1	<0.2	251	0.06	0.403	<0.2	0.859	4.05	89.5	263	12	0.043
13	M2	<30	NA	<32	<8	26.0	<0.2	239	<0.1	0.333	<0.7	0.818	3.80	82.3	237	12.4	NA
16	M2	<30	NA	<32	<8	25.0	<0.2	218	<0.1	0.402	<0.7	0.813	4.30	81.0	235	11.6	NA
42	M1	<2	38.7	<2	<1	27.1	<0.2	258	<0.06	0.411	<0.2	0.865	3.97	89	257	12.1	0.045
89	M2	<30	NA	<32	8.1	26.2	0.813	256	0.12	0.451	0.99	0.875	3.79	91.0	255	10.9	NA
90	M2	<30	NA	<32	<8	25.2	0.654	252	<0.1	0.333	<0.7	0.884	3.86	82.5	255	11.7	NA

Treated wastewater Time (day)	Method	Al μg/L	Mn μg/L	Fe μg/L	Zn μg/L	As μg/L	Se μg/L	Mo μg/L	Cd μg/L	Sb μg/L	Pb μg/L	Sr mg/L	Mg mg/L	Na mg/L	Ca mg/L	K mg/L	P mg/L
0.05	M2	76	NA	65.6	548	19.4	<0.2	115	43.92	0.543	0.79	NA	NA	NA	NA	NA	NA
0.10	M2	<30	NA	<32	76.0	23.4	<0.2	211	7.38	0.364	<0.7	NA	NA	NA	NA	NA	NA
0.13	M2	<30	NA	<32	29.8	24.2	<0.2	222	3.61	0.363	<0.7	NA	NA	NA	NA	NA	NA
0.18	M2	<30	NA	<32	26.4	22.8	<0.2	223	2.85	0.373	<0.7	NA	NA	NA	NA	NA	NA
0.30	M1	2.6	137	5.25	15.8	24.1	<0.2	239	2.06	0.421	<0.2	0.859	4.21	90.4	258	13.5	0.445
0.73	M2	<30	NA	<32	8.5	24.7	<0.2	232	0.79	0.37	<0.7	NA	NA	NA	NA	NA	NA
0.89	M2	<30	NA	<32	<8	24.7	<0.2	242	0.71	0.34	<0.7	NA	NA	NA	NA	NA	NA
1.13	M1	<2	115	5.27	8.7	24.8	<0.2	249	0.95	0.455	<0.2	0.864	4.14	90.6	261	12.5	0.329
4	M2	<30	NA	<32	<8	25.3	<0.2	227	0.11	0.381	<0.7	NA	NA	NA	NA	NA	NA
6	M2	<30	NA	<32	<8	25.9	<0.2	232	0.13	0.33	<0.7	NA	NA	NA	NA	NA	NA
8	M2	<30	NA	<32	<8	25.5	<0.2	231	0.10	0.303	<0.7	NA	NA	NA	NA	NA	NA
12	M2	<30	NA	<32	<8	21.6	0.28	196	<0.1	0.252	<0.7	NA	NA	NA	NA	NA	NA
16	M2	<30	NA	<32	<8	15.3	<0.2	125	<0.1	0.259	<0.7	NA	NA	NA	NA	NA	NA
32	M2	<30	NA	<32	<8	28.3	<0.2	232	<0.1	0.388	<0.7	NA	NA	NA	NA	NA	NA
42	M1	<2	<2	<2	2.5	29.5	<0.2	262	0.19	0.618	<0.2	0.864	3.84	90.5	259	12.4	0.070
63	M1	<2	92.7	<2	<1	26.5	<0.2	267	<0.06	0.372	<0.2	0.869	3.84	90.8	269	12.3	0.038
89	M2	<30	NA	<32	<8	28.0	0.21	256	b.d.	0.382	<0.7	NA	3.98	98.3	297	NA	NA

(figure 2). After the lowering, the blighting of shoot tops was observed and it progressed to the whole body by day 90 (figure 3). The blighting is considered to have been caused by low-temperature damage associated with water depth lowering as the result of decreased water inflow. A similar low-temperature event also occurred between days 43 and 51. At that time, growth hindrance was not observed in photos because the water level did not decrease at that time.

3.2 Chemical compositions of wastewater

The timing of sampling and the chemical compositions of the raw and treated wastewater and *E. acicularis* are listed in table 3 and graphed in figure 4. The chemical compositions of the raw wastewater show no notable variations during the experiment. They averaged 26.0±0.7 μg/L for As and 247±14 μg/L for Mo, with a ±1σ error. These concentrations of As and Mo exceed the WHO guideline values for drinking water of 10 and 70 μg/L, respectively [2]. Most of the Cd concentrations are below the detection limit, and their maximum values are 0.12 μg/L. The maximum value of Mn concentration is 38.7 μg/L. The concentrations of Cd and Mn are far lower than the guideline values of 3 and 500 μg/L, respectively. The elements Se and Pb were not detected. The other elements showed the average concentrations of 0.853±0.027 mg/L for Sr, 0.044±0.001 mg/L for P, 3.96±0.18 mg/L for

Mg, 86.2±4.1 mg/L for Na, 252±11 mg/L for Ca and 11.8±0.5 mg/L for K. The elements Al, Mn, Fe and Zn were almost below their detection thresholds (table 3).

The treated wastewater immediately after the start (~day 0.3) shows notably higher concentrations of Cd, Zn, Mn, Fe, Pb, Sb, P and Al than the concentrations in the raw wastewater (table 3 and figure 4). These enrichments could be explained by mixing of groundwater and fertilizer attached to the *E. acicularis*, which were used for growing it in sheet form before the experiment. From day 1 to day 89, there were no noticeable differences in the element concentrations between the raw and treated wastewaters, except for Cd and for temporary decreases in As, Mo and Sb concentrations at day 12 and day 16. The concentration of Cd showed a rapid decrease from day 0.05 to day 0.3, and the decrease continued gently to day 8.

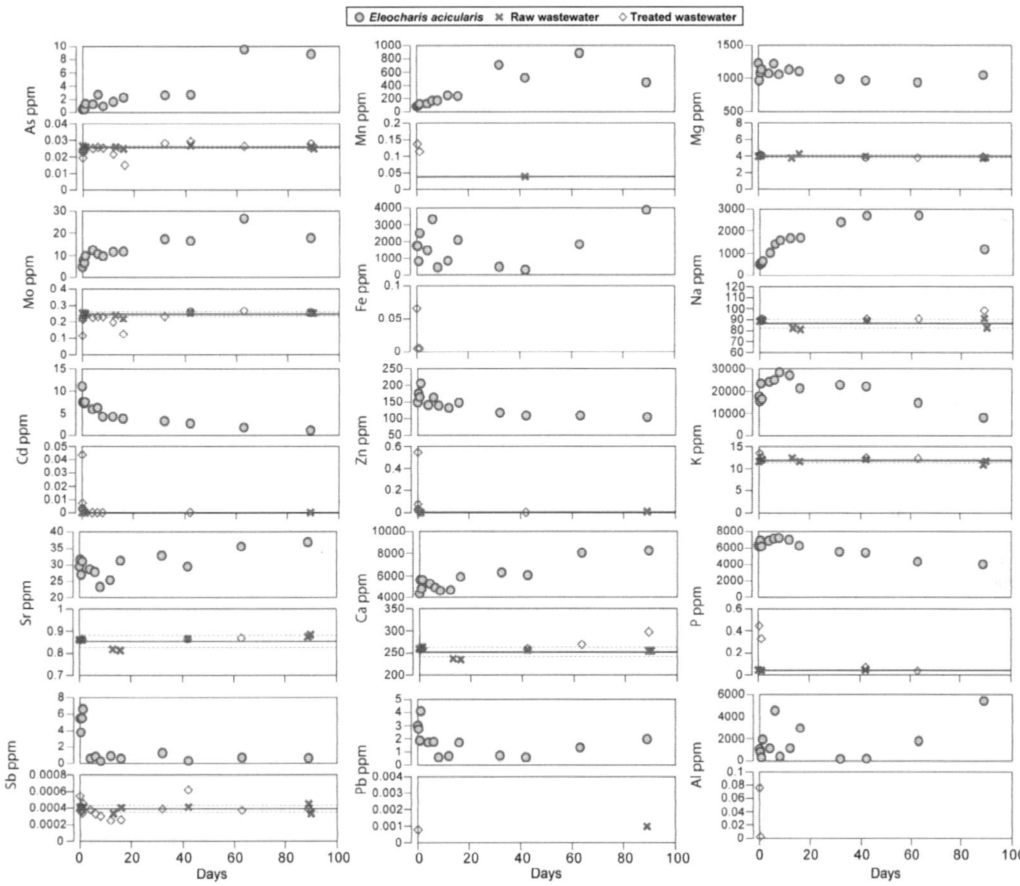

Figure 4. The graphs of measurement chemical compositions of *E. acicularis* (upper graph in each element) and raw and treated wastewaters (bottom). The unit of ppm means mg/kg for *E. acicularis* and mg/L for wastewater. Red solid and dashed lines in the bottom graphs indicate the average values and one standard diviation for the raw wastewater samples.

3.3 Chemical compositions of E. acicularis

Concentrations of As, Mo and Mn in *E. acicularis* during the experiment all show a remarkable linear increase from day 0 to day 63 and then a slight decrease from day 89 (table 3 and figure 4). From day 0 to day 63, As increased from 0.51 to 8.79 mg/kg dry wt., Mo from 4.82 to 26.62 mg/kg dry wt. and Mn from 85.9 to 884.9 mg/kg dry wt. Although Na also tended to increase from day 0 to day 63, it did not show a linear trend. The rate of increase gradually slowed toward day 63. The elements Sr and Ca increased slightly from 29.5 to 35.5 mg/kg dry wt. and from 4357 to 8058 mg/kg dry wt., respectively. The concentrations of Cd, Zn, Se and Sb decreased with time from 11.1 to 1.8 mg/kg dry wt., 150 to 110 mg/kg dry wt., 0.38 to 0.22 mg/kg dry wt. and 5.47 to 0.73 mg/kg in dry wt., respectively. The concentrations of K and P in plants are responsive to the addition or depletion of fertilizer in the growth environment. These elements increased slightly until day 8 and then decreased linearly. The variations of Fe and Al concentrations did not show any systematic tendencies.

4. Discussion

The nonsystematic decreases in As, Mo, Mn and Na concentrations observed in the sample of *E. acicularis* at day 89 suggests that the lowering of water level and flow rate at day 65 with low temperature from day 60 to day 66 caused to the plant shoots to wither in the channel. Therefore, this study evaluates the phytoremediation capability of *E. acicularis* for As, Mo and Mn, whose concentrations showed linear increases from the start to day 63 in the winter experiment.

The bio-concentration factor (BCF), a factor for estimating phytoremediation potential, is defined as the ratio of the concentration of a target element in plant tissue to that in the growth solution or soil [12]. In the most environments, the BCF for solution is higher than those for soil due to the small portion of the element responsible for uptake from soil materials and pore water to plant root. Thus, using BCFs for the comparison of phytoremediation potentials between soil and solution environments should be carefully considered. The translocation factor, which is defined as the ratio of an element's concentration in the shoot to its concentration in the root, is also a key factor in evaluating of phytoremediation ability [13]. In this study, *E. acicularis* was analyzed in terms of the whole plant, including the roots. Thus, the BCF value in this study was calculated by using the concentration in the whole plant (BCF_{whole}). The preliminary measurement for a sample with shoots separated from roots shows a translocation factor of 0.74 for As, 1.18 for Mo and 1.77 for Mn. These translocation factors are near 1. Thus, BFC_{whole} did not greatly differ from BCF for shoots and is comparable to the BCF obtained in a previous study.

4.1 Arsenic (As)

In our experiment from day 0 to 63, the As concentrations of *E. acicularis* increased from 0.51 to 8.79 mg/kg dry wt. in the wastewater flow with 0.026 mg/L and flow velocity about 0.31 L/min. The average uptake volume per day for As from the wastewater is calculated as 0.143 mg/kg dry wt./day at day 63. The BCF_{whole} for As in this study is ranges from 18-85 up to day 42 and then increases to 346 at day 69. A previous channel experiment in air temperature of 24.6-31.9°C in summer using *E. acicularis* in the same type of sloped channel with a length of 50 m and three 100-L inflows, with temperature of 26.2-34.5°C, pH 7.5, flow volume of 0.4-0.6 L/min and As concentration of 0.234 mg/L, found an increase in As concentration of *E. acicularis* from 0.14 to 4.16 mg/kg dry wt. [14]. The BCF_{whole} for solution were calculated from 0.37 to 17.8, for 2 days. The uptake rate for this 50 m channel experiment was calculated about 2 mg/kg dry wt./day, which is over 10 times higher than 0.145 mg/kg dry wt./day of this study. This corresponds to the increase in BCF_{whole} with time in our 63 days study. The lower As uptake efficiency in this study is considered as the difference of growth condition

In other case of soil environment, the two month cultivation of *E. acicularis* in the mine tailing site from September in Japan has been reported As concentration of 29.2-872 mg/kg dry wt. in shoots. These *E. acicularis* were grown in the contaminated sediment site with 188-1470 mg/kg dry wt. of As along a small river with flow velocity of 10 cm/s and As concentration of 0.039 mg/L. The BCF are estimated as 0.21-7.39 for the contaminated sediments and 1003-29966 for the river water. The BCF_{whole} for As in our flow-controlled channel experiment was lesser than 10-100 times as high as the BCF for the river water in the soil environment. This suggests that As taken up from the As and Mo contaminated wastewater in the adverse growth season is accumulated less efficiently than the As take up from the solutions in favorable growth media and season. Previous study for three other As hyperaccumulating aquatic macrophytes (*Egeria densa, Elodea canadensis* and *Lagarosiphon major*) from natural riverside with similar As concentrations in river water (0.01-0.02 mg/L) to our channel experiment reported high BCFs of 4987-24847 [15]. These BCFs are also one-two order higher than the BCF_{whole} of *E. acicularis* in our channel experiment. Laboratory batch experiments at 24°C using *E. acicularis* of 100 g as flesh weight in a solution with an As concentration of 0.057 mg/L (i.e., about twice the concentration of the present study) found an uptake rate of 0.377 mg/kg dry wt./day and BCF of 113 for 10 days [8]. The solution volume is 1L and replaced it each days. The daily water volume to the plant flesh weight was assumed as 10 L/kg fresh wt./day. This volume was smaller than the roughly estimated as 32.7 L/kg fresh wt./day for 63 days. Our uptake rate of 0.143 mg/kg dry wt./day was lower than that of the laboratory experiment. However, our BCF results were much higher, indicating the possibility that water flow would be a key to effective element uptake or that a long-term experiment would be needed in order to appropriately evaluate phytoremediation by BCF in *E. acicularis* and other aquatic macrophytes. If the contribution of transpiration from the shoots to the decrease in As concentration is not considered, then the As treatment capability through day 63 of the experiment was 3.3%, calculated as the average daily uptake mass of 0.369 mg in *E. acicularis* divided by the daily flow mass of 11.23 mg in the raw wastewater flow. This low capability of As accumulation is corresponding to the result that there is little to distinguish the As concentrations in the treated wastewater from those in inflows thought the experiment.

4.2 Molybdenum (Mo)

Our experimental results may be the first scientific report of Mo transportation from contaminated wastewater to *E. acicularis*. The average uptake ratio of Mo was 0.346 mg/kg dry wt./day, and its efficiency is evaluated as 0.8% to 106.7 mg/L/day of the daily volume in wastewater flow. The BCF for Mo increases with time from 20.2 at day 1.13 to 50.8 at day 32 and on to 88.3 at day 63. However, the rate of increase tended to slow with time. This implies that the Mo concentration of 26.62 mg/kg dry wt. in *E. acicularis* at day 63 has almost reached the maximum accumulating concentration for the wastewater with the Mo concentration of 0.247 mg/L. The Mo accumulation by *E. acicularis* in this study did not exceed the hyperaccumulation value of 1,000 mg/kg dry wt. for typical heavy metal elements [5]. However, our results demonstrate that *E. acicularis* has an uptake performance for Mo that is similar to that for As, and that *E. acicularis* has a $BCF_{whole} > 1$ for Mo accumulation. Thus, *E. acicularis* can be considered a good candidate as an accumulator of Mo from contaminated wastewater. Further experiments on accumulation and tolerance of Mo in *E. acicularis* should be carried out under various conditions.

4.3 Other element

The accumulation of Mn was 884.9 mg/kg dry wt. at day 63, which is lower than the lower limit concentration for Mn of hyperaccumulator plants at 10,000 mg/kg [15]. However, this Mn content exceeds the typical range for non-hyperaccumulators of 20-500 mg/kg dry wt. [16]. The average uptake rate was determined as 12.68 mg/kg dry wt./day. If use the maximum value of 0.0387 mg/L in the raw wastewater as the concentration in solution, then the BCF_{whole} for Mn is calculated as 240-20640. *E. acicularis* would be evaluated as having enough capability to accumulate Mn from wastewater. However, a previous report on Mn accumulation in *E. acicularis* concluded that this species is inadequate as an accumulator of Mn because the Mn concentration was higher in the roots than in shoots [8]. The preliminary analysis for separated root and shoot gives a translocation factor of 1.78, indicating that *E. acicularis* can also be a candidate as an accumulator of Mn.

4.4. Improvement point for this experimental design

The observations of the green colored shoots and increasing of As and Mo concentrations in plant body to day 63 indicate that *E. acicularis* has significant potential of As and Mo uptake from a contaminated wastewater in low air temperature condition fall below 5°C for the case that inflow water keep over 10°C in temperature and stable flow rate above 0.3 L/min. These element uptakes appear not to be caused by plant growth but by basal metabolism with stable uptake rate at equilibrium with water temperature, flow rate and element concentration of solution, because of the almost linearly increasing of As and Mo with time and As and Mo concentrations in plant tend to decrease after shoots change to non-green colored. As a future experimental design, the estimation of element uptake efficiency of *E.acicularis* should be investigated by the comparison with the water and air temperature and growth rate, with or without several types of fertilizer.

5. Conclusions

The 90-day monitoring results of weather, water temperature, flow volume and chemical compositions in water and *E. Acicularis* showed a relationship between the temperature of the wastewater flow and the uptake of heavy metal elements. While the temperatures of the inside air and the water flow in the channel were continuously kept above 0°C by the tunnel greenhouse in the channel, the bio-concentration factors of As and Mo were found exceed 1. This demonstrates that phytoremediation using *E. acicularis* is workable for the elements As and Mo in moderately contaminated wastewater flow, even in low-temperature conditions that are inhospitable to aquatic macrophytes.

6. Acknowledgements

The authors would like to thank Mr. Mito of Transportation Bureau City of Sendai for allowing us to conduct experiment in this site and Mr. Endo, Mr. Naruse and Mr. Oishi of OYO Corporation for their contributions to the setting up of the experimental channel. Mr. Sugawara of Sugawara Inc. gave us assistance with sampling.

7. References

[1] Nickson R, McArthur J, Burgess W, Ahmed K M, Ravenscroft P and Rahman M 1998 Arsenic poisoning of Bangladesh groundwater *Nature* **395** 338

[2] World Health Organization 2011 *Guidelines for drinking-water quality: fourth edition,* ed. World Health Organization p 564

[3] Ma L Q, Komar K M, Tu C, Zhang W, Cai Y and Kennelly E D 2001 A fern that hyperaccumulates arsenic *Nature* **409** 579

[4] Mohanty M, Dhai N, Patra P, Das B and Reddy P Phytoremediation: a novel approach for utilization of iron-ore wastes 2010 *Rev. Environ. Contam. Toxicol.* **206** 29-47

[5] Dhir B 2013 *Phytoremediation: Role of aquatic plants in environmental clean-up* (Springer India) p 111

[6] Sakakibara M, Harada A, Sano S, Hori R and Inoue M 2006 Phytoremediation of heavy metals contaminated by Eleocharis acicularis (in Japanese) *Proc. 12th Symp. on Soil and Groundwater Contamination and Remediation* Ed: N. Takeda (Kyoto, Japan 2006) pp. 545–548

[7] Sakakibara M, Ohmori Y, Ha N T H, Sano S, Sera K and Hori R S 2009 Practicality of phytoremediation by Eleocharis acicularis *Proc. of the 19th Symposium Geo-Environment in Geo-Technics* (Tokyo Japan) 93–96

[8] Ha N T H, Sakakibara M and Sano S 2009a Phytoremediation of Sb, As, Cu, and Zn from contminated water by the aquatic macrophyte Eleocharis acicularis *Clean (Weinh)* **27(9)** 720-725

[9] Ha N T H, Sakakibara M and Sano S 2011 Accumulation of indium and other heavy metals by Eleocharis acicularis: an ophion for phytoremediation and phytomining *Bioresour. Technol.* **102** 2228-2234

[10] Sakakibara M, Ohmori Y, Ha N T H, Sano S and Sera K 2011 Phytoremediation of heavy metal-contaminated water and sediment by Eleocharis acicularis *Clean (Weinh)* **39(8)** 735-741

[11] Ha N T H, Sakakibara M and Sano S 2009b The potential of Eleocharis acicularis for phytoremediation: case study at an abandoned mine site *Clean (Weinh)* **37(3)** 203-208

[12] Zhu Y L, Zayed A M, Qian J H, Souza M D and Terry N 1999 Phytoaccumulation of trace elements by wetland plants, II *J. Environ. Qual.* **28 (1)** 339-344

[13] Garbisu C and Alkorta I 2001 Phytoextraction: a cost-effective plant-based technology for the removal of metals from the environment *Bioresour. Technol.* **77** 3 229-236

[14] Okazaki K, Yamazaki S, Sakakibara M, Kurahashi T and Shishido M 2015 *Monthly Report of the Civil Engineering Research Institute for Cold Region* (in Japanese) **751** 11-16

[15] Robinson B, Kim N, Marchetti M, Moni C, Schroeter L, van den Dijsel C, Milne G and Clothier B 2006 Arsenic hyperaccumulation by aquatic macrophytes in the Taupo Volcanic Zone, New Zealand *Environ. Exp. Bot.* **58** 206-215.

[16] Reeves R D 2003 Tropical hyperaccumulators of metals and their potential for phytoextraction *Plant Soil* **249(1)** 57-65

Investigation about Creation Possibility of Pearl Farming in North Gorontalo, Indonesia for the Solution to Economical Poverty and Environmental Problem

Hiroki Kasamatsu[1], Mohamad Jahja[2], Masayuki Sakakibara[1]

[1] Faculty of collaborative Regional Innovation. Ehime University, Bunkyocho 3, Matsuyama, Japan
[2] International Research Collaboration. Universitas Negeri Gorontalo, Jl. Jendral Sudirman No. 6 Gorontalo, Indonesia
Email: kasamatsu.eu@gmail.com

Abstract. As a solution to economic poverty, we investigated the possibility of developing pearl farming in Kabupaten Gorontalo Utara region of Indonesia. The approximate income of farmers is 15 – 28 million rupiah/year, and that of fishers is 51 million rupiah/year. As these incomes are low and unreliable, it is important that food processing and other industries are developed and that agriculture and fisheries are promoted. Pearl farming minimize environmental load. The coastal environmental conditions are good for pearl shell growth. From 2001 till 2013, pearl farming was undertaken by a Japanese company in Kabupaten Gorontalo Utara, but the operation caused because of conflict with local people and issues with the actions of manager. To establish successful pearl farming businesses it is more important to address social issues than environmental factors.
Keywords: .poverty situation, farmer, fisher, pearl farming, Gorontalo Utara Regency

1. Introduction

Many rural areas in South Asia are dominated by economic poverty. There is an urgent need for the development of new businesses, but it is difficult for local populations to change their situation. Business opportunities appropriate to rural area are being sought to address economic poverty and develop the potential of environmental resources. Because of economic situation are undeveloped urban and industry facilities, the area retains a sustainable ecological system.

To address economic poverty we are investigating the possibility of developing pearl farming, which has much potential, and we are developing clear proposals to achieve this.

As pearl farming requires maintenance of clean water, large-scale engineering innovations are undesirable. In addition, unsustainable agricultural practices erode soils to rivers and the sea, and this could detrimentally affect the marine environment.

High quality products can generally be traded for good economic returns. Consequently, if local communities and associations are provided with appropriate technical support and management systems, good economic outcomes are possible.

We have an established relationship with pearl farmers because a pearl production facility is located in Ehime prefecture, where our university located. Technical advice is available through this relationship, making it is possible to initiate our proposal if possible results are obtained.

Indonesia and Japan differ in culture and social habits, so to initiate this new development there is a need to establish good relationship among local communities, organizations and government.

In this report, we describe the economic poverty of the study area, and we overview difficulties that overseas company face in establishing new business in this area. In particular, we describe the pearl farming that operated in Kabupaten Gorontalo Utara in the past. We outline necessity future research and multi-disciplinary approaches to establishing new business in the region.

2. General condition in the study area and method of investigation

2.1. The Gorontalo Utara Regency

We chose Kabupaten Gorontalo Utara (Propinsi Gorontalo Indonesia) as our study area (Figure 1). Kabupaten Gorontalo Utara is located in north Sulawesi, between 00°41'23" and 1°07'55" north latitude and 121°58'59" and 123°16'29" east longitude. The rainfall recorded in 2014 was approximately 6 – 321 mm/day, and the highest number of rainy days (20) occurred in November 2014. In 2014, the monthly maximum temperature ranged from 32.0 to 40.3° C and the minimum temperature ranged from 21.2 to 24.2°C. Kabupaten Gorontalo Utara has 11 districts (kecamatan), which in total cover an area of 1,777.022 km^2 [1]. The capital of kabupaten is Kwandang, which is 49 km and approximately 2 hours by car from central Kota Gorontalo.

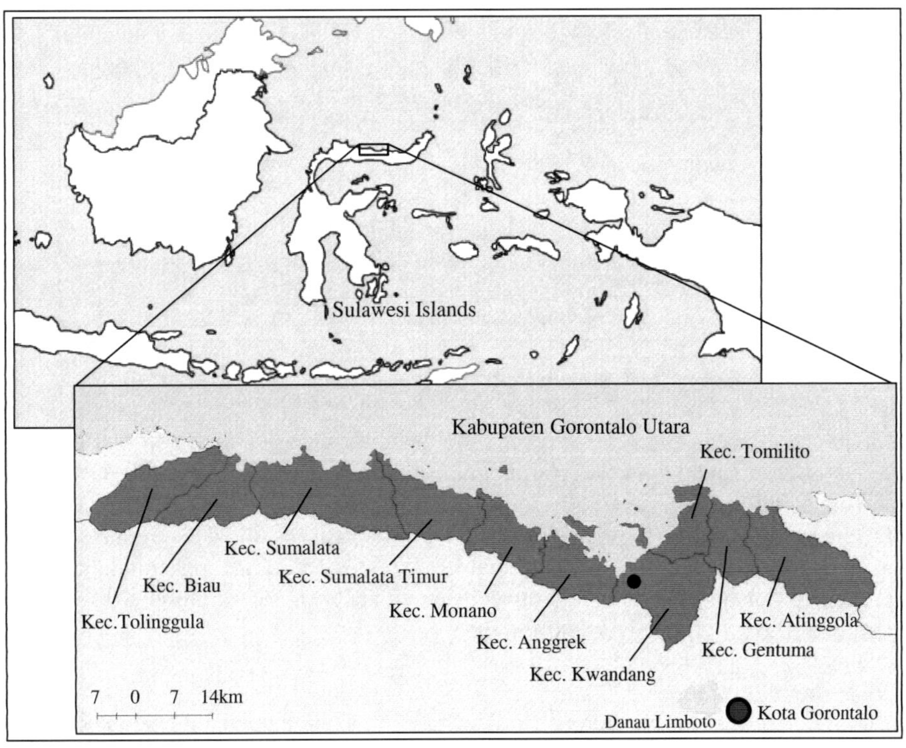

Figure 1. Map of Kabupaten Gorontalo Utara

The population in 2014 is 109,502, and the average kabupaten population growth rate from 2000 to 2010 was 1.84%. The number of household in 2010 is 24,888. As the population was 104,133 in 2010, the average of household size in this year was 4.18 [2]. The family size in this area is large compared with the average household size for Indonesia generally (3.81 in 2010) [3].

The gross regional domestic product (GRDP) from industrial activeties in Kabupaten Gorontalo Utara in 2014 was 1,989,795.43 million rupiah. The agriculture, forestry, and fishery category contributed most (951,676.84 million rupiah; 47.8% of GRDP), while the other important category was construction (213,074.79 million rupiah; 10.7% of GRDP) [4].

2.2. Method of investigation

The above shows that most people are enganged in agriculture, forestry and fisheries. Many families are supported on single incomes, but there are not always reliable. We have considered alternative options for income generation based on knowledge of the economic situation, and have identified pearl farming as one possible solution, particularly as a Japanese pearl farm operated in Kabupaten Gorontalo Utara in the past. To obtain historical details, from 12 to 16 August 2016 we undertook interviews in three kecamatan (Kwandang, Angrrek and Tomilito) involving 10 farmers and fishers, 7 desa chiefs (Tolango, Danbaro, Jembatan Merah, Bubode, Leawo, Bolango and Mutiara Laut) and government officers from Kabupaten Gorontalo Utara (Figure 2).

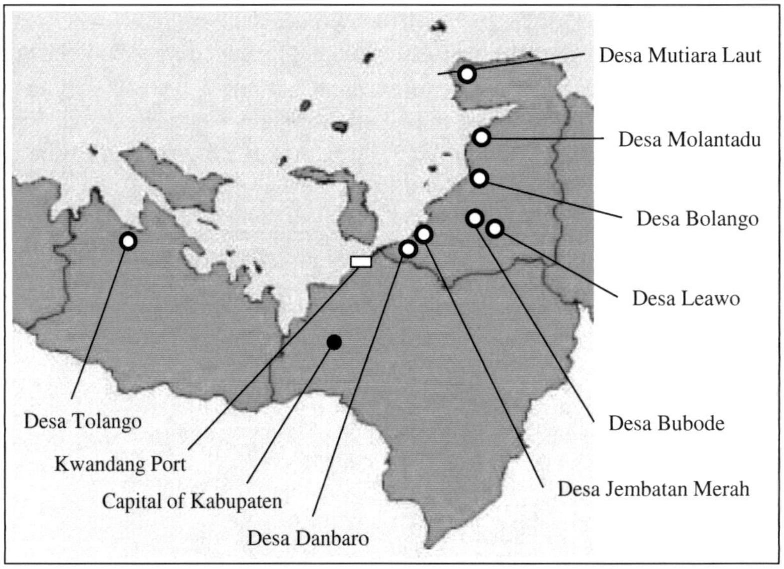

Figure 2. Research Point in Kabupaten Gorontalo Utara

To assess the best areas for pearl farming, we undertook basic investigations of environmental conditions. We obtained many details from fishers, including on wind direction and intensity, ocean currents, and the condition of fishery resources.

Pearl farming commenced in the area in 2001, but ceased in 2013 because of controversy. Understanding the issues involved is essential to the possibility of reintroducing pearl farming operations. Prior to our investigations there were many unsubstantiated rumours, so we interviewed a number of people who provided diverse opinions.

3. Rural Working Conditions

Most people are engaged in agriculture or fishieries in Kabupaten Gorontalo Utara, but there is variablity in the details among locations and seasons.

3.1. General information about farmers and fishers

The population number of households, and ratio of farmer and fisher in each desa are listed in Table 1, based on interviews with desa chiefs and local people. Fishers are common in coastal area but not in the mountainous areas. The populations of fishers and farmers varies among desa, depending on the size of the paddy fields and farms. Desa Morantabu and Desa Mutiara Laut are small so the ratio of fishers is higher than for other desas. Conversely, for desas where paddy fields and farming are possible, the number of farmers increases and the ratio of fishers is lower. Areas having a low ratio of fishers tend to have land available for cultivation, so the proportion of farmers increases.

Fishers cannot work full time in the fishery during any year. In the rainy season, many fisheries do not operate because of the strong west wind and heavy rain. During this season, fishers are engaged in agriculture.

Table 1. Basic Data of Sub Districts; Population, Household and Ratio of Farmer and Fisher (2016)

Name of Desa	Location	Population	Household	Farmer families	Fisher families
Tolango	coast, mountain	1,987	507	80%	13%
Danbaro[a]	coast, mountain	900	240	80%	20%
Jembatan Merah	coast, mountain	1,013	297	60%	40%
Bubode	mountain	1,006	294	100%	0%
Leawo[b]	mountain	600	160	90%	0%
Bolango	coast, mountain	787	205	60%	30%
Molantadu[c]	coast, mountain	1,000	300	30%	70%
Mutiara Laut	coast, mountain	500	139	10%	90%

[a, b, c] Estimation by chiefs of sub districts

3.2. Agricultural and fishery activities

3.2.1. Agriculture.

Farmers cultivate two corn crops and one rice crop each year, and all crops are cultivated on the same farm or paddy field. The cultivation period for each crop is 4 months. During the rainy season (January to April) water is available for irrigation, so rice is cultivated in the paddy fields. In the dry season, when less rainwater is available, corn crops are cultivated from May to August and from Setember to December.

However, the growing periods are variable. If farmers can get water in the dry season they cultivate rice paddies year round (Figure 3). In areas where upland rice grows well, such as Desa Jembatan Merah, cultivation commences in October or November, because upland rice needs a growth period exceeding 5 months. The farms are mostly small (less than 1 ha), and typically 1 – 3 km or more distant from the farmer's house.

Figure 3. Even if dry season, if farmer can draw enough water to paddy, they can cultivate paddy every season. Because sell price of rice is more expensive than maze, farmers want to cultivate paddy. This picture is paddy and irrigation at Desa Bubode in Dry Season, August 2016.

Farmers save rice and corn seed from their own crops, or they buy seed from neighboring farmers. Therefore, all seeds are native species, first filial generation seeds have not introduced. Farmers apply 300 kg/ha of chemical fertilizer for rice paddy cultivation and 225 kg/ha for corn cultivation. Some years ago fertilizers was provided by the kabupaten, but now farmers buy fertilizer from retail stores. The fertilizer price ranges from 200 to 300 rupiah/kg. Farmers also apply insecticides three times prior to harvest.

Each desa has a tractor provided by the kabupaten. And all farmers in the desa share its use. The desas operate on a mutial aid system for planting and harvesting of rice. Among collaborating groups of farmers, all members of the group helping in the cutting of grass.

The rice production is typically 2,000 kg/ha and for corn is 3,000 – 7,000 kg/ha. Farmers sell the crop production to brokers at a price of 8,000 – 10,000 rupiah/kg for rice and 2,000 – 3,150 rupiah/kg for corn. These prices are 200 rupiah/kg lower than themarket price. Because brokers contract with farmers, farmers have to sell to brokers.

3.2.2. Fisheries.
Most fishers use traditional fishing boats (Figure 4). In Kabupaten Gorontalo Utara there are only three vessels equipped to wrap nets; these vessels are based at Kecamatan Kwandang. Traditional style boat can carry a maxmum of three persons, but generally carry only two fishers. They leave from a bay or port late in the day or in the evening, and go offshore approximate 5 km, returning in the early morning. Fishing by hand only, nets are not used. Caught fishe are stored on ice on a small boat until they return to the bay/port. From December to January there are strong westerly winds and fishing is not possible. During this period the fishers are usually engaged in irrigated agriculture based on rainfall.

Figure 4. Fishers use traditional small boat with 2 outriggers and outboard motor. They leave from bay or sandy port at evening and come back early morning. They fish only hands with fish line. This picture is fisher's boat leaving from Kwandang bay, August 2016.

Fishers catch approximately 10 kg/boat of fish on each fishing trip. The fish are sold to brokers at 30,000 rupiah/kg, which is 20,000 rupiah/kg less than the market price. Fishers need 100,000 rupiah for each fishing trip to cover the costs of fuel (6 L), food, drink, and cigarettes. As an initial investment, 7 million rupiah is needed for the purchase of a boat, and 5 million rupiah for a 9 horsepower engine or 3.6 million rupiah for a 5 horsepower engine.

In Desa Tolango, fishers harvest and sell seaweed, which is cultivated at sea on long ropes supported by floats. After 2 weeks the seaweed is harvested, dried on nets on the coast, then sold to brokers for 12,000 rupiah/kg. The brokers sell the dried seaweed for export to Hong Kong.

In the past, fishers used dynamite fishing based on bombs made from car parts; the explosions kill the fish, which rise to surface. Some fishers from another area were killed, so it is now prohibited by government.

3.3. Economic situation of farmers and fishers

3.3.1. Income and costs.
From the above it is possible to estimate the economic situation of farmers and fishers (Tables 2 and 3). The approximate farmer income is 15 – 28 million rupiah/year. Because the price of their produce varies each year and farm size differ, farmer incomes differ. The yield is different year by year, the income of farmers is hard to be stable.

The approximate fisher income is 51 million rupiah/year. Fishers have higher incomes than farmers, but have to pay for the initial investment in boats and engines, and ongoing maintenance.

Forestry is not undertaken because the forests do not have roads, so transporting the cut timber is not possible. Consequently, most workers are farmers and or fishers, and depend on these activities for income to support their families.

Table 2. Basic Data to Calcurate farmer's Economy

Statistic data of Kabupaten Gorontalo Utara[a]		Agricultural basic units[e]	
population	109,502	amount of rice production	2,000kg/ha
average of household size[b]	4.18	amount of maize production	3,000 - 7,000kg/ha
number of household[c]	26,197	sell price of rice	8,000 - 10,000rupiah/kg
percentage of farmer	57.2%	sell price of corn	2,000 - 3,150rupiah/kg
number of farmers[d]	14,984	fertilizer (rice)	300kg/ha
rice field area	8,930ha	fertilizer (maize)	225kg/ha
farm paddy area	2,247ha	price of fertilizer	20 - 30rupiah/kg
corn area	9,054ha		

[a] Data of 2014 from "Badan Pusat Statistik Kabupaten Gorontalo Utara, 2015"[5], except average of household size.
[b] Data of 2010.
[c] Estimation by population and average of household size.
[d] Estimation by household and percentation of farmer.
[e] Data based on interview of 2016.

Table 3. Estimation of the Income and Outcome of Farmer and Fisher

Farmer	
(1) Income	
management area of rice	0.75ha
management area of maize	0.60ha
amount of rice products	1,500kg/year
amount of maize products	1,800-4,200kg/year
income of rice (minimum - maximum)	11,936,000 - 14,920,000rupiah/year
income of maize (minimum - maximum)	3,600,000 - 13,230,000rupiah/year
total income(minimum - maximum)	15,536,000 - 28,150,000rupiah/year
(2) Outcome (fertilizer)	
consumption to rice	300kg/ha
consumption to maize	225kg/ha
buy price of fertilizer	200 - 300rupiah/kg
total outcome(minimum - maximum)	72,000 - 108,000rupiah/year
(3) Deduction (minimum - maximum)	15,464,000 - 28,042,000rupiah/year
Fisher	
(1) Income	
fish catches	10kg/day
fishing days	270days/year
sell price of fish	30,000rupiah/kg
total income	81,000,000rupiah/year
(2) Outcome	
fuel, meal, drink and cigalette	100,000rupiah/day
total outcome	30,000,000rupiah/year
(3) Deduction(minimum - maximum)	about 51,000,000rupiah/year

Data from Table 3 and based on interview of 2016.

3.3.2. Support Systems.
Increasing crop harvests and fish catches, and stabilizing incomes, are the major issues for farmers and fishers. Even minor improvements are benefical, and some have already been introduced.

Support for agricultural activities has sometimes been provided by the kabupaten. For example, Kabupaten Gorontalo Utara built irrigation systems at Desa Bubode in 1974 and 1990. The system built in 1974 is near the center of desa, and **that** built in 1990 gets water from 1 km further up the catchment. These systems supply water to the rice paddies throughout the year, and are managed by the kabupaten.

Kabupaten Gorontalo Utara also provides tractors to the desas. Kecamatan Tomilito has approximately 100 desa and each desa has a tractor. Until these were provided, farmers used cows for cultivation, but this process process has become markedly easier with the availability of machinery.

Furthermore, the supply of fertilizer in the past was a kabupaten policy. The size of agricultural harvests and fish catches depends on the weather. The kabupaten also runs technical improvement programs to help stabilize incomes. Support by the kabupaten is important for farmers and for agricultural development.

Although support from the kabupaten is necessary and provided, farmer incomes vary each year. In 2015, the income situation for farmers was the worst in recent years because of a slump in the corn price; e.g., in that year a farmer could reseive 500,000 rupiah for a corn harvest of 30,000 kg, corresponding to a unit sale price of only 17 rupiah/kg, or 1/120th of that expected in an average year. Because of this crisis the local government assisted farmers to buy fertilizer in 2015, and undertook to support the purchase of corn and rice seed in 2016.

3.3.3. Other approaches.
Under conditions of income instability, the absence of support from the kabupaten, and or if an unpredicted crisis occurs, farmers and fishers face difficulties. If a bank finances 5 million rupiah over 6 months, the amount to be repaid is 5.2 million rupiah. For a farmer who makes 1.3 million rpiah/month income, repayment at this rate is not possible. Other income sources include factory work for 20,000 rupiah/day, expatriate labor to Kota Gorontalo, and load progressing duties in the large port at Kecamatan Angrrek for 70,000 rupiah/day. People can obtain employment planting and treating trees, but the income is less than that for other jobs. If a farmer's income was insufficient and he sold his 3 ha corn farm for 50 million rupiah, even if his economic situation stabilized, the price of the farm would have increased, making it is very difficult to buy back. If he wanted to work in agriculture again he would need to clear the forest and make a new farm.

The income of farmers and fishers is low and unstable, so to escape economic poverty they must break out of this situation.

3.4. Opportunities to increase incomes
To improve income it is necessary to adopt technical innovations, particularly those not requiring major environmental change.

With respect to a agriculture, it is necessary to diversify crops production beyond just rice and maize. At present If rhe one or both crops fail there is no alternative crop or jobs to supplement the family bujet. This problem is weather-dependent, and not a major issue in the rainy season. However, if water is not available in the dry season, which occupies half of the year, crops resistant to drying are needed. In the dry season the kabupaten supplies water for household use every 3 days, but there is not enough to supply water for agriculture.

With respect to fisheries, the current methods, boats, and materials are traditional, thus limiting fish catches to approximately 10 kg per fishing event. If fishery resources are abundant, the introduction of fish traps or seining will effectively increase fish catches. This may necessitate building ports, using larger boats or ships, introducing traps or nets, and learning associated techniques. How such changes cold be progressed is a future problem. However, traditional methods have only small environmental impacts,. so introducing new methods will require careful assesment to ensure their introduction does not lead to overfishing. It is essential that consensus on such changes is reached among fishers and local people. Whatever innovations are proposed, it is essencial to remember that the traditional method has maintained the environment in good condition. These considerations are necessary in deciding whether or not to introduce innovations or change methods.

A common issue in agriculture and fisheries is the need to strengthen food processing capacities. The food processing industry is adequate but local communities do not have a custom of food prosessing. There is little preserved food used in everyday meals. Rice with grilled or fried fish, chicken, egg, and vegetables is the staple food, and is available year-round because of climatic consistency. Unlike cold countries, where crop cultivation may not be possible for more than half the year, food preservation in this area is not necessary, and it is unlikely that the population of Kabupaten Gorontalo Utara has knowledge of preserving products and techniqures.

It may be possible in rural areas to develop alternative industrial economic subsystems other than agriculture and fisheries. For example, Kerawang, the traditional embroidery in North Sulawesi, is produced by female workers, usually in the houses of farmers and fishers. It is currently made and distributed locally, but there is possibility of developing a souvenir market by developing the skills associated with producing Kerawang. In Kabupaten Gorontalo Utara in 2014, the percentage of males involved in economic activities was 74.8%, but for females was only 46.8% [6]. It may be possible to promote manual industries more extensively.

The above highlights the importance of introducing innovations in food processing and otner industries, in addition to improving the conditions for farmers and fishers. The activities of women should also be promoted, as they have important potential roles in rural development.

4. Possibility of pearl farming

4.1. Environmental conditions

One potential industry to develop while minimizing environmental impacts is pearl farming. Assessing the best locations for pearl farming requires considerations of the marine environment, climatic and geographic conditions, the effects of soil erosion (especcially from agriculture), and the connections among mountainous areas, farms, paddy fields, rivers, and the sea.

A pearl farming company operated over a 12-year period at Kecamatan Anggrek and Kecamatan Tomilito (Kabupaten Gorontalo Utara),demonstrating the potential for pearl farming in this area. Interviews with fishers indicade that red tides of harmful phytoplankton, which can cause pearl shells mortality, have not been occured in the sea,suggesting it is a suitable environment for the growth of pearl shells.The water depth for pearl farming must be less than 50 m because at greater depth it is difficult to anchor the ropes and nets to the bottom. Figure 5 indicates that most of the sea area within 3 km from coast (to the red line) has appropriate topography.

Wind has the biggest influence on pearl farming. From May to October weak winds blow from the south, and the wave height is approximately 1 m. From November to April, strong winds, sometimes with rain, blow from the west and the wave height is usually 3 – 4 m, Which can impact 20 m inland from the shoreline. Fishers cannot work in December and January, when the waves are particularly high and offshore waves are higher tham on the coast. The waves near the coast are not usually high, but soil erotion and transport by rivers to the coast be significant.

As much pearl farming occurs at sea it is important to select places where the effect of west winds is least, Including areas in the lee of mountains or islands. However, the direction and force of winds can vary from place to place, so each potential site must be individual assessed.

Although little soil erosion occurs during the dry season, erotion is likely to increase during the rainy season. It is thought that there is little erosion from flat rice paddies and small areas cultivated using traditional methods, but erotion from farms on sloping land is potentially more serious. Upland rice and corn crops are commonly planted in these areas, so detailed investigations are necessory if new technical innovations are introduced in those areas.

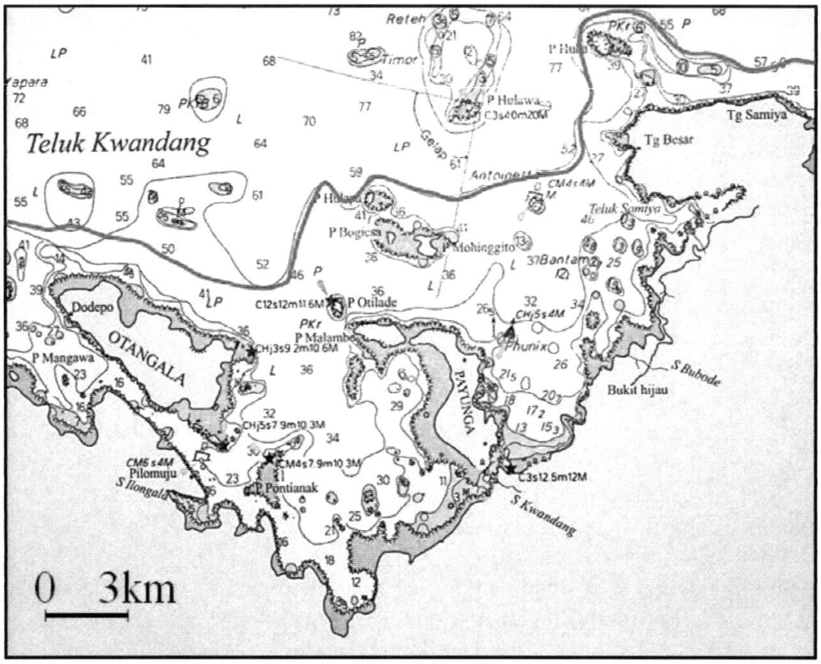

Figure 5. Depth of Sea Around Research Area [7]

4.2. History of pearl farming

Although the coastal waters of Kabupaten Gorontalo Utara are suitable for Pearl farming, the only pearl farming cased operation after 12 years. To establish the reasons, we interviewed laborers involved at that time.

4.2.1. Japanese pearl farming in Kabupaten Gorontalo Utara.

Japanese pearl farmers operated a pearl farming in Kabupaten Gorontalo Utara from 2001. They are considered the pioneers of pearl farming in this area. This case must be examined when considering the possibility of reestablishment pearl farming.

In 2001, MADOKA company established a peark farm at Desa Tolango (Kecamatan Anggrek, Kabupaten Gorontalo Utara), prior to which the company had operating at a different location. The company involved 3 Japanese people, 12 technicians from Kota Gorontalo, and 25 local fishers from Desa Tolango. The principal members were Japanese, and included the general manager, his son, and a pearl farming technician. All staff lived in a house approximately 3 m above the sea level (Figure 6). The house and workshop were built on many pillars on a very small sandy island.

From 2002 to 2003, a satellite farm knowns as MARIA at Desa Mutiara Laut. This location provided good condition for rapid shell growth. There were 15 laborers, most of whom were from Kota Gorontalo; only 3 people were locals. They floated ropes in the sea to which were tied many shell nets. Sometimes the propellers of fishers' boats cut the ropes as the boats passed the pearl farm area. Consequently, army were introduced to keep people away from the pearl farm, resulting in local fishers having to detour to reach their fishing grounds and neighboring ports. After 2 years, the guarding ceased, opposition to the activities at MARIA grew. The pearl farmers abandoned the MARIA operation; they harvested pearl products twice over 2 years, but ceased operations in 2003.

Figure 6. House and Workshop of MADOKA

The company investigated other suitable places for pearl farming prior to abandoning the MARIA operation. Farming was attenpted at Kecamatan Boroko (Kabupaten Bolaang Mongondow Utara, Propinsi Sulawesi Utara), but operations were soon discontinued as dynamite fishing was still occurring there.

In 2004, the aging general manager returned to Japan, but his son continued management of MADOKA. However, in 2013 they ceased all operations in Kabupaten Gorontalo Utara, and MADOKA moved to Kecamatan Marisa (Kabupaten Pohuwato, Propinsi Gorontalo). The first general manager and his son now live in Japan, and MADOKA is operated by another Japanese group.

4.2.2. Pearl farming process.
The general manager bought shell seedlings at Kabupaten Kupang (Propinsi Nusa Tenggara Timur), and staff bred them over 3 months. The shells were then separated to achieve 16 shells in each box, and pearl core was inserted into each shell. This was done only by Japanese staff, because it is a very delicate process and requires a high level of proficiency and technique. Next, eight shells were placed into each net, and the nets were tied to branch ropes 3 m in length. A main rope floating on the sea surface had branches to which the nets were tied. A single main rope had 250 nets attached, and the main ropes were tied to a raft in the sea. Each raft had 25 main ropes and MADOKA managed 9 rafts. Based on the above, they were growing 450,000 shells in total.

To grow the shells, he nets were taken to MARIA at Desa Mutiara Laut, Where there were 3 rafts, each of which had 10 main ropes (each 100 m long) in parallel and 10 m apart. Therefore, the pearl farming area occupied 3 ha on sea surface. Each main rope had 20 branch ropes with net which was attached to the tip of rope. The shells at MARIA were bigger than those at MADOKA, and the number of shells was different in each net. Nets were hung at various depth in the sea, including at 3, 5, and 7 m. The distance between ropes was 5 m. The distance between nets was wide because larger shells need access to more plankton.

At both MADOKA and MARIA, parasitic animals that attached to the surfaces of the shells were removed using grinding machine. This was main task, with laborers spending 20 days each month on this activity; on the other 10 days they repaired nets and ropes. One year after the core had been inserted, the shells were removed from the sea to harvest the pearls. For pearl farmers starting new operations, typically more than 1 year is required for pearls to develop, but at MADOKA pearl products were obtained in the first year, and every 3 months subsequently.

4.2.3. Situation of local laborers.

Local laborers at MADOKA were employed part-time. The wage of local laborers followed these local minimum wages. They were provided with meals and drinks in addition to wage. People from Kota Gorontalo who had training in pearl farming techniques could obtain higher wages.

The laborers at MARIA were locals from near the pearl farming area, but because the farm was some distance from their homes, they couldn't return home every day or even weekly, in addition, they were only given 3 days off every 2 months. Their wage was 400,000 rupiah/month, plus meals and drinks, which is very low. The possible reasons for the manager paying such low wage include that the work was simple and required little technique (laborers washed shells for 20 days and repaired ropes for 10 days a month), that the laborers didn't need to pay for anything when at the farm and their basic needs (shelter and foods) were supplied. However, this type of employment does not encourage development of industry because no local people learnt the techniques needed for pearl farming and pearl production, or developed management skills. In actuality, most laborers were never taught the difference between pearl products and the pearl core.

The working conditions were also very severe. Time management was strictly overseen, and laborers were not allowed to talk or smoke while working. This is very different from the customary work style of the local people.

4.2.4. Relationships with the local Communities.

Conflicts occurred between local people and MADOKA staff. Unfortunately, the MADOKA company did not provide any support for the communities or local people, and cooperated only with the desa and local government officials until the time that their operation ceased. Furthermore, the activities of fishers who made their living harvesting seaweed were disturbed by the MADOKA guards. Consequently, many local people refused to work for MADOKA.

In addition, the manager at the MARIA site did not communicate adequately with local people or members of the desa, and there was an attack on the MARIA site after the guards left the farm area, making reconcilation impossible.

Despite these problems, pearl farming has not been rejected. Local people, chiefs of the desas and mayor of the kabupaten hope that if pearl farming is re-established it may be beneficial for the local community. When MADOKA was in operation it bought 4,000 kg rice per year, and bought economic benefits to local shops, machine repairs, and schools, as well as to the laborers and their families.

While it is clear that a new pearl farming company would need to cooperate with the local people and community, the latter also benefit directly or indirectly from the activities if the company.

There are various reasons why the MADOKA and MARIA pearl farms ceased operating. Local people said that they did not get sufficient benefit, and that the new manager (the son of the first general manager) failed to cooperate with the locals, and failed to form a good relationship with local people and communities.

5. Conclusions

In Kabupaten Gorontalo Utara it is possible to develop innovate agriculture and fishery techniques or establish new business to help solve issues of economic poverty. One alternative business opportunity is pearl farming, which can provide many advantages for the area.

In tackling the problem of poverty by establishing a new industry, it is important to first address any social issues that may arise. To avoid conflicts such as those associated with the operation of the MADOKA pearl farm, a major focus must be on communication with local people, the desa, and the kecamatan and kabupaten. The company manager and community leaders are required to support the laborers, as well as the farming and fisher communities where the activities are located. This is an extremely common issue that is not restricted to pearl farming.

When we discussed with local people and government officials the possibility of reintroducing pearl farming in Kabupaten Gorontalo Utara, the reaction was generally positive; however, they do not need confliction from unilaterally way.

This study comprised only basic research. We will next need to undertake detailed investigations to establish whether pearl farming can be operated in susutainable way. To assess its potential, trans-disciplinary approach is needed to address the following issues.

Clarification of the conditions necessory for growing pearl shells, and assessment of its environmental sustainability.

Approaches to balancing pearl farming development while maintaining environment conditions.

Ensuring management of pearl farming occurs through establishment of good relationship with local people.

Ensuring laborers reseive fair wages, consistent with those reseived by farmers and fishers.

If these issures can be resolved it may lead to the development of pearl farming as one way to address poverty problems in this area.

References

[1] Badan Pusat Statistik Kabupaten Gorontalo Utara 2015 *Gorontalo Utara in Figures 2015, Geography and Climate*. p3-4

[2] Badan Pusat Statistik Kabupaten Gorontalo Utara 2015 *Gorontalo Utara in Figures 2015, Population and Employment*. p42 - 45

[3] TecKarta *Nakono. Households: Average household Size (68 countries)* https://www.nakono.com/tectcarta/databank/households-average-household-size/

[4] Badan Pusat Statistik Kabupaten Gorontalo Utara 2015 *Gorontalo Utara in Figures 2015, Regional Income*. p168 - 169

[5] Badan Pusat Statistik Kabupaten Gorontalo Utara 2015 *Gorontalo Utara in Figures 2015, Population and Employment*. p42, 45 and 50

[6] Badan Pusat Statistik Kabupaten Gorontalo Utara 2015 *Gorontalo Utara in Figures 2015, Population and Employment*. p48

[7] Tentara National Indonesia Angkatan Laut, Dinas Hidro-Oceanografi 2014 *Indonesia Sulawesi Pantai Utara Ujung Malangkah Hingga Teluk Kwandang*

IOP Conf. Series: Earth and Environmental Science **71** (2017) 012020 doi:10.1088/1755-1315/71/1/012020

Development of discrete choice model considering internal reference points and their effects in travel mode choice context

Sarif[1,2], Shinya Kurauchi[1] and Toshio Yoshii[1]

[1] Department of Civil and Environmental Engineering, Ehime University, Matsuyama, Japan
[2] Department of Civil Engineering, State Polytechnic of Ujung Pandang, Indonesia
Corresponding Author: sarifpnup09@gmail.com

Abstract. In the conventional travel behavior models such as logit and probit, decision makers are assumed to conduct the absolute evaluations on the attributes of the choice alternatives. On the other hand, many researchers in cognitive psychology and marketing science have been suggesting that the perceptions of attributes are characterized by the benchmark called "reference points" and the relative evaluations based on them are often employed in various choice situations. Therefore, this study developed a travel behavior model based on the mental accounting theory in which the internal reference points are explicitly considered. A questionnaire survey about the shopping trip to the CBD in Matsuyama city was conducted, and then the roles of reference points in travel mode choice contexts were investigated. The result showed that the goodness-of-fit of the developed model was higher than that of the conventional model, indicating that the internal reference points might play the major roles in the choice of travel mode. Also shown was that the respondents seem to utilize various reference points: some tend to adopt the lowest fuel price they have experienced, others employ fare price they feel in perceptions of the travel cost.
Keywords: Travel behavior; Reference points; Mental accounting theory; Travel mode choice; Discrete choice model

1. Introduction

In recent years, there has been a growing intersection between the transportation and the economic area of study. In microeconomic theory, there are two demand levels theory namely consumer demand and market demand. Recent behavioral research indicates that there is no simple explanation of how price might influence individual buyers' purchase decisions, even in the transportation field, such as fuel price, parking fee, and toll fee. Private car (auto) and public transport (PT) are two transportation modes that still have an interchangeable high preference to use by most of urban dweller. In urban transport field, reducing private car and increasing public transport users is challenging. Travel cost and travel time are used as reference by the traveler.

The conceptualizations of reference price rely on a substantial body of behavioral literature that examines the relationship between new point information (i.e. actual price) and established reference point information (i.e. reference price) [1]. Several theories have been proposed in dealing with reference price such as prospect theory [2], mental accounting theory [3], and strategic reference point theory [4].

According to economic theory, the price is assumed to influence buyer choice due to price serves as an indicator of the purchase cost. Thus, it is believed that the buyer has perfect information concerning price and requires the satisfaction of comparable product alternatives (Kent B. Monroe). The study of prospect theory in transport mode choice for traveling has been extensively investigated by many researchers (e.g. Stated Preference Analysis Travel Choice [5], Identification of parameters for prospect theory model for travel choice analysis [6]. Applying behavioral economics in the design of travel information system [7]. A cumulative prospect theory approach to passengers behavior modeling: waiting time paradox revisited [8]. For the transport user, sometimes travelers drive a private car and using public transport, depend on the profitable value of vehicles. The focus of this study is to find out the alternative decision frames for a short-distance trip called reference points which have two different definitions, i.e., gain and loss. The value function is statistically estimated and tested within the context of the reference price. This research is concerned only for a short-distance trip like a shopping trip from a resident (origin) to central business district (CBD). The study aims to answer what kind of the travel mode will be used and how is the influence of internal reference price in choosing the travel mode?. This research presents some hypotheses related to internal reference price across the travel mode choice, especially in the short-distance trip. The following hypotheses are;

1. The effect of internal reference price on actual travel cost are greater than perceived travel cost for public transport.
2. The effect of internal reference price on perceived travel cost are greater than reservation travel cost for public transport.
3. Actual travel time are less than perceived travel time for public transport.
4. Perceived travel time are greater than reservation travel time for public transport.
5. Public transport user tends to adopt the lowest price as reference cost.

Everyone may have different internal reference price depend on the situation of travelers. Thus, data regarding individual internal reference price was collected. This study concern on internal reference price such as fair price, perceived price and reservation price of travel cost and travel time as a reference in a short-distance trip. The purpose of this study is to analyze the influence of internal reference price on the travel behavior of transport user at a short-distance trip. In order to know the influences of internal reference price, this study analyzes the data using binomial logit model approach based on prospect theory and mental accounting theory. Another basic element of prospect theory such as weight function is beyond the scope of this study.

This study is comprised of six sections. The second section presents a brief introduction to prospect theory. The third section provides methods; that is, data collection and procedure, survey instrument and data analysis. Developed model is presented in the fourth section. The result and discussion are proposed in the fifth section. Finally, the conclusion and suggestions for future research direction are given in the last section.

2. Literature review

2.1. Prospect theory

Prospect theory is concerned with risky choices "such as whether or not to take an umbrella and whether or not to go to war, which "are made without advance knowledge of their consequence" [9]. It is like whether or not to use a private car or public transport [10]. It is found trough experiments that the choice depends on the position in which the decision maker is placed and, according to the position, whether they would have gain or lose by the perspective decision. The essence of prospect theory can be found in two scaling functions, that is, the value function and weight function, and in the reference point. The value function represents a prospect (i.e., possible outcomes and its consequences) in terms of the increase and decrease in the amount of asset from a reference point. The weight function, on the other hand, scales a prospect in term of probability associated with it [11]. The properties of the value function are showed in **Table 1**.

Table 1. Properties of the value function.

1	Defined on the deviation from the reference point.
2	Generally, concave for gains and convex for losses.
3	Steeper for losses than for gains.

Source: Kahneman and Tversky (1979), p .279. [2]

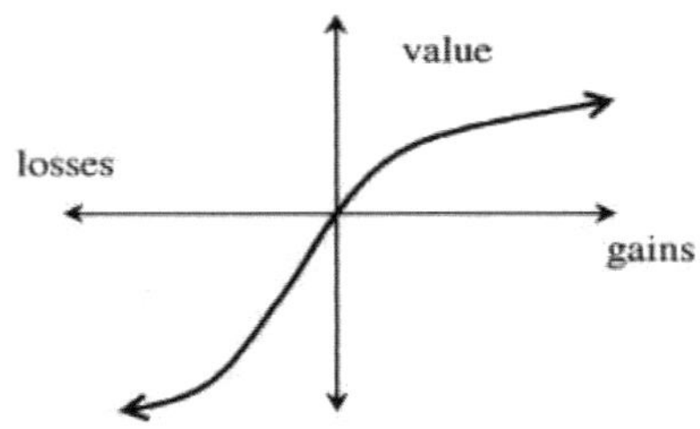

Figure 1. Value function.

Kahneman and Tversky propose the function form of the value function as

$$f(x)\begin{cases} x^{\alpha} & if\ x \geq 0 \\ \lambda(x)^{\gamma} & if\ x < 0 \end{cases} \tag{1}$$

where x refers to the outcome expressed as deviation from zero-asset position, that is, reference point, λ is the scale parameter, and α and γ are parameters of the value function in the gain region and loss region, respectively.

2.2. Reference price
The reference price is defined as the standard price considered by consumer when evaluating the actual price of the products to purchased, and it is based on the premise that consumers do not respond to prices absolutely, but only relative to the reference price [3]. The concept of a reference price asserts that consumers make decisions based on both actual and perceived price. In the transport field, it is recognized that car user behavior (travel behavior) is one of the main factors in transportation planning. Reference price might influence the perception or behavior of transport user to decide what kind of the travel mode will be used. The internal reference price and external reference price are various conceptualizations of the reference price. An internal reference price refers to a process that takes place internally, in the consumer's mind, by means of experience and judgment. Whereas, external reference price refers to information gathered from the environment, external to the person [1]. The following is detail for an internal and external reference price description.

2.2.1. Internal reference price.
The internal reference price is based on the past prices for the brand paid by consumer. It has some reference price such as fair price, reservation price, normal or standard price, aspiration price, market price, historical price, perceived price, and evoked price. The definition of some reference price has relationship each other. For instance, if someone wants to buy something/goods. They would negotiate a price to buy it. Before making a negotiation a reservation price, and fair price should be considered first. Reservation price is defined as the limit (i.e., the lowest/highest price she/he as a seller/buyer are ready to sell/buy) [12]. If the price is suitable (agree) among seller and buyer, that is a fair price. Actual price is a real price of commodity that can be sold and delivered immediately. Perceived price is the price in which a buyer can decide to buy a product. It has been claimed that the internal reference price concept has gained popularity in the literature and also provides more accuracy in

predicting consumer behavior [13]. The existence of the internal reference price is motivated by adaptation-level theory, which suggests that people adapt to the level of past stimuli on comparison with the adaptation level [14]. A consumer who concern for prices, are also more apt to pay close attention to, and stronger memory for, prices that they encountered in the past can be categorized as the internal reference price.

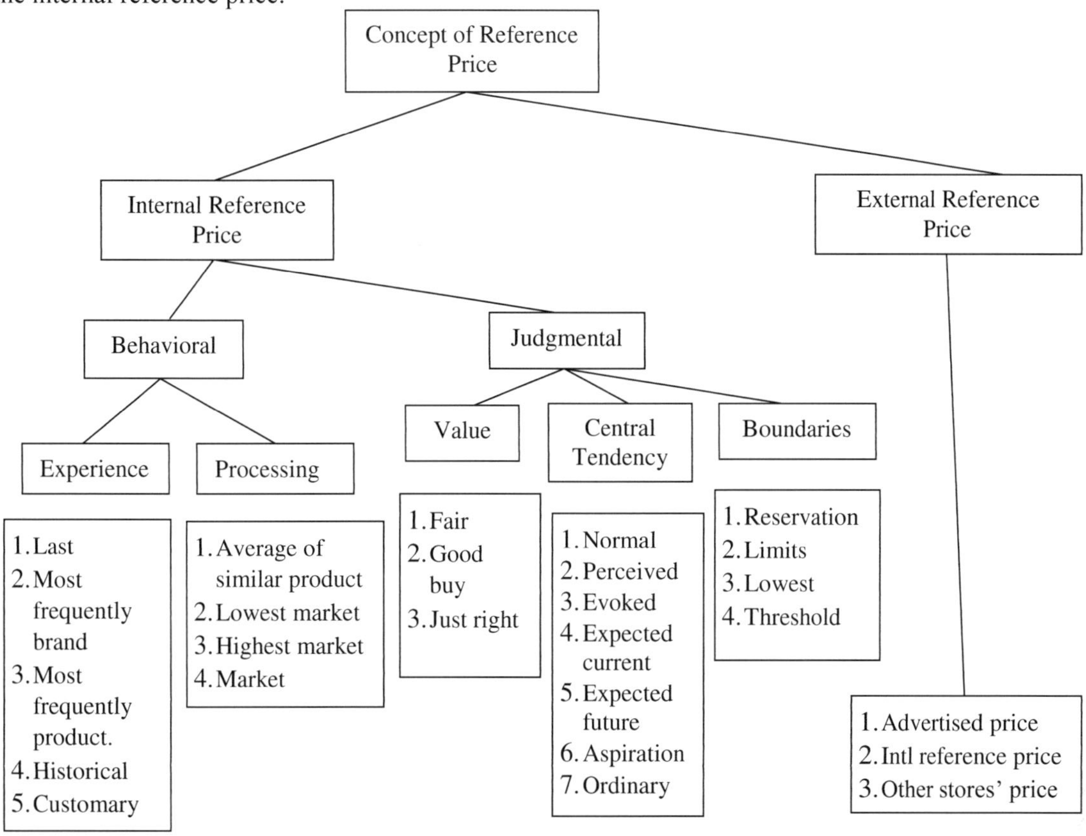

Figure 2. The concept of Reference Price.

2.2.2. *External reference price*

The external reference price is the price which consumer are not loyal to any one brand switch from brand to brand [15]. It is provided to the consumer through advertised price, international reference price and another stores' price. The advertised reference price is an offer price which can influence a consumer to buy something. Three basic external reference price formats are used in advertising: (1) comparing an advertised price with the seller's former price, (2) comparing an advertised price with the manufacturer's suggested retail price, and (3) comparing an advertised price with the price at competing [16]. Another external reference price is international reference price which reflects the price of the product in different countries. The international reference price is a part of external reference price in which the price of the product in various foreign markets using the transformable value is comparable to the local market's price. In the area of transport field, it is possible to apply the term of international reference price in a country, for example in determining the price of public transport ticket. Costumers would compare the observed point-of-purchases price with the reference price in order to implement prospect theory in the context of prices. A number of example regarding to the external reference price which related to the transport mode choice such as advertising discount fare are commonly used.

2.3. Reference point

IOP Conf. Series: Earth and Environmental Science **71** (2017) 012020 doi:10.1088/1755-1315/71/1/012020

The reference point is a zero-asset position from which the deviation measured each prospect as a gain or a loss [11]. It should be noted that the concept of a reference point has its roots in the psychology perception, and there is no formal theory available for formulating reference point [17]. Transport user will choose some alternative transport modes for traveling, such as car, bus, rail and bicycle. We assume that there are three relevant attributes of reference point: travel time, travel cost and convenience. Although the time and cost of the journey can be seen as losses in many travel situations, some travel choices could be still perceived as psychological gains if their cost and time are lower than those of other travel alternatives. At the travel cost, there are some variables associated with it, such as parking fee, toll fee and fuel price (gasoline price). Some people said that the fuel price is low and whiles others said expensive; it depends on an experience and their perception.

Thaler's (1985) model suggests that overall utility for a product can be conceptualized as function of acquisition utility (AU) and transaction utility (TU), where AU reflects the perceived value for the money or 'get' relative to 'give', and TU captures the consumer's reaction to relative price to the price compared to the expected price [18];

$$U = AU(p, \bar{p}) + \beta TU(p, p^*) \quad (2)$$

where;

U	; The utility from a purchase
$AU(p, \bar{p})$	; Acquisition utility, judged by comparing the actual price to be paid (p) to consumer's value-equivalent price ($\bar{p}$)
$\beta TU(p, p^*)$	; Transaction utility, judged by comparing the actual price to be paid (p) to the consumer's expected price p^*
p	; Actual price
$\bar{p}$	; Willingness to pay
p^*	; Reference price

3. Methods

3.1. Data collection and procedure

The data were collected in Matsuyama City, Ehime Prefecture, Japan. This study focuses on short-distance trip shopping, for example, resident to center business district such as Takashimaya, Okaido, and Gintengai. The questionnaire was addressed to transport users randomly selected in some points toward Matsuyama city officer. This questionnaire consists of 53 questions. It covers an economic term which applied in behavioral economics such as actual price, perceived price, and reserved price. By using binomial logit model, the data were analyzed to identify the reference point of travel time for public transport (PT), travel cost for public transport, and travel time for private car (auto). The respondents were asked to fill in the questionnaire sheet. The following illustrates about questionnaire sheet.

1. The questionnaire sheet describes what is the main transport mode they have used and how frequent it is. We asked them about how much is a reasonable price, fair price, reservation price and perceived price?. Moreover, reasonable price, fair price, reservation price, and perceived price serve as the reference price. Information obtained from questionnaires was analyzed by using binomial logit model in order to know the magnitude influence of internal reference price across the transport user in the short-distance trip.
2. Socio-economic and demographic characteristic, wherein surveyed respondents were asked about their gender, age, driver's license, monthly income, occupations, and zip code.

3.2. Preliminary analysis

In the preliminary analysis, data were described prior to the analysis. Analysis started with statistical analysis of the data and briefly looked at some standard techniques for description and summarization of data. Prior to application of these techniques, however, it is important to perform a data cleaning to ensure an appropriate analysis in the context of the study objectives. Matsuyama area is about 429.05 km^2, Matsuyama is the capital city of Ehime Prefecture on the island of Shikoku in Japan and also Shikoku's largest city, with a population of 515,183 as of August 1, 2016, and the number of households is 235,178, as well as the number of man and woman is 240,254 and 274,929 respectively. The ratio between data collected (sample) and Matsuyama population is shown in **Figure 3**.

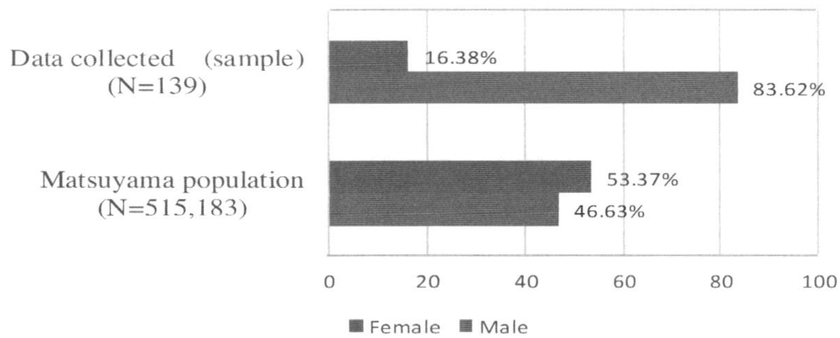

Figure 3. Data collected and Matsuyama population

Figure 3 shows the population ratio of Matsuyama for female and male are 46.63 % and 53.37 % respectively. And the population ratio of the data is 16.38 % for female and 83.62 % for male. The following figures are mode choice distribution and analyzed base on the basic information of the respondent. It can be seen in **Figure 4**, **Figure 5**, and **Figure 6**. The figures describe the mode choice (public transport and private car) based on gender, age categorize, and characteristic of travel mode choice.

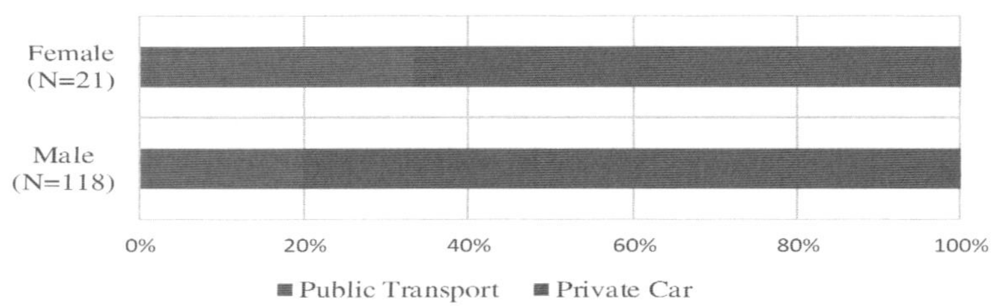

Figure 4. Mode choice based on gender.

Figure 4 shows a diagram of the relationship between travel mode based on gender. We consider the figure to know how is preference of the travel mode of choice based on gender. It can be seen that male tend to use private car compared to female.

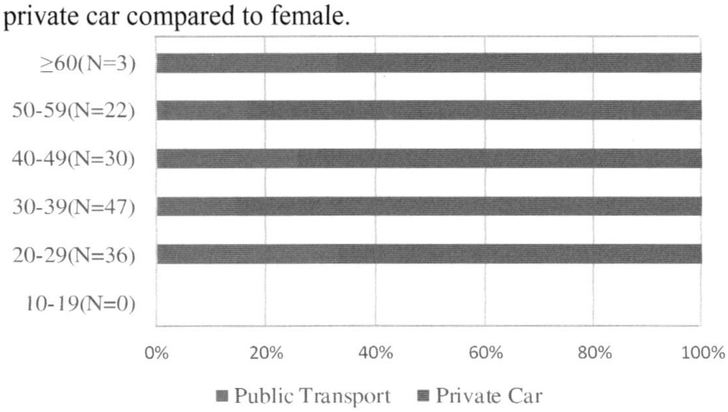

Figure 5. Mode choice based on age category.

We had 139 data and distributed in six age categories. Our data is kinds of bias in terms of age category. **Figure 5** indicates the descriptive environment between age categories and mode choice. It shows that there are not people coverage less than 20 age and there are three people who are in ≥ 60 age category. It means that transport users are mainly in the 20-29 age category to 30-39 age category. It might occur since the data which already collected in Matsuyama city office mostly use private car. Whereas, private car user must to have car license. The results also indicate that travelers mostly use private car than public transport in this data.

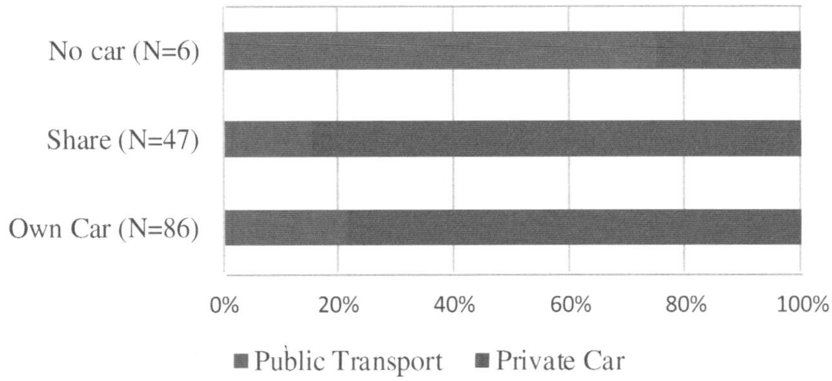

Figure 6. Travel mode share correspondent to car ownership

The characteristic of transport user on travel mode choice is presented in **Figure 6**. This figure is considered to know the effect of car ownership on use public transport or private car usage. It shows that the transport user who has a car tend to use automobile and share mobile, compared to the public transport. It might be occurred due to the private car user is over estimate to the public transport. The possibility, they do not know the actual price of a ticket the public transport.

The correlations of several variables in terms of internal reference point are shown in **Figure 7**, **Figure 8**, **Figure 9**, and **Figure 10**. Those figures are obtained from the data analysis. It is also considered to know how the transport user feels about the variables of internal reference point. The following are several variables of reference point;

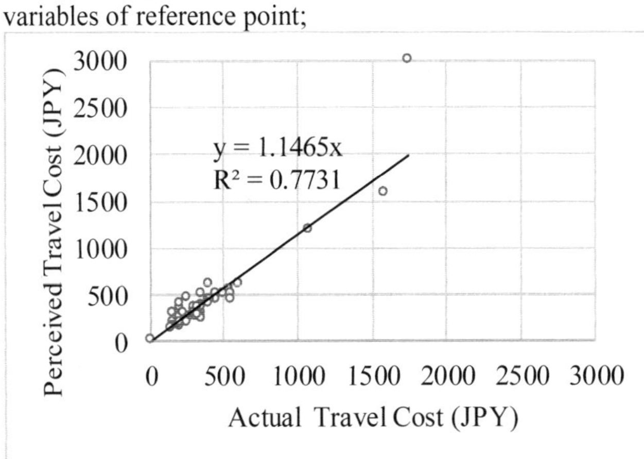

Figure 7. Actual travel cost-perceived travel cost.

Figure 7 shows the correlation between actual travel cost and perceived travel cost. It is examined in order to know the perception of public transport user to the actual travel cost toward perceived travel cost, how they feel of the cost, whether the transport user is an overestimate or underestimate. The result shows that the perceived travel cost is greater than the actual cost. Therefore, it can be said that the transport user overestimates toward perceived travel cost.

The correlation between perceived travel cost and reservation travel cost is depicted in **Figure 8**. It investigated how the public transport user feels in regard to the perceived travel cost toward reservation travel cost. The result indicates that perceived travel cost is greater than reservation travel cost. It means that the price in which a public transport user can decide to buy a ticket more than the highest price, which has a public transport.

In **Figure 9**, it shows the graph of the correlation between actual and perceived travel time. This figure is considered to know how the public transport user perception is. The figure indicates that actual travel time of public transport is quite similar to the perceive travel time. It means that the transport users feel that there is no different between actual and perceived travel time, so that it can be said that the traveler relish the condition.

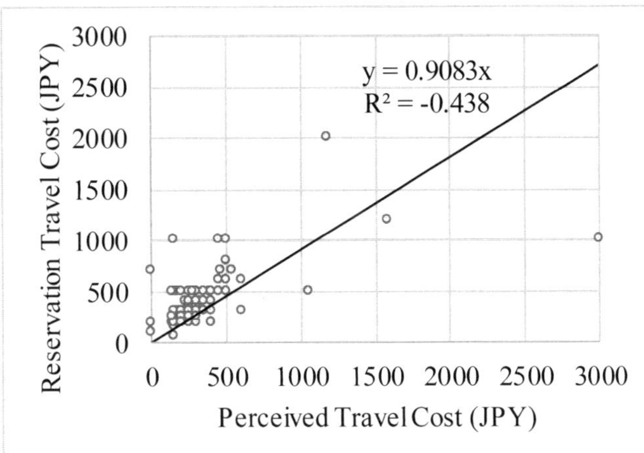

Figure 8. Perceived travel cost-reservation travel cost.

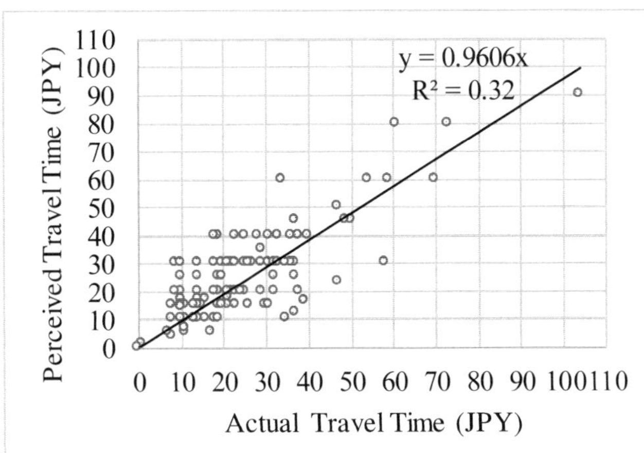

Figure 9. Actual travel time-perceived travel time.

The correlation between perceived travel time and reservation travel time as described in **Figure 10**. This analysis is performed in order to know how the public transport user feel is. The result indicates that perceived travel time is greater than reservation travel time. It can be said that the public transport users assume when they use public transport, the reservation travel time less than perceived travel time.

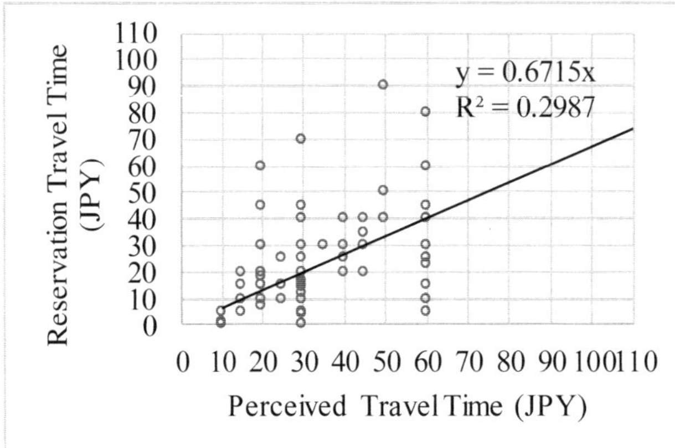

Figure 10. Perceived travel time-reservation travel time.

The description of the relationship between variables of travel time and travel cost is presented in **Table 2** and **Table 3.**

Table 2. The relationship between variables of travel cost.

Relationship between Variables	Number of samples	Share (%)
$TC_{pt}^{fair} < TC_{pt}^{percep} < TC_{pt}^{reserv}$	27	24.1
$TC_{pt}^{reserv} > TC_{pt}^{percep} = TC_{pt}^{fair}$	24	21.4
$TC_{pt}^{fair} < TC_{pt}^{reserv} < TC_{pt}^{percep}$	18	16.1
$TC_{pt}^{fair} < TC_{pt}^{percep} = TC_{pt}^{reserv}$	12	10.7
$TC_{pt}^{percep} < TC_{pt}^{fair} < TC_{pt}^{reserv}$	7	6.2
$TC_{pt}^{reserv} < TC_{pt}^{fair} < TC_{pt}^{percep}$	7	6.2
$TC_{pt}^{percep} < TC_{pt}^{reserv} < TC_{pt}^{fair}$	5	4.5
$TC_{pt}^{percep} > TC_{pt}^{fair} = TC_{pt}^{reserv}$	4	3.6
$TC_{pt}^{reserv} < TC_{pt}^{percep} < TC_{pt}^{fair}$	4	3.6
$TC_{pt}^{percep} = TC_{pt}^{fair} = TC_{pt}^{reserv}$	3	2.7
$TC_{pt}^{percep} < TC_{pt}^{fair} = TC_{pt}^{reserv}$	1	0.9
Total	112	100

Table 2 shows the relationship between variables of travel cost. In the first row, it shows that the public transport fair is relatively expensive. On the other hand, some possibilities for using public transport might be occurred due to perceived travel cost for public transport is lower than reservation cost. The second case, fair cost of public transport is equal to the perceive cost, but lower than reservation cost. In this case, people also can use public transport. In the third, the fair travel cost is less than reservation cost, while perceived travel cost is greater than fair travel cost and reservation travel cost. It means that the perceived of public transport user is the highest. So, it is difficult to use public transport. At fourth case, the reservation travel cost of public transport is equal to the perceive travel cost but greater than fair cost. In the fifth row, fair travel cost exists between perceived travel cost and reservation travel cost but perceived travel cost less than reservation travel cost. It means that public transport user is an underestimate. The similar situation to the fifth case, in the sixth situation, but perceived travel cost is greater than reservation travel cost, so it can be said the public transport user is over estimate. The seventh case indicates that fair travel cost is greater than two conceptual internal reference point namely reservation travel cost and perceived travel cost. It can be said that the public transport users are dominant or still consistent to use it. Moreover, at the eighth case describes the relationship which reservation travel cost is equal to fair travel cost, but less than perceive travel

IOP Conf. Series: Earth and Environmental Science **71** (2017) 012020 doi:10.1088/1755-1315/71/1/012020

cost. It means that it is difficult to use the public transport. In the ninth case, fair travel cost is greater than two conceptual reference point. It means that people accurately use public transport. At the tenth, perceive travel cost and fair travel cost as well as reservation travel cost is equal. It can be said public transport users assume same for all attributes. The last case, perceived travel cost less than fair travel cost but same with reservation travel cost. It means that people have the possibility to use public transport

Table 3. The relationship between variables of travel time.

Relationship between Variables	Number of samples	Share (%)
$^{fair}_{pt} < {}^{percep}_{pt} < {}^{reserv}_{pt}$	41	36.6
$^{reserv}_{pt} > {}^{percep}_{pt} = TC^{fair}_{pt}$	31	27.7
$^{fair}_{pt} < {}^{percep}_{pt} = {}^{reserv}_{pt}$	17	15.2
$^{percep}_{pt} < {}^{fair}_{pt} < {}^{reserv}_{pt}$	9	8.0
$^{fair}_{pt} < {}^{reserv}_{pt} < {}^{percep}_{pt}$	7	6.2
$^{percep}_{pt} > {}^{fair}_{pt} = {}^{reserv}_{pt}$	5	4.5
$^{percep}_{pt} = {}^{fair}_{pt} = {}^{reserv}_{pt}$	1	0.9
$^{percep}_{pt} < {}^{reserv}_{pt} < {}^{fair}_{pt}$	1	0.9
Total	112	100

Table 3 shows the relationship between variables of travel time for public transport. In the first row, it describes that the time of public transport fair is relatively long. In this situation, public transport users feel the perceived travel time is longer than fair travel time. On the other hand, they have some possibilities to use public transport because perceived travel time for public transport is shorter than reservation travel cost. The second case, the fair travel time of public transport is equal to the perceived travel time but lower than reservation time. In this case, people also can use public transport. At third case, the reservation travel time of public transport is equal to the perceived travel time but longer than fair time. It means the public transport user is over estimate to using public transport. In the fourth row, fair travel time existed between perceived travel time and reservation travel time but perceived travel time less than reservation travel time. It means that public transport user is an underestimate about travel time. In the fifth situation describes the fair travel time is less than reservation time, while perceived travel time is longer than them. It means public transport user consider to use public transport. The sixth case describes the relationship which reservation travel time is equal to the fair travel time but less than perceived travel time. It might be said that it is difficult to use the public transport user. At the seventh, perceived travel time and fair travel time as well as reservation travel time is equal. It can be said public transport user feels that they can choose any transport modes they need. The eighth row shows fair travel time is longer than two conceptual internal reference point namely reservation travel time and perceive travel time. It is also can be said the public transport users is dominant or still consistent to use it.

4. Developed model.

The data were analyzed by using the binomial logit model. It is developed base on mental accounting theory and prospect theory. The following functional form for the value function fits the prospect theory in Equation **(3)**;

$$f(x)\begin{cases} x^{\alpha} & if\ x \geq 0 \\ \lambda(x)^{\gamma} & if\ x < 0 \end{cases} \tag{3}$$

The binary logit model can be seen in Equation **(4)** [19].

$$P_n\,(i) = \frac{e^{Vin}}{e^{V_{in}} + e^{V_{jn}}} \tag{4}$$

$$U_c = \beta_0^c + \beta_1^c X_1^c + \beta_2^c X_2^c + \cdots + \beta_k^c X_k^c \quad (5)$$
$$U_{pt} = \beta_0^{pt} + \beta_1^{pt} X_1^{pt} + \beta_2^{pt} X_2^{pt} + \cdots + \beta_k^{pt} X_k^{pt} \quad (6)$$
$$\text{Car (c)} = \exp(\, b[1] + b[3]^* x[.,] + b[4]^* x[.,] + b[5]^* x[.,] + b[7]^* x[.,] \quad (7)$$
$$\text{Public Transport (pt)} = \exp(\, b[2] + b[3]^* x[.,] + b[4]^* x[.,] + b[5]^* x[.,] + b[6]^* x[.,] \quad (8)$$

The parameters are defined for each utility which used to the model such as b [1] = constant utility of car, b [2] = constant utility of PT, b [3] = coefficient parameter of cost, b [4] = coefficient parameter of time, b [5] = coefficient parameter of age, b [6] = coefficient parameter of gender, b [7] = coefficient parameter of license and U_c is utility of car as well as U_{pt} is utility of public transport.
The mental accounting theory is showed in Equation **(9)**;

$$U_c = AU_c(p, \bar{p}) + \beta TU_c(p, p^*) \quad (9)$$
$$AU_c = \beta_2^c (TC_c - WTP_c) \quad (10)$$
$$TU_c = \beta_3^c (TC_c - RC_c) \quad (11)$$
$$U_c = \beta_0^c + \beta_1^c X_1^c + \beta_2^c (TC_c - WTP_c) + \beta_3^c (TC_c - RC_c) + \beta_5^c TT, \quad (12)$$
$$U_{pt} = \beta_0^{pt} + \beta_1^{pt} X_1^{pt} + \beta_2^{pt} (TC_{pt} - WTP_{pt}) + \beta_4^{pt} (TC_{pt} - RC_{pt}) + \beta_5^{pt} TT \quad (13)$$

The following is some cases of the probabilistic choice framework.
Case 1 (linear-in-parameter)

$$U_c = \beta_0^c + \beta_1^c X_1^c + \beta_2^c X_2^c + \beta_3^c X_3^c + \beta_4^c X_4^c \ldots + \beta_k^c X_k^c \quad (14)$$
$$U_{pt} = \beta_0^{pt} + \beta_1^{pt} X_1^{pt} + \beta_2^{pt} X_2^{pt} + \beta_3^{pt} X_3^{pt} + \beta_4^{pt} X_4^{pt} \ldots + \beta_k^{pt} X_k^{pt} \quad (15)$$

In this case, the utility function for all attributes of travel behavior is a linear parameter. The travel cost, travel time and the other attributes work in each variable. It is called conventional model.
Case 2 (non-linear parameter)
The utility function of linear-in-parameter to be a utility function nonlinear-in-parameter is developed by incorporating between prospect theory and binomial logit model using power function. The utility function can be expressed as follows;

$$U_{pt} = \alpha^{pt} + \beta_t^{pt} (TT_{per} - TT_{rp})^{\exp(\eta t)} + \beta_c^{pt} (TC_{per} - TC_{rp})^{\exp(\eta c)} + \gamma_t + \varepsilon, \text{ if } TT_{per} \geq TT_{rp} \text{ and } TC_{per} \geq TC_{rp} \quad (16)$$
$$U_{pt} = \alpha^{pt} + \beta_t^{pt} (TT_{rp} - TT_{per})^{\exp(\eta t)} + \beta_c^{pt} (TC_{rp} - TC_{perc})^{\exp(\eta c)} + \gamma_t + \varepsilon, \text{ if } TT_{rp} \geq TT_{per} \text{ and } TC_{rp} \geq TC_{perc} \quad (17)$$

Model A (Travel time for public transport, there are four models).
The utility function of the public transport U_{pt} of the loss frame which the fair travel time and reservation travel time as a reference point are;

$$U_{pt} = \alpha^{pt} + \beta_t^{ptloss} (TT_{per} - TT_{fair})^{\exp(\eta t)loss} + \beta_c^{pt} (TC_{per} - TC_{fair})^{\exp(\eta c)loss} + \gamma_t + \varepsilon, \text{ if } TT_{per} \geq TT_{fair} \text{ and } TC_{per} \geq TC_{fair} \quad (18)$$
$$U_{pt} = \alpha^{pt} + \beta_t^{ptloss} (TT_{per} - TT_{reserv})^{\exp(\eta t)loss} + \beta_c^{pt} (TC_{per} - TC_{reserv})^{\exp(\eta c)loss} + \gamma_t + \varepsilon, \text{ if } TT_{per} \geq TT_{reserv} \text{ and } TC_{per} \geq TC_{reserv} \quad (19)$$

The utility function of the public transport U_{pt} of the gain frame which the fair travel time and reservation travel time as a reference point are;

$$U_{pt} = \alpha^{pt} + \beta_t^{ptgain} (TT_{fair} - TT_{per})^{\exp(\eta t)gain} + \beta_c^{pt} (TC_{fair} - TC_{per})^{\exp(\eta c)gain} + \gamma_t + \varepsilon, \text{ if } TT_{fair} \geq TT_{per} \text{ and } TC_{fair} \geq TC_{per} \quad (20)$$
$$U_{pt} = \alpha^{pt} + \beta_t^{ptloss} (TT_{per} - TT_{reserv})^{\exp(\eta t)gain} + \beta_t^{pt} (TC_{per} - TC_{reserv})^{\exp(\eta c)gain} + \gamma_t + \varepsilon, \text{ if } TT_{per} \geq TT_{reserv} \text{ and } TC_{per} \geq TC_{reserv} \quad (21)$$

Model B (Travel cost for public transport, there are four models).

The utility function of the public transport U_{pt} of the loss frame which the fair travel cost and reservation travel cost as a reference point are;

$$U_{pt} = \alpha^{pt} + \beta_t^{ptloss} (TT_{per} - TT_{fair})^{\exp(\eta t) loss} + \beta_c^{pt} (TC_{per} - TC_{fair})^{\exp(\eta c) loss} + \gamma_t + \varepsilon, \text{ if } TT_{per} \geq TT_{fair} \text{ and } TC_{per} \geq TC_{fair} \quad (22)$$

$$U_{pt} = \alpha^{pt} + \beta_t^{ptloss} (TT_{per} - TT_{reserv})^{\exp(\eta t) loss} + \beta_c^{pt} (TC_{per} - TC_{reserv})^{\exp(\eta c) loss} + \gamma_t + \varepsilon, \text{ if } TT_{per} \geq TT_{reserv} \text{ and } TC_{per} \geq TC_{reserv} \quad (23)$$

The utility function of the public transport U_{pt} of gain frame which the fair travel time and reservation travel time as a reference point are;

$$U_{pt} = \alpha^{pt} + \beta_t^{ptgain} (TT_{fair} - TT_{per})^{\exp(\eta t) gain} + \beta_c^{pt} (TC_{fair} - TC_{per})^{\exp(\eta c) gain} + \gamma_t + \varepsilon, \text{ if } TT_{fair} \geq TT_{per} \text{ and } TC_{fair} \geq TC_{per} \quad (24)$$

$$U_{pt} = \alpha^{pt} + \beta_t^{ptloss} (TT_{per} - TT_{reserv})^{\exp(\eta t) gain} + \beta_t^{pt} (TC_{per} - TC_{reserv})^{\exp(\eta c) gain} + \gamma_t + \varepsilon, \text{ if } TT_{per} \geq TT_{reserv} \text{ and } TC_{per} \geq TC_{reserv} \quad (25)$$

In the existing application of utility function, it is recognized that it has several attributes such as travel time, travel cost and conventional model. The travelers are affected by travel time or travel cost to choose the travel mode. This research is grounded by tracing the private car behavior and public transport behavior deeply, where TC_c is travel cost of the car, RC_c is reference cost of the car, TT_c is travel time of the car, RTT_c is reference travel time of the car, ε_c is Gumble distribution of the car. Also, α, β, η and γ are variables of the value function in the gain and loss area. From the eight models, it will be obtained which is the better model or goodness-of-fit model. In the conventional model, the utility of a risky prospect is linear in outcome probabilities. By developing the model, the mental accounting theory can be employed into the transport field (travel behavior), which both gain frame and loss frame is separately created, where the reference point serves as asset position. The following is a figure of the reference point (non-linear parameter)

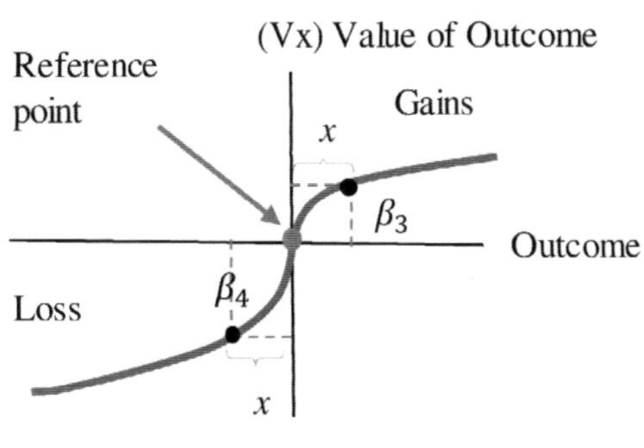

Figure 11. A Reference point (non-linear parameter).

5. Result and discussion

This study developed a travel behavior model base on the mental accounting theory. In conventional model usually employ the linear-in-parameter. In the mental accounting theory, researchers use the non-linear parameter. Travel time for public transport is presented in **Table 4**. By using Gaussian Software, we obtain the **Table 4**. The $\theta_1, \theta_2, \theta_3, \theta_4$ are the satisfaction threshold of public transport. β AU loss is variable of acquisition utility in the loss area, η AU loss is transaction utility in the gain area, AIC is an Akaike's Information Criteria and ρ^{-2} is an informal goodness-of-fit measure.

For a binary choice model, the value of goodness-of-fit (ρ^{-2}) must lie between 0 and 1 as well as the closer to one is a better. As we can see in **Table 4** that the better of goodness-of-fit is fair travel time because of its value closer to 1.

IOP Conf. Series: Earth and Environmental Science **71** (2017) 012020 doi:10.1088/1755-1315/71/1/012020

Table 4. Travel time for public transport.

Variables	Objective TT		Perceived TT		Reference Point			
					Fair TT		Reservation TT	
	Estimates	t-statistic	Estimates	t-statistic	Estimates	t-statistic	Estimates	t-statistic
θ_1	-2.20	-7.70	-2.29	-7.86	-2.26	-8.36	-0.441	-1.79
θ_2	-1.40	-5.45	-1.48	-5.67	-1.33	-5.71	0.352	1.45
θ_3	-0.76	-3.14	-0.82	-3.39	-0.59	-2.85	0.96	3.77
θ_4	0.59	2.32	0.55	2.14	1.24	4.66	2.36	7.59
β AU loss	-2.47	-4.97	-2.45	-5.29	-2.94	-4.96	-0.17	-0.22
η AU loss					-0.92	-3.19	-0.75	-0.13
β TU gain					1.09	2.43	1.93	3.58
η TU gain					-10.1	-0.08	-0.65	-1.47
Number of observations	110.00		115.00		115.00		115.00	
AIC	310.04		314.36		298.67		338.74	
ρ^{-2}	0.12		0.12		0.19		0.08	

By performing similar analysis at **Table 4**, we gain goodness-of-fit ρ^{-2} in **Table 5**. It is shown that the goodness-of-fit of travel cost for public transport is fair travel cost, because the highest value of ρ^{-2} is in fair travel cost, the value of ρ^{-2} is 0.21.

Travel time for the private car is shown in **Table 6,** it has three models. The First model is analyzed for average travel – minimum travel time, model 2 is analyzed for maximum-minimum travel time, and model 3 is analyzed for maximum-minimum travel time.

Table 5. Travel cost for public transport.

Variables	Objective TT		Perceived TC		Reference Point			
					Fair TC		Reservation TC	
	Estimates	t-statistic	Estimates	t-statistic	Estimates	t-statistic	Estimates	t-statistic
θ_1	-1.73	-7.36	-1.83	-7.46	-2.43	-8.27	-0.33	-1.48
θ_2	-0.80	-4.04	-0.87	-4.26	-1.32	-5.40	0.66	2.89
θ_3	-0.22	-1.15	-0.27	-1.41	-0.60	-2.67	1.30	5.32
θ_4	1.47	5.26	1.43	5.10	1.46	4.54	3.08	8.75
β AU loss	-2.00	-3.77	-2.20	-4.01	-3.51	-4.97	0.00	0.00
η AU Loss					-0.93	-4.02	2.54	0.04
β TU gain					0.97	1.49	1.75	3.84
η TU gain					-10.76	-0.03	-1.64	-2.63
Number of observations	110.00		110.00		112.00		112.00	
AIC	307.75		303.84		283.53		305.80	
ρ^{-2}	0.11		0.12		0.21		0.08	

There are three attributes in **Table 6**; average travel time, maximum travel time and minimum travel time. Then these attributes are formulated each other, thus it obtained three models. β_5 and β_6 are the variables of travel time for private car. By using Gaussian Software, it can be obtained the value of goodness-of-fit of the model. As can be seen that the better model of goodness-of-fit is model 1 since the value is 0.24, which is closer to one.

Table 6. Travel time for private car

Variables	Reference Point					
	Model 1		Model 2		Model 3	
	AV-Min		Max-AV		Max-Min	
	Estimates	t-statistic	Estimates	t-statistic	Estimates	t-statistic
θ_1	-2.79	-5.82	-2.07	-2.98	-36.16	-0.23
θ_2	-1.79	-4.05	-1.15	-1.73	-35.21	-0.22
θ_3	-1.00	-2.30	-0.40	-0.61	-34.44	-0.22
θ_4	1.00	2.38	1.45	2.13	-32.56	-0.21
5	-6.05	-1.36	-0.50	-0.62	-34.96	-0.22
6	-0.20	-0.33	-1.03	-0.30	-4.30	-0.94
N observations	113.00		113.00		113.00	
AIC	276.16		290.17		285.82	
ρ^{-2}	0.24		0.20		0.21	

From the data analysis, it found a reference point for each model in travel time for public transport model, travel cost for public transport model, and travel time for the private car model. Which then the models were integrated. The integrated model is increasingly recognized as a tool to support transport policy. This study also identify which variable have the most influence on these models should be integrated. Furthermore, it is important to develop the multi choice mode, which kind will be predicted. Integrated model constitutes three combined attribute models; travel time for public transport, travel cost for public transport, and travel time for the private car. The Integrated models were presented in **Table 7**.

Table 7. Integrated Model

No	Variables	Estimates	t-statistic
1	θ_1	-2.339	-8.105
2	θ_2	-1.3705	-5.556
3	θ_3	-0.5119	-2.404
4	θ_4	1.2195	4.54
5	AU loss	-5.2442	-3.406
6	η AU loss	-0.3595	-1.27
7	TU gain	1.0703	2.375
8	η TU gain	-10.0936	-0.075
9	θ_1	-2.5374	-7.891
10	θ_2	-1.3196	-5.133
11	θ_3	-0.4916	-2.146
12	θ_4	1.4964	4.549
13	AU loss	-5.6977	-4.63
14	η AU Loss	-0.5352	-2.718
15	TU gain	1.2326	1.982
16	η TU gain	-10.76	-0.041
17	θ_1	-2.7178	-5.383

18	θ_2	-1.7617	-3.735
19	θ_3	-0.921	-1.994
20	θ_4	1.0103	2.262
21	5	-7.2779	-1.155
22	6	-0.064	-0.101
23	constants	-0.0235	-0.05
24	travel cost pt	0.363	1.974
25	travel time	0.0863	0.561
26	gender	0.2532	0.598
27	age30	0.5982	1.815
28	age50	0.3983	0.897
29	own pc	0.0009	0.003
30	frequency	0.165	0.367

Table 7 shows the result of the integrated models. The variables were analyzed among each other. The value of variables obtained by analyzing the data trial and error for all variables until found best value. As can be seen in **Table 7** that the variables number 1 to 4 are satisfaction threshold of travel time for public transport, variables number 5 to 8 are acquisition utility and transaction utility in gains and loss area. Then, variables number 9 to 12 are satisfaction threshold of travel cost for public transport, variables number 13 to 16 are acquisition utility and transaction utility in gains and loss respectively. Hereafter, variables number 17 to 20 are satisfaction threshold of travel time for private car and variables number 21 to 22 are variable of travel time for private car. Variables number 23 is constant of the model. Finally, the variables number 24 to 30 are dummy variables of the integrated model. In this model travel time for public transport is not too significant but travel cost for public transport significantly affect the model because its value is greater than the level of significant (1.974).

6. Conclusion

This study investigated the decision-making process of travel mode choice based on the prospect theory and mental accounting theory. The conventional model of the utility function is affected by travel time, travel cost, and convenience as well as other attributes. Three attributes are utilized as a reference point such as reservation travel time, perceive travel time and the fair travel cost. According to the AIC (Akaike's Information Criteria) index and ρ^{-2} (Mac Fadden pseudo) which found the value of AIC and ρ^{-2} is 298,69 and 0.19 for fair travel time for public transport, the value of AIC and ρ^{-2} is 283.53 and 0.21 for fair travel cost for public transport, and the value of AIC and ρ^{-2} is 276.16 and 0.24 for average-minimum travel time for private car. By developing a model using mental accounting theory and prospect theory, the results showed that the fair travel time and fair travel cost are utilized as reference point for public transport user seem to be diminishing according to the gain and loss. The results also showed that the goodness-of-fit of the developed model is higher than that of the conventional model which means that the internal reference point might play the major roles in the choice of travel mode. Several limitations of this study provide research for the future study. First, given the importance of the reference point, the development of mental accounting theory in the transport field is recommended. Second, how to identify what kind of reference point is utilized for the evaluation of travel cost such as perceived transit fare and cost of the other travel mode. Third, how the effect of the reference point after implementing the transport policy whether it is stable, decrease or increase?.

7. Acknowledgements

This research was financially supported by Scholarship of Ministry of Research, Technology and Higher Education of Indonesia and is conducted under collaboration with Ehime University.

IOP Conf. Series: Earth and Environmental Science **71** (2017) 012020 doi:10.1088/1755-1315/71/1/012020

8. Reference

[1] Oded lowengart 2002 Reference Price Conceptualisations : An Integrative Framework of Analysis *J. Mark. Manag.* **18** 145–71

[2] Kahneman D and Tversky A 1979 Prospect theory: An analysis of decision under risk *Econom. J. Econom. Soc.* **47** 263–91

[3] Thaler R 1985 Mental Accounting and Consumer Choice *Mark. Sci.* **4** 199–214

[4] Fiegenbaum A V I and Schendel D A N 1996 Strategic reference point theory **17** 219–35

[5] Hensher D 1994 Stated preference analysis of travel choices : the state of practice *Transportation (Amst).* **21** 107–33

[6] Avineri E and Bovy P H L 2008 Identification of Parameters for a Prospect Theory Model for Travel Choice Analysis *Transp. Res. Rec.* **2082** 141–7

[7] Avineri E 2011 Applying behavioral economic in the design of travel information system *UTSG*

[8] Avineri E 2004 A Cumulative Prospect Theory Approach to Passengers Approach to Passengers Behavior Modeling : Waiting Time Paradox *Intell. Transp. Syst.* 37–41

[9] Kahneman D and Tversky A 2000 *Choices, values, and frames*

[10] Neumann J V and Morgenstern O 1944 *Theory of and Economic* (Princeton: Princeton University Press)

[11] Senbil M and Kitamura R 2004 Reference Points in Commuter Departure Time Choice : A Prospect Theoretic Test of Alternative Decision Fames *J. Intell. Transp. Syst.* **8** 19–31

[12] Poucke D Van and Buelens M 2002 Predicting the outcome of a two-party price negotiation : Contribution of reservation price , aspiration price and opening offer *J. Econ. Psychol.* **23** 67–76

[13] Briesch R, Krishnamurthi L, Mazumdar T and Raj S P 1997 A Comparative Analysis of Reference Price Price Models *J. Consum. Res.*

[14] Kumar V, Karande Kiran and Reinartz W J 1998 The Impact of Internal and External Reference Prices on Brand Choice : The Moderating Role of Contextual Variables *J. Retail.* **74** 401–26

[15] Mazumdar T and Papatla P 1995 Loyalty Differences in the Use of Internal and External Reference Prices *Mark. Lett.* **6** 111–22

[16] Albert J, Bitta D, Monroe K B and Ginnes J M 1981 Consumer Perceptions of Comparative Price Advertisements Consumer Perceptions of Comparative Price Advertisements

[17] Tversky, Amos, and Daniel Kahneman. 1986 Rational choice and the framing of decisions.*Journal of business* S251-S278

[18] Urbany J E, Bearden W O and Borrero M S 1997 Transaction Utility Effects When Quality Is Uncertain *J. Acad. Mark. Sci.* **25** 45–55

[19] Ben-Akiva, M; Lerman S R 1985 *Discrete Choice Analysis: Theory and Application to Travel Demand.* (Amerika: MIT Press)

IOP Conf. Series: Earth and Environmental Science 71 (2017) 012021 doi:10.1088/1755-1315/71/1/012021

Identification of Altered Minerals Based on Synthetic Aperture Radar (SAR) For Mineral Exploration in a Tropical Area

Panggea Ghiyats Sabrian[1,4], Asep Saepuloh[2], Syafrizal[3], Arie Naftali Hawu Hede[3]

[1] Graduate Program of Mining Engineering, Faculty of Mining and Petroleum Engineering, Institut Teknologi Bandung (ITB), Bandung, Indonesia 40132
[2] Research Group of Geology, Faculty of Earth Sciences and Technology, Institut Teknologi Bandung (ITB), Bandung, Indonesia 40132
[3] Research Group of Earth Resources Exploration, Faculty of Mining and Petroleum Engineering, Institut Teknologi Bandung (ITB), Bandung, Indonesia 40132
[4] Department of Civil Engineering, Faculty of Engineering, Universitas Borneo Tarakan (UBT), Tarakan, Indonesia 77123
E-mail: panggea@gmail.com

Abstract: Geological investigation by remote sensing using surface physical properties in tropical regions is challenging. To minimize the effects of atmosphere and vegetation in the obtained optical images, we used the synthetic aperture radar (SAR) images. The phased array L-band synthetic aperture radar (PALSAR) onboard the advanced land observing satellite (ALOS) was selected due to ability of the L-band to penetrate clouds and canopy vegetation. The polarimetric decomposition method based on Cloude–Pottier classification was used as the basis for backscattering analyses. The Ciseuti area in West Java, Indonesia was selected as the study site due to the existence of mining activities, including gold and galena mines. The identification is focused on the spatial distribution of prospected minerals regardless of cloud and vegetation canopy. The classified prospect zones could be extended using the Cloude–Pottier polarimetric decomposition into moderate random entropy and alpha double bounce scattering at argillic alterations, moderate random entropy and alpha surface scattering at intermediate argillic alterations, and highly moderate random entropy and alpha volume diffusion at advanced argillic alterations. The entropy and alpha extracted from ALOS PALSAR data based on Cloude–Pottier decomposition were very useful for identifying alteration zones.
Keywords: mineral exploration, synthetic aperture radar, alteration, polarimetric decomposition, entropy, alpha.

1. Introduction

Various remote sensing methods have been developed to obtain the most accurate geological information possible. The optical remote sensing is used frequently for mineral mapping because it can provides large amounts of spectral information in the visible, near-infrared, and shortwave infrared regions. This method effectively identifies minerals based on their reflectance spectra. However, optical images are affected by atmospheric conditions and the presence of vegetation can interfere with the reflectance spectra used to analyze the target materials (soils, rocks, or minerals). Synthetic aperture radar (SAR) is advanced to detect geological targets on the ground based on physical properties of the soil surface in the tropics because of its ability to minimize the atmospheric and vegetation effects [1].

Full polarimetric SAR data can be used to discriminate the target from the earth's surface by measuring the type of scattering generated by the surface [2]. While decomposition techniques have been widely used in the environmental field to classify polarimetric radar data of target surface scattering, its application to geosciences is still rare [3]. Saepuloh et al. (2012) [4] used the polarimetric decomposition techniques of Cloude and Poittier to extract the geomorphology and structure of active volcanoes by using surface roughness at the volcanoes. From the surface roughness they were able to distinguish between surface alterations, hot mud, and hot springs. The level of acidity (pH) in the surface will cause its roughness [5].

Hydrothermal alteration is a complex process that includes changes in mineralogy, chemistry, and textures of rocks. Alterations are generated from the interaction of hydrothermal solutions with rocks in its path under certain chemical-physical condition [6]. Studying alterations is one way to identify a mineral deposit in an area. Using polarimetric Cloude–Poittier decomposition technique and surface roughness enables the classification of alteration zones in the study area.

2. Study Area

The research area focused on the mineral prospects in Ciseuti, Purwakarta. At the coordinates 756345-765375 Em and 9275275-9266245 Nm UTM Zone 48 S (Figure 1). The island of Java was affected by the subduction zone between the Indo-Australian plate and the Eurasia plate, called the Sunda-Banda Magmatic Arc [7]. The presence of volcanic activity in the subduction zone resulted in the mineralization of precious metals and base metals. Geologically, Ciseuti is located in the Bogor zone, where the morphology is generally hilly with rocks predominantly volcanic rocks and sedimentary rocks [8]. The oldest rock formation in this area is the Rajamandala formation. A sedimentary rock formation, Jatiluhur, is also well exposed in the study area. Younger igneous rocks such as andesite hornblende and porphyry diorite hornblende intrude into this formation [9].

3. Data

Data used in this study was collected using the phased array type L-band synthetic aperture radar (PALSAR) onboard the advanced land observing satellite (ALOS) in ascending orbit mode with full polarization (HH, HV, VV, VH) level 1.0. Fieldwork was carried out to collect alteration from ten different locations around the study area.

4. Methodology

Decomposition methods have been introduced in the analysis of two or full polarimetric radar images to interpret the physical characteristics of an object through reflection. One of the latest techniques in decomposition of polarization data is to use eigenvalue analysis to obtain the physical attributes. One of the decomposition algorithms that widely used is Cloude–Pottier decomposition [2] for polarimetric data (covariance matrix or coherence shape). Cloude–Pottier decomposition can be used to convert the highly complex full polarimetric data into three simpler units of analysis, namely entropy, alpha angle, and anisotropy.

$$H = \sum_{i=1}^{3} -P_i \log n\ P_i; \quad P_i = \frac{\lambda_i}{\sum \lambda_i} \tag{1}$$

Entropy (H) has a range of values between 0 and 1; it shows the degree of randomness object scattering on an object in the Earth's surface and can be referred to as the relative intensity of the scattering i-process. $H = 0$ indicates a single object, a unique, or an object with a low surface roughness. $H = 1$ indicates a random process of backscattering with no dominant object or a surface with high roughness in the coverage area. Thus, when an object on the Earth's surface is single, unique, or has a low surface roughness, the entropy values will tend to be close to zero.

$$\bar{\alpha} = \sum_{j=1}^{3} \alpha_j P_J \tag{2}$$

Another important parameter is the angle alpha, which shows the type of backscattering (of any type of object) that is the most dominant in the observed pixels. The value varies between 0° and 90°. Zero degrees indicates odd-bounce scattering from a flat surface. Even-bounce scattering can be observed around α = 90°, while α = 45° indicates dipole scattering.

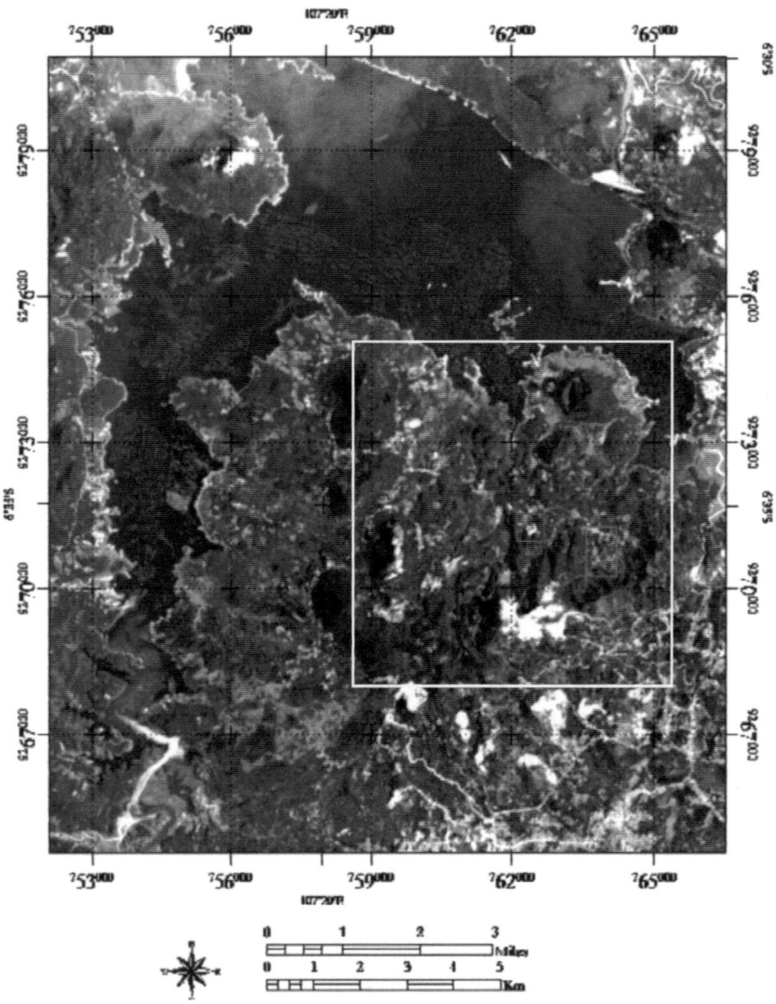

Figure 1. The yellow box denotes the study area, the image was captured from the Landsat 8 (band 4, 5, and 6 in RGB composition).

5. Result and discussion

Before the decomposition classification the data must pass through several processes including focusing and multilooking, followed by geocoding and terrain correction. The process of geocoding and terrain correction requires digital elevation model (DEM) data, in this case we used shuttle radar topographic mission which is available online The geocoding process, the projection map is entered according to the location of the research area, Ciseuti areas is located in UTM 48S zone in the WGS84 datum.

IOP Conf. Series: Earth and Environmental Science 71 (2017) 012021 doi:10.1088/1755-1315/71/1/012021

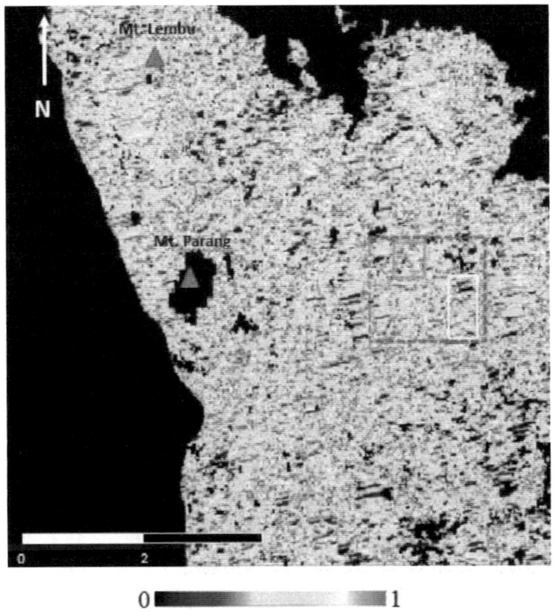

Figure 2. Entropy map; Ciseuti area, gold mining site, and galena mining site presented by purple, yellow, and blue rectangles, respectively.

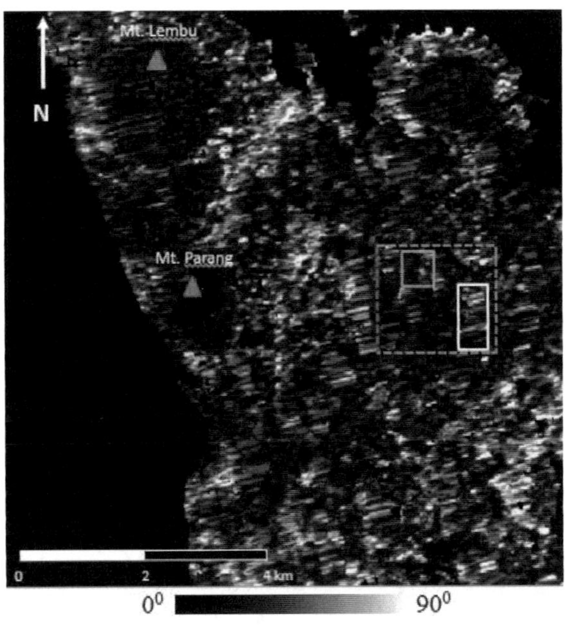

Figure 3. Alpha map; Ciseuti area, gold mining site, and galena mining site presented by purple, yellow, and blue rectangles, respectively

The next process is *H*/alpha decomposition, which produces the entropy (*H*) and alpha values. After obtaining these two of parameters, the next process is polarimetric classification based on *H* / alpha. In the results shown in Figure 2, the minimum alpha value of 24 is shown with a dark blue and the maximum value of 87 is shown in light blue, while the black color represents areas for which there is no data. The minimum entropy value 0.06 is indicated by blue and the maximum value of 0.96 is shown in red; purple represents areas for which there is no data (Figure 3).

Table 1. Values of entropy and alpha, and hydrothermal alteration zone at the observation point.

Code	Easting (Em)	Norting (Nm)	alpha	entropy	Hydrothermal Alteration Zone [10]
AL01	763856	9270696	66.5936	0.6217	"Adularia" — sericite, sericitic, argillic
AL02	763955	9270531	58.6789	0.7537	"Adularia" — sericite, sericitic, argillic
AL03	763855	9270456	51.8378	0.759	Argillic, intermediate argillic
AL04	763749	9270311	53.8481	0.7811	Argillic, intermediate argillic
AL05	762601	9271548	39.9066	0.792	Argillic, intermediate argillic
AL06	762535	9271247	50.3473	0.7735	Argillic, intermediate argillic
AL07	762531	9271296	44.7217	0.8624	Argillic, intermediate argillic
AL08	762475	9270860	41.4975	0.8146	Advanced argillic — acid sulphate
AL09	762392	9270689	41.1937	0.8085	Advanced argillic — acid sulphate
AL10	762484	9271120	51.251	0.8563	Advanced argillic — acid sulphate

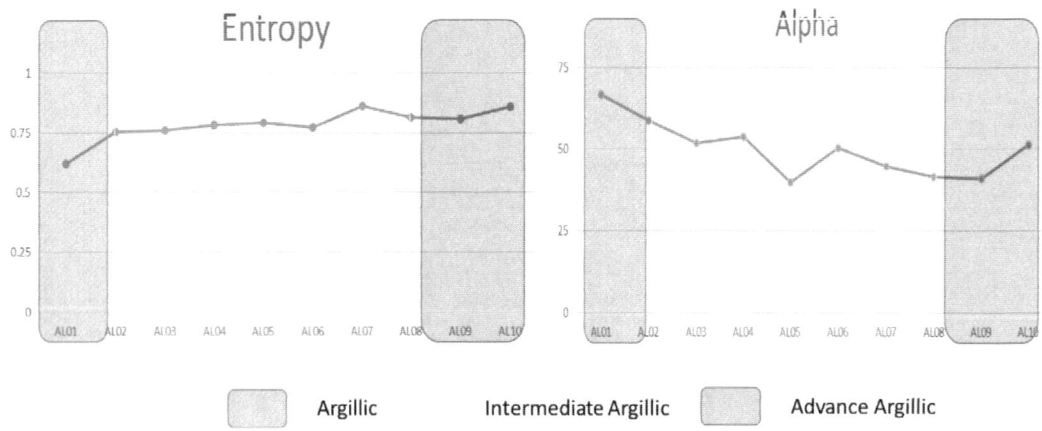

Figure 4. Graph entropy value (left) and alpha value (right) at the observation point.

When the entropy value obtained at a point of observation (show in Figure 5) is paired with alteration zones identified by Thompson (1996) [10], then argillic zones, intermediate argillic zones, and advanced argillic zones show different patterns (Figure 4). AL01 and AL02 are argillic zones. AL03, AL04, AL05, AL06 and AL07 are intermediate argillic zones. AL08, AL09, and AL10 are advance argillic zones.

The Cloude–Pottier polarimetry decomposition analysis generates the H (entropy) and α (alpha) values. The value of H, as previously noted, indicates the level of randomness of surface scattering of an object, where the surface roughness is a function of H. The surface of altered rock will be rougher than the surface of unaltered rocks. More acidic rock surface, with a pH value less than 7, will have a higher surface roughness than rocks with a neutral pH or pH close to 7 [5].

The argillic alteration zone is characterized by the presence of clay minerals such as illite, kaolinite, and montmorillonite. This alteration is caused by the process of cations are widely leached in acidic conditions, transforming plagioclase into kaolinite and into montmorillonite at the outer areas. The intermediate argillic zone shows presence of minerals such as montmorillonite, illite, chlorite, hydromicas, and locally kaolinite. The formation of these minerals is related to the availability of limited amounts of K, Ca, and Mg. Advanced argillic alteration refers to complete acid attack with the formation of the mineral kaolinite-dickite and a number of diaspore, alunite, amorphous silica, slightly corundum, and pyrophyllite minerals [6].

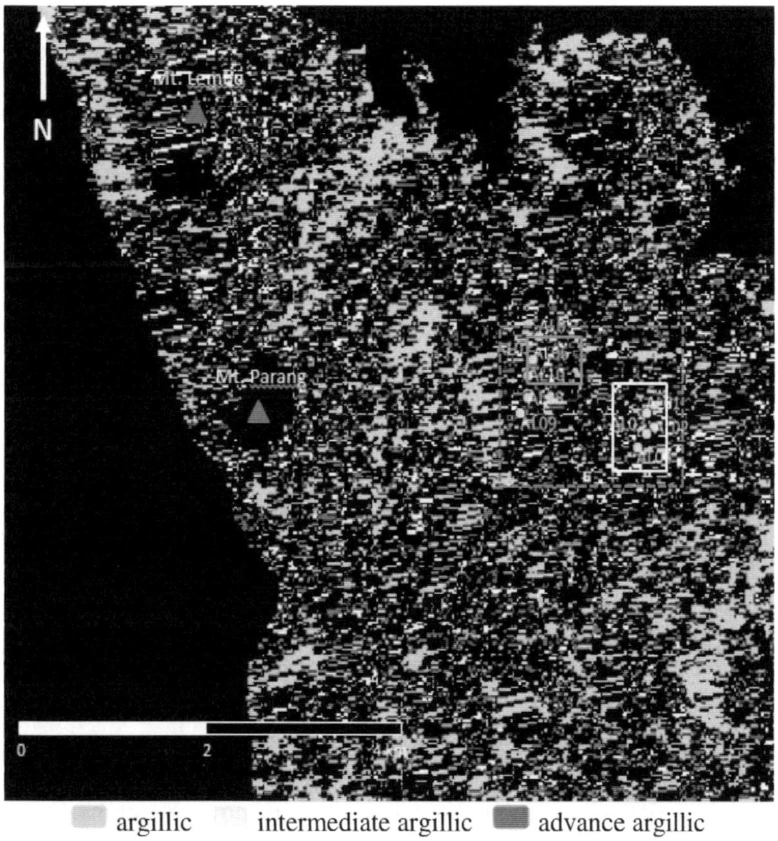

Figure 5. Map of the distribution of alteration zones; Ciseuti area, gold mining site, and galena mining site presented by purple, yellow, and blue rectangles, respectively; blue dots are observation point.

The entropy value of argillic zones AL01 0.6217 and AL02 0.7537 are moderate random entropy. Intermediate argillic zones AL03 0.759, AL04 0.7811, AL05 0.792, AL06 0.7735 and AL07 0.8624 are moderate random entropy. Advanced argillic zones AL08 0.8146, AL09 0.8085, and AL10 0.8563 are highly random entropy. This is because advanced argillic zones have the highest acidity (lowest pH), which causes the surface of altered rocks to be more rugged or rough. Argillic and intermediate argillic zones have lower acidity so that the surface are more smooth then advance argillic. Because higher entropy value causes more rough the surface.

Alpha value based on the number of reflections before the signal received by the sensor. The alpha value of argillic zones AL01 66.59 and AL02 58.68 are double-bounce scattering. Intermediate argillic zones AL03 51.84, AL04 53.85, AL05 39.91, AL06 50.35 and AL07 44.72 are volume diffusion. Advanced argillic zones AL08 41.50, AL09 41.19, and AL10 51.25 are volume diffusion. It's because on the surface argillic zones little backscatter specular reflection, whereas on intermediate argillic and advance argillic zones moderate backscatter. The backscatter of the signal influence the number of signal reflection before reach sensor.

From the map of the distribution of alteration zones shown in Figure 5, it appears that the area around Ciseuti area is dominated by advanced argillic alteration zones. This is in agreement with the results of research conducted by Dewangga (2014) [8] which also showed that the Ciseuti region is dominated by advanced argillic alteration.

6. Conclusion

Full polarization data obtained from ALOS PALSAR was deconstructed to the distribution of entropy and alpha values. With entropy and alpha values can describe alteration zones base on physical properties of the surface. The advance argillic with strong acid have rougher surface than argillic and intermediate argillic zone, cause rougher surface higher entropy value. Alpha value base on backscatter of the signal from the surface.

Areas of argillic alteration have moderate random entropy and alpha double-bounce scattering. Intermediate argillic zones have moderate random entropy and alpha volume scattering. Advanced argillic zones have highly moderate random entropy and alpha volume diffusion. Ciseuti area is dominated by advanced argillic alteration zones.

7. References

[1] Saepuloh A, Koike K, Omura M, Iguchi M, and Setiawan A 2010 *SAR- and gravity change-based characterization of the distribution pattern of pyroclastic flow deposits at Mt. Merapi during the past ten years* (Bulletin of Volcanology) 72(2) pp 221-232.

[2] Cloude S R and Pottier E 1996 *A review of target decomposition theorems in radar polarimetry* (IEEE Trans. Geosci. Remote Sens) vol. 34 no. 4, pp. 498–518.

[3] Silva A Q, Paradella W R, Freitas C C, and Oliveira CG 2013 *Evaluation of Digital Classification of Polarimetric SAR Data for Iron-Mineralized Laterites Mapping in the Amazon Region* (Remote Sens) vol 5, pp 3101-3122.

[4] Saepuloh A, Koike K, and Omura M 2012 *Applying Bayesian decision classification to Pi-SAR polarimetric data for detailed extraction of the geomorphologic and structural features of an active volcano* (IEEE Geoscience and Remote Sensing Letters) 99(4) pp 554-558.

[5] Saepuloh A, Koike K, Heriawan M N, and Kubo T 2016 *Quantifying Surface Roughness to Detect Geothermal Manifestations from Polarimetric Synthetic Aperture Radar (PolSAR) Data* (Stanford: PROCEEDINGS Fourtieth Workshop on Geothermal Reservoir Engineering Stanford University pp 22-24.

[6] Pirajno, F 2009 *Hydrothermal Processes and Mineral Systems* (East Perth – Australia: Springer Geological Survey of Western Australia) 1250 hal.

[7] Garwin S, Hall R, and Watanabe Y 2005 *Tectonic Setting Geology Gold and Copper Mineralization in Cenozoic Magmatic Arcs of Southeast Asia and the West Pasific* Economic Geology 100th Anniversary pp 891 – 930.

[8] Dewangga, A H 2014 *Geology and Hydrothermal Alteration Study in Ciseuti and Surrounding Area Purwakarta District West Java* (Bandung: Bandung Institute of Technology Press).

[9] Sudjatmiko 1972 *Geological map*, (Bandung: Center of geological survey Indonesia).

[10] Thompson A J B and Thompson J F H 1996 *Atlas of alteration: A Field and Petrographic Guide to Hydrothermal Alteration Minerals* (Geological Association of Canada Mineral Deposits Division) pp 119.

The removal of heavy metals by iron mine drainage sludge and *Phragmites australis*

Nguyen Thi Hoang Ha[1] and Bui Thi Kim Anh[2]

[1]VNU University of Science, 334 Nguyen Trai, Hanoi, Vietnam
[2]Institute of Environmental Technology, Vietnam Academy of Science and Technology, 18 Hoang Quoc Viet, Hanoi, Vietnam
Email: hoanghantvnu@gmail.com; hoangha.nt@vnu.edu.vn

Abstract. This study was conducted to assess the removal of heavy metals from solutions by the combination of modified iron mine drainage sludge (sorbent column) and surface and subsurface flow constructed wetlands using the common reed (*Phragmites australis*) during 30 days of experiment. The results of this study demonstrated that the average removal rates of Zn, Pb, Mn, and As by sorbent column were 59.0, 55.1, 38.7, and 42.4%, respectively. The decreasing trend of removal rates of metals by sorbent column was obtained during the experiment. The average removal rates of Zn, Pb, Mn, and As by sorbent column-surface constructed wetland were 78.9, 73.5, 91.2, and 80.5%, respectively; those by sorbent column-subsurface flow constructed wetland were 81.7, 81.1, 94.1, and 83.1% which reflected that subsurface flow constructed wetland showed higher removal rate than the surface system. Concentrations of heavy metals in the outlet water were lower than the Vietnamese standard limits regulated for industrial wastewater. The results indicate the feasibility of integration of iron mine drainage sludge and constructed wetlands for wastewater treatment.
Keywords: constructed wetland, heavy metal, iron mine drainage sludge, *Phragmites australis*, removal

1. Introduction

Mining and smelting activities of metal ores have generated considerable amounts of waste materials to the surrounding environment [1]. Production of Cu, Pb, and Zn causes the greatest environmental degradation as the result of releasing large quantities of heavy metals such as Cu, Cd, Pb, and As [1]. A variety of chemical-physical technologies have been applied to deal with the removal of these metals from water, including precipitation [2], ion exchange [3], electrocoagulation [4], and biological processes [5]. Each treatment technology possesses both advantages and disadvantages. In low income countries, the remediation cost appears to be one of the most important criteria for choosing treatment technologies.

Several natural sorbents have been reported for wastewater treatment, including kaolinite [6], sludge [7,8], diatomite [9,10], and laterite [11,12]. Adsorption of heavy metals is considered an effective and low cost method which offers flexible design and operation, rapid removal, and small space [2,13]. However, adsorption can be reversible, energy consumption and removal efficiency depends on the type of adsorbents and substances to be removed [13]. Therefore, the combination with other treatment technologies may be required for effective removal of water contaminated with multiple heavy metals.

Phytoremediation have emerged as cost-effective, minimal energy consumption, and environment-friendly technologies for metal removal from solutions [14]. Aquatic plant species (e.g., *Lemna minor*

L., *Eichhornia crassipes*, *Polygonum amphibium* L.) have been widely used [15]. A number of studies and applications of constructed wetlands have been conducted for wastewater treatment for more than five decades [16]. It involves biotechnological process in which plants are important part of the treatment system [16]. Common reed (*Phragmites australis*) is one of the most widely distributed species in the world [17]. In addition, this plant species is well adapted with contaminated environments and is a potential candidate for phytoremediation of heavy metals [16,18]. Different substrate materials in constructed wetlands for *P. australis* growth have been reported such as zeolite and series of limestone and cocopeat [19], blast furnace slag, caramsite, vermiculite, gravel, paddy soil, red soil, and turf [20], coke and gravel [21].

This study was conducted to assess the removal rates of heavy metals from solutions by the combination of modified iron mine drainage sludge and constructed wetland using the *P. australis* in a pilot greenhouse scale experiment.

2. Materials and methods

2.1. Preparation of plant, sorbent, and inlet water for experiment

Iron mine drainage sludge from an iron processing mine in northern Vietnam was used for modification of sorbent in the experiment. Sorbent were prepared by addition of 10% sodium silicate solution with deionized water. The mixed samples were pressed using a machine to generate sorbent of 2-mm diameter and 1-2 cm length. Sorbent was then heated at 400 ^{0}C in 3 hours by an oven. The specific gravity of sodium silicate solution was 1.46 ± 0.01g/ml; the contents of Na_2O and SiO_2 were 11.5% ~ 12.2%, and 27.5% ~ 29.5%, respectively. The percentage of quartz, muscovite, illite, kaolinite, hematite, talc, illite, goethite, and magnetite in the sorbent was 43, 13, 13, 12, 7, 7, 4, and 1%, respectively.

Phragmites australis was collected along a riverside of the Red River in Hanoi, Vietnam and washed with tap water. Plant was transplanted in two constructed tanks filled with soil and gravel/pebbles/sand for growth stabilization during 3 months by clean water and fertilized in a certain time. Plant density was one clumps of 3 stems per 15 cm x 15 cm. Average temperature was 22 ^{0}C and daily sunlight hours were approximately 12 hours.

Four metals (Pb, Zn, Mn, and As) were prepared as metal salts $Mn(NO_3)_2{\cdot}4H_2O$, $Zn(NO_3)_2{\cdot}4H_2O$, $Pb(NO_3)_2$, and Na_2HAsO_4 provided by Merck (Germany). The reagents were dissolved in Milli-Q water to obtain the desired contamination levels. Concentrations of Mn, Zn, Pb, and As in the inlet water in the experiment were 4.0, 1.5, 0.6, and 0.4 mg/L, respectively. Initial concentrations of heavy metals in water solutions were set the same as those in outlet wastewater of a processing ore area in the largest Pb-Zn mine in northern Vietnam after running through two settling reservoirs. The resulting solutions were adjusted to pH 7 with NaOH and HNO_3 using the Horiba D54 pH meter.

2.2. Experiment design

In the present experiment, pre-filtration column (sorbent column) and subsequent two types of constructed wetlands were set (Figure 1). After passing through sorbent column with constant water velocity of 100 L water per day, water was separated evenly into surface and subsurface constructed wetlands. Retention time of water in the sorbent column was approximately 30 minutes; that in constructed wetlands was 48 hours. The retention times were set based on optimal retention time for heavy metals removal at the previous laboratory experiments [22].

Sorbent column: modified iron mine drainage sludge of 1400 g was kept in a 1.8 L transparent acrylic tube.

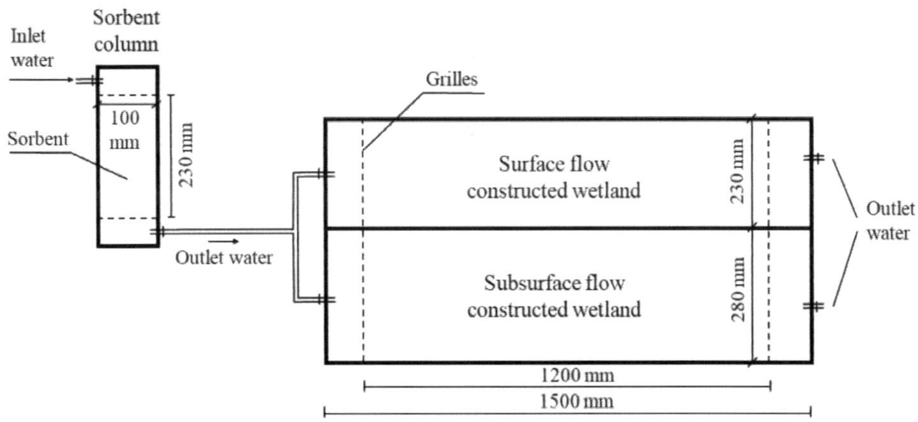

Figure 1. Experiment design of sorbent column and constructed wetland

Surface flow constructed wetland: Natural soil was used as a substrate. Porosity of substrate was 55% and safety factor was 2%. The height and length of constructed wetland system were designed to be 500 x 1500 mm, given the optimal conditions of anaerobic environment for bacteria activities and retention time of water in constructed wetland system. Design of surface flow constructed wetland was calculated by equations as follows:

$$\text{Vsfcw} = Q\times t = \frac{50}{24\times 1000}\times 48 = 0.1 \text{ m}^3$$

$$V = \frac{\text{Vsfcw}}{\varphi} = \frac{0.1}{57\%} = 0.175 \text{ m}^3$$

$$S = \frac{0.175}{0.5} = 0.35 \text{ m}^2$$

$$W = \frac{0.42}{1.5} = 0.23 \text{ m}$$

where Vsfcw is active volume (m^3), Q is water flow (m^3/hour), t is retention time (hour), V is constructed volume (m^3), φ is porosity of substrate added with safety factor (%), S is surface area of constructed wetland, and W is width of constructed wetland.

Subsurface flow CW: active volume, height and length of this system were set the same as that of the surface flow. Water ran in a horizontal pathway. Gravel of 2-4 cm grain size was used as a substrate. Pore volume of substrate was 45% and safety factor was 2%. Design of subsurface flow constructed wetland was calculated by equations as follows:

$$\text{Vsfcw} = Q\times t = \frac{50}{24\times 1000}\times 48 = 0.1 \text{ m}^3$$

$$V = \frac{\text{Vsfcw}}{\varphi} = \frac{0.1}{47\%} = 0.21 \text{ m}^3$$

$$S = \frac{0.21}{0.5} = 0.42 \text{ m}^2$$

$$W = \frac{0.42}{1.5} = 0.28 \text{ m}$$

Additional length of 150 mm was constructed between sorbent column and constructed wetlands for evenly and constantly allocating input water.

2.3. Sampling

Input and output water samples of sorbent column, surface and subsurface constructed wetland systems were collected before and at 1, 3, 5, 7, 10, 13, 16, 19, 22, 25, and 30 days of the experiment. Sorbent and plant samples were taken at the start and the end of the experiment.

2.4. Analytical methods

Eh and pH parameters were measured by HORIBA pH/COND METER D-54 equipment. Biological Oxygen Demand (BOD_5) was measured by Winkler titration and Chemical Oxygen Demand (COD) was determined by dichromate reflux methods using potassium permanganate. The concentration of heavy metals in water was determined by Atomic Absorption Spectrometer (AAS; Agilent 240FS with hydride generation accessory VGA77) at VNU University of Science.

3. Result and Discussion

3.1. pH, BOD_5 and COD in water environment

pH of the water samples in the outlet of sorbent column, surface flow constructed wetland and subsurface flow constructed wetland varied within 7.03-7.85; 7.33-7.83 and 7.22-7.71, respectively. The results demonstrated that pH was stable during experiment. These pH values were within favourable ranges for *P. australis* growth [23, 24].

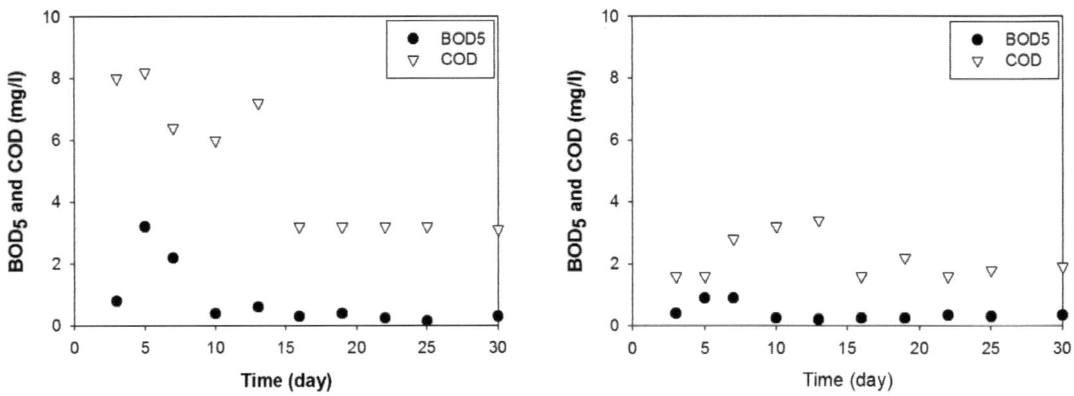

Figure 2. BOD_5 and COD in surface flow wetland (left) and subsurface flow wetland (right).

Biochemical oxygen demand (BOD_5) and chemical oxygen demand (COD) in outlet water of surface flow constructed wetland ranged within 0.2-3.2 and 3.1-8.2 (Figure 2). Those values of subsurface flow constructed wetland ranged within 0.2-0.9 and 1.6-3.2. Higher values of COD and BOD_5 of surface flow constructed wetland than those in the subsurface system were possibly due to higher amount of organic materials leaching from soil substrate. Variation of COD and BOD_5 values during the first 13 days of experiment may reflect the residue of organic fertilizer supply for plant growth before the experiment (Figure 2).

3.2. Concentration of heavy metals in water

The concentrations of heavy metals in outlet water of sorbent column and constructed wetlands decreased during the experiment in comparison with the inlet flow (Figure 3). The average removal rates of Zn, Pb, Mn, and As by sorbent column were 59.0, 55.1, 38.7, and 42.4%, respectively. The average removal rates of Zn, Pb, Mn, and As by sorbent column-surface flow constructed wetland were 78.9, 73.5, 91.2, and 80.5%, respectively; those by sorbent column-subsurface flow constructed wetland were 81.7, 81.1, 94.1, and 83.1%. These results demonstrated that the removal rates of heavy metals in subsurface flow constructed wetland were higher than those in the surface flow system. The removal rate of Pb by sorbent column was higher than that of As, Mn, and Zn. Subsurface and surface flow constructed wetlands demonstrated better removal of Mn than the other metals (Figures 3 and 4).

Concentration of Mn in outlet of sorbent column decreased dramatically during the first 10 days and remained almost constantly other days (Figure 3). Concentration of Mn in inlet of sorbent column was 4 mg/l which decreased to 0.92 mg/L on the first day and 2.78 mg/L on the 30th day of the experiment. Concentration of Mn in outlet of sorbent column exceeded the allowable limits on national technical regulation for industrial wastewater type A (QCVN40/2011-A: water source for domestic water supply) and type B (QCVN40/2011-B: water source not for domestic water supply) [25] (Figure 3). The removal rate of Mn was 77.1% on the first day, which decreased to 30.6% at the last day of the experiment. Removal rates of Mn in sorbent column-surface flow constructed wetland and sorbent column-subsurface flow constructed wetland were 96.0 and 99.6%, respectively after 3 days, which decreased to 81.2 and 84.8% on the last day (Figure 4). Mn concentration in outlet water of constructed wetlands met the regulation QCVN40/2011-A during 19-25 days and QCVN40/2011-B [25] during the experiment. However, the decreasing trend of removal rate reflected treatment limit of Mn by the system which highlights the need for further setting of constructed wetlands for longer remediation.

Concentration of Zn was steadily decreased during 30 days of the experiment by sorbent column (Figure 3). The removal rate of Zn was 74.6% during the first day which decreased to 51.1% after 30 days of the experiment (Figure 4). However, the removal rates of constructed wetlands were relatively stable, indicating the long term remediation of Zn by these systems. Concentrations of Zn in both inlet and outlet water of 3 systems were lower than the allowable limit which was set to be 3 mg/L [25] (Figure 3). This fact required studies to assess the possibility of wastewater treatment with higher concentrations of Zn by constructed wetlands using *P. australis*.

Concentration of Pb in outflow of sorbent column gradually decreased during 30 days of the experiment (Figure 3). The removal rate of Pb was higher than that of other metals. The removal rate of Pb was 96.7% at the 1st day which decreased to 22.8% after 30 days of the experiment (Figure 4). The removal rate of Pb by sorbent column-surface flow constructed wetland and sorbent column-subsurface flow constructed wetland was 93.1 and 98.4%, respectively after 3 days which decreased to 54.7 and 64% after 30 days of the experiment. Concentrations of Pb in outlet water of sorbent column, sorbent column-surface flow constructed wetland and sorbent column-subsurface flow system at the 3rd day were 0.06, 0.03, and 0.01 mg/L, respectively which increased to 0.46, 0.27, and 0.22 mg/L after 30 days. Pb concentrations in outlet of 3 systems were lower than the regulation QCVN40/2011-B [25] which was set to be 0.5 mg/L (Figure 3). However, it is noted that concentrations of Pb in inlet water was slightly higher than values regulated at QCVN40/2011-B [25]. Pb concentrations in outlet water of constructed wetlands after 10 days of experiment met the QCVN40/2011-B limit [25].

Concentration of As in water decreased dramatically by sorbent column during 30 days of the experiment (Figure 3). Removal rates of As by surface and subsurface flow constructed wetlands were almost the same. The removal rates of As by sorbent column was 75% at the 1st day which decreased steadily (average 42.4 %) at the end of the experiment (Figure 4). Average of removal rates of As in sorbent column-surface flow and sorbent column-subsurface flow constructed wetlands were 80.5 and 83.1%, respectively. Concentrations of As in outlet water of sorbent column, sorbent column-surface flow constructed wetland and sorbent column-subsurface flow system varied within 0.10-0.28, 0.06-0.09, and 0.05-0.08 mg/l, respectively. Concentration of As in outlet of sorbent column exceeded the allowable limits regulated at QCVN40/2011-B [25]; however, those of sorbent column-constructed wetland were lower than the regulation QCVN40/2011-B during the experiment. Constant concentrations of As in outlet water indicated the stability of As removal by the systems; however, the near-allowable limit required longer contact duration of water with sorbent or constructed wetlands, higher volume of sorbent or larger scale of constructed wetlands.

Using the same sorbent as that in the present study, Chinh et al. [26] conducted the laboratory column experiment in which initial concentrations of Mn, Pb, Zn, and As in water solutions were 20, 20, 6, and 1 mg/L, respectively. The removal rates of Mn, Pb, Zn, and As during 25 days of the experiment were 27.9-97.6, 73.9-97.4, 51.0-53.0, and 96.4-98.2%, respectively [26]. In the present study, removal rates of Mn, Pb, Zn, and As by sorbent column varied within 30.0-77.1, 22.8-96.7, 51.1-74.6, and 31.2-75.0%, respectively. It is noted that concentrations of heavy metals in inlet water conducted by Chinh et al. [26] were higher than those in the present study; however, reverse trend of removal rates of Mn, Pb, and As was obtained. It is possibly due to different experimental conditions (e.g., temperature, sunlight) and initial metal concentrations. This is in agreement with the effects of temperature and metal concentrations on metal adsorption by sorbents [27, 28].

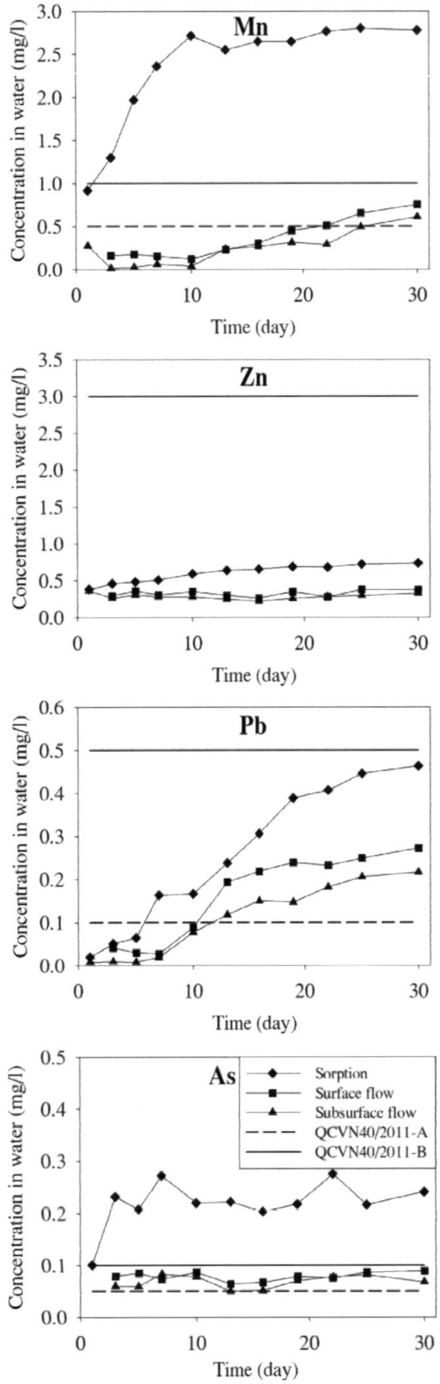

Figure 3. Concentrations of Mn, Zn, Pb, and As in outlet water from sorbent column, surface flow and subsurface flow constructed wetland systems.

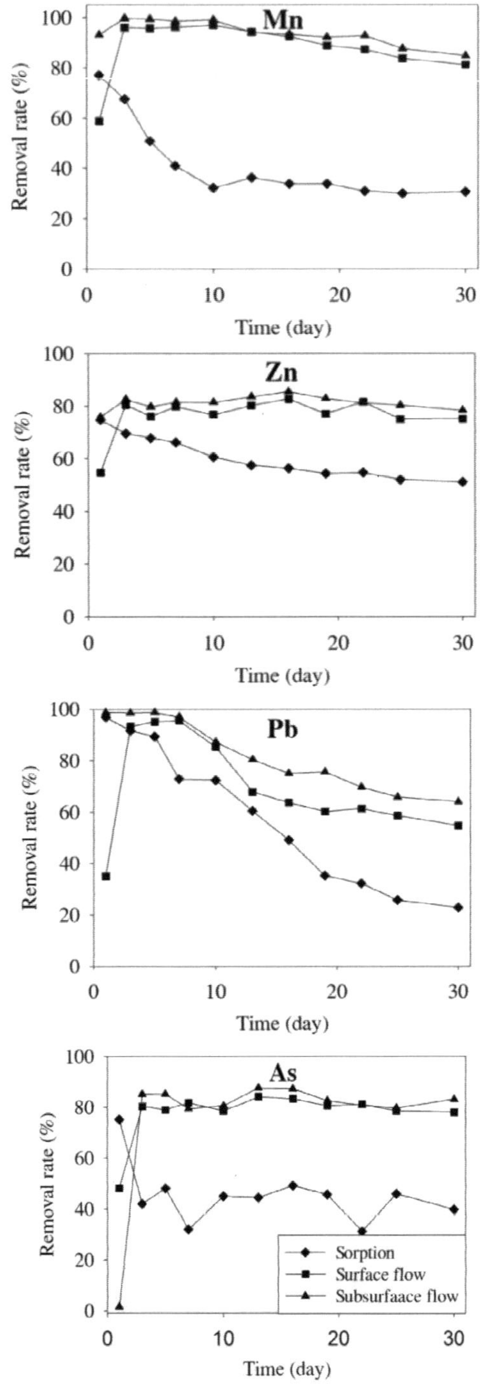

Figure 4. Removal rates of Mn, Zn, Pb, and As by sorbent column, surface flow and subsurface flow constructed wetland systems.

In sorbent column, heavy metals were possibly removed from water by both sorption and precipitation processes [26]. In constructed wetlands, heavy metals were removed or retained by both aerobic and anoxic/anaerobic processes [16]. Under anaerobic conditions, highly insoluble metal sulphides are formed; under anoxic conditions, metals can also be removed from water due to formation of carbonates which are less stable than sulphides and co-precipitation to roots. Other amount of metal removal from constructed wetlands may be adsorbed to substrates and organic matters and could be retained by filtration and deposition in the rhizosphere [29]. *Phragmites autralis* roots can also uptake and accumulate heavy metals in roots or aboveground part of the plant [16,22,30,31]. Highest concentrations of Mn, Zn, Pb, and As in the plant roots were 3920, 1020, 1350, and 183 mg/kg-DW, respectively; those in the stems were 465, 108, 227, and 74.0 mg/kg-DW; and those in the leaves were 716, 150, 157, and 88.3 mg/kg-DW [22]. Concentrations of Zn and Pb in in the frond biomass of *P. autralis* were reported to be 217 and 264 mg/m^2, respectively [16].

4. Conclusions

The integration of modified iron mine drainage sludge and surface flow or subsurface flow constructed wetlands showed potential of heavy metals treatment in the water. Concentrations of Mn, Zn, Pb, and As in the outlet of sorbent column-surface flow constructed wetland and sorbent column-subsurface flow constructed wetland were within Vietnamese regulation for industrial wastewater with the exception of Pb after 10 days of experiment. Sorbent column-subsurface flow constructed wetland demonstrated better heavy metals removal rate than sorbent column-surface flow system. Further studies need to be conducted for higher initial concentrations of inflow, longer experimental duration, and larger pilot scales of these integration systems before it can be applied in field conditions.

5. Acknowledgments

The authors are grateful for the support of the National Project coded KHCN-TB.02C/13-18 of the National Science and Technology Program for Sustainable Development of Vietnam' Northwest Region (KHCN-TB/13-18).

6. References

[1] Dudka S and Adriano D C 1997 Environmental impacts of metal ore mining and processing: a review *J. Environ. Qual.* **26** 590-602
[2] Gunatilake S.K 2015 Methods of removing heavy metals from industrial wastewater *J. Multidiscip. Eng. Stu.* **1** 12-18
[3] Gailkwad R W, Sapkal V S and Sapkal R S 2010 Ion exchange system design for removal of heavy metals from acid mine drainage wastewater *Acta Montan. Slovaca* **15** 298-304
[4] Turdean G L and Popescu I C 2013 Electrochemical Method for Heavy Metals Detection by Inhibition of Acetylcholinesterase Immobilized on Pt-nanoparticles Modified Graphite Electrode *E3S Web of Conferences* vol 1 p 05007
[5] Sonali R D and Jayant P K 2013 Biological methods for heavy metal removal - A review *International J. Eng. Sci. Innovative Technol.* **2**(5) 304-09
[6] Adebowale K O, Unuabonah I E and Olu-Owolabi B I 2005 The effect of some operating variables on the adsorption of lead and cadmium ions on kaolinite clay *J. Hazard. Mater.* **134** 130-39
[7] Hung D D and Minh N T 2011 Sorption of Zn(II) on Bao Loc modified- granular bauxite mine drainage sludge by column experiment *Vietnam J. Earth Sci.* **33** 591-98
[8] Hülya G, Christian T, David M and Olaf S 2003 Adsorption of arsenate from water using neutralized red mud *J. Colloid Interface Sci.* **264** 327-34
[9] Al-degs Y, Khraisheh M A M and Tutunji F 2001 Sorption of lead ions on diatomite and manganese oxides modified diatomite *Water Res.* **15** 3724-28
[10] Li E, Zeng X and Fan Y 2009 Removal of chromium ion (III) from aqueous solution by manganese oxide and microemulsion modified diatomite *Desalination* **238**(1-3) 158-65
[11] Thao N H P, Ha N T H, Thuy P T, Khai N M, Nga T T H 2016 Study the environmental pollution treatment ability of laterite *VNU J. Science* **32**(1S), 321-26
[12] Yu X, Zhu L, Guo B and He S 2008 Adsorption of mercury on laterite from Guizhou Province,

China *J. Environ. Sci.* **20** 1328-34

[13] Fu F and Wang Q 2011 Removal of heavy metal ions from wastewaters: A review *J. Environ. Manage.* **92** 407-18

[14] Ha N T H, Sakakibara M and Sano S 2011 Accumulation of Indium and other heavy metals by Eleocharis acicularis: An option for phytoremediation and phytomining *Bioresour. Technol.* **102** 2228-34

[15] Wang Q, Cui Y and Dong Y 2002 Phytoremediation of polluted waters: potentials and prospects of wetland plants *Biotechnol. Acta* **22**(1-2) 199-208

[16] Vymazal J and Brezinova T 2016 Accumulation of heavy metals in aboveground biomass of Phragmites australis in horizontal flow constructed wetlands for wastewater treatment: a review *Chem. Eng. J.* **290** 232-42

[17] Hocking P J, Finlayson C M and Chick A J 1983 The biology of Australian weeds. 12. Phragmites australis (Cav.) Trin. ex Steudel *J. The Australian Institute of Agricultural Science* **49** 123-32

[18] Jiang X and Wanga C 2008 Zinc distribution and zinc-binding forms in *Phragmites australis* under zinc pollution *J. Plant Physiol.* **l165**(7) 697-704

[19] Allende L K, McCarthy D T and Fletcher T D 2014 The influence of media type on removal of arsenic, iron and boron from acidic wastewater in horizontal flow wetland microcosms planted with Phragmites australis *Chem. Eng. J.* **246** 217-28

[20] Zhu W L, Cui L H, Ouyang Y, Long C F and Tang X D 2011 Kinetic adsorption of ammonium nitrogen by substrate materials for constructed wetlands *Pedosphere* **21**(4) 454-63

[21] Chen M Z, Tang Y Y, Li X P and Yu Z X 2009 Study on the heavy metals removal efficiencies of constructed wetlands with Different Substrates *J. Water Resource Prot.* **1** 1-57

[22] Nga T T H and Ha N T H 2016 Simultaneous removal of some heavy metals and arsenic from aqueous solutions by *Phragmites australisVietnam J. Sci. Technol.* **54**(2A) 259-64

[23] Chabreck R H 1972 *Vegetation, water and soil characteristics of the Louisiana coastal region Bulletin 664* (Louisiana State University, Louisiana Agricultural Experiment Station. Baton Rouge, LA) p 72

[24] USDA NRCS 2012 The PLANTS database [Online] *Available. http://plants.usda.gov*

[25] QCVN40:2011-BTNMT National Technical Regulation on Industrial Wastewater Quality

[26] Chinh L S, Nhuan M T, Hai N X, Hai N T, Thang D N, Giang N T, Quy T D and Ha N T H 2016 The potential of modified iron mine drainage sludge for treatment of water contaminated with heavy metals *VNU J. Sci.* **32**(1S) 38-44

[27] Johnson B B 1990 Effect of pH, temperature, and concentration on the adsorption of cadmium on goethite *Environ Sci. Technol.***24** 112-18

[28] Dzombak D A and Morel F M M J 1986 *Colloid Interface Sci.* **112** 588-98

[29] Marchand L, Mench M, Jacob D L and Otte M L 2010 Metal and metalloid removal in constructed wetlands, with emphasis on the importance of plants and standardized measurements: a review *Environ. Pollut.* **158** 3447-61

[30] Afrous A, Manshouri M, Liaghat A, Pazira E and Sedghi H 2011 Mercury and arsenic accumulation by three species of aquatic plants in Dezful, Iran, Afr *J. Agric. Res.* **6**(24) 5391-97

[31] Vaiekonyt R, Kiviat E, Nsenga F and Ostfeld A 2014 An exploration of common reed (*Phragmites australis*) bioenergy potential in North America *Mires. and Peat* **13**(12) 1-9

Mercury Distribution in the Processing of Jatiroto Gold Mine Wonogiri Central Java Indonesia

Dwi Fitri Yudiantoro[1], Muhammad Nurcholis[1], Dewi Sri Sayudi[2], Mirzam Abdurrachman[3], Intan Paramita Haty[1], Wiryan Pambudi[4], Arum Subroborini[5]

[1] Geological Engineering University of Pembangunan Nasional "Veteran" Yogyakarta Indonesia
[2] Geological Agency Indonesia
[3] Geological Engineering Institute of Technology Bandung Indonesia
[4] Undergraduate Student University of Pembangunan Nasional "Veteran" Yogyakarta Indonesia
[5] Magister Bioscience Student University of Sebelas Maret Surakarta Indonesia
Email: d_fitri4012@yahoo.com

Abstract. The research area is one of the Wonogiri gold producer. In this region there are nearly 30 gold processing locations. This area has a steep morphology which is part of Mt. Mas. The work of the gold processing is a part time job besides for the local farmer population. To get the gold bearing rocks, are by digging holes manually around Mt. Mas, while gold processing is carried out in their homes. As a result of these activities, then identified the distribution of mercury in the surrounding settlements. Analytical methods used in this study is the measurement mercury content using Hg meter on altered rocks, soil and using XRF (X-Ray Fluorescence) for plant samples. This results of research shows that there are conducted on mercury contents in the altered rocks, soil and plants showed significant mercury contents in altered rocks, soil and plants. This proves that mercury has polluted the environment surrounding residents, both of people living in the hill down on the lower plain areas. The results of this study are expected to be used as reference to help overcome the pollution of the area.

1. Introduction

Mining waste materials containing heavy metals Hg, Fe, Pb, As, Cu, Cd, Zn, Cr and S which are toxic [1, 2, 3]. It is very dangerous to humans, plants and animals. This mine waste can pollute the environment through the media water and air [4], so that it can pollute the environment.

Mercury is a trace element in the earth's crust which is only about 0.08 mg/kg [5]. Cinnabar (HgS) is the most common natural source of mercury [6, 7]. In addition, sulphide minerals, such sphalerit (ZnS), wurtzite (ZnS), chalcopyrite (CuFeS) and galena (PbS), can also contain trace amounts of mercury. Cinnabar is soluble in water, but due to weathering of rocks and soil erosion, then cinnabar can release mercury into the aquatic environment [8]. Mercury is a heavy metal that is very dangerous if taken into the body of living beings such as humans, animals, and plants [5]. The previous study [9] states that the type of mercury found in the organs of living beings, such as fish, is methyl mercury (CH_3Hg^+) or dimethyl mercury ($(CH_3)_2Hg$). Cases of mercury poisoning were quite famous [10, 11] was the case that occurred in Minamata Bay, Japan in the 1950s. The chemical industry which operates around the Minamata Bay dumping mercury waste into the waters of the bay and caused mothers who consume seafood obtained from Minamata Bay gave birth to children with congenital defects. In that case, of the 111 poisoning cases, 43 people died [12].

The spread of mercury in the research area is quite extensive because the deployment location in the population area. This is supported by research site is an alteration zone which is the source of the metal element and also the processing and waste disposal is in the settlements each resident.

The research location is in the hamlet of Mesu, Boto, District Jatiroto, Wonogiri Central Java Province. This location is a gold mine that whole processing of gold was at each house (see Figure 6).

2. Data and Method

Mercury is a heavy metal that is a liquid and gas type. So the method of analysis has to do with geological mapping. To study of mercury levels in the study area and its effect on plants, mercury measurements were performed in alteration rocks, soil and some plants. Measurements of mercury in alteration rocks and soil use Hg meter, while the measurement of mercury in plants using X-Ray Fluorescence (XRF). The pH meter used to study the content of the acidity of soil and alteration rocks. The results of this research will be presented on geological and mercury distribution map.

3. Result and Discussion

3.1. Geological Setting

The research area was an area of hills morphologies were composed by volcanic breccias, andesite intrusion and andesite lava that have undergone hydrothermal alteration (Figure 1, 2). The geology based on [13] research areas compiled by Nglanggran volcanic breccias (Oligocene). Volcanic breccias interbedded with tuff, volcanic sandstone, andesite lava and andesite intrusion. Characterize of volcanic breccias were massive, textured grains supported texture, andesite fragments, matrix of volcanic material and silica cement. The andesite intrusion generally has undergone a process of alteration. The altered andesite intrusion was colored red-gray, fresh color dark gray, hypocrystaline, faneric-afanitic, subhedral minerals, inequigranular, mineral composition are pyroxene, hornblende, plagioclase in groundmass volcanic glass. The secondary minerals were present in andesite intrusion was chlorite, quartz and hematite.

Figure 1. The hilly morphology of research area.

IOP Conf. Series: Earth and Environmental Science **71** (2017) 012023 doi:10.1088/1755-1315/71/1/012023

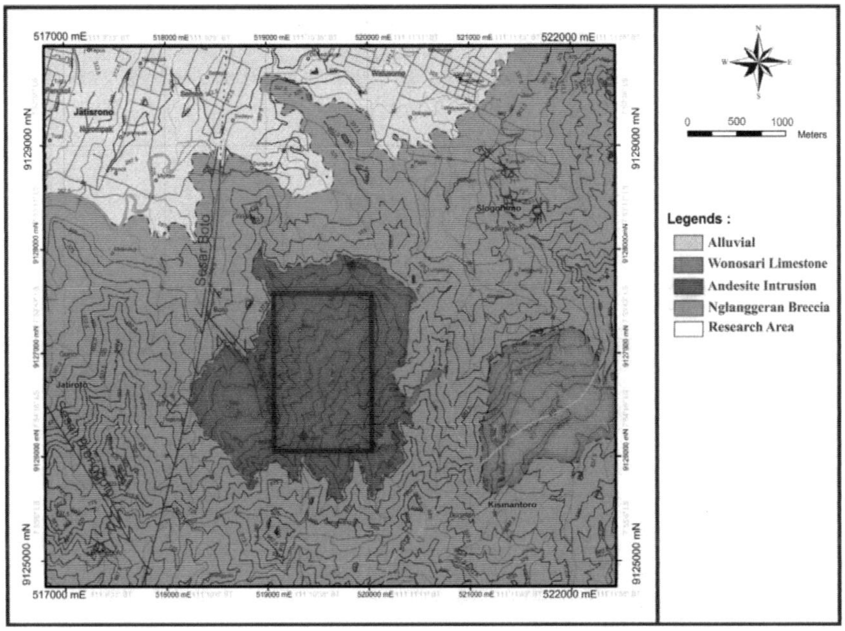

Figure 2. Geological map (black box is a research area).

The type zone of surface alteration that has been found in the study area was advanced argillic. In this type of alteration minerals of advanced argillic were found some of minerals as clay mineral, quartz, alunite, hematite, pyrite and fine cinnabar (Figure 3).

Figure 3. Outcrop of advance argillic alteration in Mesu area.

3.2. Mercury distribution

Measurements of mercury using Hg-meter (Figure 4) conducted on several locations both on alteration rock and gold processing area. The levels of mercury in Mesu area measured by Hg-meter were 0-0.06 mg/cm^3, while the green zone has a content of 0-0.01 mg/cm^3 mercury. The yellow zone has a content of 0.01-0.02 mg/cm^3, and the orange zone has a content of 0.021-0.04 mg/cm^3. The red zone has a content of mercury around 0.041-0.06 mg/cm^3 (Figure 5).

Figure 4. Measurement of mercury in soil by using Hg-meter.

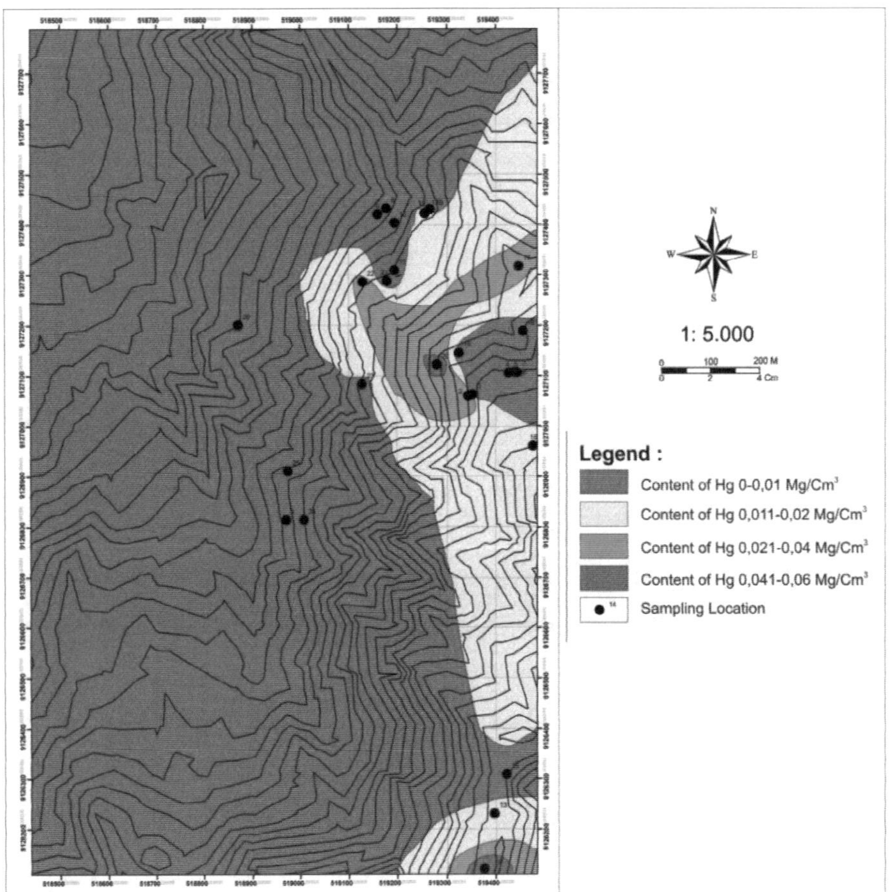

Figure 5. The distribution mercury in Mesu area.

The highest distributing of Hg was at site 20 and 14. Site 20 was a gold processing with high intensity was at residential locations (Figure 6).

Figure 6. Gold processing that was in the human settlement and mercury measurement using Hg-meter in mine waste

Site 14 was the site of mining and mercury measurements done at the rocks undergo advanced argillic alteration. It was to know that the alteration rocks has naturally contains mercury. The results of mercury measurements in alteration rocks and soils were presented in Table 1.

Table 1. Measurement mercury concentration on altered rocks and soil

Sample Name	Hg (mg/cm^3)	pH	Sample Name	Hg (mg/cm^3)	pH
80	0.001	7.0	A 14	0.060	5.2
81	0.021	5.0	A 15	0.000	0.0
82	0.001	4.4	A 16	0.013	5.8
83	0.006	4.6	85	0.036	4.5
84	0.001	6.0	86	0.003	5.0
A 6	0.000	5.0	87	0.002	6.6
A 7	0.001	4.8	88	0.044	5.2
A 8	0.002	6.4	89	0.001	5.8
A 9	0.001	4.8	90	0.036	4.4
A 10	0.002	6.6	91	0.001	4.8
A 11	0.001	5.7	92	0.004	5.7
A 12	0.019	6.8	93	0.003	6.2
A 12 b	0.001	5.0	94	0.001	5.2
A 13	0.012	5.8	95	0.006	5.6
A 14	0.060	5.2	96	0.000	4.0

Results analysis of Hg content in plants was the range of 0.0041-0.0168%. Samples were taken of plant that was consumed by the community in this area. The results of this analysis shows that the plants were consumed by people have contained Hg quite high. For example, mercury levels contained by kale leaves and bay leaves containing 0.0168% mercury (Figure 7).

(a) (b)

Figure 7. Type of plants containing 0.0168% mercury concentrations of plant; (a) kale leave and (b) bay leave

4. Conclusion

Hg element derived from the gold processing using the amalgamation process and also formed naturally from the alteration rocks. The spread of Hg was controlled by the topography, bedrock and water flow. Mercury in the soil was already polluting the local people as well as animals and plants.

5. Acknowledgments

The author would like to thank : UPN "Veteran " Yogyakarta, ESDM Wonogiri, Education Minister (DIKTI), Boto-Mesu village, in giving permission and the opportunity to conduct this research. As well as colleagues involved and assist in research.

6. References

[1] Al-Jaloud AA, Hussain G, Al-Saati AJ, Karimulla S 1995 Effects of Wastewater Irrigation and Sludges in Kuwait. *Journal of Environment Science and Health.* **39.** 397-407

[2] El Khalil H, El Hamani O 2008 Heavy Metal Contamination from Mining Sites in South Morocco: Monitoring Metal Content and Toxicity of Soil Runoff and Groundwater. *Journal Environ Monit Asses.* **136**. 147-160

[3] Galavi M, Jalali A, Ramroodi M 2010 Effects of Treated Municipal Wastewater on Soil Chemical Properties and Heavy Metal Uptake by Sorghum (*Sorghum bicolor* L.). *Journal of Agricultural Science.* **2** (**3**). 235-241

[4] Faruruw DM, Yauri UA, Dangoggo SM 2013 Cadmium, Copper, Lead and Zinc Levels in Sorghum and Millet Grown in The City of Kano and Its Environs, *Global Advanced Research Journal of Environtment Science and Taxicology.* **2(3)** 82-85

[5] Moore JW 1991 Inorganic Contaminations of Surface Water. *Springer-Verlag. New York.* 334 p

[6] Novotny V and Olem H 1994 Biological Oceanographyc Proceses. Third edition. Pergamon Press. *New York.* 330 p

[7] Soemirat 2007 Toksikologi Lingkungan *Yogyakarta* Gadjah Mada University Press

[8] McNeely RN, Nelmanis VP and Dwyer L 1997 Water Quality Source Book, A Guide to Water Quality Parameter, Inland Water Directorate, Water Quality Branch, *Ottawa Canada.* 89 p

[9] Dugan PR 1972 Biochemical Ecology of Water Pollution Plenum Press New York. 159 p

[10] Harada M 1995 Minamata Disease: Methylmercury Poisoning in Japan Caused by Environmental Pollution, Critical Reviewer in Toxicology **2591.**1-24

[11] Lestari, Edward 2004 Dampak Pencemaran Logam Berat Terhadap Sumberdaya Laut Dan Sumberdaya Perikanan (Studi Khusus Kematian Massal Ikan-Ikan Di Teluk Jakarta) Jakarta

Makara
[12] Sawyer, CN and McCarty PL 1972 Chemistry for Environmental Engineering, Thrid edition, McGraw-Hill Book Company Tokyo 532 p
[13] Sampurno and Samodra H 1991 Geologi Lembar Ponorogo, Jawa, Pusat Penelitian dan Pengembangan Geologi Indonesia.

Regional geochemistry Bandung Quadrangle West Java: for environmental and resources studies

Purnama Sendjaja and Baharuddin

Centre for Geological Survey, Geological Agency, The Ministry of Energy and Mineral Resources, Republic of Indonesia, Jln. Diponegoro no. 57 Bandung.
Email: purnama.sendjaja@esdm.go.id, pppurs@gmail.com

Abstract. Geochemical mapping based on the stream sediment method has been carried out in the whole of Java Region by the Centre for Geological Survey. The Regional Geochemistry Bandung Quadrangle as part of West Java Region has been mapped in 1:100.000 scale map, base on the Geological Map of Bandung Quadrangle. About 82 stream sediment samples collected and sieved in the 80 mesh sieve fraction during the field work session at 2011. This fraction was prepared and analysed for 30 elements by X-ray fluorescence spectrometry at the Centre for Geological Survey Laboratory. There are some elements indicating significant anomaly in this region, and it is important to determine the present abundance and spatial distribution of the elements for presuming result from natural product or derived from human activities. The volcanic products (Tangkuban Perahu Volcano, Volcanic Rock Complex and Quarternary Volcanic-Alluvial Deposit) are clearly identified on the distribution of As, Ba, Cl, Cu, Zr and La elements. However Mn, Zn, V and Sr are related to precipitation in the Tertiary Sediments, while the influence of human activities are showing from a geochemical map of Cl, Cr, Cu, Pb and Zn that show scattered anomalies localized close to the cities, farming and industries.
Keywords: Geochemical map; Stream Sediment; Bandung Quadrangle.

1. Introduction

Systematic regional geochemical mapping in Indonesia has been done for several island regions i.e.: Sumatera [1], Sulawesi [2], East and West Nusa Tenggara [3,4], East and West Kalimantan [5]. For the Java Island, Geochemical field sampling carried out during the period of 2011-2013 and already published the Geochemical Atlas of West Java Region [6] within the scale of 1:1.250.000 in 2015. This atlas is a compilation from the 22 regional geochemistry maps of 1:100.000 scale basis on the geological map quadrangle published by the Centre for Geological Survey. In this paper, shows the Regional Geochemistry Bandung Quadrangle as part of West Java Region for more detail explanation in interpretation of anomalies and purposes to provide data for basic geology, mineral resources and the environment.

The geochemical sampling was using stream sediment method. Stream sediment sample represents of upstream material from the sampling site for recognizing geochemical or mineralogical anomalies for follow-up work and reconnaissance studies in drainage basins.

The Regional Geochemistry of Bandung Quadrangle area is bounded by Longitude 107°30'E and 108°00'E; Latitude 6°30' S and 7°00'S. this region is included into the Central Depression and Tertiary Sedimentary Rocks [7][8]. Lithology of the region is mainly covered by alluvial, sedimentary formation and volcanic deposit materials as product of several volcanic centre that surroundings the region (Figure. 1).

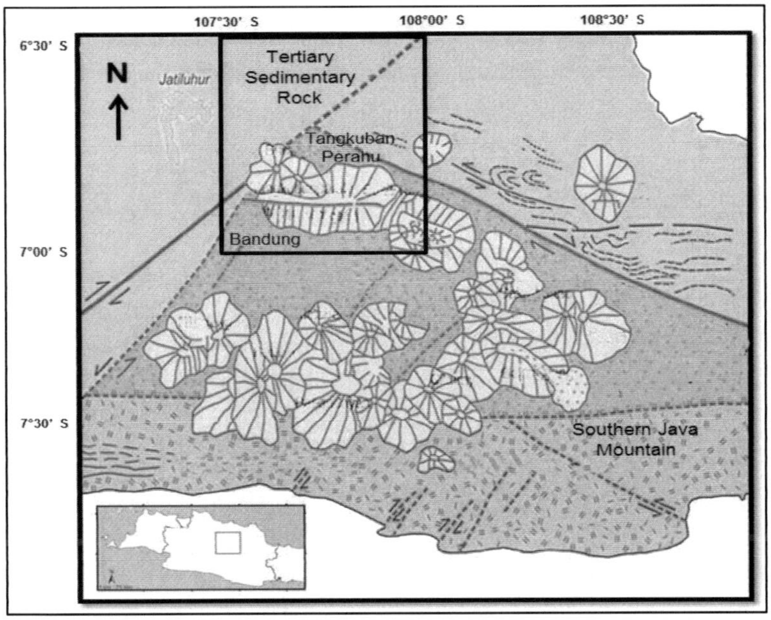

Figure 1. Study area and regional geology of the Eastern part West Java [7]

2. Data and Method

The methods applied in this geochemical mapping were fieldwork and laboratory examined. During the fieldwork, eighty-two stream sediment samples from the active river were taken, with the density of samples represent one sample for about 25 km^2. A grid with 5 x 5 km cells was drawn on 1:100.000 topographic and geological maps of the Bandung quadrangle [9] (Figure.2). The target of sampling was to collect one stream sediment sample from each of the grid cells. Stream sediment were chosen because they represent a composite of the drainage basin lithology [10]. The samples collected by geological students as a collaboration project between the Geological Agency and universities.

The 80 mesh stream sediment samples from field were dried for about 24 hours, and then were crushed by jaw crusher to powder with obtaining particle size of 200 mesh by using a ball mill. The sample powder then pressed by pellet machine with previously adding polyvinyl alcohol and borate acid.

The samples pellet were then analyses using an ARL 9900 Thermo Scientific X-Ray Fluorescence instrument at the Geological Laboratory of Centre for Geological Survey in Bandung. The number of chemical elements that can be determined by XRF are 30 elements, as follows: Ag, Al, As, Ba, Ca, Ce, Cl, Co, Cr, Cu, Fe, Hg, K, La, Mg, Mn, Mo, Na, Ni, P, Pb, Rb, Sc, Si, Sr, Te, Ti, V, Zn and Zr. The certified reference material (CRM) used to prepare the calibration and quality control standards. Five CRM (STSD1, STSD2, STSD3, STSD4 and LKSD1-4) for calibration were included in every interval samples analysis. The accuracy of the method for all determinants is better than ±5%.

The result of the geochemical analysis was then subjected to configure geochemical maps with an interpolated value used ArcGIS and Geosoft Oasis Montaj software. Contouring was using the kriging method on the ArcGis software with the colour for class intervals was based on a cold to hot trend with increasing elemental concentration. The kriging method is recommended for gridding of irregularly spaced data points, such as element concentrations measured in stream sediment samples [11]. Geosoft Oasis Montaj was used to create frequency distribution diagrams and determine correlation coefficients for the elements.

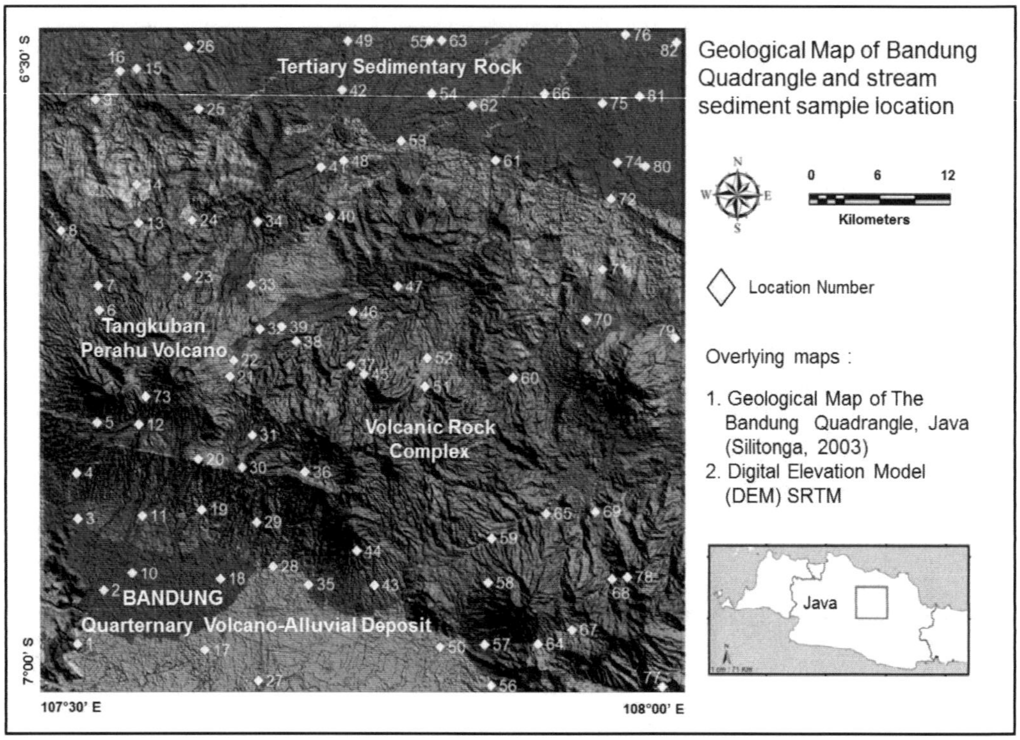

Figure 2.Digital elevation model, Geological map and Sample Location.

3. Result and Discussion

Result of geochemical analysis of stream sediment samples is shown in Table 1, displayed in alphabetical order. From 30 elements which were could be analyzed by XRF method, there were 5 elements concentration (Ag, Hg, Mo, Ni and P) below detection limit in all samples location. Fourteen (14) elements representing major and trace elements were selected for discussion to any possible adverse environmental problem associated with the element and/or possibility for mineral resources.

From the Table 1, As, Ba, and Mn have their highest concentration values in sample 14-P130 (Loc No 80). The location at Indramayu district, taken from the field rice and close to the village. High concentration of As, only occurred in sample 14-P011, 14-P043, 14-P116, 14-P119, 14-P130 and 14-P132 (Loc. No. 9, 32, 11, 75, 80 and 82 respectively) while the rest of the other samples below detection limit. Possibility of high content of As in location 32 and 11 are volcanic activity of Tangkuban Perahu volcano while the remain samples are located in agricultural area may cause by utilization of pesticides or fertilizer.

The result of the correlation analysis of some selected elements (Table 2) shows that Ba has strong correlation with Mn (0.77), a moderate correlation with Pb and very weak with Cl (0.30). These correlation may correspond to the distribution of Ba on the stream sediment map is similar to that of Mn in the east part of the map, and similar to that Cl in the Tangkuban Perahu volcano. Elevated Ba values may indicate the presence of volcanic felsic rocks surrounding volcano, in association with Cl and Zr, however in the west part of map indicate Tertiary Sediment may contains calcareous in association with Sr [11].

Other group of elements with strong correlation is Zn-V (0.73), the distribution of Zn on the stream sediment map is similar to that of V. The pattern may representing co-precipitation of these elements in the Volcanic and Tertiary Sediments formation. Zinc also has a weak correlation with Cl, Mg and Sr, whereas V in stream sediment shows a weak correlation with Cl and Cr. Negative correlation coefficient are recorded between Cr and As, Ba, Ce; Sr and Ce, Cu, La.

Table 1. Stream sediment geochemical data (ppm)

Location No.	Sample No.	As	Ba	Ce	Cl	Cr	Cu	La	Mg	Mn	Pb	Sr	V	Zn	Zr
1	14-P002	n.d	n.d	n.d	102	66	39	22	7950	1180	47	66	176	97	n.d
2	14-P003	n.d	n.d	n.d	56	64	46	17	7410	1250	n.d	42	167	113	n.d
3	14-P004	n.d	240	n.d	72	65	66	22	8590	1390	32	107	397	176	n.d
4	14-P005	n.d	260	n.d	n.d	31	51	27	1360	1090	n.d	0	212	58	n.d
5	14-P006	n.d	332	n.d	n.d	0	58	22	5020	1590	n.d	167	137	84	55
6	14-P007	n.d	164	n.d	133	36	34	25	14500	2340	n.d	116	876	181	105
7	14-P008	n.d	180	n.d	166	44	54	21	8350	2390	n.d	123	828	194	132
8	14-P009	n.d	267	n.d	74	n.d	50	29	3760	1820	n.d	87	500	139	201
9	14-P011	121	n.d	n.d	n.d	49	n.d	19	9090	814	n.d	68	155	88	252
10	14-P015	n.d	245	n.d	n.d	40	86	26	6200	1260	n.d	123	241	126	59
11	14-P016	n.d	295	72	52	33	90	25	6780	1210	n.d	118	221	121	54
12	14-P017	n.d	356	n.d	242	32	109	22	5220	1020	n.d	78	185	77	69
13	14-P021	n.d	128	n.d	45	16	n.d	29	3790	437	n.d	35	388	n.d	155
14	14-P022	n.d	206	n.d	40	n.d	n.d	22	5340	285	n.d	62	122	n.d	136
15	14-P023	n.d	165	n.d	n.d	33	n.d	22	4130	312	n.d	54	189	n.d	262
16	14-P024	n.d	220	n.d	71	125	20	22	10300	1880	n.d	134	494	140	217
17	14-P026	n.d	236	76	160	84	61	21	12300	1700	101	125	425	156	n.d
18	14-P027	n.d	n.d	n.d	n.d	19	50	30	1100	402	n.d	0	172	44	53
19	14-P028	n.d	243	n.d	n.d	34	42	21	4990	2580	n.d	36	201	101	n.d
20	14-P029	n.d	370	n.d	n.d	n.d	97	28	2880	1410	n.d	0	198	106	132
21	14-P030	n.d	498	n.d	87	n.d	416	26	2280	1460	n.d	37	325	76	359
22	14-P031	n.d	199	n.d	123	187	22	19	9190	1360	n.d	231	399	105	386
23	14-P032	n.d	196	n.d	124	164	18	13	8360	1150	n.d	193	358	82	335
24	14-P033	n.d	279	n.d	100	n.d	73	32	2870	2050	n.d	63	686	235	210
25	14-P035	n.d	n.d	n.d	64	90	46	28	2920	1890	n.d	30	744	129	121
26	14-P036	n.d	162	n.d	101	123	41	28	5710	2480	n.d	59	1340	190	191
27	14-P037	n.d	298	n.d	n.d	44	66	30	2490	1500	n.d	0	225	88	72
28	14-P039	n.d	n.d	n.d	75	68	85	24	9620	1370	32	158	261	178	n.d
29	14-P040	n.d	368	n.d	83	55	52	26	4000	1400	n.d	93	190	90	n.d
30	14-P041	n.d	192	n.d	71	57	79	23	11500	1830	n.d	110	373	129	n.d
31	14-P042	n.d	261	n.d	154	60	81	22	6320	2190	n.d	118	690	165	183
32	14-P043	98	958	58	837	62	75	34	4250	999	81	112	481	84	332
33	14-P044	n.d	248	n.d	227	96	87	22	4510	1580	n.d	126	579	165	227
34	14-P045	n.d	232	n.d	84	31	48	28	4170	2050	n.d	66	745	165	192
35	14-P051	n.d	255	n.d	128	47	50	21	10700	1440	n.d	148	202	100	n.d
36	14-P053	n.d	246	n.d	149	121	41	22	6200	1570	n.d	101	528	143	160
37	14-P054	n.d	221	n.d	86	36	26	20	8730	1530	n.d	151	413	125	138
38	14-P055	n.d	221	n.d	149	62	55	20	5420	2300	n.d	92	837	170	216
39	14-P056	n.d	567	n.d	554	155	77	32	11800	1360	50	92	1360	145	223
40	14-P057	n.d	178	n.d	131	70	50	28	2680	2100	n.d	58	659	179	355
41	14-P058	n.d	193	n.d	145	76	19	17	7780	1580	n.d	141	277	115	226
42	14-P059	n.d	282	n.d	192	24	38	22	7350	1430	n.d	134	229	116	247
43	14-P063	n.d	178	n.d	n.d	59	42	24	5990	2200	n.d	n.d	478	107	n.d
44	14-P064	n.d	277	n.d	n.d	43	34	25	7460	1400	n.d	101	204	83	n.d
45	14-P066	n.d	293	n.d	108	61	29	25	5050	1530	n.d	218	307	104	232
46	14-P067	n.d	275	n.d	159	33	38	23	5370	1550	n.d	181	334	118	224
47	14-P068	n.d	237	n.d	111	92	42	28	3430	2230	n.d	68	715	154	210
48	14-P070	n.d	210	n.d	107	n.d	38	20	4330	571	n.d	73	251	88	267
49	14-P072	n.d	264	n.d	139	40	30	25	7000	1710	n.d	146	395	141	227
50	14-P074	n.d	278	n.d	n.d	64	49	27	2840	1460	n.d	n.d	244	79	n.d
51	14-P078	n.d	261	n.d	86	33	28	22	7580	1510	n.d	244	293	110	116
52	14-P079	n.d	322	n.d	103	77	44	26	4570	1610	n.d	164	305	115	226
53	14-P082	n.d	212	n.d	104	136	0	20	12300	1730	n.d	208	393	125	198
54	14-P083	n.d	274	n.d	198	105	45	22	5790	4150	n.d	105	731	160	269
55	14-P084	n.d	228	n.d	170	170	26	21	7110	1530	n.d	154	609	172	331
56	14-P085	n.d	211	n.d	n.d	67	68	29	3950	1470	n.d	35	259	96	n.d
57	14-P086	n.d	249	n.d	n.d	70	63	25	5850	1650	n.d	25	417	84	n.d
58	14-P087	n.d	371	n.d	n.d	29	69	28	2030	1520	n.d	n.d	210	80	50
59	14-P088	n.d	314	n.d	n.d	76	51	22	4790	1850	n.d	63	218	84	n.d
60	14-P091	n.d	228	n.d	66	216	42	28	5060	2220	n.d	108	534	123	159
61	14-P094	n.d	265	n.d	67	60	23	20	8080	1640	n.d	143	177	95	190
62	14-P095	n.d	273	n.d	80	80	18	23	10900	1730	n.d	288	217	102	167
63	14-P096	n.d	286	n.d	183	n.d	26	24	4550	421	n.d	96	85	100	268
64	14-P098	n.d	246	n.d	n.d	112	59	28	3720	2350	n.d	n.d	430	121	46

Location No.	Sample No.	As	Ba	Ce	Cl	Cr	Cu	La	Mg	Mn	Pb	Sr	V	Zn	Zr
65	14-P100	n.d	258	n.d	n.d	55	47	24	4520	1630	n.d	54	195	76	n.d
66	14-P107	n.d	n.d	n.d	142	328	34	21	13900	2680	n.d	128	1330	270	281
67	14-P110	n.d	n.d	n.d	49	127	34	20	12800	1600	n.d	126	377	96	n.d
68	14-P111	n.d	228	n.d	n.d	35	43	20	10500	1170	n.d	94	244	70	n.d
69	14-P112	n.d	209	n.d	n.d	53	44	22	11400	1370	n.d	67	250	74	n.d
70	14-P115	n.d	261	n.d	242	166	38	24	16000	2080	n.d	211	483	156	102
71	14-P116	75	n.d	n.d	58	83	n.d	17	9510	779	n.d	94	118	96	294
72	14-P117	n.d	265	n.d	79	101	18	16	13600	1300	n.d	214	262	118	171
73	14-P118	n.d	310	n.d	209	35	106	25	5560	963	n.d	88	190	79	53
74	14-P118	n.d	321	n.d	122	378	26	16	16600	2120	n.d	190	675	172	174
75	14-P119	104	n.d	n.d	34	86	18	16	4310	1530	n.d	51	180	57	281
76	14-P120	n.d	320	n.d	49	66	20	19	8010	1340	n.d	74	165	76	224
77	14-P121	n.d	230	n.d	n.d	51	47	22	2830	1780	n.d	n.d	292	93	n.d
78	14-P123	n.d	247	73	n.d	112	35	28	4680	1440	n.d	n.d	403	96	101
79	14-P127	n.d	224	n.d	201	303	46	16	17600	3160	n.d	180	691	174	90
80	14-P130	88	1850	n.d	60	n.d	39	n.d	6140	25500	133	196	208	126	138
81	14-P131	n.d	n.d	n.d	128	124	26	21	8140	1180	n.d	290	261	82	232
82	14-P132	167	n.d	n.d	230	57	18	19	5990	1950	82	115	173	75	245

Table 2: Correlation Coefficient Elements

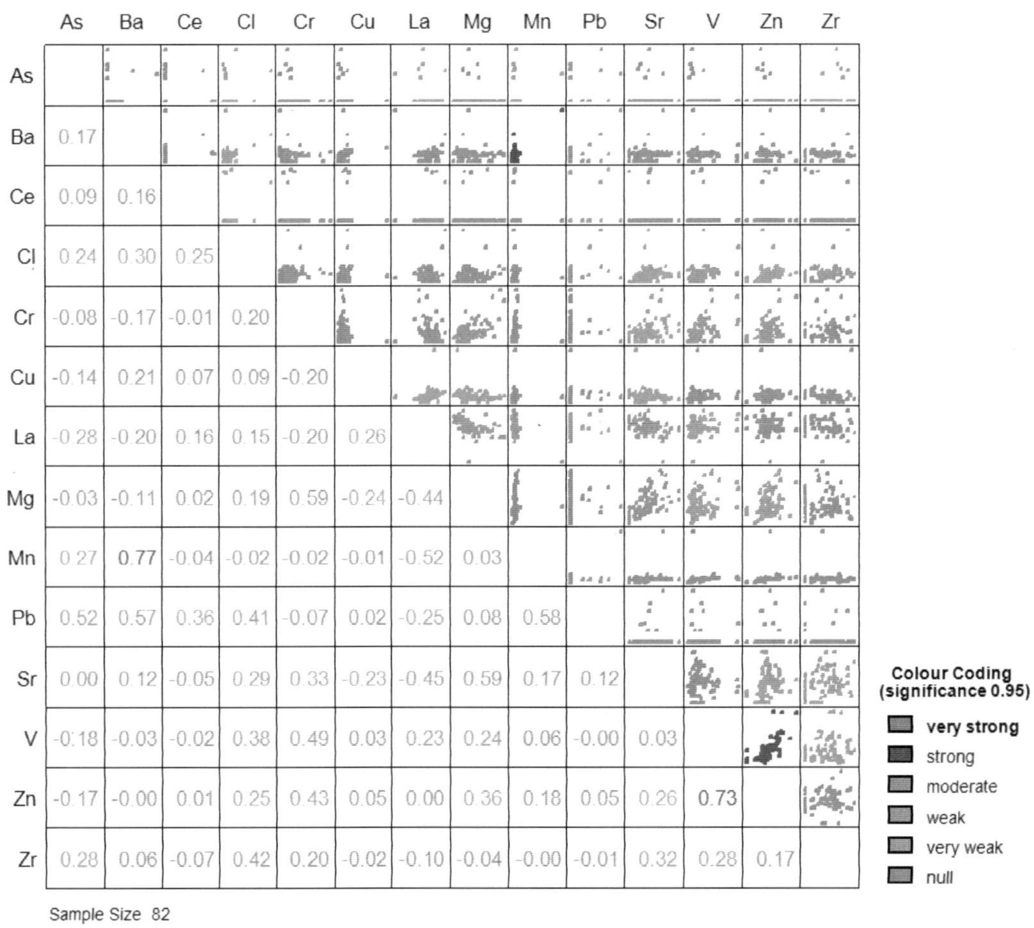

Areas of anomalous elements concentration and statistical analysis are described in the regional geochemical map of Bandung (Figure 3 – Figure 16). Geochemical maps are an important tool in the exploration of mineral deposits and in the last decade it is also become increasingly important for environmental issues [12]. In purpose of geochemical prospecting of mineralization resources, Cu, Pb and Zn are used as indicator and pathfinder for many types of Cu and gold deposit. However there are no correlation between these elements (Cu-Pb (0.02), Cu-Zn (0.05) and Pb-Zn (0.05)) for this area. Only iron deposit as lateritic type reported as economic mineralization near Subang area [13].

The largest scale regional pattern observed on the maps is formed by elements such as Ba, Cl and Cu. All these elements show significantly higher concentrations in the Tangkuban Perahu volcano area than in the south and north part. The element patterns coincident with volcanic rocks distribution of the Tangkuban Perahu volcano [9] and typical enriched elements from volcanic and igneous rocks source [14,15].

Other noticeable feature is the distinctive distribution elements pattern between north and south part of the map. Some element show low concentration (Cu and La) in portions of the northern part of the Tangkuban Perahu volcano and Volcanic Rock Complex. This area includes the district of Purwakarta, Subang and Indramayu. This feature is likely due to a Tertiary Sedimentary Rock formation. The only exceptions among the mapped elements to this trend of low concentration are Zr (Figure 16). This element show relatively high concentrations along the Tangkuban Perahu volcano to the north. This trend is believed to reflect the accumulation of heavy minerals, such as zircon for their resistance to weathering [11].

The Bandung city containing elevated concentrations of Pb (101 ppm) and Ce (76 ppm) in Location No. 17. The location is a region of urban and industrial complex, lies on the Quarternary Volcano-alluvial Deposit formation. These increasing of anthropogenic elements may correspond to industrial and urban pollutions. Some possible element pollutant such as As, Cu, Zn, Sr and La [11], shows also in Purwakarta, Indramayu and Subang, characterized by rather high concentration anomalies and the distribution pattern scattered locally close to industries, field rice area and cities. These are example of geochemical map of element that can be related to known natural features or associated with anthropogenic pollution.

Other patterns have no obvious explanation; for example, the striking differences of La, Cu, and Mg patterns between the north-eastern part area of the map (Indramayu district) as compared to the south- western part (Bandung city) and north-western part (Purwakarta and Subang district). Possible causes for that pattern is the strike-slip fault on these locations.

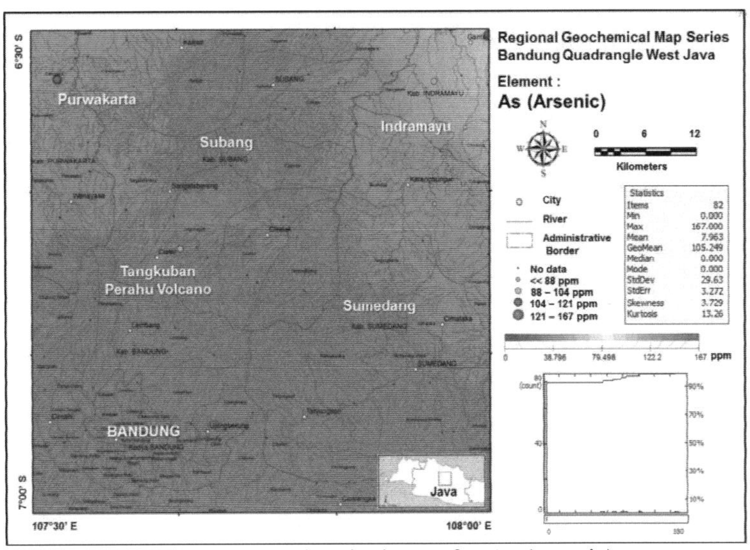

Figure 3. Geochemical Map for As (arsenic)

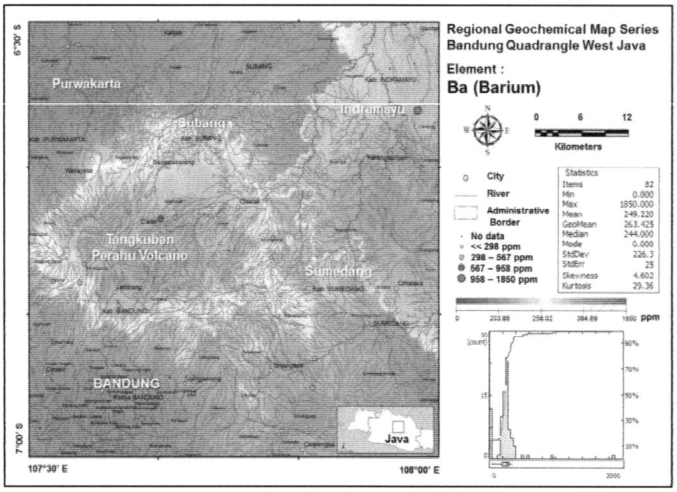

Figure 4. Geochemical Map for Ba (barium)

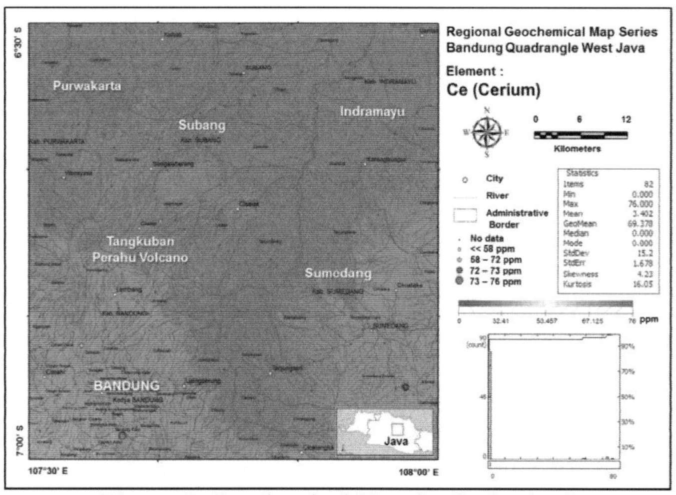

Figure 5. Geochemical Map for Ce (cerium)

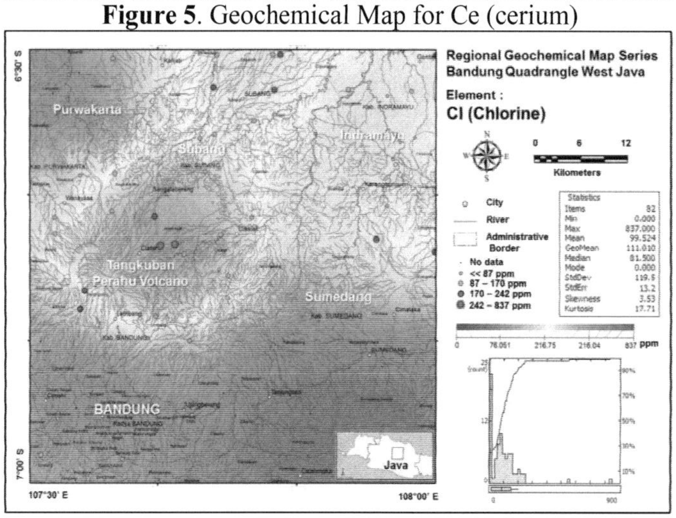

Figure 6. Geochemical Map for Cl (chlorine)

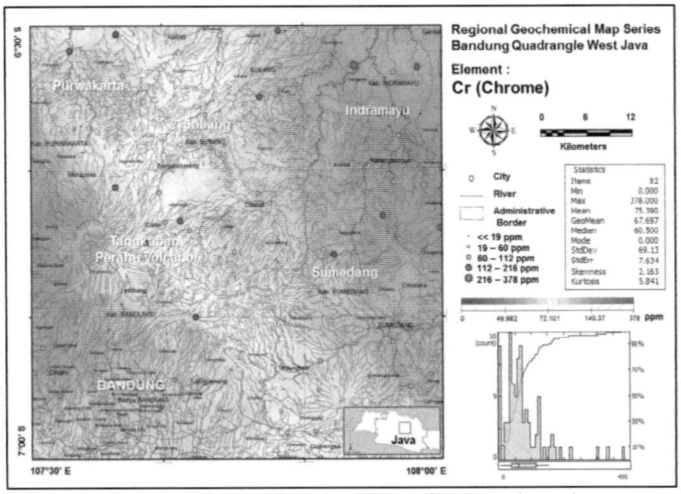

Figure 7. Geochemical Map for Cr (chrom)

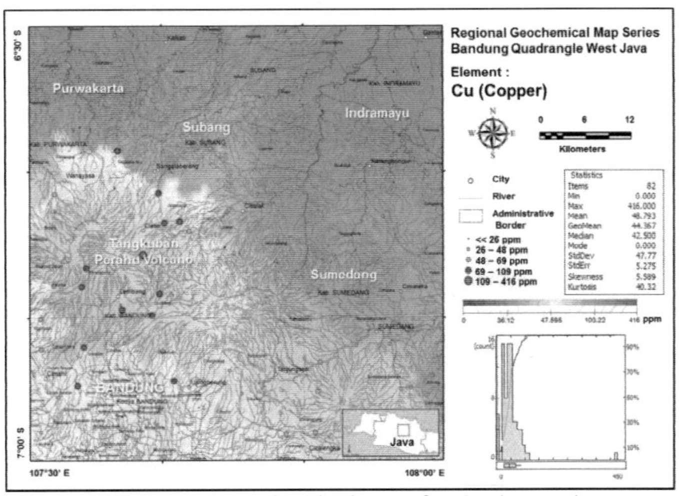

Figure 8. Geochemical Map for Cu (copper)

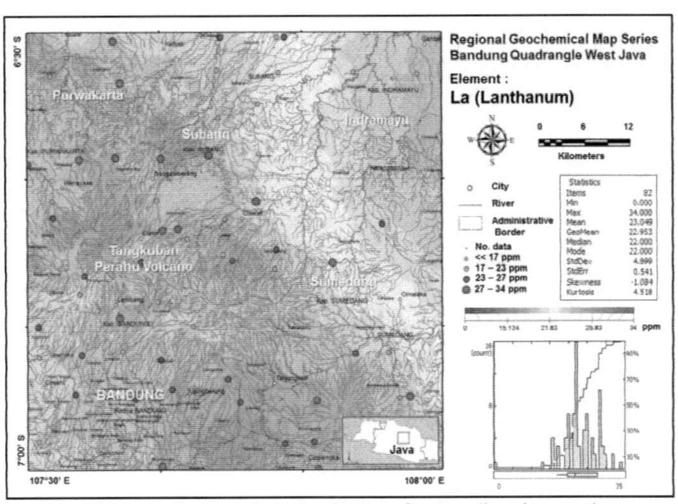

Figure 9. Geochemical Map for La (lanthanum)

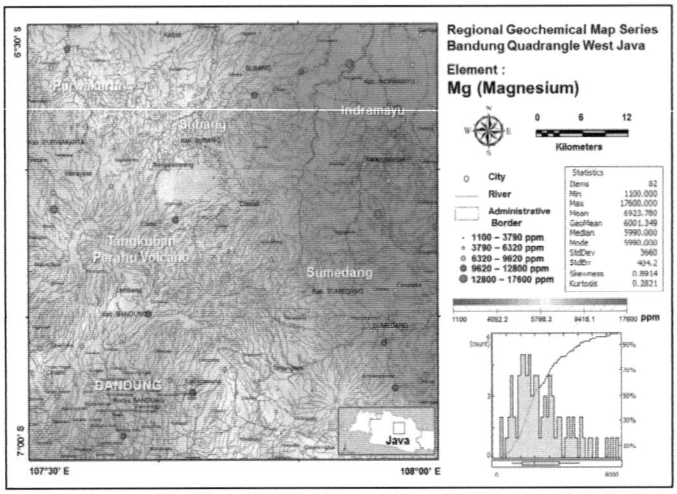

Figure 10. Geochemical Map for Mg (magnesium)

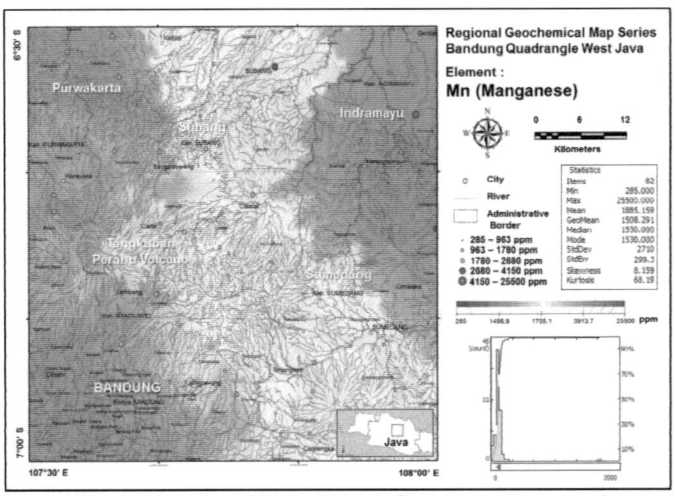

Figure 11. Geochemical Map for Mn (manganese)

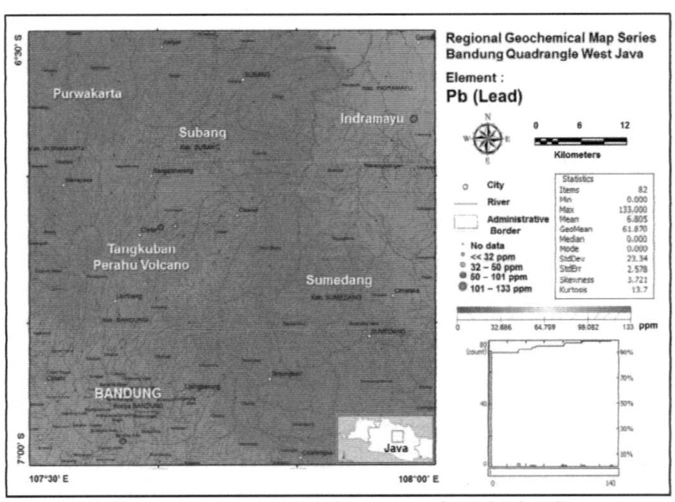

Figure 12. Geochemical Map for Pb (lead)

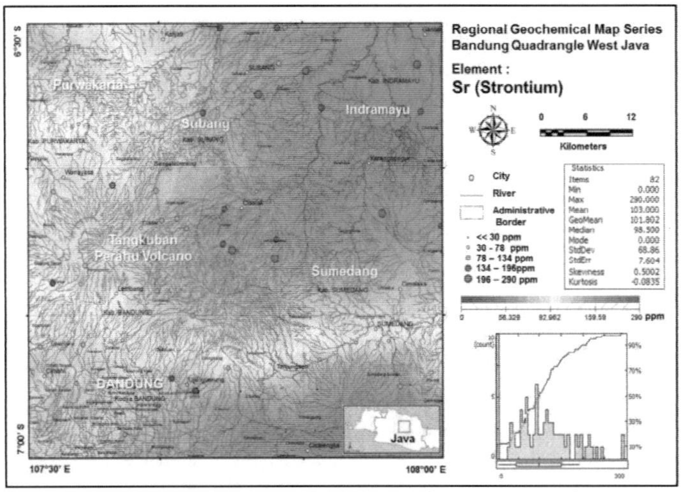

Figure 13. Geochemical Map for Sr (strontium)

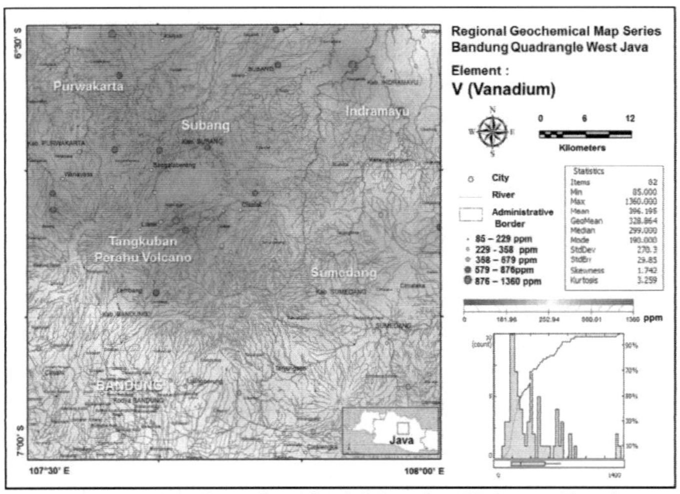

Figure 14. Geochemical Map for V (vanadium)

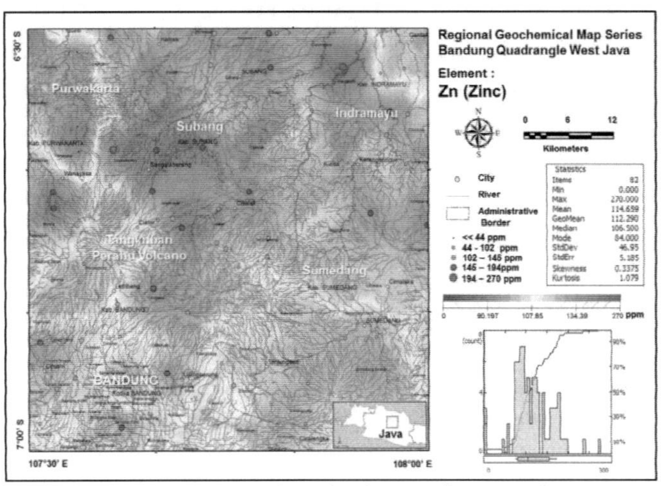

Figure 15. Geochemical Map for Zn (zinc)

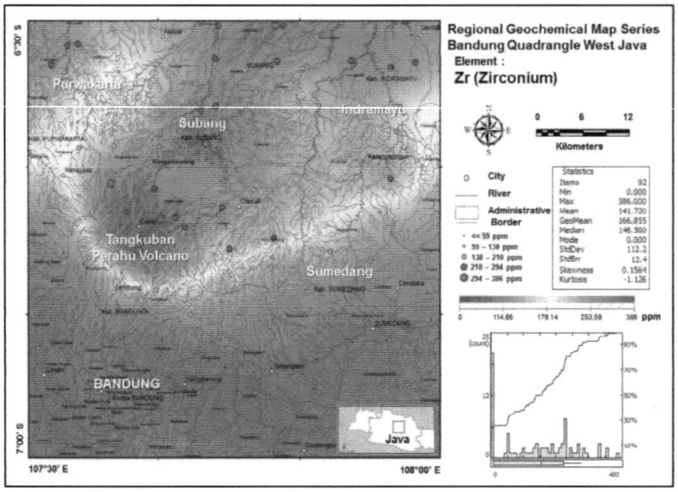

Figure 16. Geochemical Map for Zr (zirconium)

4. Conclusion

As a conclusion from the integration of geochemical result and statistical analysis is that the areas have significant anomaly indications. The volcanic products (Tangkuban Perahu Volcano, Volcanic Rock Complex and Quarternary-Alluvial Deposit formation) are clearly identified on the distribution of As, Ba, Cl, Cu, Zr and La elements, meanwhile the elements related to precipitation in the Tertiary Sediments are Mn, Zn, V and Sr. The Possibility influences of human activities are showing in a geochemical map of Cl, Cr, Cu, Pb and Zn that scattered locally close to the city, farming, and industry areas.

5. Acknowledgements

The authors would like to thank Agus Hendratno M.T., and his geological students from Gadjah Mada University for the invaluable help during the field work. Dwi P. Novitasari and Verry E. Setiawan for composing the spatial database in the GIS.

6. References

[1] Suprapto S J 2008 Geokimia regional Pulau Sumatera conto endapan sungai aktif fraksi -80# Buletin Sumber Daya Geologi 3-3 1
[2] Suprapto S J 2006 Geokimia regional Sulawesi bagian Utara percontoh endapan sungai aktif -80 mesh Jurnal Geologi Indonesia 1-2 73
[3] Gurniwa A 2005 Penyelidikan geokimia bersistem Lembar Ruteng Barat Kabupaten Manggarai dan Manggarai Barat Provinsi Nusa Tenggara Timur Proc. Hasil Lapangan – DIM Bandung 6-1
[4] Sumartono 2005 Penyelidikan geokimia regional sistematik Lembar Denpasar dan Mataram Provinsi Bali Proc. Hasil Lapangan – DIM Bandung 12-1
[5] Baharuddin 2012 Kandungan unsur Th, U dan Ce dalam stream sedimen di wilayah Pegunungan Schwaner Proc. Seminar pemanfaatan peta geokimia untuk sumber daya mineral dan lingkungan Yogyakarta
[6] Sendjaja P, Kusniadi A, Partoyo E, Baharuddin, Hikmat A, Suyono, Andriani S, Rosana M, Safri I and Hendratno A 2015 Atlas Geokimia Jawa Bagian Barat Pusat Survei Geologi Bandung ISBN: 978-602-6923-06-6 39 p
[7] Katili J A and Sudrajat A 1984 Volcanological Survey of Indonesia, Bandung, 102 p
[8] Dam M A C, Suparan P, Nossin J J, Voskuil R P G A and GTL Group 1996 A chronology for geomorphological developments in the greater Bandung area, West-Java, Indonesia J. SE. Asia Sci. 14-112 pp 101-105
[9] Silitonga P H 2003 Geological Map of The Bandung Quadrangle Geological Research and

Development Centre, Bandung
[10] Darnley A G, Björklund A, Bølviken B, Gustavsson N, Koval P V, Plant J A, Steenfelt A, Tauchid M, Xuejing X, Garrett R G and Hall G E M 1995 A global geochemical database for environmental and resource management: Final report of IGCP Project 259 Earth Sciences 19 UNESCO Pub. Paris 122 p
[11] Steenfelt A 2001 Geochemical atlas of Greenland – West and South Greenland Geological Survey of Denmark and Greenland Report 2001/46 39 p
[12] Bogen J, Walling D E and Day T J 1992 Proc. Int. Sym. Erosion and Sediment Transport Monitoring Programmes in River Basins Oslo IAHS Publ. 210 527 p
[13] Rosana M F, Widhiyatna D and Kartawa W 2013 Potensi Sumber Daya Mineral Jawa Barat: menuju pembangunan Jawa Barat yang Berkelanjutan UNPAD open Repository Bandung e print 16907 13 p
[14] Ohta A, Imai N, Terashima S and Tachibana Y 2005 Influence of surface geology and mineral deposits on spatial distributions of elemental concentrations in stream sediments of Hokkaido Japan J. Geochem. Explor. 86 86-103
[15] Eilenberg S, Carr M J 1981 Copper contents of lavas from active volcanoes in El Salvador and adjacent regions in Central America J. Economic Geology Society of Economic Geologist 76-8 p 2246-2248

Heavy metals accumulation by *Athyrium yokoscence* in a mine area, Southwestern Japan

Hendra Prasetia[1], Masayuki Sakakibara[1,2], Akinari Takehara[1,3], Yuri Sueoka[1].

[1] Graduate School of Science and Engineering, Ehime University, Matsuyama 790-8577, Ehime, Japan
[2] Faculty of Collaborative Regional Innovation, Ehime University, Matsuyama, Ehime, Japan
[3] Fukuizumi Junior High School, Sakai City, Osaka, Japan
Corresponding Author: hendra@sci.ehime-u.ac.jp

Abstract. Mine dumps pose environmental problems worldwide. Phytoremediation is a remediation technique that uses plants to clean polluted environments. *Athyrium yokoscence*, a fern of the family Aspidiaceae, is a well-known Cd hyperaccumulator used in phytoremediation. The aim of the current study is to determine the extent of heavy metal absorption by *A. yokoscence* and evaluate its potential use in phytoremediation. The shoots, stems, and roots of *A. yokoscence* were analysed by inductively coupled plasma–optical emission spectrometry (ICP–OES) to determine the concentrations of Cd, As, Pb, Cu, and Zn, yielding maximum concentrations in the shoots of 851, 215, 192, 60.9, and 769 mg/kg-DW, in the stems of 481, 140, 881, 28.9, and 495 mg/kg-DW, and in the roots of 1210, 1868, 6473, 2484, and 5446 mg/kg-DW, respectively. The results indicate that *A. yokoscence* is a hyperaccumulator of Cd and can be used for the translocation of As, Pb, Cu, and Zn within the shoot tissue. The roots translocate heavy metals through the stems and they finally reside within the shoot tissue. The results suggest that *A. yokoscence* has considerable potential use as an environmental assessment tool for Cd, As, Pb, Cu, and Zn, and in helping to remediate mine dumps.
Keywords: hyperaccumulator; phytoremediation; translocation, maximum concentration

1. Introduction

The toxicity of heavy metals is a serious problem for the integrity of environmental, ecological, and nutritional systems [1]. Heavy metal pollution has been recognized as a global issue for decades. The pollutants are introduced to the environment directly as a result of accidental spills during transportation, leakage from waste disposal sites, and industrial contamination [2]. The metals contaminate air, water, and soil. The most common heavy metal contaminants are Cd, Cr, Cu, Hg, Pb, and Zn [3].

Various techniques and chemical methods are used to remediate heavy metal contamination, such as soil washing, soil vapor extraction, land farming, soil flushing, solidification or stabilization, thermal desorption, biopiles, bioslurry systems, bioventing, encapsulation, and aeration [4].

Biological methods, utilizing living organisms can be applied to assess the degree of heavy metal contamination within an environment [5,6]. Phytoremediation is a well-known, cost effective, green technology that uses plants to remediate heavy metal contamination in water, air, and soil. This technology consists of two components: root-colonizing microbes and the plants themselves [7].

Hyperaccumulator plants are commonly used in phytoremediation. A hyperaccumulator is defined as a plant in which the concentrations of heavy metals in the above-ground parts are 50–100 times higher than in plants from non-polluted environments [8]. A plant's translocation factor (TF) is the ratio of a metal or metalloid's concentration in the shoots to that in the roots [9], and this value can be used to determine if a species is a hyperaccumulator, as well as being an important parameter in studies of heavy metal uptake by plants [10,11].

A plant's species and genotype determines its tolerance to heavy metals [12,13] *Athyrium yokoscence* is a fern of the family of Asipidaceae [14] that is used as an indicator of metal-contaminated soils in Japan [15]. Nishizono *et al.* [16] reported that *A. yokoscence* grows predominantly on metal-contaminated soils in Japan and is a non-habitual condenser. The fern accumulates high levels of Cu and Zn in the roots, and Cd in the shoots, when grown on soils contaminated with heavy metals [16]. Previous studies have shown that *A. yokoscence* has potential for use in the phytoremediation of soils contaminated by heavy metals. In this context, the aim of this study is to determine the potential use of *A. yokoscence* for phytoremediation in mine dumps.

2. Materials and Methods

This is a preliminary study that investigates the use of *A. yokoscence* to assess the degree of contamination in an abandoned mine site and evaluates its potential use in phytoremediation and as an indicator of environmental toxicology. The organic and mineral contents (%vol) of soil from the site were measured to a depth of 5 cm, using a soil moisture kit, to determine the relationship between the composition of the soil, in which the plant takes root, and heavy metal uptake by the plant tissue.

2.1. Sampling plots

A field survey was completed at an abandoned mine dump in Southwestern Japan on 10–11 May 2014. The dump is located at an altitude of 400–500 m above sea level, in a warm-temperate zone. Copper and Pb were the principal metals to have been smelted at the abandoned mine site, between AD 808 and 1972. The surface of the waste dump consists mainly of slag fragments and tailings composed of andesite, mudstone, and limestone. Solidified and coherent slag has become partially exposed on the slag and tailings dump. *A. yokoscence* grows naturally on this mine dump, and 12 live *A. yokoscence* plants were collected for analysis from randomly selected sites across several areas of the mine dump (Figure 1).

2.2. Analytical methods

The fern samples were first washed with tap water and separated into shoots, stems, and roots. After separation they were rinsed in an ultrasonic frequency cleaner with ultrapure water for 5 min intervals until clean. The samples were then dried at ~80 °C for two days in a ventilated oven. Once dry, the samples were pulverized using a powder mill (Varian PM-2005 m, Osaka Chemical). Each 20 mg sample of shoots, stems, or roots was digested with a solution of H_2O_2, HF, and HNO_3 at a ratio of 2:5:10. Heavy metal concentrations were determined by inductively coupled plasma–optical emission spectrometry (ICP–OES) (Varian Optima^{TM} 8300 series, PerkinElmer, Kanagawa, Japan) at the Integrated Centre for Sciences, Ehime University, Matsuyama, Japan. The different operating wavelengths of metals were used to identify the element in the sample [17,18].

2.3. Translocation Factor

The TF is defined as the ratio of the heavy metal concentration in the shoots to that in the roots [19,20] as follows [21,22]:

$$TF = C_{shoot}/C_{root} \quad (1)$$

where C_{shoot} is the metal concentration in the shoots and C_{root} is that in the roots. The TF is calculated using the heavy metal concentrations and is used to evaluate the phytoextraction ability of a plant, in particular the plant's ability to translocate heavy metals from the roots to the shoots [22].

2.4. Statistical analysis

Statistical analyses were performed using OriginPro 9.0 for Windows. The Shapiro–Wilk test was used to check the normality of the heavy metal concentrations in the shoots, stems, and roots. The data were log normally distributed, so the Kruskal–Wallis ANOVA test was used to test for significant differences, with $\rho < 0.05$ considered statistically significant.

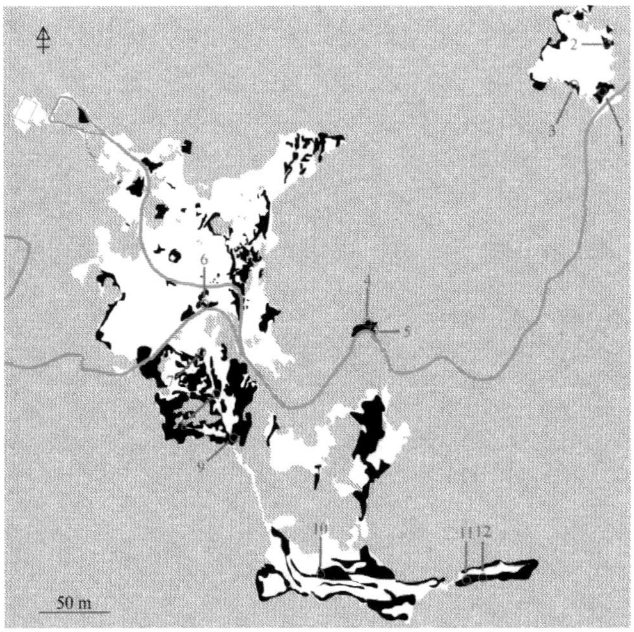

Figure 1. Distribution map of *A. yokoscence* and sampling point in an abandoned mine site

3. Results

The organic and mineral contents of the soil samples ranged from 3.80 to 31.6 and 0.20 to 25.3 (%vol) (Table 2), respectively. These results indicate that there is no significant relationship between the organic and mineral contents of the soil and heavy metal uptake into the plant tissue. The total concentrations of Cd in the plant shoots ranged from 60.9 to 851 mg/kg dry weight (DW) (Table 1). The total concentrations of Cd in the stems and roots ranged from 111 to 481 and from 184 to 1210 mg/kg DW (Table 1), respectively. These results suggest that *A. yokoscence* can accumulate high concentrations of Cd, at levels that exceed the standard concentration for hyperaccmulators of Cd, at this abandoned mine site.

Table 1. The Cd, As, and Pb concentrations and organics and minerals content of soil of *A. yokoscence*.

Sample	Cd				As				Pb			
	Shoot (mg/kg-DW)±SD	Stem (mg/kg-DW)±SD	Root (mg/kg-DW)±SD	TF	Shoot (mg/kg-DW)±SD	Stem (mg/kg-DW)±SD	Root (mg/kg-DW)±SD	TF	Shoot (mg/kg-DW)±SD	Stem (mg/kg-DW)±SD	Root (mg/kg-DW)±SD	TF
1	487±87.1	434±71.5	1210±280	0.40	51.8±9.27	61.2±10.1	84.2±19.5	0.61	92.1±16.5	257±42.3	1147±265	0.08
2	230±40.6	181±32.4	184±31.3	1.25	ND	55.3±9.91	833±142	ND	133±23.4	441±79.0	3876±658	0.03
3	301±54.3	240±41.2	652±120	0.46	215±38.8	ND	ND	ND	64.7±11.7	55.0±9.44	445±82.0	0.15
4	425±75.2	390±69.2	982±167	0.43	161±28.4	ND	156±26.5	1.03	40.4±7.14	454±80.6	158±26.8	0.26
5	280±48.2	234±39.1	714±124	0.39	47.6±8.18	26.0±4.32	2.31±0.40	20.58	32.1±5.53	246±41.1	740±129	0.04
6	381±65.9	210±39.1	917±159	0.42	104±17.9	57.7±10.7	ND	ND	24.3±4.21	99.1±18.4	422±72.8	0.06
7	448±78.4	375±64.9	448±79.4	1.00	ND	ND	1868±331	ND	59.6±10.4	150±26.0	6473±1148	0.01
8	492±87.1	333±62.3	545±96.4	0.90	92.8±16.4	ND	112±19.7	0.83	114±20.3	49.6±9.30	3182±563	0.04
9	350±64.1	305±51.7	792±169	0.44	102±18.7	47.1±7.98	127±27.0	0.81	46.2±8.45	204±34.7	1091±232	0.04
10	60.9±10.6	111±19.2	232±40.8	0.26	161±28.1	ND	545±95.7	0.30	191.7±33.5	880±152	3361±590	0.06
11	851±154	481±81.7	953±171	0.89	144±26.1	119±20.2	363±65.2	0.40	160±29.0	373±63.2	2236±401	0.07
12	458±81.8	354±59.2	760±135	0.60	ND	140±23.5	388±68.9	ND	62.8±11.2	175±29.3	1032±184	0.06

DW: Dry Weight; SD: Standard Deviation; ND: Not Detected.

The shoots of *A. yokoscence* accumulated As and Pb at total concentrations ranging from not detected (ND) to 215 and from 24.3 to 192 mg/kg DW, respectively Table 1). Ranges of Cu and Zn total concentrations in the shoots were from ND to 60.9 and from ND to 770 mg/kg DW, respectively Table 2). These results suggest that *A. yokoscence* cannot accumulate high concentrations of As, Pb, Cu, and Zn from the roots into the shoots. Ranges of As, Pb, Cu, and Zn in the stems were from ND to 140, 49.6 to 881, ND to 21.5, and 63.9 to 494.7 mg/kg DW, respectively, and in the roots were from ND to 1868, 158 to 6473, ND to 2484, and 508 to 5446 mg/kg DW, respectively (Table 1 and 2). The amount of heavy metals transported from roots to stem show a positive correlation at a logarithmic scale, with the root concentrations (Figure 2). The results indicate that *A. yokoscence* has the ability to uptake heavy metals into its root tissue, but is generally unable to transport them to its stem tissue.

Table 2. The Cu and Zn concentrations and organics and minerals content of soil of *A. yokoscence.*

Sample	Cu				Zn					
	Shoot (mg/kg-DW)±SD	Stem (mg/kg-DW)±SD	Root (mg/kg-DW)±SD	TF	Shoot (mg/kg-DW)±SD	Stem (mg/kg-DW)±SD	Root (mg/kg-DW)±SD	TF	Organic (%vol)	Mineral (%vol)
1	ND	ND	33.2±7.67	ND	497±89.0	254±42.0	1456±336	0.34	6.40	2.43
2	ND	1.04±0.18	741±126	ND	180±31.8	151±27.0	2459±418	0.07	16.6	11.6
3	ND	ND	ND	ND	367±66.3	103±17.7	833±153	0.44	4.80	1.00
4	60.9±10.8	28.9±5.13	ND	ND	583±103	415±73.6	1650±280	0.35	15.3	10.40
5	ND	ND	ND	ND	225±38.7	151±25.3	1109±193	0.20	18.9	13.7
6	ND	15.1±2.80	ND	ND	213±37.0	157±29.2	946±164	0.23	5.60	1.6
7	ND	18.7±3.24	1403±250	ND	694±122	495±85.6	5447±966	0.13	23.7	18.2
8	20.1±3.56	21.5±4.03	2484±439	0.01	659±117	332±62.3	5026±888	0.13	15.5	10.7
9	ND	ND	7.21±1.53	ND	200±36.6	145±24.6	509±108	0.39	18.1	12.7
10	ND	ND	682±120	ND	ND	63.9±11.0	790±140	ND	15.6	9.40
11	37.2±6.74	6.67±1.13	904±162	0.04	770±139	368±62.5	3275±588	0.23	12.2	7.47
12	ND	ND	486±86.5	ND	456±81.6	242±40.5	1904±340	0.24	17.3	12.4

DW: Dry Weight; SD: Standard Deviation; ND: Not Detected.

4. Discussion

4.1. Heavy metal absorption and transportation from the roots to stem of *A. yokoscence*

Heavy metals are absorbed into plant tissue through the phloem, in conjunction with nutrient absorption. The heavy metals are absorbed in the form of essential macro- and micronutrients, induced by the selective uptake of ions by roots, or by the diffusion of elements in the soil [23,24]. The ability of a plant to accumulate heavy metals is considered a detrimental trait in the long term [25], because even micronutrients can become toxic for plants when they exceed threshold values [24]. The composition of the soil substrate and its water content may influence a plant's ability to uptake heavy metals [26]. The separation of metals from their source, by uptake to a root, is complicated by the mediating effect of soil properties and plant processes [23]. Heavy metal uptake by plants depends on several plant factors, such as root intrusion, water, and ion flux, the kinetics of metal solubilization in soils, biological parameters, and a plant's ability to adapt metabolically to metal stresses in the environment [23]. This study includes some observations of soil composition, but more detailed analysis would be required to fully understand the role of the soil substrate in the absorption and transport of heavy metals by *A. yokoscence*.

Cadmium represents a serious environmental hazard because it can be absorbed through the alimentary tract, it penetrates through the placenta, and damages membranes and DNA [27]. The preferential uptake of Cd, as observed in this study, is of particular concern because it demonstrates that Cd can be present at elevated levels in plant tissues not considered to be phytotoxic. This could pose a threat to humans and animals if the plants form part of their food chain [24]. Iron deficiency promotes the uptake of Cd in *Thlaspi caerulescens* Ganges ecotype [28]. In *Arabidopsis halleri*, Cd transfer from the growing medium to the root xylem is partially shared with Zn and/or Fe transport [29]. On the other hand, a manganese-enriched medium promotes the uptake of Cd in *Lactuca sativa* [30].

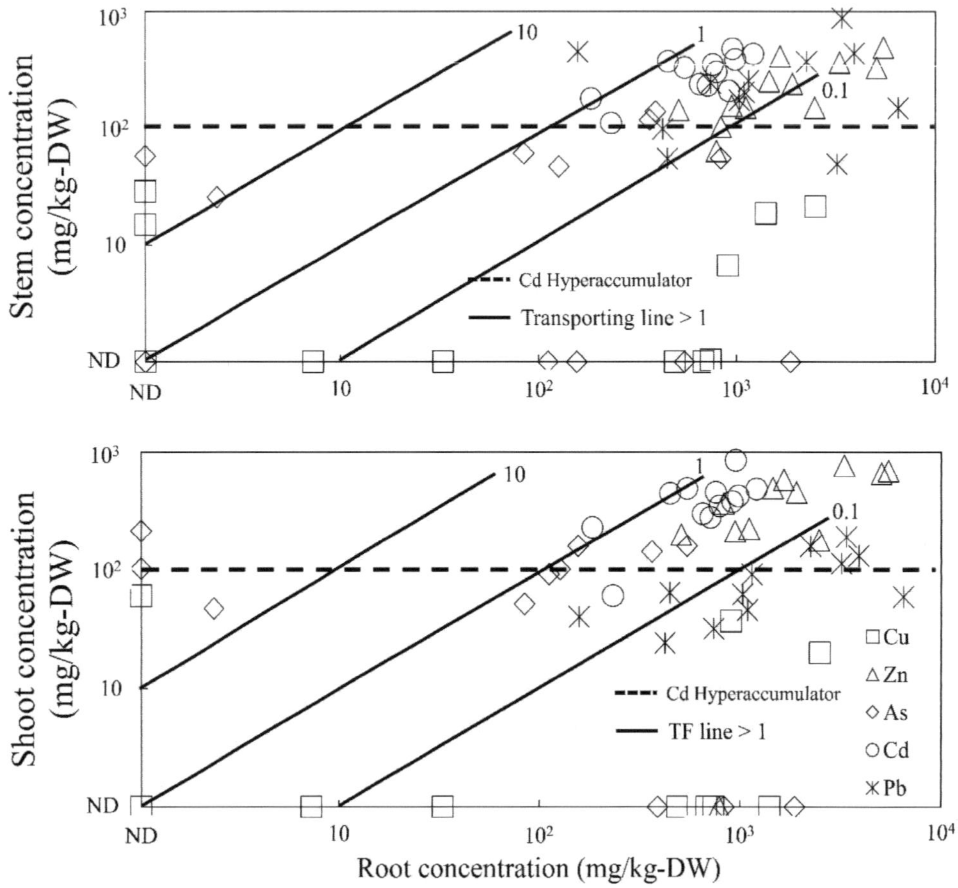

Figure 2. TF and transporting of heavy metals of *A. yokoscence*; ND: Not Detected; DW: Dry Weight.

Cd concentrations in the shoots of *A. yokoscence* may exceed 100 mg/kg DW, indicating this is a hyperaccumulator plant [31]. Total Cd concentrations observed in shoots, in previous studies by Van *et al.* [15] and Morishita and Boratynski [32], were <1200 and 996 mg/kg DW, respectively. In the present study, Cd uptake to the stem tended to increase and stabilize when Cd was present at high concentrations in the roots (Figure 2). A positive correlation between Cd transport to the stem and root concentrations indicates that *A. yokoscence* has the potential to accumulate high concentrations of Cd in its shoot tissue (Figure 2). This suggests in turn that *A. yokoscence* can be used for phytoremediation and as a preliminary step in environmental assessments of Cd contamination.

In the present study, *A. yokoscence* also accumulated As, Pb, Cu, and Zn (Figure 2). Other studies have reported that As accumulation is influenced by soil properties [33]. Plants with concentrations of As in their shoots exceeding 1000 mg/kg DW have been classified as As hyperaccumulators [34].

Lead is highly mobile in the soil and accumulates at depths of up to 20.32 cm [3]. The amount of Pb in a plant is controlled by the pH, organic matter content, and phosphorus concentration of the soil [3]. In this study, no relationship was observed between Pb concentrations and the organic content of the soil, which suggests that Pb uptake and accumulation are not controlled by organic matter content in this case.

In root tissue, Cu is almost always present in complexed forms; however, it is most likely that the metal enters the root cells in dissociated forms [35]. The rate of Cu uptake by plants differs widely according to speciation [35]. The concentration of Cu in root tissue affects a plant's capacity to retain Cu against transport to the shoots [35].

Zinc uptake is controlled by plant metabolism [34]. When Zn concentrations in contaminated soil exceed the level at which Zn is required as a nutrient, it may cause phytoextraction [1]. The accumulation and absorption of Zn into the shoots is necessary, given its role as a micronutrient [26].

The positive correlation between the concentration of metals in the stem and roots, as observed in this study (Figure 2), indicates that *A. yokoscence* has potential both as a metal-tolerant plant and in preliminary environmental assessments of As, Pb, Cu, and Zn at the level of an individual park. The uptake of these metals and their transport from roots to stem tend to increase and become stabilized at high concentrations in the root tissue. In the present study it is possible that leaching occurred during washing and sample preparation, which may have caused a bias in the data and the very wide range in concentrations observed in this study. A logarithmic scale was used to determine the correlations between the concentrations of Cd, As, Pb, Cu, and Zn in the roots and stems, and in roots and shoots, given the wide range in values.

4.2. Translocation Factor

The TF is a measure of a plant's ability to translocate heavy metals from the roots to the shoots. While some species are able to accumulate heavy metals in their shoots, most plants have a low TF value, which indicates they have limited ability to translocate heavy metals. In this study, the TF values of Cd, As, Pb, Cu and Zn range from 0.26 to 1.25, ND to 20.6, 0.01 to 0.25, ND to 0.04, and ND to 0.44, respectively. When a TF value exceeds 1, it indicates the plant is a hyperaccumulator. Mycorrhizal fungi may influence TF values [9]. The TF values in this study indicate the hyperaccumulation of Cd and As by *A. yokoscence* (Figure 2). However, the TF results differ from the concentration results, especially for As (Figures 2). This finding indicates that a plant should not be classified as a hyperaccumulator based solely on TF values, but that concentration data for the metal of interest in the shoots should also be considered in the classification, and vice versa.

5. Conclusions

The present results show that *A. yokoscence* steadily accumulates Cd, As, Pb, Cu, and Zn, and translocates high concentrations of Cd to its shoots. The positive correlation between heavy metals in the shoots, stem, and roots indicates that *A. yokoscence* can be used to determine contamination by these metals in a preliminary environmental assessment. The results indicate that the hyperaccumulator classification should reflect a combination of shoot accumulation and TF. These factors indicate that *A. yokoscence* has potential for the phytoextraction of Cd at abandoned mine sites.

6. Acknowledgements

One author (Hendra Prasetia) wishes to thank the Japanese Government for providing a Monbukagakusho Scholarship for graduate study at Ehime University. This research was supported by Feasibility Study "Social Acceptance of Regional Innovation for Reducing High-Impact Environmental Pollution", Research Institute for Humanity and Nature (RIHN) and JSPS KAKENHI Grant Number 16H02706 to M. Sakakibara.

7. References

[1] Nagajyoti P C, Lee K D and Sreekanth T V M 2010 Heavy metals, occurrence and toxicity for plants: a review *Environ. Chem. Lett.* **8** 199–216
[2] Riser-Roberts E 1992 *Bioremediation of petroleum contaminated sites* (CK Smoley)
[3] Tangahu B V, Sheikh Abdullah S R, Basri H, Idris M, Anuar N and Mukhlisin M 2011 A review on heavy metals (As, Pb, and Hg) uptake by plants through phytoremediation *Int. J. Chem. Eng.*
[4] Khan F I, Husain T and Hejazi R 2004 An overview and analysis of site remediation technologies. *J. Environ. Manage.* **71** 95–122

[5] Peng K J, Luo C L, Chen Y H, Wang G P, Li X D and Shen Z G 2009 Cadmium and other metal uptake by Lobelia chinensis and Solanum nigrum from contaminated soils *Bull. Environ. Contam. Toxicol.* **83** 260–4

[6] Brookes P C 1995 The use of microbial parameters in monitoring soil pollution by heavy metals *Biol. Fertil. Soils* **19** 269–79

[7] Sarma H 2011 Metal hyperaccumulation in plants: a review focusing on phytoremediation technology *J. Environ. Sci. Technol.*

[8] Shen Z-G and Liu Y-L 1998 Progress in the study on the plants that hyperaccumulate heavy metal *Plant Physiol. Commun.* **34** 133–9

[9] Trotta a., Falaschi P, Cornara L, Minganti V, Fusconi a., Drava G and Berta G 2006 Arbuscular mycorrhizae increase the arsenic translocation factor in the As hyperaccumulating fern Pteris vittata L. *Chemosphere* **65** 74–81

[10] Zayed A, Gowthaman S and Terry N 1998 Phytoaccumulation of trace elements by wetland plants: I. Duckweed *J. Environ. Qual.* **27** 715–21

[11] Marchiol L, Assolari S, Sacco P and Zerbi G 2004 Phytoextraction of heavy metals by canola (Brassica napus) and radish (Raphanus sativus) grown on multicontaminated soil *Environ. Pollut.* **132** 21–7

[12] Adriano 2001 *Trace Elements in Terrestrial Environments* vol 32 (Springer-Verlag)

[13] Tlustoš P, Pavlíková D, Száková J and Balík J 2006 Phytoremediation of Metal-Contaminated Soils ed J-L Morel, G Echevarria and N Goncharova (Dordrecht: Springer Netherlands) pp 53–84

[14] Kitagishi K, Yamane I and others 1981 *Heavy metal pollution in soils of Japan.* (Japan Scientific Societies Press)

[15] Van T K, Kang Y M, Fukui T, Sakurai K, Iwasaki K, Aikawa Y and Phuong N M 2006 Arsenic and heavy metal accumulation by Athyrium yokoscense from contaminated soils *Soil Sci. Plant Nutr.* **52** 701–10

[16] Nishizono H, Ichikawa H, Suziki S and Ishii F 1987 The role of the root cell wall in the heavy metal tolerance of Athyrium yokoscense *Plant Soil* **101** 15–20

[17] Bettinelli M, Beone G M, Spezia S and Baffi C 2000 Determination of heavy metals in soils and sediments by microwave-assisted digestion and inductively coupled plasma optical emission spectrometry analysis *Anal. Chim. Acta* **424** 289–96

[18] Wang J, Nakazato T, Sakanishi K, Yamada O, Tao H and Saito I 2004 Microwave digestion with HNO 3/H 2 O 2 mixture at high temperatures for determination of trace elements in coal by ICP-OES and ICP-MS *Anal. Chim. Acta* **514** 115–24

[19] Yoon J, Cao X, Zhou Q and Ma L Q 2006 Accumulation of Pb, Cu, and Zn in native plants growing on a contaminated Florida site *Sci. Total Environ.* **368** 456–64

[20] McGrath S P and Zhao F-J 2003 Phytoextraction of metals and metalloids from contaminated soils *Curr. Opin. Biotechnol.* **14** 277–82

[21] Ladislas S, El-Mufleh A, Gérente C, Chazarenc F, Andrès Y and Béchet B 2012 Potential of Aquatic Macrophytes as Bioindicators of Heavy Metal Pollution in Urban Stormwater Runoff *Water, Air, {&} Soil Pollut.* **223** 877–88

[22] Mattina M I, Lannucci-Berger W, Musante C and White J C 2003 Concurrent plant uptake of heavy metals and persistent organic pollutants from soil *Environ. Pollut.* **124** 375–8

[23] Cataldo and Wildung 1978 Soil and plant factors influencing the accumulation of heavy metals by plants *Environ. Health Perspect.* **Vol. 27** 149–59

[24] Peralta-Videa J R, Lopez M L, Narayan M, Saupe G and Gardea-Torresdey J 2009 The biochemistry of environmental heavy metal uptake by plants: implications for the food chain *Int. J. Biochem. Cell Biol.* **41** 1665–77

[25] Kumar P B A N, Dushenkov V, Motto H and Raskin I 1995 Phytoextraction: the use of plants to remove heavy metals from soils *Environ. Sci. Technol.* **29** 1232–8

[26] Prasetia H, Sakakibara M, Sueoka Y and Sera K 2016 Pteris cretica as a Potential Biomarker

and Hyperaccumulator in an Abandoned Mine Site, Southwest Japan *Environments* **3** 15

[27] Kabata-Pendias 2011 2011 *Trace elements in soils and plants,* pp. 287-304.

[28] Lombi E, Tearall K L, Howarth J R, Zhao F-J, Hawkesford M J and McGrath S P 2002 Influence of iron status on cadmium and zinc uptake by different ecotypes of the hyperaccumulator Thlaspi caerulescens *Plant Physiol.* **128** 1359–67

[29] Ueno D, Iwashita T, Zhao F-J and Ma J F 2008 Characterization of Cd translocation and identification of the Cd form in xylem sap of the Cd-hyperaccumulator Arabidopsis halleri *Plant Cell Physiol.* **49** 540–8

[30] Ramos I, Esteban E, Lucena J J and Gárate A 2002 Cadmium uptake and subcellular distribution in plants of Lactuca sp. Cd--Mn interaction *Plant Sci.* **162** 761–7

[31] Baker A J M 1987 Metal tolerance *New Phytol.* **106** 93–111

[32] Morishita T and Boratynski J K 1992 Accumulation of cadmium and other metals in organs of plants growing around metal smelters in Japan *Soil Sci. Plant Nutr.* **38** 781–5

[33] Wei C-Y and Chen T-B 2006 Arsenic accumulation by two brake ferns growing on an arsenic mine and their potential in phytoremediation *Chemosphere* **63** 1048–53

[34] Bondada B R, Tu S and Ma L Q 2004 Absorption of foliar-applied arsenic by the arsenic hyperaccumulating fern (Pteris vittata L.) *Sci. Total Environ.* **332** 61–70

[35] Kabata-Pendias, A.; Pendias, H., 2011, *Trace elements in soils and plants*, pp. 253-268

An Artificial Channel Experiment for Purifying Drainage Water Containing Arsenic by Using *Eleocharis acicularis*

Kenji Okazaki[1], Shusaku Yamazaki[1], Toshiyuki Kurahashi[1], and Masayuki Sakakibara[2].

[1] Civil Engineering Research Institute for cold region, Public Work Research Institute, Sapporo, Japan.
[2] Department of Earth Science, Graduate School of Science and Engineering, Ehime University, Matsuyama, Japan.
Email: 90185@ceri.go.jp

Abstract: This paper reports the results of an artificial channel experiment in which water containing arsenic was purified by using *Eleocharis acicularis*. The experiment was conducted to investigate the feasibility of phytoremediation by *Eleocharis acicularis* in civil engineering projects. In the experiment, 15 m^2 of *Eleocharis acicularis* mats were laid in an artificial channel. Three sessions of artificial flow were implemented by leading 100.0 L of river water containing 0.234 mg/L of arsenic into the channel each time. The arsenic concentration of the leachate from the channel was analyzed. As the results of experiment, the arsenic concentrations of the leachate for the three sessions were 0.045 mg/L, 0.133 mg/L, and 0.249 mg/L. This shows that the arsenic concentration decreased during the first two sessions, whose flow totaled 200 L. The arsenic concentrations in the *Eleocharis acicularis* were 0.87 mg/kg, 1.01 mg/kg, and 4.16 mg/kg, which show that the plant absorbs arsenic. Moreover, it was found that the amount of sample water was reduced through evapotranspiration from the plant and the artificial channel.
Keywords: Artificial Channel; Purifying Drainage Water; *Eleocharis acicularis*; Arsenic; Evapotranspiration.

1. Introduction

Concentrations of heavy metals such as arsenic, selenium, lead and zinc that leach from soil and rock excavated at road construction sites occasionally exceed the criteria values prescribed for each such metal. When excavated soil and rock are temporarily stored at road construction sites, rainwater or snowmelt that seeps from such soil and rock may be contaminated by heavy metals. The Manual for Controlling Heavy Metal Pollution Caused by Natural Rocks and Soil at Construction Sites [1] proposes leachate treatment as a remedial measure. In this method, by exposing excavated soil and rock to the air at a temporary storage place, heavy metals contained in the leachate are properly treated. However, if heavy metal concentrations in the leachate exceed the corresponding criteria values, treatment of the heavy metals can be expensive, which poses a problem.

The experiments reported herein were conducted for the purpose of evaluating the applicability of phytoremediation to civil engineering work. Phytoremediation, a technique for cleaning up contaminated environments, has been used for purifying mine effluent. This technique has advantages and disadvantages. Its effectiveness is more dependent on weather conditions than is the effectiveness of other techniques for cleaning up environmental contaminants. Additionally, contaminants are removed only in the rhizosphere, and it takes time to grow plants large enough to be useful for phytoremediation. However, this technique is applicable at a relatively low cost, consumes no energy, and is environmentally sound. This study used *Eleocharis acicularis* (hereinafter: *E. acicularis*), a plant in the Cyperaceae family that absorbs and accumulates heavy metals, because this plant has been proven to be effective in mine effluent treatment [2-4]. *E. acicularis* is tolerant of a wide variety of heavy metals, and it absorbs these metals. The plant is a perennial that grows in clumps at ponds, marshes, reservoirs, and paddy fields, and it is distributed throughout Japan. *E. acicularis* propagates with runners that branch off from the root area. In the experiments reported below, *E. acicularis* was used for purifying river water that contained arsenic and flowed in an artificial channel. The concentrations of arsenic in the river water and in *E. acicularis* were measured.

2. Outline of the experiments

2.1. Experiment site and materials

Experiments were conducted by using an artificial channel that was installed on the roof of a laboratory building (Figure 1). The 50-m-long channel is made of PVC and consists of 5 sections. Each section is 30 cm wide and 10 m long. Short pipes were used for assembling the channel with a hydraulic gradient of 0.5 permillage (‰). In the channel, 15 m^2 of *E. acicularis* mats with a wet weight of 54.5 kg were placed (Photo 1). The *E. acicularis* used in the experiments was hydroponically cultured, and the roots of adjacent plants were intertwined. The plants were shaped to form mats. Water taken from a river near a hot spring in the city of Sapporo was used for the experiments. The river water includes hot spring water that contains arsenic welling up from the bedrock of the river [5].

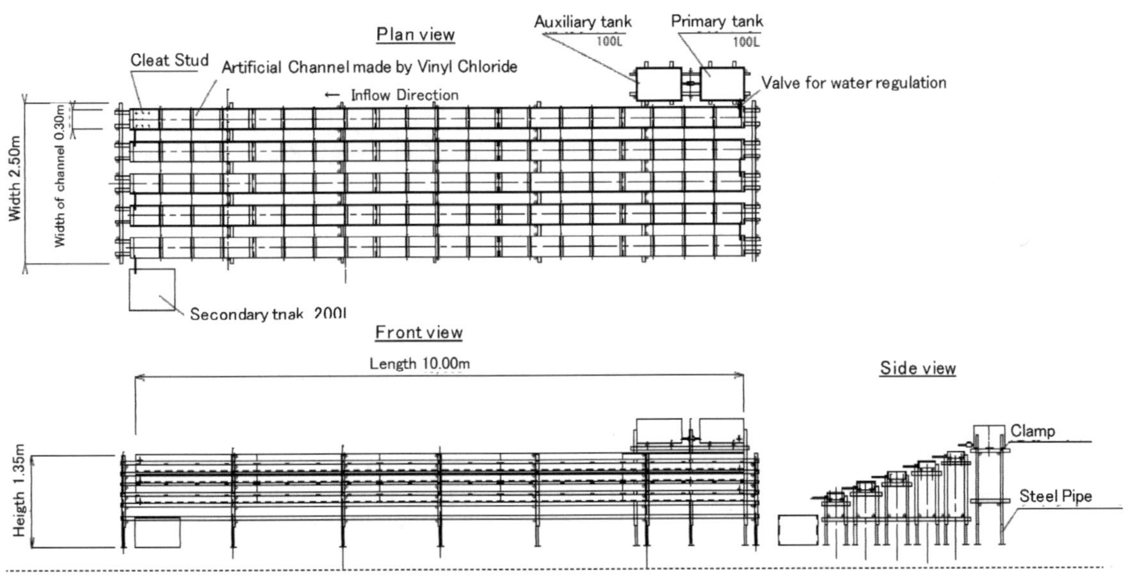

Figure 1. Outline of Artificial Channel

2.2. Experimental method

2.2.1. *Arsenic removal*

After pouring river water into the artificial channel in which *E. acicularis* mats had been put, *E. acicularis* and seepage water were sampled to measure the arsenic content and concentration. A total of 300.0 L of river water was used, and 100.0 L of river water was poured into the channel three times over 2 days. The inflow rate was 0.4 - 0.6 L/min. The experiment regarding arsenic removal was conducted between late July and early August, when the average air temperature was 25 °C or higher. At the onset of the experiment and each time 100.0 L of river water was poured into the channel, 50 ml of seepage water was sampled at the outflow end, 50 m from the inflow end. Values of air temperature, river water temperature, pH and electrical conductivity (EC) were measured when the experiment was started and every time the seepage water was sampled. At three points (0 m, 25 m and 50 m from the inflow end) in the channel, 50 g of *E. acicularis* was sampled at the onset of the experiment and each time 100.0 L of river water was poured into the channel. Specimens were weighed after excess moisture was wiped off. The wet weight of all specimens was 1,050.6 g. After drying, the weight of these specimens was 195.7 g, 19 % of the wet weight.

Photo 1. *Eleocharis acicularis* in Artificial Channel

2.2.2. *Experiment to determine the rate of evapotranspiration*

To determine the evaporation rate of water from *E. acicularis* (i.e., the transpiration rate) and from the artificial channel, tap water was poured into the channel five times over 5 days. A total of 500.0 L of tap water was used for this purpose. The inflow rate was 0.4 - 0.6 L/min. The experiment was conducted at the beginning of October. At that time of year, the average air temperature is usually 15 °C or lower. The wet weight of *E. acicularis* used in the experiment was 68.6 kg. When 100.0 L of tap water was poured into the channel without the *E. acicularis*, 3 % of the tap water remained in the channel and the rest ran off from the channel. Although the ratio of the evapotranspiration and the evaporation is obtained as the ratio of the evapotranspiration and the open water evaporation [6]. In the experiment, the transpiration rate for *E. acicularis* and the evaporation rate of water from the channel were estimated by subtracting the total amount of seepage water from the total amount of tap water poured into the channel. The evapotranspiration rate estimated by using river water and that estimated by using tap water were analyzed in light of the difference in the average air temperature at the time when the experiments were conducted by using river water or tap water.

2.2.3. *Analysis method*

The river water used for measuring arsenic content and concentration was sampled, filtered, mixed with nitric acid to yield a 1 wt.% nitric acid solution, and preserved. The preserved river water was analyzed for arsenic concentration by using an Inductively Coupled Plasma Mass Spectrometry; Agilent 7500cx, Agilent Technologies, (hereinafter: ICP-MS). For the purpose of measuring the arsenic content of the *E. acicularis*, sampled *E. acicularis* was rinsed thoroughly with ultrapure water, dried at 80 ºC in a dryer for 2 days, and pulverized. The fine powder from pulverization was mixed with 30 wt.% hydrogen peroxide, 61 wt.% nitric acid and 38 wt.% hydrofluoric acid, and the mixture

IOP Conf. Series: Earth and Environmental Science **71** (2017) 012026 doi:10.1088/1755-1315/71/1/012026

was evaporated at 98 °C until dry. Then, 61 wt.% nitric acid was added, and the mixture was re-evaporated at 98 °C until dry. The nitric acid solution was prepared by adding 30 wt.% nitric acid to the *E. acicularis* specimen and analyzing for arsenic content by using an ICP-MS.

3. Results

3.1. Changes in the arsenic concentration of the seepage water

The initial concentration of arsenic in river water was 0.234 mg/L. After the first, second, and third inputs of river water into the channel, the arsenic concentration in seepage water was 0.045 mg/L, 0.133 mg/L, and 0.249 mg/L, respectively (Table 1). The arsenic concentration after the third input of river water was 0.249 mg/L, a value close to the initial concentration of 0.234 mg/L. This suggests that *E. acicularis* has a limited arsenic absorption capacity under the experiment conditions, but the reductions in the arsenic concentration indicate that *E. acicularis* helped lower the arsenic concentration in the river water. As the average pH value of the river water was 7.5 before it was poured into the channel and was 7.4 when the river water seeped out, the pH remained neutral during the experiment. Regarding the EC of the river water, the average value was 0.62 mS/m before the river water was poured into the channel, and was 0.64 mS/m when the river water seeped out. These results show that there was no significant change in water quality.

Table 1. Results of experiments for river water.

Number of inflow	Test Water (*: At the start/end of experiment –: Unmeasured)							*Eleocharis acicularis*		
	Arsenic Leachate (mg/L)	Temperature* (°C)	Water Temperature* (°C)	pH*	EC* (mS/m)	Inflow Time (Hour)	Quantify of Water* (L)	Arsenic Concentration (mg/kg)		
								0m point	25m point	50m point
Initial	0.234	–						0.14	0.07	0.12
1	0.045	24.6/31.9	26.2/31.0	7.51/6.72	0.52/0.38	2.5	100/8.9	0.87	0.41	0.15
2	0.133	31.9/30.4	31.9/34.5	7.57/7.87	0.64/0.63	4.5	100/66.8	1.01	0.71	0.31
3	0.249	27.5/ -	31.0/26.5	7.50/7.52	0.70/0.90	4.5	100/32.8	4.16	2.46	1.10
Total	-	-	-	-	-	11.5	108.5	-	-	-

3.2. Changes in the arsenic content of the *E. acicularis*

At the point 0 m from the inflow end of the channel, the initial arsenic content in *E. acicularis* was 0.14 mg/kg. After the first, second and third inputs of river water into the channel, the arsenic content was 0.87 mg/kg, 1.01 mg/kg, and 4.16 mg/kg, respectively. The arsenic content in the *E. acicularis* was the highest at the point 0 m from the inflow end and the lowest at the point 50 m from the inflow end. This suggests that the *E. acicularis* at the points 25 m and 50 m from the inflow end had the capacity to absorb more arsenic at the end of the experiment. The experiment results show that *E. acicularis* has the capacity to absorb and accumulate arsenic. However, because this capacity varies depending on the arsenic concentration in water and the inflow rate of water, it is necessary to examine conditions under which *E. acicularis* absorbs arsenic more efficiently.

Table 2. Results of experiments for tap water.

Number of Inflow	Test Water (*: At the start/end of experiment –: Unmeasured)				
	Temperature* (°C)	Humidity* (%)	Water Temperature* (°C)	Inflow Time (Hour)	Quantify of Water* (L)
1	16.7/11.5	41.7/70.7	16.5/10.9	17.0	100/56.2
2	11.5/8.1	64.1/72.1	16.3/9.2	21.0	100/54.4
3	11.3/13.6	58.4/59.0	16.2/12.5	24.0	100/69.0
4	13.6/10.2	59.0/73.7	16.2/10.0	22.0	100/58.5
5	10.2/ -	73.7/ -	16.0/ -	9.0	100/56.9
Total	–	–	–	93.0	295.0

3.3. The quantity of water used, and the quantity of seepage water

When the experiment using river water was conducted, the air temperature was between 27.3 °C and 29.8 °C, and the average air temperature was 28.3 °C. Because the quantity of river water used for the experiment was 300.0 L and the total volume of seepage water was 108.5 L, 64 % of the river water was lost during the experiment. When the experiment using tap water was conducted, the air temperature was between 9.5 °C and 17.6 °C, and the average air temperature was 12.6 °C (Table 2). The quantity of tap water used was 500.0 L and the total volume of seepage water was 295.0 L; thus, 41 % of the tap water was lost. The difference between the volume of water poured into the channel and the volume of seepage water was understood as the volume of evapotranspiration. It was confirmed that the volume of evapotranspiration varies depending on the average air temperature.

4. Discussion

4.1. Total content of arsenic in seepage water

The total arsenic content in the seepage water after it seeped out of the *E. acicularis* was estimated on the basis of the total content of arsenic in the river water used in the experiment. The total content of arsenic in the river water was defined as the product of the volume of river water and its initial arsenic concentration. The total arsenic content in the seepage water after it seeped out of the *E. acicularis* was obtained by subtracting the arsenic content of the water remaining in the channel and the product of the seepage water volume and the arsenic concentration of the seepage water from the total content of arsenic in the river water. The total content of arsenic in the river water is calculated as follows:

Total content of arsenic in the river water
= 0.234 mg/L (initial arsenic concentration) × 300.0 L (total volume of river water)
= 70.2 mg

The arsenic content remaining in the channel is calculated as shown below by using 1) the volume of water remaining in the channel before *E. acicularis* was placed in it, and 2) the volume of water retained within or on the surface of the *E. acicularis* (i.e., a value obtained by subtracting the weight of *E. acicularis* after drying from the wet weight of *E. acicularis*).

The volume of water remaining in the channel before the *E. acicularis* was placed in it
= 3.0 L (3 % of total inflow water)
The volume of water retained within or on the surface of the *E. acicularis*
= 54.5 kg (wet weight of *E. acicularis* placed in the channel) × (100.0 - 19.0 %)
= 44.2 L

The arsenic content of the water remaining in the channel is obtained by multiplying the volume of water remaining in the channel by the arsenic concentration measured after the third water input (0.249 mg/L).

The arsenic content of the water remaining in the channel
= 0.249 mg/L × (3.0 L + 44.2 L)
= 11.8 mg

The total arsenic content in the seepage water before it seeped out of the *E. acicularis* can be obtained by subtracting the arsenic content of the water remaining in the channel from the total arsenic content of the river water poured into the channel.

The total arsenic content in the seepage water before it seeped out of the *E. acicularis*
= 70.2 mg - 11.8 mg
= 58.4 mg

The seepage water volume is multiplied by the arsenic concentration of the seepage water for the first, second, and third water inputs, and the three products are summed as shown below:

The sum of the products:
= 8.9 L × 0.045 mg/L + 66.8 L × 0.133 mg/L + 32.8 L × 0.249 mg/L
= 17.5 mg

Because the arsenic content for the total amount of the seepage water before it seeped out of the *E. acicularis* is 58.4 mg and because that after it seeped out of the *E. acicularis* is 17.5 mg, we conclude that 40.9 mg of arsenic was absorbed by the *E. acicularis*. This reduction in arsenic content is equivalent to a 58 % reduction of the total content of arsenic in the river water (i.e., 70.2 mg) used for the experiment.

4.2. Rate of evapotranspiration
The rate of evapotranspiration was estimated per unit area of *E. acicularis* mats per hour. The total area of *E. acicularis* mats was 15 m^2, the total duration of river water input was 11.5 hours (Table 1) and the total duration of tap water input was 93.0 hours (Table 2). Regarding the experiment using river water, the rate of evapotranspiration is obtained by dividing the amount of water loss by the total area of *E. acicularis* mats and by the total duration of water input.

The rate of evapotranspiration in the experiment using river water
= (300.0 - 108.5 L) / 15 m^2 / 11.5 h
= 1.1 L/m^2/h

Regarding the experiment using tap water, the rate of evapotranspiration is obtained by dividing the amount of water loss by the total area of *E. acicularis* mats and by the total duration of water input.

The rate of evapotranspiration in the experiment using tap water
= (500.0 L - 295.0 L) / 15 m^2 / 93.0 h
= 0.2 L/m^2/h

When the experiment using river water was conducted, the air temperature was 27.3 - 29.8 °C (28.3 °C on average). The air temperature was 9.5 -7.6 °C (12.6 °C on average) when tap water was used for the experiment. The difference in the rate of evapotranspiration per unit area per hour is understood to be attributable to the difference in the average temperature. In the experiments conducted by using an artificial channel, *E. acicularis* was found to absorb arsenic. At the same time, it was found that *E. acicularis* and the channel were useful for reducing the volume of water poured into the channel through evapotranspiration.

5. Conclusions
The findings of the experiments explained above are as follows:

1. The results of the experiment conducted on using *E. acicularis* to remove arsenic in an artificial channel show that *E. acicularis* helps lower the arsenic concentration in seepage water and that the capacity of *E. acicularis* to absorb arsenic depends on the volume of water. The arsenic content in the *E. acicularis* increased with increase in the number of water inputs, and the arsenic content in the *E. acicularis* was higher at the upper reach of the channel and lower at the lower reach.
2. The total arsenic content in the seepage water was equivalent to 58 % of the total content of arsenic in the river water used for the experiment.
3. The volume of water was reduced due to evapotranspiration by 64 % when the average air temperature was 25 °C or higher, and by 41 % when the average air temperature was 15 °C or lower. The reduction rate of the water volume per unit area per hour was 1.1 L/m^2/h when the average air temperature was 25 °C or higher, and it was 0.2 L/m^2/h when the average air temperature was 15 °C or lower.

In the future, research and analysis will be continued for the following purposes at a site that is subjected to the constant inflow of drainage water: confirming arsenic removal by *E. acicularis* and the arsenic absorption rates of *E. acicularis* under various weather conditions; and identifying elements that are contained in water evaporated from channels.

6. References

[1] Committee of the Manual for Controlling Heavy Metal Pollution Caused by Natural Rocks and Soil on Construction Sites., 2010. The Manual for Controlling Heavy Metal Pollution Caused by Natural Rocks and Soil on Construction Sites (in Japanese). pp. 72-84.

[2] Ha, N. T. H., Sakakibara, M., Sano, S., Hori, R. S. and Sera, K., 2009. The potential of Eleocharis acicularis for phytoremediation: case study at an abandoned mine site, *CLEAN - Soil, Air, Water*, Vol.37, pp.203-208.

[3] Ha, N. T. H., Sakakibara, M. and Sano, S., 2009. Phytoremediation of Sb, As, Cu and Zn from Contaminated Water by the Aquatic Macrophyte Eleocharis acicularis, *CLEAN - Soil, Air, Water*, Vol.37, pp.720-725.

[4] Sakakibara, M., Harada, A., Sano, S. and Hori, R. S., 2009. Heavy Metal Tolerance and Accumulation in Eleocharis acicularis, a Heavy Metal Hyper Accumulating Aquatic Plant Species, *Geo-pollution Science, Medical Geology and Urban Geology*, Vol.5, pp.1-8.

[5] Kubota, Y., Sakakibara, M. and Sano, S., 2010. Arsenic cocentration of hot spring and river waters at Jyozankei hot spring area in Sapporo city, western Hokkaido, Japan (in Japanese). *Ehime University Faculty of Science Bulletin*, Vol.16, pp.7-13.

[6] Claude, E. Boyd., 1987. Evapotranspiration/Evaporation ratios for aaquatic plants, *J. Aquat. Plant Manage*, Vol.25, pp.1-3.

Socio-demographic characteristics of traditional gold smelters in Makassar, south Sulawesi, Indonesia

Hasriwiani Habo Abbas[1,4], Masayuki Sakakibara[1,2], Lukmanul Hakim Arma[3], Iva Hardi Yanti[4]

[1]Departement of Earth Science, Graduate School of Science and Engineering, Ehime University, 2-5 Bunkyo-cho, Matsuyama, Japan
[2]Faculty of Collaborative Regional Innovation, Ehime University, 348-58 Tomegamori, Takizawa, Iwate 020-0137Japan
[3]Faculty of Engineering, Hasanuddin University, Jl. Perintis Kemerdekaan km. 10, Tamalanrea, Makassar, Indonesia
[4]Faculty of Public Health, Muslim Indonesia of University, Gedung Menara Lt.9, Jl. Urip Sumahorjo Km 5, Makassar, Indonesia Indonesia
Email: haboabbashasriwiani@yahoo.com

Abstract. The traditional gold smelting in Makassar, South Sulawesi, Indonesia, is an informal work with the manufacture of gold jewelry as the core activity. Stages of the gold processing include panning, smelting, and refining with mercury. In the current study, we used a social demography analysis to classify the traditional gold smelter workers in this region. Data (e.g. sex, age, education level, time working, and income) were obtained from a questionnaire survey of 58 smelter workers in the Wajo and Tallo Sub-districts of Makassar. Results showed that 84.5% of the workers were males aged from 21 to 50 years with on the average 15 year of work. The gold smelter were last educated in elementary school (31.0%), junior high school (36.2%), and senior high school (27.6%) levels whereas 5.1% have no education. We found that the monthly income of an un-skilled worker was ~Rp. 2 million (USD 147.0) whereas that of a skilled worker was between Rp. 2.5 million (USD 183.76) and Rp. 5 million (USD 367.51). An owner could earn over Rp. 5 million (USD 367.51) per month. The result suggested that the traditional gold smelting used rudimentary technique and attracted young people with a low education level. This business continues to exist because the worker earn sufficient income and may higher through mastering gold smelter proficiency.
Keywords: Socio-demographic, Goldsmith, UAGM

1. Introduction

Makassar, in the southern part of Sulawesi, is the fifth largest city in Indonesia in terms of population. According to the Statistics Center of Indonesia, the main business fields are trade, processing industries, services, transportation and communication, banking, and finance [1]. One of the traded commodities is gold jewelry. Since a gold shopping center was formally established in the 1970's, Makassar has become one of the main gold trading centers in Indonesia. Shops sell various styles of both local and imported gold items.

In Makassar, jewelry is produced from local gold in traditional gold smelters. The job of those working smelters is to make gold jewelry and to recover fine gold particles from the waste generated during goldsmith activities. Other smelters in the urban part of Makassar recover gold from waste using a recovery process that is quite similar to the artisanal gold process used at mining sites. Herein, those who recover gold from waste are referred to as urban artisanal gold mining (UAGM) workers.

The goldsmiths make gold jewelry by refining, casting, forming, and shaping a gold bar or gold doré. Jewelry trader buy gold doré from gold shop in artisanal and small-scale gold mining (ASGM) sites. The goldsmith is requested by the gold trader to make jewelry. In order to get a malleable composition, the gold doré is refined by burning using a torch flame in an open clay bowl, and then alloyed with silver (Ag) or Cupper (Cu) (e.g., alloy material comprising 65% Au and 35% Cu). The gold alloy is then reduced to granules of diameter 2–3 mm to facilitate re-melting and subsequent jewelry fabrication such as forming, hammering, bending, cleaning, and polishing. Generally, the gold doré still contains small amount of mercury, which may potentially contaminate the surrounding environment during smelting or refining [2,3]. The goldsmith activities generate waste containing fine gold particles. This waste is processed by the UAGM worker to recover the gold particles, which are then returned to the goldsmith. The working relationship among the gold shops, goldsmith workers, and UAGM workers is shown in Figure 1.

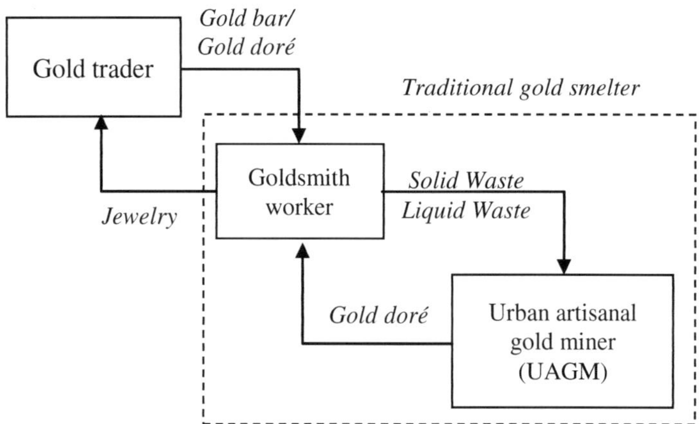

Figure 1. The work relations between gold shop, goldsmith, and UAGM.

The UAGM does not excavate primary or secondary ore but collects the waste from goldsmith activities and recovers the gold incorporated in the waste. The recovery process relies on rudimentary amalgamation and smelting techniques during which mercury is released to environment. The waste is put into a flooded pond. The pond occupies an area of about 1–2 m^2 and is 30–40 cm deep. These processes are mostly done in the open yard of the urban miner's house. Few workers use shallow dishes to pan the slurry, while others use a carpeted sluice box in which the gold particles settle because of the density of the gold. Most workers use low-capacity rotating trommel because it is more efficient for obtaining amalgam gold than the dishes or the carpeted sluice boxes. Amalgamation is applied in panning or in rotating trommel. Liquid mercury was added and mixed in rotating trommel for 3–4 hours. To obtain gold amalgam resulted from rotating trommel, the excess mercury is squeezed through a fabric and may be re-used 2–3 times.

While there have been many studies of ASGM activities, few researchers have examined the traditional gold smelting industry. Public health research has focused mainly on the impact of mercury exposure on the health of people living near to ASGM and gold shop activities [4–7]. Previous studies have shown that after skin contact with mercury during the amalgamation process, workers in gold shops developed physical and mental health problems [8]. Mercury vapor from smelting in gold shops is released to urban air and dispersed by wind. An integrated socio-demographic study, in which socio-demographic factors (e.g. population density, sex, income, and level of education) are characterized, complements a public health assessment. However, these factors have not been assessed in traditional gold smelting communities. To our knowledge, no previous studies have reported the sociodemography of traditional gold smelters who operate in urban areas of Makassar. In this study, therefore, we present socio-demographic information about the smelters, such as the number of workers, age, sex, educational level, income, and length of employment. The traditional gold smelting industry in Makassar, which has been active for more than 40 years, is informal and dominated by small businesses. In this study, we also present background information that might help to explain why this sector persists.

2. Data and Methods

Data were collected in period March 2015 – March 2016. There were 33 UAGM workers and 25 goldsmiths participated in this research. Data for the socio-demographic information of the gold smelters namely sex, age, period of employment, education, income, qualification such as owner, skilled and un-skilled workers, were obtained using a questionnaire. The gold smelters were living in sub-district of Wajo and Tallo, Makassar city shown in Figure 2. An observation were conducted to figure out the process of making jewelry and urban mining.

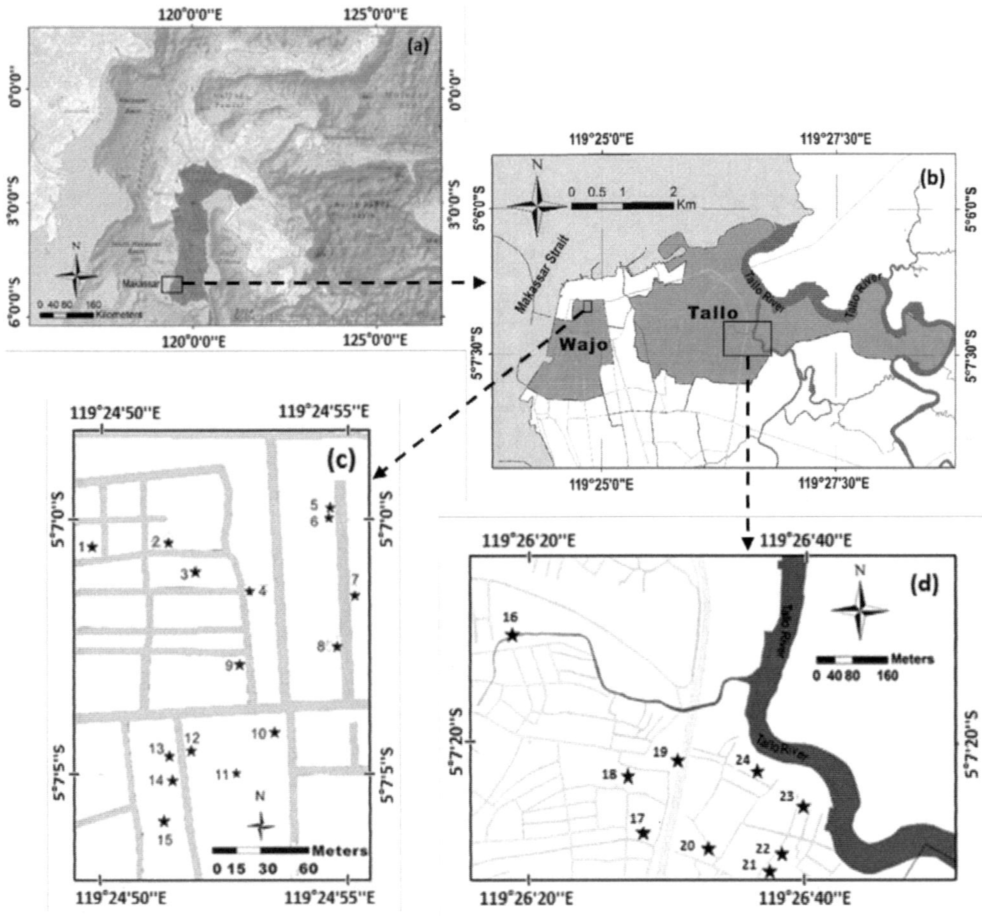

Figure 2. Maps of sampling sites at traditional gold smelters. (a) Sulawesi island with the location of Makassar city (square box) in south-Sulawesi (grey area). (b) Sampling sites (two square boxes) in Wajo and Tallo sub-district. (c) and (d) sampling points at Wajo and Tallo sub-district, respectively.

3. Results and Discussion

The socio-demographic information of the smelters is summarized in Table 1. Additional table recording socio-demographic infomation of the gold smelters in every sites shown in Figure 2(c) and

(d) was provided in the appendix. The majority are male, aged between 15–65 years old. However, women may be engaged in this work, as the results show that 15.5% of the smelters are women [15,16]. Figure 3 shows that workers of productive age, 25–45 years old, dominate these occupations. In this range of age, they have high working hours. The workers comprise skilled workers, who play an important role in this industry, and un-skilled workers. Many of the skilled UAGM workers are in the 25–39 year old age group (44%) as shown in Figure 3a while skilled goldsmith are mostly in the 30–44 year old age group (36%) as shown in Figure 3(b).

Table 1. Characteristics of socio- demographic traditional gold smelters

Socio-Demographic characteristics	UAGM (n=33)		Goldsmith (n=25)	
	n	%	n	%
Sex :				
Male	28	84.8	17	68
Female	5	15.2	8	32
Age :				
18 – 19	2	6	0	0
20 – 24	5	15	2	8
25 – 29	4	12	5	20
30 – 34	5	15	5	20
35 – 39	7	22	6	24
40 – 44	5	15	3	12
45 – 49	3	9	2	8
50 – 54	1	3	1	4
55 – 59	1	3	0	0
60 – 65	0	0	1	4
Owner	3	9.1	5	20
Skilled worker	19	58	14	56
Un-skilled worker	11	33	6	24
Length of service,				
Mean (Year)	14.8		15.5	
max. – min.	(3 – 40)		(3 – 40)	
Level of education :				
No education	2	6	1	4
Elementary school	13	39	5	20
Junior high school	9	27	12	48
Senior high school	9	27	7	28
Income :				
< 2.5 million rupiah	11	33.3	6	24
2.5–5 million rupiah	19	57.6	14	56
5–8 million rupiah	3	9.1	5	20

Figure 3 also demonstrates that the business owners are either between 35 and 49 for UAGM or between 30 and 65 years old for goldsmith. During our interviews we found that owners between 60 and 65 years old were first generation traditional smelters whereas owners between 30 and 50 years old were second generation smelters who are continuing the business after retirement of their parents or elders. We also noted that children were not involved in the work but were living in the areas around the smelters. This contrasts with the ASGM industry, in which many children work at the mining sites [17,18].

Education level is an important socio-demographic factor that can be used to study the characteristics of this community of traditional gold smelters. The education levels of the un-skilled and skilled workers, and the owners of both the goldsmith and UAGM activities, are presented in Figures 4 and 5. The majority of the skilled goldsmith workers have a low level of education, having only attended elementary and junior high school. The education level of the UAGM workers, however, is relatively lower than that of the goldsmiths. This indicates that simple, traditional technologies and management practices are used in this industry. In fact, this entire business depends on the technical skills of the workers and there are no formal education requirements.

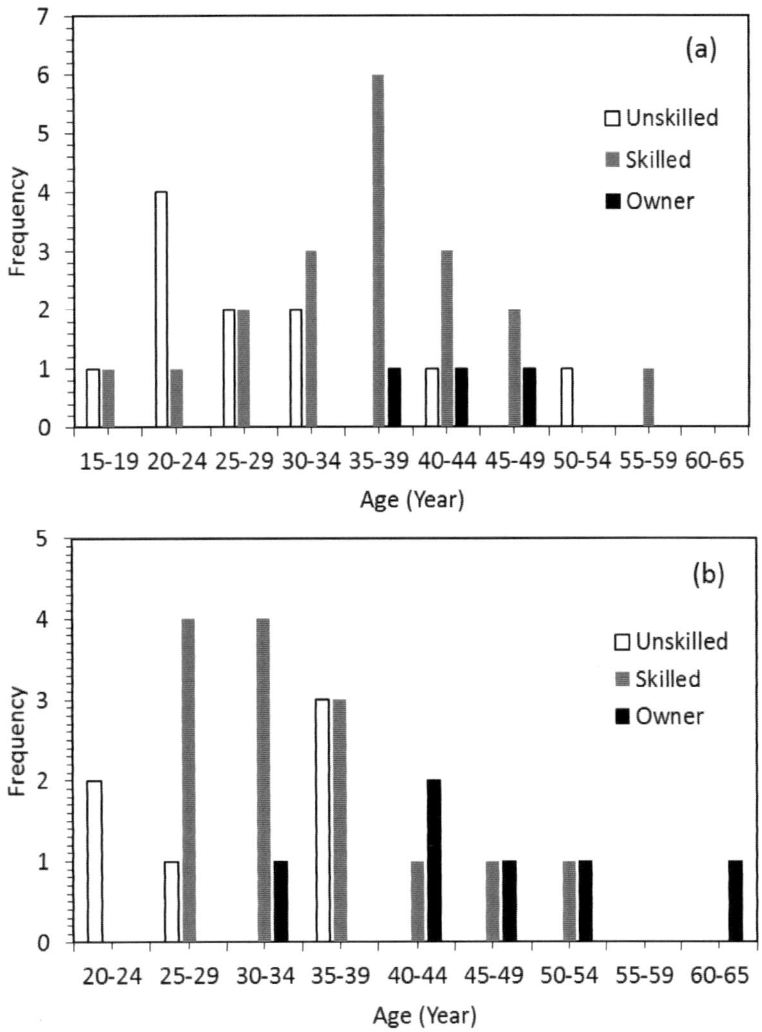

Figure 3. Distribution of traditional gold smelter classified by owner, skilled, and un-skilled workers in (a) UAGM and (b) goldsmith.

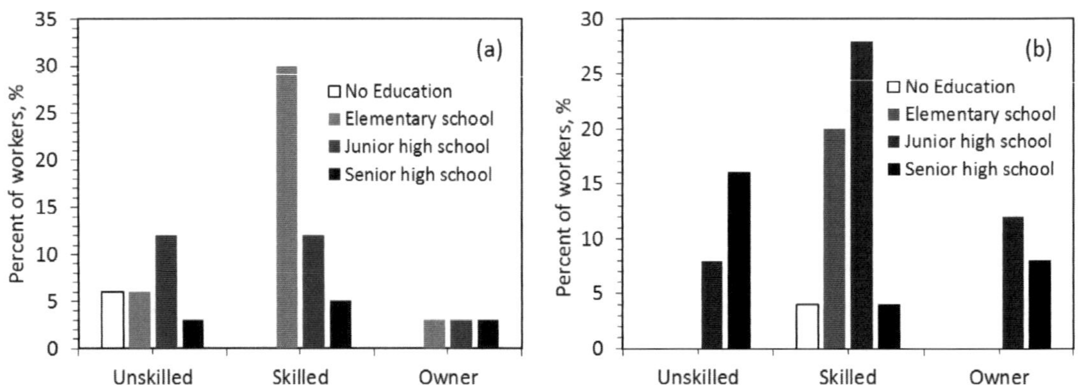

Figure 4. Distribution of unskilled and skilled worker, and owner classified by educational level. (a) UAGM worker. (b) goldsmith.

The workers may improve their skills by serving as apprentices. Therefore, workers are classified as un-skilled and skilled according to their experience. As shown in Figures 5(a) and (b), the majority of both UAGM and goldsmith skilled workers have been working for between 10 and 19 years. The goldsmith workers need to have higher skills because they have to successfully master the gold composition (Au–Ag) process, and the jewelry manufacturing, design, and decoration processes. To maintain their businesses, the owners hire new workers, mostly their own relatives.

The new workers are trained to be skilled workers and to become proficient goldsmiths. The workers are introduced to this work at between 18 and 20 years of age to ensure they are able to master the skills required by the ages of 25–45 years of age. Most of the 24% un-skilled workers are new workers and receiving 'on-the-job' training (Figure 5).

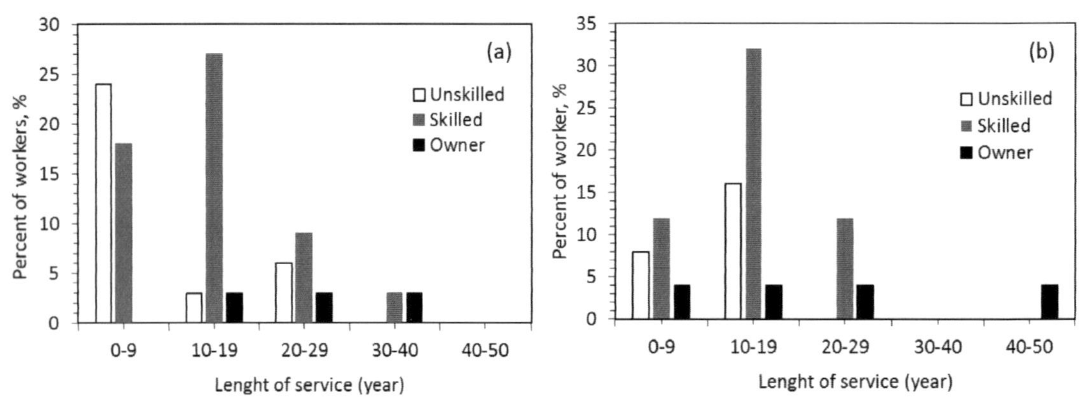

Figure 5. Distribution length of service of the unskilled and skilled workers, and owner. (a) UAGM worker. (b) goldsmith.

The workers' income is determined by their skill level. Hence, unskilled workers earn Rp. ~2.5 million per month, while skilled workers can earn Rp. 2.5–5 million. The minimum wage of employees in Makassar is Rp. ~2.3 million. The income of the owner, at Rp. 5–8 million per month, is considerably higher. Second-generation owners who have held their own business for 3–5 years are already earning around this amount.

These socio-demographic characteristics suggest that the traditional gold smelting industry is attractive to urban inhabitants in Makassar, and especially to young people with a low education level. Although it is an informal industry, traditional gold smelting continues to exist, mainly because the income is sufficient for daily life. There is also the opportunity to upskill and become a skilled worker, thereby increasing one's income.

The UAGM industry applies a recovery process to waste generated by the manufacture of gold jewelry. The stages of the recovery process are similar to those of the ASGM industry in Indonesia, and they include panning, amalgamation, and smelting [7,9–13,19]. Although the recovery process is relatively straightforward, the number of UAGM workers shows an increase over time. There are no official records of the number of traditional smelters in the Makassar government, but our interviews indicate a slight increase in the number of UAGM workers and a decrease in the number of goldsmiths in this research period. This contrasts with the ASGM industry in rural Indonesia, which is showing rapid growth.

4. Conclusions

Traditional gold smelters in Makassar have unique demographic characteristics. They either work as goldsmiths or urban artisanal gold miners. Most of them are in productive age and have only a low level of education, yet they can improve their skills by serving an apprenticeship while working, thereby increasing their income. Mercury released into the environment during the traditional gold smelting process raises the risks of mercury pollution among the inhabitants of Makassar. We recommend a follow-up epidemiological study to assess the degree of mercury pollution in urban areas of Makassar.

5. Acknowledgment

The authors would like to thank the Moslem Indonesia of University and South Sulawesi of Indonesia government that has given permission to conduct research. One author (Hasriwiani Habo Abbas) would like also to thank to Indonesian government, Dikti Scholarship for their funding support of doctoral studies in Ehime University. I would like also to thank to Japanese government to support this research by (conducted as a part of) feasibility study "Social Acceptance of Regional Innovation for Reducing High Impact Environmental Pollution", Research Institute for Humanity and Nature (RIHN) and JSPS KAKENHI Grant Number 16H02706 to M. Sakakibara.

6. References

[1] BPS 2014 *Badan Pusat Statistik-Statistics Indonesisa, Makassar City*

[2] Pirrone N, Cinnirella S, Feng X, Finkelman R B, Friedli H R, Leaner J, Mason R, Mukherjee A B, Stracher G B, Streets D G and Telmer K 2010 Global mercury emissions to the atmosphere from anthropogenic and natural sources *Atmos. Chem. Phys.* **10** 5951–64

[3] Salameh Azimi and Sadeghi Moghaddam M 2013 Effect of Mercury Pollution on the Urban Environment and Human Health *Environ. Ecol. Res.* **1** 12–20

[4] Wip D, Warneke T, Petersen A K, Notholt J, Temme C, Kock H and Cordy P 2013 Urban mercury pollution in the City of Paramaribo, Suriname *Air Qual. Atmos. Heal.* **6** 205–13

[5] Nakazawa K, Nagafuchi O, Kawakami T, Inoue T, Yokota K, Serikawa Y, Cyio B and Elvince R 2016 Human health risk assessment of mercury vapor around artisanal small-scale gold mining area, Palu city, Central Sulawesi, Indonesia *Ecotoxicol. Environ. Saf.* **124** 155–62

[6] Kristensen A K B, Thomsen J F and Mikkelsen S 2014 A review of mercury exposure among artisanal small-scale gold miners in developing countries *Int. Arch. Occup. Environ. Health* **87** 579–90

[7] Bose-O'Reilly S, Drasch G, Beinhoff C, Rodrigues-Filho S, Roider G, Lettmeier B, Maydl A, Maydl S and Siebert U 2010 Health assessment of artisanal gold miners in Indonesia *Sci. Total Environ.* **408** 713–25

[8] Gul N, Khan S, Khan A and Ahmad S S 2015 Mercury health effects among the workers extracting gold from carpets and dusted clays through amalgamation and roasting processes *Environ. Sci. Pollut. Res.* **22** 17965–74

[9] Stapper D 2011 Artisanal Gold Mining, Mercury and Sediment in Central Kalimantan,

Indonesia *ProQuest Diss. Theses* 159

[10] Krisnayanti B D, Anderson C W N, Utomo W H, Feng X B, Handayanto E, Mudarisna N, Ikram H and Khususiah 2012 Assessment of environmental mercury discharge at a four-year-old artisanal gold mining area on Lombok Island, Indonesia *J. Environ. Monit.* **14** 2598–607

[11] Arifin Y, Sakakibara M and Sera K 2015 Impacts of Artisanal and Small-Scale Gold Mining (ASGM) on Environment and Human Health of Gorontalo Utara Regency, Gorontalo Province, Indonesia *Geosciences* **5** 160–76

[12] Tomiyasu T, Kono Y, Kodamatani H, Hidayati N and Rahajoe J S 2013 The distribution of mercury around the small-scale gold mining area along the Cikaniki river, Bogor, Indonesia *Environ. Res.* **125** 12–9

[13] Limbong D, Kumampung J, Rimper J, Arai T and Miyazaki N 2003 Emissions and environmental implications of mercury from artisanal gold mining in north Sulawesi, Indonesia *Sci. Total Environ.* **302** 227–36

[14] Veiga M M, Baker R F, Fried M B and Withers D 2004 *Protocols for environmental and health assessment of mercury released by artisanal and small-scale gold miners* (United Nations Publications)

[15] Yakovleva N 2007 Perspectives on female participation in artisanal and small-scale mining: A case study of Birim North District of Ghana *Resour. Policy* **32** 29–41

[16] Armah F A, Boamah S A, Quansah R, Obiri S and Luginaah I 2016 Working conditions of male and female artisanal and small-scale goldminers in Ghana: Examining existing disparities *Extr. Ind. Soc.* **3** 464–74

[17] Bose-O'Reilly S, Lettmeier B, Matteucci Gothe R, Beinhoff C, Siebert U and Drasch G 2008 Mercury as a serious health hazard for children in gold mining areas *Environ. Res.* **107** 89–97

[18] Bose-O'Reilly S, McCarty K M, Steckling N and Lettmeier B 2010 Mercury exposure and children's health *Curr. Probl. Pediatr. Adolesc. Health Care* **40** 186–215

[19] IPEN 2013 Mercury Hotspots in Indonesia ASGM sites : Poboya and Sekotong in Indonesia 1–8

IOP Conf. Series: Earth and Environmental Science **71** (2017) 012028 doi:10.1088/1755-1315/71/1/012028

Heavy metals concentrations in scalp hairs of ASGM miners and inhabitants of the Gorontalo Utara regency

Yayu Indriati Arifin[1], Masayuki Sakakibara[2], and Koichiro Sera[3]

[1]Department of Geology, State University of Gorontalo, Jl. Jend. Sudirman No. 6 Kota Gorontalo, 96128 Indonesia
[2]Faculty of Collaborative Regional Innovation, Ehime University, 2-5 Bunkyo-cho, Matusyama 7908577, Japan
[3]Cyclotron Research Center, Iwate Medical University, 348-58 Tomegamori, Takizawa, Iwate 0200173, Japan
E-mail: yayu_arifin@ung.ac.id

Abstract. We performed the Particle Induced X-ray Emission (PIXE) analysis on scalp hair samples of 115 ASGM miners and inhabitants of Gorontalo Utara regency. Along with mercury (Hg), we presented other trace elements such as Copper (Cu) and Manganese (Mn). Concentrations of Cu, Mn and Hg in the scalp hairs of ASGM miners are higher non miners. Significant and positive correlations coefficients between Cu and Hg concentration with Mn concentration may indicate that there are still unknown metabolism process related with ASGM activities.

1. Introduction

The Gorontalo Utara regency is home of the historical gold mining site Hulawa (formerly known as Buladu), the Hulawa reopen again at 1970 by local inhabitants and continues operates as ASGM. Since 2000 several new ASGM sites in the regency has been start operating. Recent estimation that ASGM sites in the regency produces 286 kg of gold and emits 572 kg of Hg per year [1]. Elevated mercury concentrations on the environments and human hairs of ASGM miners and inhabitants have been reported [2]. Arsenic (As) and lead (Pb) concentrations in human scalp hairs have been detected and positively correlate with mercury [3].

In this report, we presented the evidence of other trace elements (Cu and Mn) in scalp hair of miners and inhabitants of the Gorontalo Utara regency. Correlations of Cu and Mn with Hg in scalp hair may add information on source of contamination or fate of elements in human body.

2. Experimental

2.1. Study Area

Samples were collected from five districts in the Gorontalo Utara regency: Anggrek, Kwandang, Monano, Sumalata, and Tolinggula (Figure 1). Geographically, Tolinggula, Sumalata, Monano, Anggrek, and Kwandang are situated on hills and mountains along the coastline of the Gorontalo Utara Regency. Inhabitants of the Gorontalo Utara regency mainly work as farmers and fishermen. Marine fish is commonly a part of their diets, along with rice, corn and vegetables, which are also produced on the nearby hills alongside the coastline.

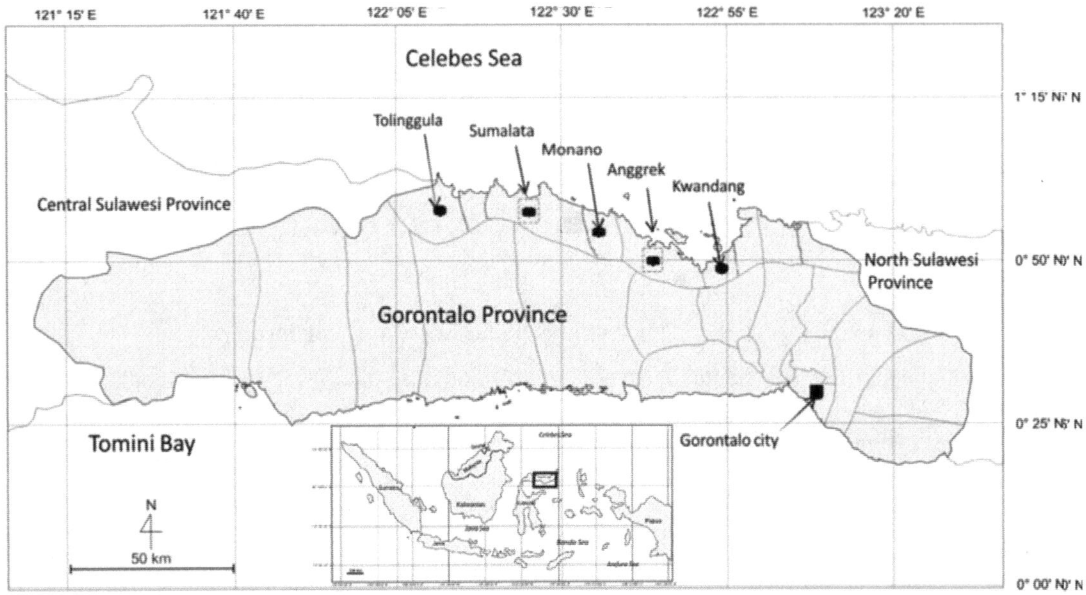

Figure 1. Sampling locations of human hair from the Gorontalo Utara regency are indicated by (•) on Gorontalo Province map, Tolinggula, Sumalata, Anggrek, Monano and Kwandang are name of subdistricts.

The ASGM activities in the Sumalata and Anggrek districts are located along the Wubudu and Anggrek riverbanks, respectively. The bioaccumulation of Hg, which may occur in living organisms such as paddy rice, corn, and marine fish, becomes agents that spread Hg contamination through the food web of inhabitants in the Gorontalo Utara regency. The Hg concentration in river sediments and fish will be used as background information about the Hg in the biotic and abiotic environments.

The Sumalata and Anggrek districts are locations with ASGM activities, while Kwandang, Monano, and Tolinggula are districts without mining activities. The residents of Anggrek and Sumalata are considered the ASGM miners group, while the residents of Kwandang, Monano, and Tolinggula are considered the control group.

2.2. Sampling

Human scalp hair samples were taken from 95 participants from inhabitants of Anggrek (n=25), Sumalata (n=24) Kwandang (n=15), Monano (n=39), and Tolinggula (n=12) between 2012 and 2013. Of the 115 participants, 66 were female, and the mean age was 23 years (range: 8 months – 63 years). Among the 115 participants, 21 were ASGM workers, 22 were housewives, 7 were officers of the regency, 6 were unemployed, 1 was a teacher, 1 was a university student and 38 were children (participants with ages below 18 years old).

The Cu, Mn, and Hg concentrations in the hair samples from Anggrek, Kwandang, Monano, Sumalata, and Tolinggula were determined to understand the status of contamination. The distribution of participants according to sex, location, and occupation is summarized in the table 1. Approximately 10-20 strands of hair were cut close to skin from the right backside (mastoidal region of the temporal bone) and then labelled and stored in a sample plastic bag [5].

The Hg concentration in hair samples will be used to characterize the risk through a comparison with reference values published by the German Human Biomonitoring Commission in 1999 (Commission Human – Biomonitoring of the Federal Environmental Agency Berlin, 1999) [6]. The German Human Bio-Monitoring (HBM) commission established toxicology threshold limits, which can be put into three categories. The first category is below normal or HBM I, where the Hg level in hair is below 1 μg/g. The above normal category is an alert level between HBMI and HBM II, where the Hg hair content is from 1 to 5 µg/g. Meanwhile, above 5 µg/g is categorized into the high level or over HBM II. Cu and Mn are essential elements for human body, however at certain conditions their excess concentrations are related to disturbance in organ functionalities [7-9].

Table 1. Number of scalp hair donor distribution related to their job, sex and living place in Gorontalo Utara regency.

Residence	Sex	N	Job	N
Anggrek	F	13	Miner	8
	M	12	Non Miner	17
	Total	25	Total	25
Kwandang	F	7	Miner	0
	M	8	Non Miner	15
	Total	15	Total	15
Monano	F	24	Miner	1
	M	15	Non Miner	38
	Total	39	Total	39
Sumalata	F	12	Miner	12
	M	12	Non Miner	12
	Total	24	Total	24
Tolinggula	F	10	Miner	0
	M	2	Non Miner	12
	Total	12	Total	12
Total	F	66	Miner	21
	M	49	Non Miners	94
	Total	**115**	**Total**	**115**

2.3. Analytical Procedure

Elemental analysis for the scalp hair samples was performed by particle induced x-ray emission (PIXE) in the Cyclotron Research Center, Iwate Medical University, Japan. The precision and accuracy of this method have been reported elsewhere [10-12] and will be described as follows. We used Zinc with known concentration (256 ppm) as internal standard; PIXE detects peak count of K_{α} about 2,325.2 ±50 (2% error). Hair samples were washed using Milli-Q water and shaken in an ultrasonic bath for 1 minute. Then, the samples were dried by wiping them with a tissue. The dried hair samples were washed again by being stirred in acetone for 5 minutes. Then, they were washed again using Milli-Q water, wiped well with tissue and left to dry at room temperature. The hair samples (approximately seven hairs per person) were stuck on a target holder. A 2.9 MeV-proton beam hit the target after passing through a beam collimator of graphite, whose diameter was 6 mm. X-rays of energy higher than that of the K- K_{α} line were detected by a Si(Li) detector (25.4 mm thick Be window; 6 mm active diameter) with a 300 mm-thick Mylar absorber. For measurements of X-rays lower than the K-K_{α} line, a Si(Li) detector (80 mm Be; 4 mm active diameter), which has a large detection efficiency for low energy X-rays, was used. Descriptions of the data acquisition system and the measuring conditions are reported elsewhere [8]. The typical beam current and integrated beam charge were 100 nA and 25 µC, respectively. The procedure for the standard-free method for untreated hairs is almost the same as that reported in the previous studies [6].

3. Result and discussion

Table 2 shows the summary of statistical results of Cu, Mn, and Hg concentrations in hairs of the inhabitants in the Gorontalo Utara regency. There is clear indication that means concentrations of Cu, Mn and Hg for ASGM miners are higher than miners. But there were no clear trends of mean concentrations of Cu, Mn, and Hg as functions of distance of living place to ASGM locations.

We perform the Spearman correlation on the total sample, and we found the tree coefficients (see the Table 3). There are strong and significant correlations of Hg and Mn concentrations to Cu concentration in hairs of inhabitants of the Gorontalo Utara regency. While weak and non-significant correlation is found between Cu and Hg.

Table 2. Cu, Mn, and Hg concentrations (μg/g) in hairs of inhabitants in the Gorontalo Utara regency.

Region	Cu		Mn		Hg	
	Mean ± SD	range	Mean ± SD	range	Mean ± SD	range
Anggrek	32.2±2.1	14.9-136.1	14.0±2.6	1.3 – 56.7	16.9 ± 29.2	2.1 – 144.8
Kwandang	15.0±1.7	6.6 - 34.8	3.9±5.7	0.4 - 103.4	6.5 ± 1.5	3.5 – 14.6
Monano	11.9±1.6	1.5 - 26.9	10.3±2.7	0.9 - 63.8	5.6 ± 2.8	2.8 – 28.1
Sumalata	16.6±1.5	6.2 - 45.7	4.5±4.8	1.0 – 58.6	8.2 ± 1.9	2.5 – 69.8
Tolinggula	11.4±1.3	1.3 - 16.1	12.1±3.0	1.8 - 87.9	5.1 ± 0.8	4.3 – 6.0
Miners	22.6 ± 1.9	6.2 – 89.3	13.6 ± 1.7	1.6 – 136.1	8.2 ± 1.7	2.7 – 17.9
Non miners	13.6 ± 1.7	1.5 – 136.7	8.5 ± 3.4	0.4 – 88.0	6.3 ± 2.0	1.1 – 144.8
Total	**1.3±2.7**	**0.0 - 19.9**	**8.7±10.4**	**0.0 - 67.6**	**10.2 ± 16.9**	**2.1-144.8**

Table 3. Correlation coefficient between elements.

Element	Mn		Hg	
	Coeff.	Sig.	Coeff.	Sig.
Cu	0.12	0.06	0.14	0.15
Hg	0.25	0.007		

4. Conclusion

Concentration of Cu, Mn, and Hg in average of miners and inhabitants of exposed group (Anggrek and Sumalata) are less elevated than control group (Kwandang, Monano, and Tolinggula). While according to HBM: inhabitants of the Gorontalo Utara regency have elevated Hg concentration (already in the alert level), while mean value for inhabitants exposed groups are in the danger level. The amount of Pb in hair of inhabitants of the Gorontalo Utara regency are still in the safe level, while As are already higher than the limit.

We also found the significant and strong correlation between concentrations of Cu and Hg to Mn in scalp hairs of inhabitants of the Gorontalo Utara regency. Such correlation may reveal the unknown sources of contamination, routes of contamination, and impact on health status of individual hair donors.

5. References

[1] Arifin, Y.I., Sakakibara, M., Takakura, S., Jahja, M., Lihawa, F., and Machmud, M., 2013 Artisanal and small sale Gold mining in Gorontalo Utara regency, Indonesia, *23rd Symmposium on Geo-Environments and Geo-Technics*, Tsukuba, Japan.

[2] Arifin, Y.I., Sakakibara, M., and Sera, K. 2015 Impacts of Artisanal and Small Scale Gold Mining (ASGM) in Gorontal Utara Regency, North Sulawesi, Indonesia, *Geosciences*.

[3] Bose-O'Reilly, S., Mc.Carthy, K.M., Steckling, N., and Lettmeier, B., 2010, Mercury Exposure and Children's health, *Curr. Probl.Pediatr. Adolesc. Health Care*, **40**,186-215.

[4] Cordy, P.; Veiga, M.; Crawford, B.; Garcia, O.; Gonzalez, V.; Moraga, D.; Roeser, M.; and Wip, D. 2013 Characterization, mapping, and mitigation of mercury vapour emissions from

artisanal mining gold shops. *Environ. Res*, **125**, 82-91.

[5] Foo, S.C., and Tan, T.C., 1998 Elements in the hair of South-east Asian islanders. *The Sci. of the Total Environ.*, **209**, 185-192.

[6] Schulz, C.; Angerer, J.; Ewers, U.; Kolossa-Gehring, M. 2007 The german human biomonitoring commission. *Int. J. Hyg. Environ. Health*, **210**, 373-382.

[7] Grandjean, P., 1978 Lead concentration in single hairs as a monitor of occupational lead exposure, *Int. Arch. Occup. Environ. Health*, **42**, 69-81.

[8] Niculescu, T., Dumitru, R., Botha, V., Alexandrescu, R., and Manolescu, N., 1983 Relationship between the lead concentration in hair and occupational exposure, *Br. J. Ind. Med.*, Feb: **40(1)**:67-70.

[9] Hinwood, A. L., Sim, M. R., Jolley, D., de Klerk, N., Bastone, E. B., Gerostamoulos, J., & Drummer, O. H. 2003 Hair and toenail arsenic concentrations of residents living in areas with high environmental arsenic concentrations. *Environmental health perspectives*, **111(2),** 187.

[10] Kempson I.; Skinner W. M. 2006 Advanced Analysis of metal distribution in human hair, *Environ. Sci. Technol.*, **40**, 3423.

[11] Esteban, M. and Castano A. 2009 Non-invasive matrices in human biomonitoring: *A review Environ. International*, **35**, 438-449.

[12] Sera, K.; Futatsugawa, S. and Matsuda, K. 1999 Quantitative analysis of untreated bio-samples. *Nuclear Instrumen and Method B.*,**150**, 226.

Corporate social responsibility for regional sustainability after mine closure: a case study of mining company in Indonesia

Andi Erwin Syarif, Tsuyoshi Hatori

Department of Civil and Environment Engineering, Ehime University, 3 Bunkyo-Cho, Matsuyama 790-8577, Japan
Email: erwin.sustainable.development@gmail.com

Abstract. Creating a soft-landing path for mine closure is key to the sustainability of the mining region. In this research, we presents a case of mine closure in Soroako, a small mining town in the north-east of South Sulawesi province, in the center of Sulawesi Island in Indonesia. Especially we investigates corporate social responsibility (CSR) programs of a mining company, PT Vale Indonesia Tbk (PTVI), towards a soft-landing of mine closure in this region. The data of the CSR programs are gathered from in-depth interviews, the annual reports and managerial reports. Furthermore we presents an integrated view of CSR to close mining in a sustainable manner. We then evaluate CSR strategies of the company and its performance from this viewpoint. Based on these steps, the way to improve the CSR mine closure scenario for enhancing the regional sustainability is discussed and recommended.
Keyword: Soft-Landing Mine Closure, Corporate Social Responsibility, Structuration Theory, Regional Sustainability, Sustainable Development, PT. Vale Indonesia Tbk (The Indonesian Mining Company)

1. Background

Over the past 48 years, PT Vale Indonesia Tbk (PTVI) has been operating under the Contract of Work (CoW) agreement with the government of Indonesia to open pit nickel mines and a processing plant facilities near Sorowako on the island of Sulawesi. As the world's leading nickel producer, PTVI has played a central role in regional's economy through revenues, employment, business empowerment and social development programs. In the community life, the presence of PTVI has dominated many aspects of the community's socioeconomic activities. There are most of soroako's residents working as employees for the mining company PTVI, and others become contractors, traders, civil servants, etc. Given the circumstance, the number of sorowako' population has been increasing dramatically and uncontrolled. Sorowako in the past was just a small and isolated settlement where occupied by less than one thousand people, but nowadays the number of population has reached around 24,738 people [1]. This stemmed from regional migration and the growing local population. There are many issues that has been created such as poverty, unemployment, health care issues, agrarian conflict, public order crimes, degradation of protected forest in-and-around the CoW, the slump area on the bank of the lake matano, environmental pollution threatens water resources and the quality of air, etc.

All of these problems above have been attributed to the PTVI's presence both direct and indirect. So that, PTVI always face some business interruptions or social movements from the local communities and non-profit organization (figure 1). They ask and criticize PTVI about the negative impact on the environment, as well as in relation to cultural and socio-economic issues. The multifarious responses and opinions depend on the respective viewpoint, and PTVI continues to operate under the shadow of perpetuating debate to the present day. Actually, the case of PTVI provides an example of a general situation that faces many mining industries in today's developing country, including Indonesia

Figure 1. Social Movement

For reducing and solving social movement, PTVI needs to consider a new approach in its strategy development in order to maintain a good social relationship, which are critical aspects influencing corporate success continuously. One of the important things that the company has to be considered is the implementation of the Corporate Social Responsibility (CSR) programs. It must be accepted and supported by the local community as a key stakeholder and involved them in decision making. In the practice, the maximum realization of the CSR program can't be reached if the base of relationships between the company and the local community are not relatively harmonious. Other words, if a company ignores its stakeholder relations and the social environment in which it operates. Therefore, PTVI has conducted a strategic CSR program within long-term basis called *CSR for soft-landing mine closure scenario* [2] (figure 2). The CSR program which encompasses two dimensions: socioeconomic and environmental concern. As shown in the diagram below, there are important things where environmental commitment set-up need to be implemented such as: (a) the recovery of environmental functions and supporting capacity, i.e., activities leading to the recovery of protected forests and production forests; (b) nature conservation, which consists of endemic conservation and preservation; (c) wastewater management and emission management, which consist of sediment control, lake conservation, the control of Cr^{+6} (heavy metal) and dust emissions. Meanwhile, the socioeconomic commitment consists of activities such as (a) community empowerment, which includes activities such as the industrial training program, and create a mine closure foundation; (b) providing support to regional development, which provide of public infrastructures, and community-based forest management; (c) Alternative sustainable livelihood which includes institutional creation of employment opportunities outside the mine through sustainable agriculture and Industrial tourism.

From schematic and its explanation, PTVI has realized that greater corporate social responsibility (CSR) can contribute to the betterment of the community, human prosperity, and environmental sustainability, particular in the phase of mining closure. However, CSR for soft-landing mine closure always finds significant constraints to translate into on-the-ground CSR practice especially community development program. So that, CSR programs should be re-identified to create a local self-reliance, and also to adapt the regional sustainability. That is why the purpose of this paper is:

- To evaluate the effectiveness of CSR implementation based on the concept of accountability norms. Specifically, the CSR program that have a multi-years mechanism, and then directly connected to the soft-landing mine closure program
- To recommend the CSR model towards soft-landing mine closure towards the regional sustainability. Technically, to explore what should be done by CSR PTVI in supporting the achievement of regional sustainability after mine closure

In turn, CSR program becomes more successful in a sustainable manner, follow the best mining practice and bring a bright legacy for the next generation

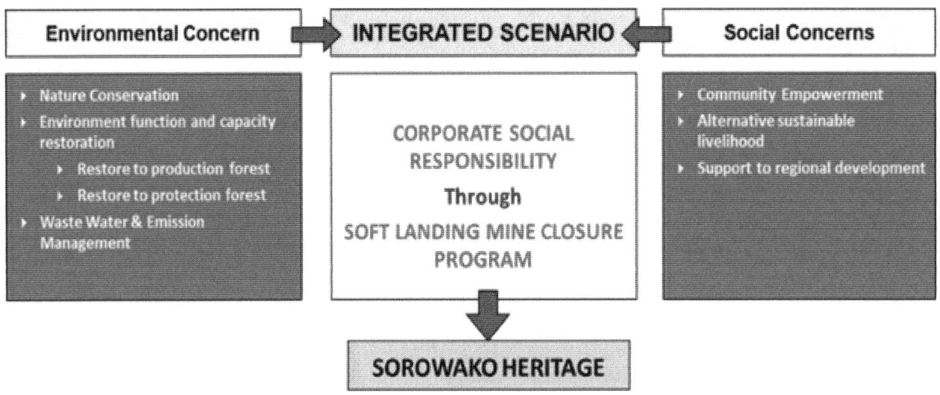

Figure 2. Schematic of PTVI's soft-landing mine closure

2. Material and Methods

CSR concepts and its evaluation and CSR for sustainable development were gathered through a literature review. Firstly, several journals and paper of CSR were identified through a content analysis. Based on this, Researchers conducted a specific analysis of methods, approaches, and strong point messages. Secondly, the table information was developed and applied to obtain comprehensive overviews. In which case, a specific theory for the evaluation of CSR will be decided and concluded. Thirdly, on-the-ground CSR practice, examining a total of 21 reports published by PTVI. The data set includes 6 sustainability reporting, 11 community development reports and progress, 3 annual reports published by PTVI and 1 document of a memorandum of understanding, and another data is 13 Verbeek Tabloid from local government (total data set, 34 reports). From the reports, we focus on CSR program both on the community development programs and mine closure (rehabilitation). Fourthly, the impact of CSR programs and the degree of social relationship between the company and local community are identified, listed, analysed and interpreted based on norm accountabilities. Finally, we developed the concept of CSR for regional sustainability after mine closure, which is the framework of sustainable development as a research basis

3. Result

3.1. Gathering CSR concepts and its evaluation

There have been many attempts to establish a better understanding of CSR from initial concept, planning stage, executing a process, until the stage of monitoring and evaluation (table 1). According to the phrase "Corporate Social Responsibility" originates with H. Bowen [3], the concept of CSR is used to describe businesses' integration of social and environmental issues into decisions, goals, and operations. The sole purpose of CSR is to identify and improve a company's impact on society and the environment while driving stronger business results such as brand enhancement, market differentiation, and employee satisfaction. Another perspective from World Bank [4], CSR is the

IOP Conf. Series: Earth and Environmental Science **71** (2017) 012029 doi:10.1088/1755-1315/71/1/012029

commitment of business to behave ethically and to contribute sustainable economic development by working with all relevant stakeholders to improve their lives in ways that are good for business, the sustainable development agenda, and society at large. Refer to Milton Friedman [5] in his book "A Stakeholder Approach" stated that CSR is interpreted as a means of increasing or maximizing the profits of the company where the shareholders are the focal point in pursuit of profit maximization. So that, stakeholder includes local community doesn't engage in business operation as key concern. For the evaluations of CSR stages, these literature reviews are indeed necessary in order to provide an overview of the concept CSR and its procedures. According to the table is showed some published papers and journals that have been reviewed and analysed.

Table 1. Journals and Papers on the evaluation of CSR programs

Title of Papers	Approaches (a), and key aspect of studies (b)
The Pyramid of Corporate Social Responsibility: Toward the Moral Management of Organizational Stakeholders by Archie B. Carroll [6]	a. Stakeholder Management b. Constructed the model of CSR as a pyramid formed in four-parts; 1st-economic responsibility, 2nd-legal, 3rd-ethical and 4th-discretionary responsibility= philanthropy
Managing corporate sustainability and CSR: A framework combining values, strategies, and instruments by Rupert J. Baumgartner [7]	a. Stakeholder Management b. Identified the relevance of CSR and the significant sustainability aspects related to the different tasks and action levels (Normative Management, Strategic Management, and Operational Management)
How Well Do Social Ratings Actually Measure Corporate Social esponsibility? by Aaron K. Chatterji [8]	a. Corporate Social Performance b. CSR assessed on the transparency of a well-established social and environmental performance rating system and provides a score (through the indication of the number of strengths and concerns)
Analysis of CSR Towards Mining and Tobacco Companies Listed IDX Based on Global Reporting Initiative by Hendy Indra Sasmita [9]	a. Strategic Competitive Advantages b. CSR disclosure with likers scale, and descriptive Analysis according to Annual Report and Global Report Initiative (GRI)Sustainability Reporting Standard
How Corporate Social Responsibility is Defined: An Analysis of 37 Definitions' by Alexander Dahlsrud, [10]	a. Communicative approach b. CSR provided a guidance on how the CSR should be defined and evaluated, and five dimensions of CSR are developed through a content analysis of existing CSR definitions
Corporate Social Responsibility (CSR)	a. CSR evaluation Based on the method of

And Project Management by Dave Nielsen [11]	project management b. CSR rationalized through the organization's demands according to my project's objectives. This paper uses actual examples to point out what to avoid and offers action in the CSR programs
A proposed methodological framework for the evaluation of corporate social responsibility by Konstantin G. Aravossis, Nikolaos A. Panayiotou and Katerina Tsousi [12]	a. Corporate Social Performance b. The measurement of CSR uses the weighted performance indicators relating to a company's impact on different areas of activity such as the environment, the community, the human capital, the shareholders and the marketplace

By analyzing the concept CSR and its evaluation, a substantial number of journals and papers are a lack of explaining the complicated relationship between local community concerns and company socially responsible behavior. Hence, it is required to build a framework which can explain or disclosure their relationship. In response to this matter, the evaluation of CSR can be analyzed on the basis of the structuration theory of Anthony Giddens [13].

3.2. Proposing Structuration Theory as The Frameworks
As formulated by Anthony Giddens [13], Structuration theory is centrally concerned with the order as "the transcending of time and space in human social relationships (figure 3). In other words, Structuration is about conditions governing the continuity or transformation of structures, and therefore the reproduction of social system. Principally, Structuration theory contains three structures that examined 1) Structure of Signification, 2) Structure of Legitimacy, and 3) Structure of Domination [14]. The means by which structures are translated into actions are called modalities, which are interpretive schemes, facilities, and norms. These modalities can explain why and how interaction is affected. The three dimensions of interaction are described as communication, power, and sanctions. Giddens calls to this interaction as the "duality of structure". It is the essential to understand the complexities that exist in the relationship between the two elements. The table below will be drawing the structuration theory.

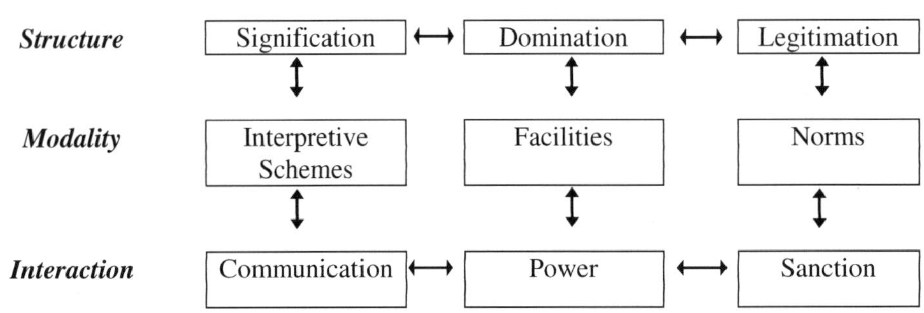

Figure 3. Dimensions of the "Duality of Structure" (Adapted from [13])

Based on the schema, the structuration theory provided a meta-framework, which is to accomplish a complicated task. Structuration theory can be summarized what is important as follow.

- The structure of signification; representing the constitution of meaning, which provide individuals with interpretive schemes by which they make sense of what other say and do. They are a set of categories or systems of relevance applied in the sustaining of day-to-day communication [14]. On the one hand, interpretive schemes are enabling as they guide organizational action, allow interpretation of ambiguous situations, and reduce uncertainty in conditions of complexity and change [15]. According Giddens [13], the human being is an intentional agent when admitting that the agents use their linguistic skills to talk, write and reflect on the rules and resources used in social interaction. The Interpretive act can results in multiple and mistake meanings.
- The structure of legitimation; providing a moral order via naturalization of societal norms, values, and standards. When individual agents interact, they exhibit consciously, subconsciously, or unconsciously meanings of their behaviour (Giddens refers to these as sanctions). The literature distinguishes broadly among three types of legitimacy; the first is normative, or what moral legitimacy. The second is pragmatic legitimacy. The third is cognitive legitimacy [16] (table 2).
- The structure of domination; Understanding the power relationship as a form of interaction between the actor and the structure. In this interaction, resources can be used as a form of authority resources, which allow agents to control persons. Resources can also be used in the form of allocative resources, which allow agents to control material objects or property [17]. In Giddens' view, the resources are the means by which power is exercised, and also to promote facilities in social relations. According to Weber [14] defines power as the probability that one actor, within a social relationship, is able to carry out his own will despite resistance.

Table 2. The Type of legitimacies

Type	Definition	Relationship with constituents
Normative legitimacy	Organization reflects acceptable and desirable norms, standards, and values.	Organization meets normative judgments about outputs/results, procedures and technologies, structures, personal, etc.
Pragmatic legitimacy	Organization fulfills needs and interests of its stakeholders and constituents.	Organization exchanges goods and services that constituent's want, and receives support and legitimacy.
Cognitive legitimacy	Organization pursues goals and activities that fit with broad social understandings of what is appropriate, proper, and desirable.	Organization "makes sense" and/or is "taken for granted" according to socially construct "realities."

Each structure has the above aspects, involving different combinations of rules and resources. These rules and resources the agents rest on in the production and reproduction of social systems are the products of social action [18]. For Giddens, the structure is more specific and detailed than a system, referring to structured practices.

3.3. CSR activities based on The Structuration Theory

In the CSR implementation, the relationship between the local community (truster) and company (Trustee) is of central to the use of structuration theory. Researchers have analysed some data from sustainability reporting, management reports, and the result of the field observation including semi-structured interviews and open conversations. The use of the structuration theory is concluded at three main objectives in the evaluation of CSR program. They are 1) CSR communication for maintaining a good stakeholder relationship (Structure of Signification), CSR for Good Corporate Governance (Structure of Legitimation) and The Impact of CSR program and Procedural Fairness (Structure of Domination).

3.3.1. CSR communication for maintaining good relationship as a Structure of Signification

In practice, the initial basis for CSR communication is constructed from engagement with members of the local community and providing information on the CSR programs, and what may happen in the future (Mine Closure). A CSR communication to be credible is largely based on providing consistent, factually correct and clear information and by complying with commitments made to the community.

There is two CSR communication connecting with the local communities.

- Direct communication, The Company conducts an open communication with the local community directly. It emphasizes that the company is there to listen, is highly concerned and would like to be followed up. For example, Public consultation, participatory need assessment, etc.
- Indirect communication, The Company provides a media information to the community about CSR program and its progress such as a newsletter, annual reports, information sessions, charitable donations and employee volunteering, information on social and environmental performance through a sustainability report, etc.

Refer to structuration theory, it represents the constitution of meaning, which provide interpretive schemes as an analytic lens for examining community assumptions and facilitate communication. It may be identified the use of common language and the disparity in language structure within CSR communication strategy. In the practice, a common language could be applied to the local community when facing different interpretation about special CSR program, for example, CSR for renewable energy with the advanced technology. While the disparity in language structure can be analyzed when individual stakeholders conduct an interpretation with their own background perceptions. For example, "Mapping" , it could be interpreted such as drawing a land contour by general people, but a CSR facilitator (Company) means conducting a social survey in participatory need assessment, etc. The disparity in language structure can generate a highly different understanding between company and community. Therefore, the disparity of language structure can be reduced or found out solution with a communication regularly both of them in order to make the same vision or a mutual understanding.

In Corporate Social Responsibility communication, there are some strategies that are adapted individually, case by case to the needs and standards of the company and its environment. All these Corporate Social Responsibility communication strategies are also separated into internal and external measures and are adjusted to the status of the program (CSR Company). According to Jen Boynton [19], CSR communication has some barriers to be effective, such as 1) Scattershot CSR communications, Company want to caution against disclosing too much information. At the same time, it is important to be strategic about how the information is shared in order to maximize the positive impact of the communication, 2) Communications operates separately from operations. There's a harsher term for the case where the communications message is not backed up by the actions of other departments: greenwashing. When a communications team runs around telling stories about the great things a company does without the input of the X, bad things can happen, 3) Avoiding the negative. Many CSR communications focus on the positive and that's great, but it's also important to be honest when things aren't going so well, otherwise, it's hard to believe the good things. In the practice, CSR communication to be credible may consist in reporting the change and reassuring stakeholders by informing them on the program's progress, as well as on misconceptions in relation to the CSR implementation process. Therefore, Communicating on CSR initiatives and performance is a critical competency for successful companies.

Refer to structuration theory, CSR communication needs to be followed the constitution of meaning, which provide interpretive schemes as an analytic lens for examining community assumptions and facilitate communication. At the local community, there is a wide variety of interpretation and expectation associated with the performance of CSR programs including mine closure. The difference of the interpretation on CSR program is a moderate way or logic (common sense), but sometimes it touches into the diverse structure of language. Therefore, CSR communication may be identified the use of common language and the disparity in language structure. In the practice, a common language could be applied to the local community when facing misinterpretation about special CSR program, which is required a technical knowledge. For example, the method of participatory rural appraisal (PRA), a manager or CSR facilitator can explain the meaning of PRA and its approaches to the local community. Hence, the local community can be more understand about PRA in a simple and unambiguous. While the disparity in language structure can be established when the local community makes an interpretation with their own background experience and conditions. For example, "Community Development", it could be interpreted by the local community as charity program through donation for the local community continuously, but a CSR facilitator (Company) means conducting CSR programs in order to the local community could be self-sufficient based on their local resources, not dependent on the company, etc. The disparity in language structure can generate a highly different understanding between company and community. Therefore, the disparity of language structure can be reduced or found out solution with a communication regularly both of them in order to make the same vision or a mutual understanding

3.3.2. CSR for Good Corporate Governance as a Structure of Legitimation

Mining companies must be aware and understand the norms of the local community and be able to collaborate with them, as they represent the local 'rules of the game'. One of the rules of games is "The right thing to do" as a moral basis. If failure to do that the mining operation can be stopped or rejected. As a multinational mining company, PTVI is still struggling on how to manage the risk on its business and getting a legitimacy from the local community as well as complying with government regulation. As one of the instruments for getting the legitimation, Corporate Social Responsibility can be seen as a manifestation of good corporate governance models, which are pertinent issues within transparency, responsibility, accountability, and fairness. It should be a part of company's values. For understanding company values, It could be seen its vision and mission as an entry point to measure the legitimation. The development of a genuine CSR vision and mission is often triggered by shareholder interest and accommodated by local stakeholders needs. According to PTVI's vision and mission stated: "to transform natural resources into prosperity and sustainable development, and then to be the number one global natural resources company in creating long-term value, trough excellence and passion for people and the planet" [20]. From this statement, it reflects that PTVI has a global standard to implement the best mining practice and the framework of sustainable development in the higher scale priority. These are organizational values or norms that are likely to have implication for getting the legitimation. Therefore, consistent with company's values, embedded in the company culture, and ideally integrated into the company business operations. In other words, the mission and vision should be transferred as well as its implementation from Board of Directors, Strategic Management until Operational Level. At the certain level, the effectiveness of vision and mission may be beyond normative legal compliance and the liability of the company. At this point, CSR for good corporate governance could achieve the "Cognitive Legitimacy" in the structure of legitimation. This means a company pursues goals that society deems to be proper and desirable. Other words, Constituency support for the company is not due to self-interest, but rather due to its taken-for-granted character.

IOP Conf. Series: Earth and Environmental Science **71** (2017) 012029 doi:10.1088/1755-1315/71/1/012029

Technically, a legitimation should not only judge on their economic success but on non-economic criteria, such as comply with the law (what is permissible and what is not), ethical aspect (be moral, fair, respect people's rights, avoid harm or social injury and prevent harm caused by others). Therefore, PT Vale Indonesia (PTVI) declared to comply with all laws both formal based on government regulation and informal based on local culture [21]. Based on report data [22], we assume that PTVI is generally accepted by the public at large because of its role in the advancement of society's needs and well-being, and the economic growth for regional sustainability. Specifically, there are two key factors that PTVI can be accepted at the phase of mining operation: 1) Because of the need for labor, PTVI can absorb a large body of manpower and provide many opportunities for work, and 2) Local government income directly or indirectly is obtained from PTVI and has helped provide more positions in the local government bureaucracy. However, this acceptance is neither automatic nor unconditional, one thing that is very important to remember that the quality of legitimacy factors is dynamic, intangible and responsive to changes in perceptions regarding the company and is also susceptible to outside influences.

3.3.3. The Impact of CSR program and Procedural fairness as a Structure of Domination

In this paper, CSR activities according to Structure of domination could be influenced by the impact of CSR program and the implementation of procedural fairness. Here are summarize of them

- The impact of CSR program - Based on data of soft-landing mine closure [2], CSR program has been divided into two approaches namely socioeconomic concerns and environmental concern. In the practice, socioeconomic concerns explain about the alternative sustainable livelihood, support to regional development, Community Empowerment, while environmental concerns focus on waste and emission management issues, impact on biodiversity and habitat, and Environment and capacity restoration. In the practice, there are two impacts of CSR programs, which will be increasing local's trust, they are "Independency and Sustainability". In detailed, Independency means the community should be independent enough to empower its own resources and to decide how best to explore these resources without depending on the existence of the company. While, Sustainability refers to the process of building equitable, productive and participatory structures to increase the economic empowerment of communities and their surrounding regions. Therefore, The main challenge for PTVI is to improve CSR programs that maintain goodwill for the company and address the long-term developmental needs of communities in a sustainable way (sustainable livelihood), without creating a culture of dependency [23]. CSR program for soft-landing mine closure can be accepted or unaccepted by local people depend on the CSR impacts. If CSR for socioeconomic and environmental impacts may be worse than expected, hence, the local community will reject the CSR program. However, if CSR for socioeconomic and environmental may be positively experienced than expected level, hence, the perception of CSR impact will be found a trust and acceptance by local community.
- Procedural fairness – This approach is used to evaluate on how local community perceive the CSR procedures are designed. In mine closure program, the implementation of CSR program have been designed with two model of decision makings: 1) Following the schema of CSR for community development, It uses a participatory need assessment that the program is designed through focus group discussions and in-depth interviews with local community, and partnership with local government. For example, Sustainable agriculture programs, grant for public infrastructures, community-based forest management, and Industrial training program, and 2) Involving specific expert, community representative and NGO for technical assessment at the first stage, After that making on public consultation at the second stage, and finally, Returning to Executive Management for making a final decision . For example, constructing Ecological Park and Mining Park. Based on these approaches, PTVI has tried to demonstrate procedural fairness when designed CSR program. However, it needs a detailed assessment from both on the

decision-making process and the decision outcomes. From this finding, we can assume that local community infers the procedural fairness of an authority based on the extent to which they feel involved in the decision-making process. The basic question for it, whether local community regards the procedure as fair or unfair, whether decision makers treat local community in a respectful way, or disrespectful way, and whether the local community has actively participated in decision-making processes, or as casual to be participated.

Both The Impact of CSR program and Procedural fairness is in form of allocative resources, which the company provides a way to examine how the evaluation of CSR comprehensively and structurally

4. Discussion

4.1. Regional Sustainability based on CSR Mining Closure

After the evaluation of CSR program through Structuration Theory, next step, the improvement of CSR mine closure scenario for enhancing the regional sustainability can be analysed generally. In this paper, CSR mine closure based on the frameworks of sustainable development is a basic measurement for the successful implementation of regional sustainability after mine closure. This implies that the CSR for mine closure can be designed, in comprehensive terms, as the creation of sustainable outcomes through the minimization of environmental, social and economic impacts on host communities and the maximization of positive legacies that contribute positively to future generations. Based on this, there are the three the framework of sustainable development could be analysed for regional sustainability [24].

Economic development – Regional sustainability suggests the investment of generated revenues, to ensure the future development and long-term livelihood of the local communities. CSR for mining closure is the commitment of businesses to behave ethically and to contribute to sustainable economic development by working with all relevant stakeholders to improve sustainable livelihood. Especially, PTVI has to support the local community & government to explore replacement renewable resources as economic activities within PTVI operation area, to replace nickel as the main source for economic growth

Environmental protection – Regional sustainability recognizes that all life has intrinsic value, is interconnected and that environmental protection and ecological integrity are part of the irreplaceable life support systems. In the mining industry, minimizing the environmental impact of mining resource exploitation and land rehabilitated to allow successive use [25]. Specifically, PTVI as mining companies has an obligation because their business deals a lot with extracting minerals from below the ground. In order to the company must give a specialized CSR programs dedicated to restoring the land through reclamation, which means restoring the disturbed areas in returning the land to its original state to the maximum extent possible, so that it could still be of use to the community, and to conserving the biodiversity spots from mainland until fresh water lake.

Culture and Social Cohesion – Regional sustainability promotes the reality and diversity of community and regions for increasing social cohesion, and the critical importance of 'sense of place' and heritage (buildings, townscapes, landscapes, and culture) in any plans for the future. Specifically, PTVI has to fully support the welfare and quality of life of local community especially indigenous people, preserving the culture of local community, and reducing the social and cultural disruption to communities, maintenance of stakeholder dialogue and transparency of operation

In the practice, CSR mining closure for regional sustainability requires a strategic alliance collaboration among Government, Community, and Private Sector, and then discover the power in the concept for providing a coherent, integrated each program and the CSR's long-term vision-mission. As the multi-international mining company, PTVI should show its moral obligation or responsibility for supporting regional sustainability. The decision to create a regional sustainability after mine closure can be also connected to its vision and mission on "the framework of sustainable development" [26]. It becomes a valuable vision for exploring ways to bring up internal changes in culture and structure. In turn, a commitment to the environmental protection and the development of socioeconomic may be beyond normative legal compliance and the liability of individuals. At this point, CSR for soft-mine closure could be categorized as "Cognitive Legitimacy" in the structure of legitimation. This means a company pursues goals that society deems to be proper and desirable. Other words, Constituency support for the company is not due to self-interest, but rather due to its taken-for-granted character. At the end, Public trust can be presented in form of social license to operate for the company.

5. Conclusion

The implementation of CSR towards mine closure is not an instantaneous process, and requires a sustainable manner. Therefore, the evaluation of CSR program must be conducted in order to review the mine closure program and its progress. In turn, CSR for mine closure should be soft-landing that will guarantee the sustainable livelihood (socioeconomic) for the local community, and also to maintain carrying capacity of the environment. Based on this, Regional sustainability can be achieved and maintained in order to face the future in the phase of mining closure

In this paper, the evaluation of CSR for soft-landing mine closure program is examined with using Giddens' structuration theory as a framework analysis. At this point, the understanding social relationship between the company and the local community is disclosed and analyzed. In providing the analysis, Researchers analyzed the CSR for soft-landing mine closure, which is referred to three structures: 1) structures of signification, 2) structures of legitimation, and 3) structures of domination. For the structure of signification to be implemented, CSR communication both direct and indirect must be designed with “credible”. It must honestly and transparently explain all the implications, and benefits of CSR program, including unsuccessful CSR program to be the lesson learned. As a language, CSR communication tends to be accommodative in form of “common language” to avoid misinterpretation between company and local community. The structure of Legitimation, the company can identify, and then implements strategies to maintain the social relationship with the local community. Therefore, the company (PTVI) needs to know its legitimacy level (normative, pragmatic, and cognitive) and what has to be done to maintain and/or improve. CSR program can be seen as media to sanction social relation (approval or not). It depends on company behavior. The structure of Domination, Trust and Corporate Social Responsibility (CSR) are two interrelated terms that need to be understood how they influence each other. Trust is the main thing. If there is no trust between the company and local people, nothing good will come out. In the practice, CSR should make a resource measurably in “trust” based on the impact of CSR program, and conduct a procedural fairness in CSR program. Actually, the impact of CSR programs positively is the most important to get "trust" from the local community.

The framework of sustainable development is required to construct the CSR programs for supporting regional sustainability. In the practice, a trilateral process of consultation among mining company (PTVI), local government, and the local community is conducted. To be fully effective, the process of consultation for mine closure should start at the mine design stage, not in the phase of mining operation or phase of closure planning. For the future research, the development of logic model for CSR mining closure can be designed according to the process of dialogue or consultation among them (Local Community, Government, and Company). The logic model describes logical linkages among program resources, activities, outputs, and audiences, and highlights different orders of outcomes related to a specific problem or situation [27]. In this way, logic models can be seen to support CSR mine closure programs both planning and evaluation

6. References

[1] Central Agency on Statistics 2016 The population administration services for Nuha Subdistrict,

Luwu Timur. p1
[2] Erwin A S and Aris P A 2007 A Roadmap to Soft-Landing Mine Closure Scenario 2007– 2025
[3] Bowen H R 1953 Social Responsibilities of the Businessman
[4] World Commission on Environment and Development 1987 Report of the World Commission on Environment and Development: Our Common Future (The Brundtland Report)
[5] Freeman E R 1984 A Stakeholder Approach pp 40-43
[6] Carroll A B 1991The Pyramid of Corporate Social Responsibiiity: Toward the Moral Management of Organizational Stakeholders *Business Horizons* **34** pp.39–48
[7] Baumgartner R J and Ebner D 2006 The relationship between Sustainable Development and Corporate Social Responsibility *Corporate Responsibility Research Conference* **17**
[8] Aaron K, David I L, and Michael W 2008 How Well Do Social Ratings Actually Measure Corporate Social Responsibility Journal of Economy & Management Strategy 125–69
[9] Hendy I S 2014 Analysis of CSR Towards Mining and Tobacco Companies Listed IDX Based on Global Reporting Initiative 1–12
[10] Dahlsrud A 2006 How Corporate Social Responsibility is Defined:an Analysis of 37 Definition Corporate Social Responsibility and Environmental Management pp.1–13
[11] Dave Nielsen 2010 Corporate Social Responsibility (CSR) and Project Management
[12] Aravossis K G, Panayiotou and Tsousi K 2006 A Proposed Methodological Framework for the Evaluation of Corporate Social Responsibility Environmental Economics and Investment 87–95.
[13] Anthony Giddens 1984 the Constitution of Society Outline of the Theory of Structuration', Cognitive Therapy and Research Cognitive Therapy and Research **12(4)**
[14] Kobayashi K, Hatori T and Jeong H 2011 Public Reviews and Trus Formation in Disaster Risk Management *Journal of Natural Disaster Science* **32.2** 89–97
[15] Yates J, Orlikowski A and Takeshi Y 2002 Community-based Interpretive Schemes Exploring the Use of Cyber Meetings within a Global organization Proceedings of the Annual Hawaii International Conference on System Sciences 3576–85 p1
[16] Suchman M C 1995 Managing Legitimacy: Strategic and Institutional Approaches The Academy of Management Review **Vol. 20** pp. 571-610
[17] Weber M 1947 the Theory of Social and Economic Organizations
[18] Englund H, Gerdin J & Burns J 2011 25 Years of Giddens in accounting research: Achievements, limitations and the future Accounting, *Organizations and Society* **36** pp 494-513
[19] Boynton J 2012 Top 10 Mistakes in CSR Communications
[20] PT Vale Indonesia 2013 Sustainability Reporting Improving Sustainability Performance for Our Future
[21] PT Vale Indonesia 2014 Sustainability Reporting Creating Added Value Throught Performance Excellence for Sustainability
[22] PT Vale Indonesia 2011 Sustainability Reporting Responds to Challenges
[23] PT Vale Indonesia 2015 Menghadapi tantangan: Rising to the Challengge Annual Report
[24] Murphy K 2012 The social pillar of sustainable development : a literature review and framework for policy analysis Sustainability Science, Practice, & Policy
[25] Sánchez, L.E. 1998. Industry response to the challenge of sustainability: the case of the Canadian nonferrous mining sector, *Environmental Management*, **22(4)**: 521-531
[26] PT Vale Indonesia 2012 Annual Report A Strategy for the Long Term
[27] W.K. Kellog Foundation 2004 Logic Model Guide

IOP Conf. Series: Earth and Environmental Science **71** (2017) 012030 doi:10.1088/1755-1315/71/1/012030

Hydrothermal alteration and timing of gold mineralisation in the Rumbia Complex, Southeast Arm of Sulawesi, Indonesia

Musri Mawaleda[1], Emmy Suparka[2], Chalid Idham Abdullah[2], Nurcahyo Indro Basuki[2], Marnie Forster[3], Jamal[1], Kaharuddin[1]

[1]Department of Geology, Hasanuddin University, Makassar, Indonesia
[2]Department of Geology, Institute of Technology Bandung, Indonesia
[3]Research School of Earth Sciences, Australian National University, Canberra, Australia
Corresponding author's: musri.mawaleda@unhas.ac.id

Abstract. The Rumbia Mountains, which in this study named Rumbia schist Complex is an east-west oriented, composed by a high-pressure/low-temperature, and a medium-pressure/low-temperature metamorphic rocks. Identified as mica schist, glauchopane schist, and green schist. Rumbia complex known as the location of gold deposits prospects discovered by local communities since 2007. The results of research showed that the metamorphic rocks are as hosts. There are two phase of gold mineralization that occurs in this area, namely: 1) Associated with tectonic deformation and metamorphic rocks exhumation, and 2) gold-related hydrothermal deposits. Radiometric age dating used $^{40}Ar/^{39}Ar$ geochronology, indicate that the first of gold mineralisation in the Rumbia Complex occurred ~23 million years ago, and the second gold mineralisation were subsequently overprinting at 7 million years ago.
Keywords: Hydrothermal, Gold, Deposits, $^{40}Ar/^{39}Ar$, Timing, Rumbia, Complex, Southeast, Sulawesi

1. Introduction

Rumbia Complex or Rumbia Mountains is located in the southern part of the Southeast Arm of Sulawesi (Figure 1). Those are a folded mountain composed of metamorphic rocks. The highest peak has reaches 1,100 m above sealevel. This area is located ±1,800 km east of Jakarta or ±150 km to the south of Kendari City, the capital of Southeast Sulawesi Province (Figure 1).

2. Background

Geological investigation in the Southeast Arm of Sulawesi, especially in the Rumbia Complex, and Mengkongga Complex around Kolaka was first performed by; **(i)** Wunderlin (1913); Bothe and Hetzel (1932) in Bemmlen (1949). Petrology Studied and mineralogy has been done by Gisolf, 1924 in Bemmlen (1949); De Roever (1947, 1950, 1956), suggests that the metamorphic rocks in the Rumbia Complex, and Mendoke Mountains are epidote amphibolite and glaucophane schist. Unlike in the Central Sulawesi metamorphic rocks are characterized by the presence of jadeite-aegyrine, crrosite, lawsonite and ferrocarpholite; **(ii)** Helmers et al. (1989) describe the Rumbia Complex as mica-graphite schist and quartzite mica with marble and metabasite intercalation; **(iii)** Simandjuntak et al. (1994) identifies the Pompangeo Complex in Central Sulawesi describe of mica schist, chlorite schist, mica-graphite schist, mica-quartz schist, glaucophane schist, amphibolite-ruby schist, gneiss, hornfels, and eclogite is the Late Cretaceous-Paleocene. The same thing illustrated that the Rumbia Complex and Mendoke in Southeast Arm of Sulawesi are Cretaceous-Paleocene. While the Mengkongga Complex, consisting of schist, quartzite and gneiss is the Permo-Carbon; **(iv)** Parkinson (1998) and Kadarusman et al. (2004), assume that the Rumbia Complex as the same metamorphic path with Pompangeo Complex in Central Sulawesi, Bantimala accretion Complex in the South Sulawesi and Mekongga Complex in the Southeast Arm of Sulawesi is the Early Cretaceous in age.

These last few years, discovered placer gold deposits in the Rumbia Complex, allegedly derived from schist and metabasite debris in the Rumbia Complex, and Mendoke Mountains (Surono and Tang, 2009). Idrus et al, (2010) proposed that the placer gold deposits in the Langkowala Valley (Rumbia Complex) as orogenic gold deposits are associated with the formation of metamorphic rocks that are completely unrelated to the volcanic activity.

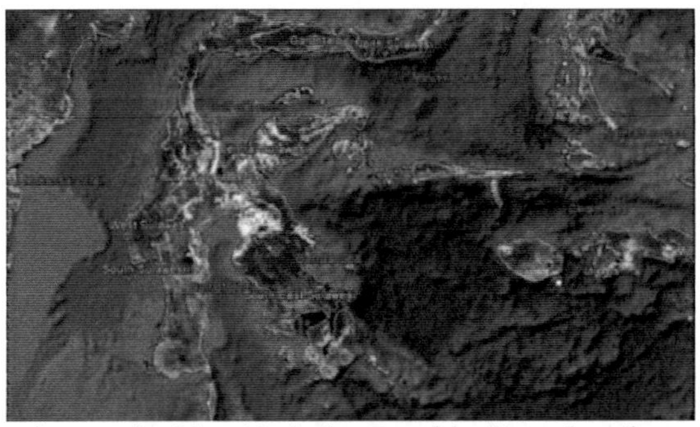

Figure 1. Location of study area 'red box' is it the Rumbia Complex (https//earth.com: Download January 9, 2016).

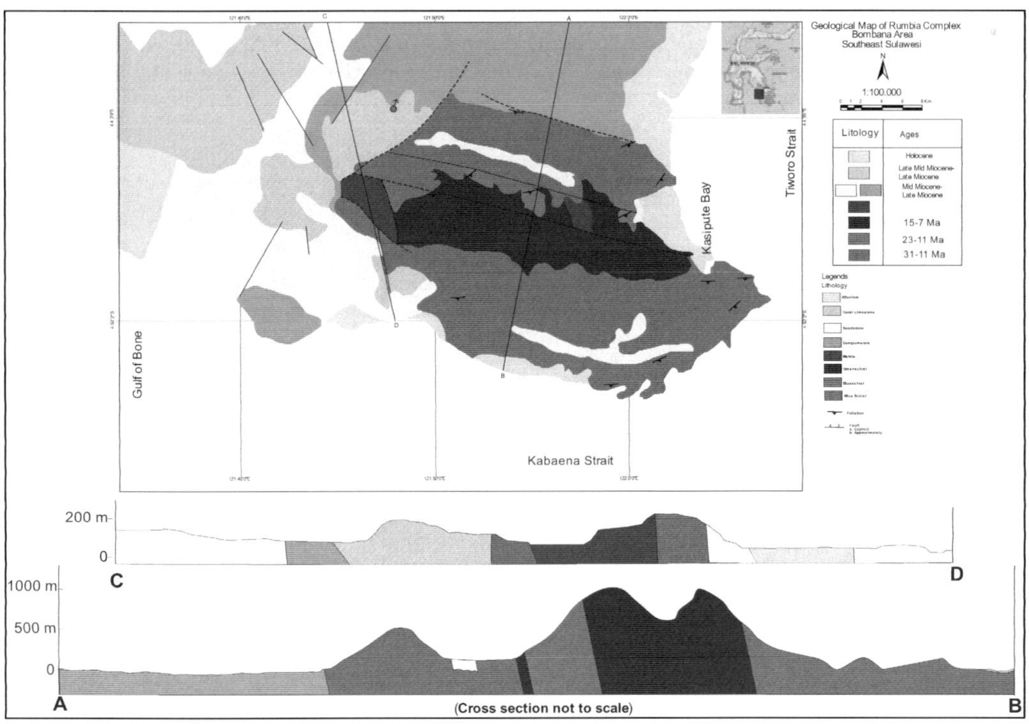

Figure 2. Geological map of Rumbia Complex, Southeast Arm of Sulawesi, Indonesia (Musri, 2015).

3. Geology of Rumbia Complex

Based on the relief, the study area can be distinguished into three geomorphology units, namely: 1) Folded mountain geomorphology unit; 2) Hilly geomorphology; and 3) Plain geomorphology unit. Folded mountains geomorphology unit includes Rumbia Mountains consist of North Block and South Block, with its highest peak 1,100 m above sea level. This unit occupied ±35% of the study area. North and South Blocks are separated by "strike slip fault zone" is relatively east-west trending. Folded mountain geomorphology unit, the overall occupied by metamorphic rocks.

The hilly geomorphology unit, occupies the west, southwest along the south of the Rumbia Mountains, approximately ±25% of the study area. The highest point is 60-100 meters above sealevel. Relief relatively coarse, partly in the form of barren hills with scrub vegetation and grasslands. Common local secondary trees consisting of teak and other wood cultivated plants. This area composed by coral reef limestones, conglomerates, and sandstone. Plain geomorphology unit occupies about ±40% of the study area. These units are spread out in the north in the direction of the Langkowala Valley, the west and the east and south that follows the coastline.

Plain geomorphology unit, generally occupied by sandstones, conglomerates, alluvial and beach swampy. The northern part of the plain, sedimentary rocks are dominated by conglomerates and sandstones. Both of these lithological units, known as Sulawesi Molasse (Sarasin and Sarasin 1901 in Bemmelen 1949). While Kartadiputra (1983) refer to it as Langkowala Formation and Alangga Formation. In the central part of the plain geomorphology unit, drained by the Langkowala river and Tahiite River.

Almost all locations in the Langkowala Formation and Alangga Formation is a placer gold mining area, as a traditional gold mine, organized by local communities.

4. Research Methodology

Research methodology includes field survey and laboratory analysis: 1) field survey cover of rocks sampling of alteration and mineralization. 2) laboratory analysis include petrographic analysis by thin section for hydrothermal alteration minerals, and ore petrography analysis by polish section for ore minerals. While the age analysis using micro fossil data and radiometric age dating of $^{40}Ar/^{39}Ar$.

5. Result and Discussion

5.1. Hydrothermal Alteration of Rumbia Complex

Hydrothermal minerals analysis by petrography analysis results of the Glaucophane Schist altered, Chlorite Schist altered and Mica Schist altered. The ranges of alteration from 10% to 75%, and divided into two type, i.e. propylitic alteration and phylic alteration type.

5.1.1 Propylitic alteration type

Propylitic type developed in almost all lithological units, with the alteration percentage between 10% until 75%. Figure 3 shows a thin section of a propylitic chlorite schist altered, and alteration in Figure 4, shows the percentage of alteration up to 50-75%. Alteration minerals assemblages are chlorite, epidote, sericite, quartz and opaque minerals. This sample shows propylitic alteration overprinting with phyllic alteration type.

5.1.2 Phylic alteration type

Phylic alteration type developed in all lithological units of the Rumbia Complex. Developing pervasive alteration in mica schist and chlorite schist. A number of samples are mineralized with sulfide minerals (pyrite, arsenopyrite, cinnabar and antimony), and oxide minerals (hematite and goethite). The following figure is a microphotograph (Figure 5) of metamorphic rocks of the Rumbia Complex with phylic alteration.

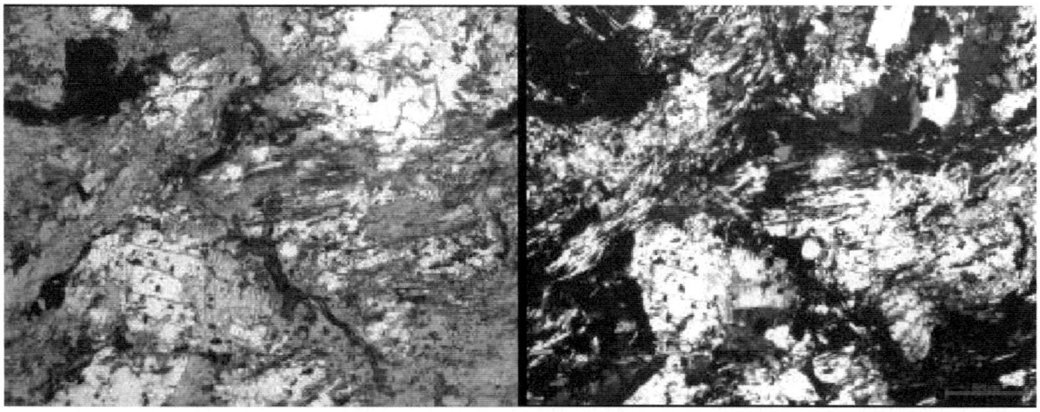

Figure 3 Microphotograph of thin section of Chlorite Schist Altered with alteration minerals assemblage are chlorite, epidote, quartz, sericite, carbonate and opaque minerals representing a propylitic overprinting phyllic alteration type.

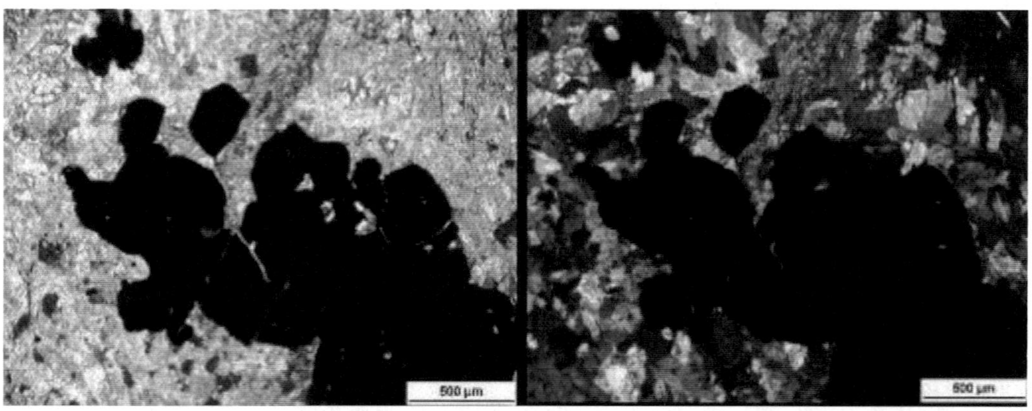

Figure 4 Microphotograph of thin section of Glaucophane Schist Altered, where alteration is up to 75%. Assemblage of alteration minerals, i.e. epidote, chlorite, quartz and opaque minerals.

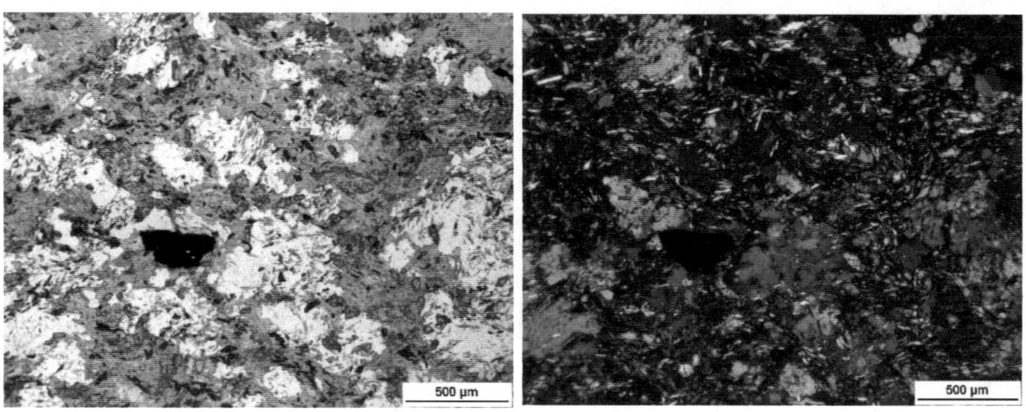

Figure 5. Microphotograph of thin section of chlorite schist (RS-38-11), alteration up to 60%. (propylitic overprinting with phyllic alteration).

The alteration in the Sample of Glaucophane Albite Schist altered (Figure 6) has alteration up to 75% which is accompanied by mineralization, this is also characterized by the presence of opaque minerals. Secondary sericite is also present that fills cracks in albite crystals and replaces unknown porphyroblast as pseudomorphs.

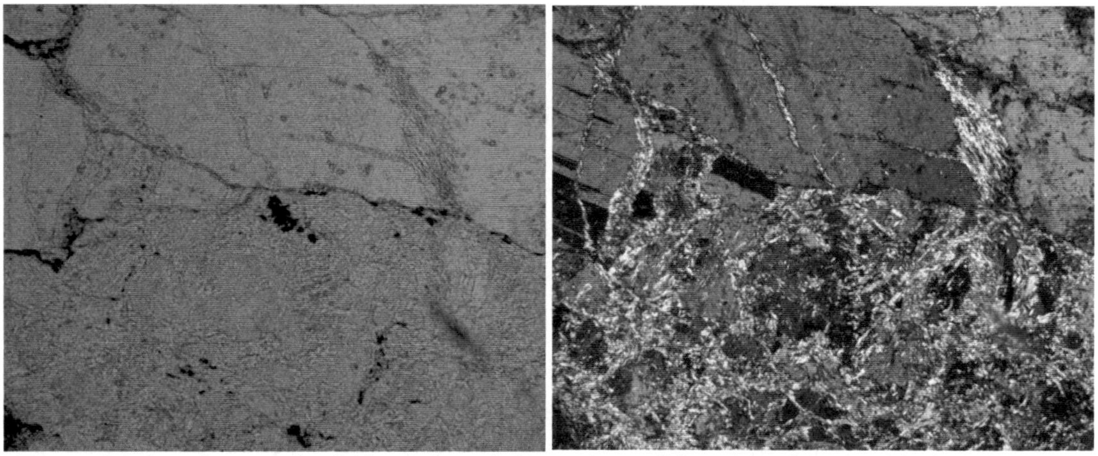

Figure 6. Microphotograph of thin section of glaucophane albite schist altered, with alteration up to 75% (phylic type).

The microphotograph of alteration in Figure 7 is suggested to have been a glaucophane albite schist, although this is difficult to be sure. This sample has only been found only exposed on the East Side of Rumbia Complex, near the Kasipute City. This alteration at this location is accompanied by sulfide minerals (pyrite and arsenopyrite, and galena). Sulfides can be easily been seen with the naked eye as disseminated pyrite minerals in the rocks, fissure veins and filling fractures associated with quartz veining. The presence of chlorite and sericite minerals, indicating that this rock which suffered propylitic alteration overprinting by phylic alteration type

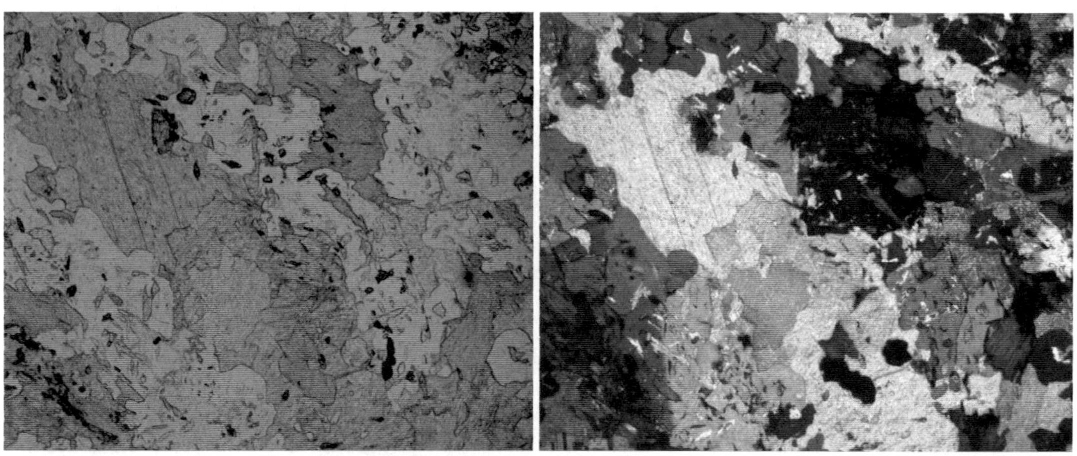

Figure 7 Microphotograph of thin section of altered rock (RS-30C-11), it is suggested to be an altered glaucophane albite schist. The alteration percentage is up to 75%, with assemblage minerals: sericite, chlorite, calcite, quartz, opaque minerals.

5.2. Ore petrography

Ore petrography was undertaken of nine rocks altered sample of Rumbia schist Complex. These samples had significant ore sulfide-content. The analyses included identification of ore, texture, association and ore paragenesis. Result showed that ore minerals are associated with sulfide minerals (pyrite, arsenopyrite, antimony, galena and cinnabar). There is also an associated with oxide minerals (hematite, and goethite). See **Table 1** Ore mineral paragenesis (growth phase).

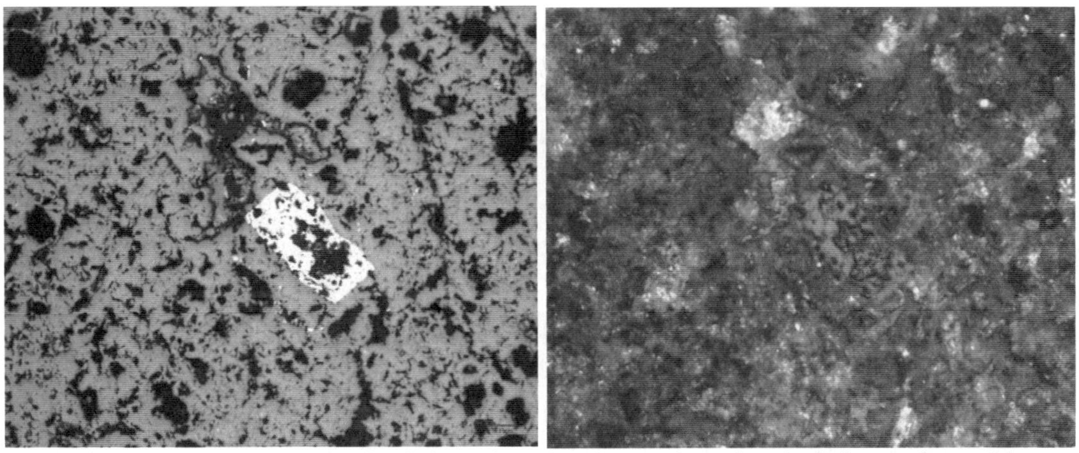

Figure 8 Microphotograph of polish section, showing a prismatic euhedral crystal, possibly hornblende that has been replaced by pyrite (pale yellow).

In this this sample (Figure 8), showing antimony associated with pyrite, and partially replaced by iron oxide (hematite) which sporadically occurring within gangue of quartz. Antimony interpreted as ore mineral that formed in the initial period of mineralization in the Rumbia Complex.

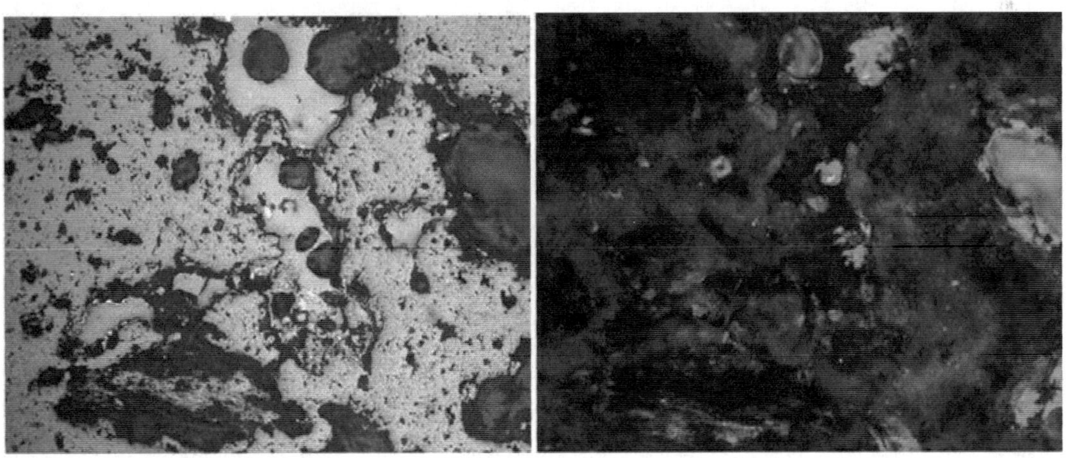

Figure 9 Microphotograph (RS-23A-11) showing antimony (fibrous), irregular, gray color. 5Iron oxide which is brown in cross polar (XPL), is filling the fractures

Figure 8 and Figure 9 showing that mineralization occurs in two period, namely pre-tectonic period and post-tectonic period. In the early period or pre-tectonic mineralization, hydrothermal solutions filled cracks within the rock and rock foliation and thus were incorporated into the exhumation of the metamorphic rocks by inclusion in the fabric. In contrast, the second period occurs after this deformation and it is associated with fluid that caused by alteration in metamorphic rocks (look Figure

10). In Figure 10, appears pyrite comes later, spread in the rocks formation, partially filling fissures and cracks in antimony, and anhedral crystal, and indicated as a post-tectonic mineralization. Very likely, that it presence with regard to the exhumation phase, or extensional regime. While the second phase, seems to be presented after deformation, it is most likely associated with hydrothermal solutions that generate alteration in metamorphic rocks.

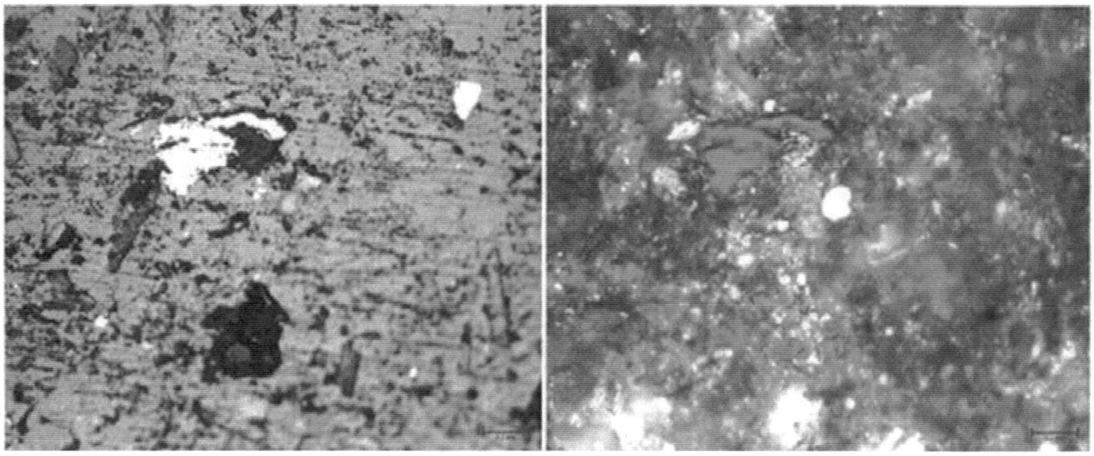

Figure 10. Microphotograph of polish section: showing pyrite, white color (*ppl*), to be disseminated within the rock. The pyrite is seen as anhedral grains and is classified as post-deformation mineralization. Antimony (gray color) was dominant in the rock, and is fibrous, anhedral and deformed. This suggests mineralization as a pre-tectonic or early syn-tectonic mineralization

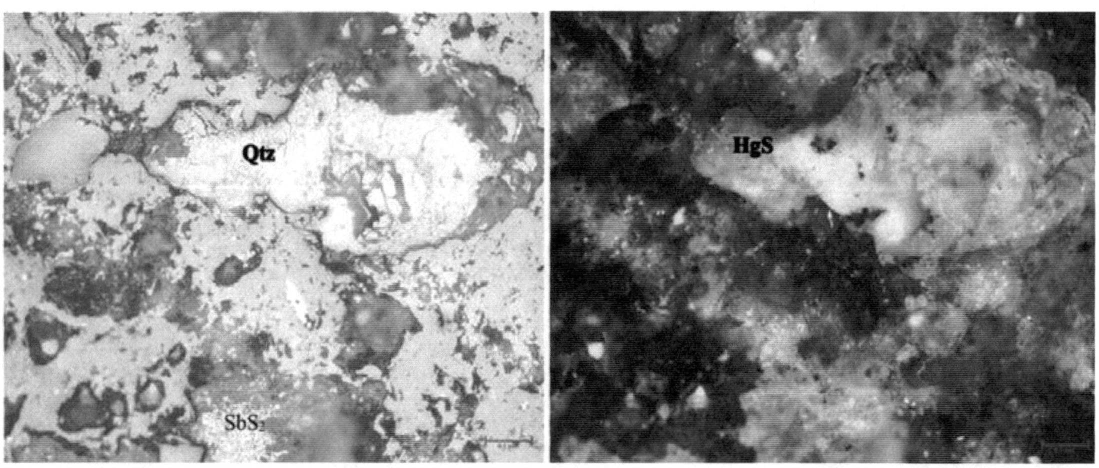

Figure 11 Microphotograph of polish section (RS-23B2-11) shows Mercury ore (Cinnabar 'red color') overprinting with quartz. Mercury Ore (Cinnabar) replaces existing minerals, or filling cavities.

In Figure 11, appears Cinnabar (Mercury Ore), filling fissures and or replace the existing mineral, overprinting with quartz. However the presence of Cinnabar, indicating that the carrier of hydrothermal ore minerals associated with magmatism activity. In this case, understood that Mercury 'HgS' as heavy metals, so it is interpreted that the source of the solution is relatively close to the location found today. Interesting to note, that in Figure 11, it seems antimony (stibnite ore 'SbS_2'), visible presence early, before the formation of the cinnabar (Mercury Ore 'HgS'). Antimony is deformed, forming cracks were very intense, and partly replaced by Cinnabar.

In **Figure 12**, shows the ore mineral pyrite cavities filling presented in gangue quartz. Is the result of hydrothermal ore deposits and as a post-tectonic deposit are included in the phase three on the minerals paragenesis in the Rumbia Complex.

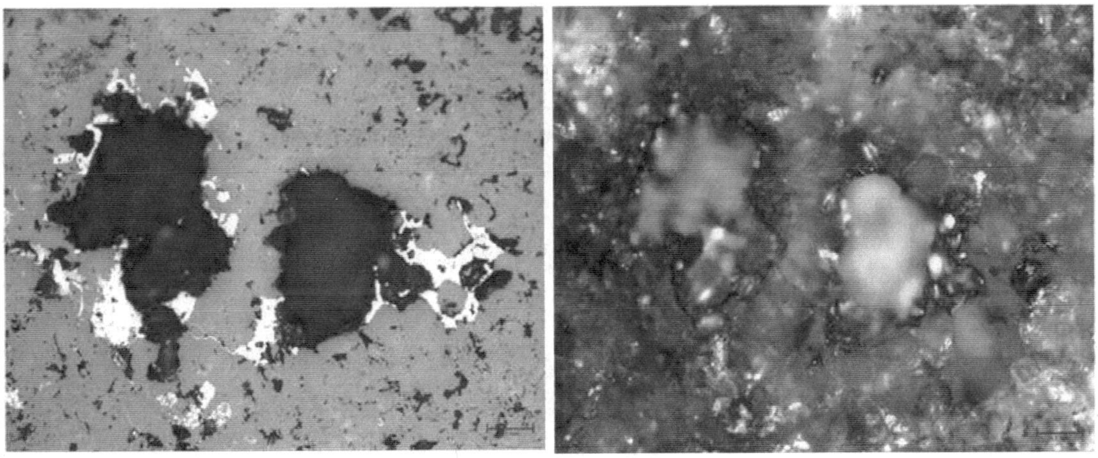

Figure 12. Microphotograph of polish section (RS-07-12); Showed, pyrite present to filled the cavity (as a post-tectonic deposits) in a gangue of quartz.

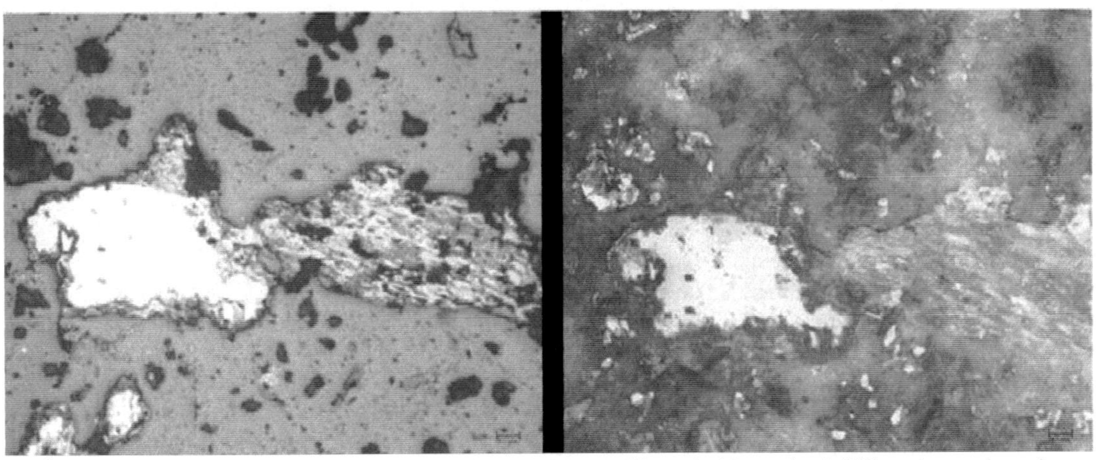

Figure 13. Microphotograph, sample code: RS-23B1; Silver (white color) filling the cavity of a stibnite (antimony). At the edge of the crystals appears to corrosion as a result of weathering.

Figure 13, shown the silver minerals, which is irregular, cavities filling in a gangue of quartz along with antimony (stibnite). The presence of silver in this sample are as post-tectonic deposits or phase three in ore mineral paragenesis in the Rumbia Complex. Figure 14, showing electrum (yellowish white), irregular, as inclusions in gangue of quartz, disseminated sporadically in the rock. Electrum presence in the rock samples it is a hydrothermal deposits are post-tectonic (phase 3). While in Figure 15, shows a oxide mineralization of ore deposits related to the final phase of weathering and oxidation. This phase is the period at the end of the mineralization in the Rumbia Complex. Occurs when the host rocks in contact with ground water or surface water or the atmosphere.

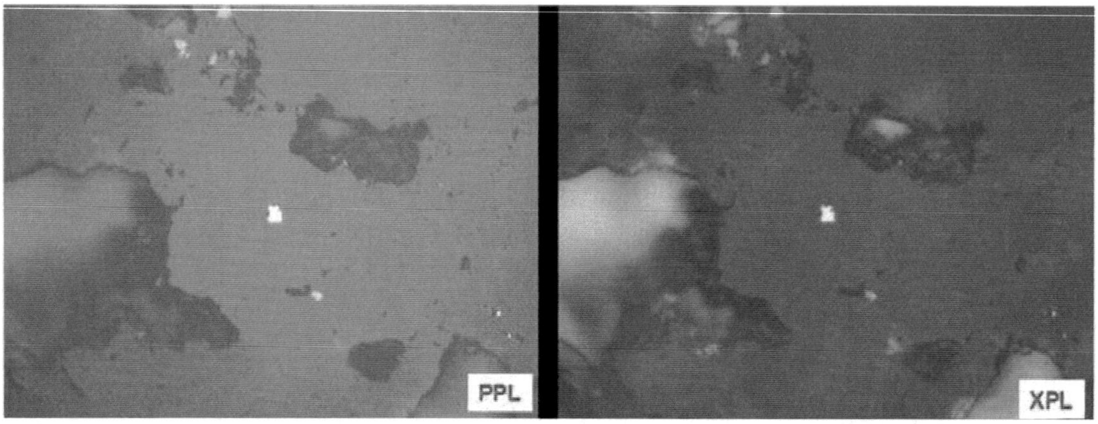

Figure 14 Microphotograph, sample code: RS-27B-12; Showing electrum (yellowish white), irregular, as an electrum crystals disseminated in a gangue of quartz.

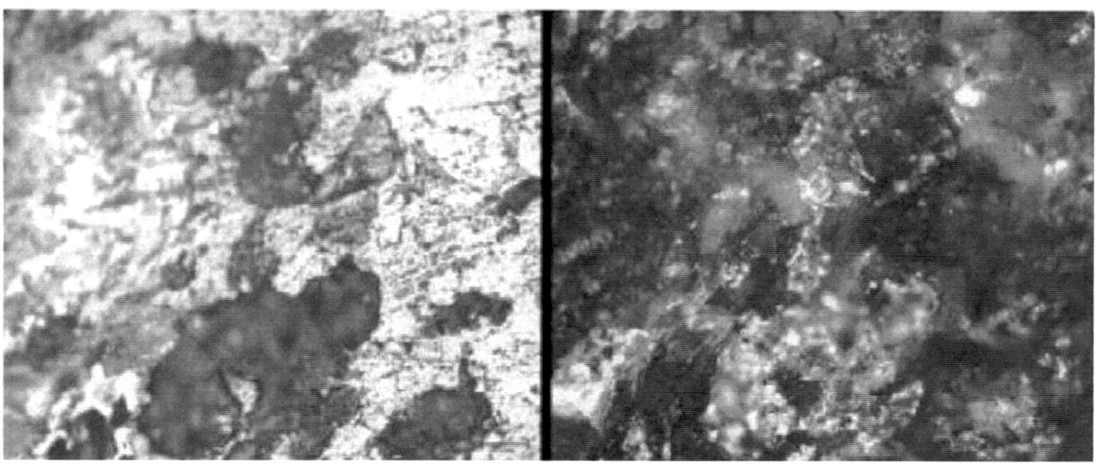

Figure 15 Microphotograph of polish section, showing chalcopyrite ores formed at phase 1. Chalcopyrite was partially replaced Fe-oxide mineralization which grows in the final phase (phase 4).

5.3. Gold deposits

Gold deposits in Rumbia Complex, discovered by the local community, in 2007. There after geological investigations conducted by Surono and Tang (2009) concluded that the placer gold deposits in the Langkowala Valley (Bombana), derived from metamorphic rocks debris in the Rumbia Mountains and Mendoke Mountains (metmamorphic complex). Furthermore, Idrus (2010) concluded as a primary gold deposits in Bombana (Rumbia Complex) as orogenic gold deposits, and not relation with volcanism. While Musri (2011, and 2015), argues that the primary gold deposits in Bombana (Rumbia Complex), associated with hydrothermal alteration in the Early Miocene magmatism. So the metamorphic rocks in the Rumbia Complex, and limestones as a host. Musri (2015) also of the view that the same thing also happens in the metamorphic complex of Mendoke Mountains, and Mekongga to the Northern part of the Southeast Arm of Sulawesi. Early Miocene magmatism occurred as a result of thickening of the crust after the collisional between Rumbia micro-continent and Mekongga, following the amalgamation of Southeast Arm of Sulawesi, Indonesia.

5.4. Location of gold deposits

Primary gold deposits in the Rumbia Complex spread at least 4 major locations, i.e. the Northern, Northwest, Central, and Western part of the Rumbia Complex (See: Figure 16).

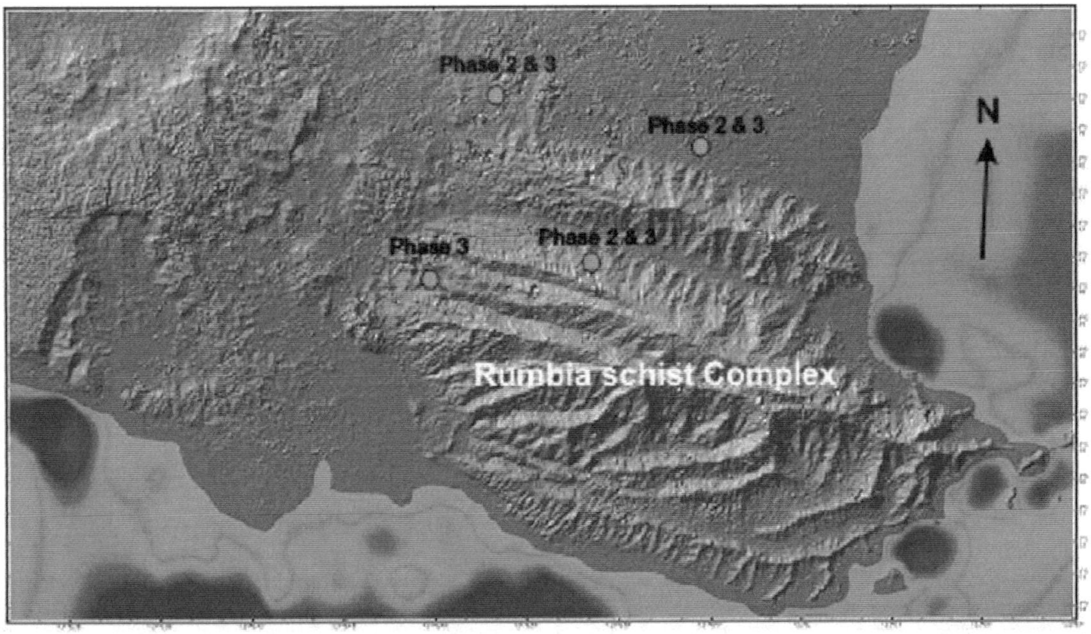

Figure 16 Map location of gold mineralization in the Rumbia Complex. Phase 2 is the first timing of gold mineralization, and phase 3 is the second timing of gold mineralization (look: Table 1, the phase of ore paragenesis in the Rumbia Complex).

Based on the petrographic analysis, ore petrography and $^{40}Ar/^{39}Ar$ ages spectrum, it is known that the primary gold deposits in the Rumbia Complex are occur in two phases; (i) The initial phase relates to the exhumation of *HP* metamorphic rocks. Mineralization is characterized by presence of gold, silver, stibnite, chalcopyrite, galena, and pyrite) within the foliation defined by *HP* minerals, where it deposits in fractures, and low stress zones during an intense period of deformation (Figure 17). This mineralization phase is classified as syn-tectonic mineralization. In addition, gold has been found in the mica-schist. These locations occur mostly in the northern part and north-west of the Rumbia Complex (Figure 16). In the mica schist gold deposited after or late in the formation of the white-mica defined foliation, in low strain zones, fractures and cracks (Figure 18); (ii) Mineralization is associated with an extensional phase. In this phase, the ore minerals pseudomorph mafic minerals, infill cavities and fractures. In general, gold mineralized bearing quartz veins occur during deformation or syn-tectonic to post-tectonics (Figure 19). Some of ore deposits (e.g. electrum) is found as an inclusion, or are disseminated in the rocks, frequently in a gangue of quartz (Figure 20).

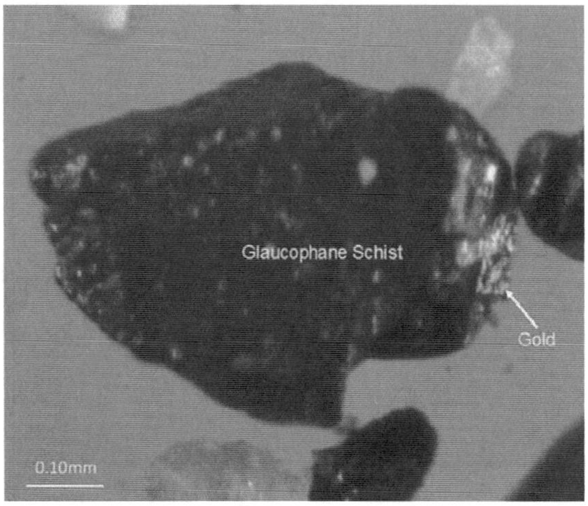

Figure 17 Fractures filled by gold within glaucophane schist fragment (molasses fragment of Amitopa River, Notrh Poleang, Bombana).

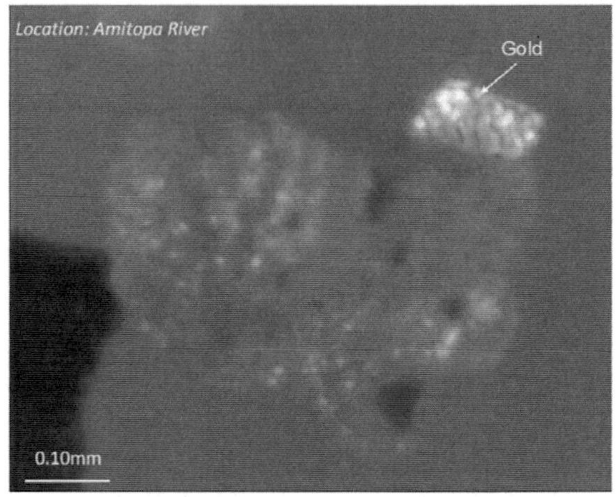

Figure 18. Fissure veins and or foliation filled by gold in the mica schist fragment (placer deposits) from Amitopa River, North Poleang, and Bombana. The first of gold mineralization period.

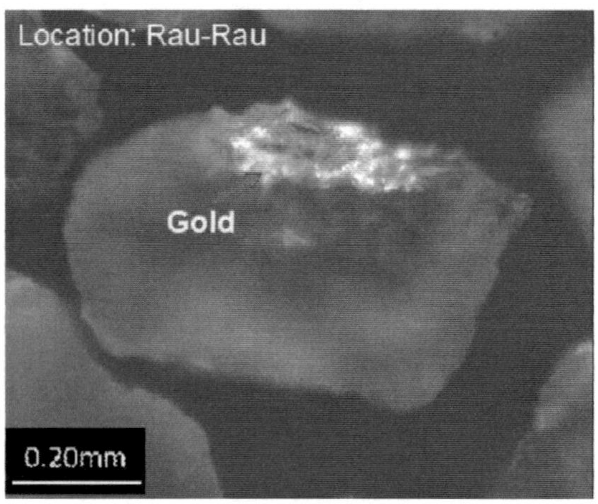

Figure 19 Gold deposits bearing quartz (placer gold deposits). The second primary gold deposits. Quartz grain comes from metamorphic rocks debris (mica schist, gaucophane schist, and chlorite schist) of Rumbia Mountains.

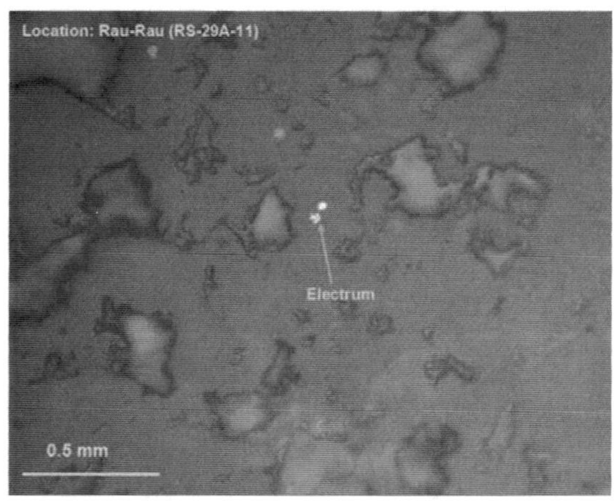

Figure 20 Electrum (fine grains) disseminated in the gangue of quartz, location from Rau-Rau (close to the host spring. The second stage of gold mineralization in the Rumbia Complex.

5.5. $^{40}Ar/^{39}Ar$ Age dating and timing of gold mineralization in the Rumbia Complex

$^{40}Ar/^{39}Ar$ dating has been done on the metamorphic rock with gold content from the key mineralization locations. An array of ages have been determined showing that mineralization has occurred at different times. Results of $^{40}Ar/^{39}Ar$ dating of the *HP* metamorphic foliation is from ~23 Ma to ~11 Ma suggesting that mineralization may have occurred in this zone during that time. Because the gold mineralization occurred during the foliation formation it is suggested that the Phase 1 was deposited at ~23 Ma. This phase of gold mineralization has been found in situ in the *HP* metamorphic rocks as well as Placer gold after the erosion.

5.6. Ore paragenesis in the Rumbia Complex
Ore mineral texture analysis results through a polish section, and integrated analysis of the hydrothermal alteration minerals, known that the ore mineral deposits in the study area was divided into four (4) phase, i.e: **Phase1:** At this phase of ore deposits formed before the tectonic deformation or commonly referred to as pre-tectonic deposits. Ore minerals were formed are antimony, galena, chalcopyrite, cinnabar, magnetite and pyrite. **Phase 2:** In phase 2, which formed the ore deposits associated with hydrothermal flow through rock fractures associated with metamorphic exhumation and upthrusting. Furthermore ore deposits or filling in the fractures and in the metamorphic rocks foliation. Ore deposits such as these occur as syn-tectonic deposits. Minerals that form the antimony, galena, pyrite, chalcopyrite, cinnabar, silver, gold, and magnetite. **Phase 3:** In phase three, mineral deposits formed after tectonic deformation, and in this study is referred to as post-tectonic deposits. Minerals that form the chalcopyrite, pyrite, cinnabar, silver and gold. At this phase with respect to the ore deposits hydrothermal alteration. **Phase 4:** Phase 4 is the ore deposits were formed at the end of the period of mineralization in the Rumbia Complex. This event with respect to oxidation as a result of the formation is influenced, groundwater, atmosphere, and or weathering. Also referred to as oxide ore deposits. Table1, showing ore minerals paragenesis in the Rumbia Complex.

Table 1. Ore minerals paragenesis in the Rumbia Complex

Ore minerals	**Phase 1**	**Phase 2**	**Phase 3**	**Phase 4**
Antimony				
Galena				
Chalcopyrite				
Arsenopyrire				
Pyrite				
Silver				
Cinnabar				
Electrum/gold				
Magnetite				
Fe-Oxide				

Source: Musri (2015)

5.7. Geology model of gold deposits in the Rumbia Complex
The geological model of gold deposits in the Rumbia Complex is based on ore petrography, ore minerals paragenesis, the evolution of metamorphism and tectonic evolution in space and time. Approach model to integrate the age data of metamorphic rocks, tectonic evolution, and geological formation of gold deposits in the Rumbia Complex. As shown in Figure 21 and Figure 22, are the

IOP Conf. Series: Earth and Environmental Science **71** (2017) 012030 doi:10.1088/1755-1315/71/1/012030

alternative scenarios of the geological model of gold deposits in the Rumbia Complex. The first alternative scenario (Figure 21), on the understanding that Rumbia Complex as a micro-continent that is different from other continental in the Southeast Arm of Sulawesi (Musri,2015). In this context, the Proto-Rumbia Microcontinent suffered subduction with Proto-Mekongga Microcontinent in the Nortn part of Southeast, which ended with a collision and caused by emplacement of ultramafic rocks that is currently known as Mendoke Mountains (Musri, 2015). This scenario, the case relates to the end of metamorphism. Hot solution can be derived from Uppermost Mantle, which carries out ore minerals, mixed with meteoric water, and precipitation in the rock mass. In this case, the fracture, and foliation, could act as a canal way. Based on the radiometric age data, at least occurred after 23Ma. As the initial phase of glaucophane schist formation. Allegedly occurred during the 23-11 Ma.

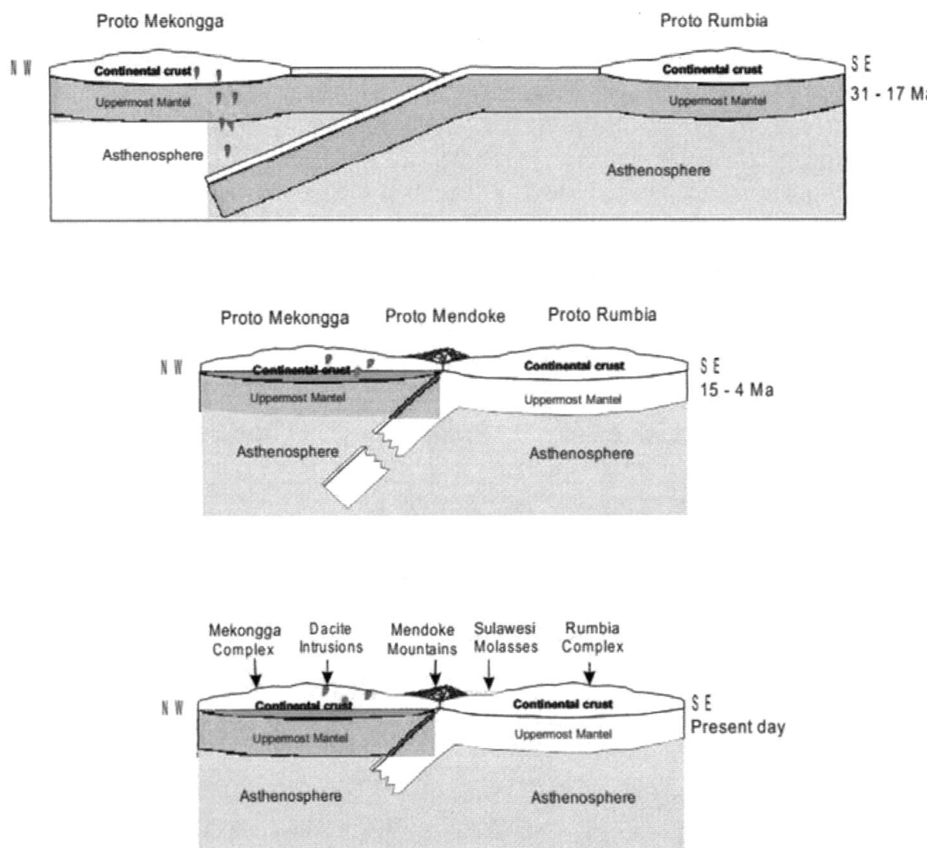

Figure 21 The first scenario, formation of primary gold deposits in the Rumbia Complex, which is interpreted that the hydrothermal solution moves up began in Figure 5.6 (A). In this model the Proto Rumbia Complex as micro-continent, before experiencing subduction that began 31 Ma, with Proto Mekongga Micro continent, as well as collision at 17 Ma.

The second alternative scenario it is possible the contribution of new subduction by others micro-continent in the South or Southeast of the Rumbia Complex (Figure 22). Very likely a micro-continent Buton as hypothesized by Ali et al., (1996). In this scenario, gold mineralization occurs in conjunction with extensional phase. Based on data from radiometric ages of the metamorphic rocks in the Rumbia Complex, the events of this mineralization occurred on 15 Ma to 7 Ma. Gold mineralization of this phase, interpreted as a second phase of gold mineralization in the Rumbia Complex, which can be observed in the textural of ore minerals as electrum, anhedral, and disseminated in the rocks, and cavities filling.

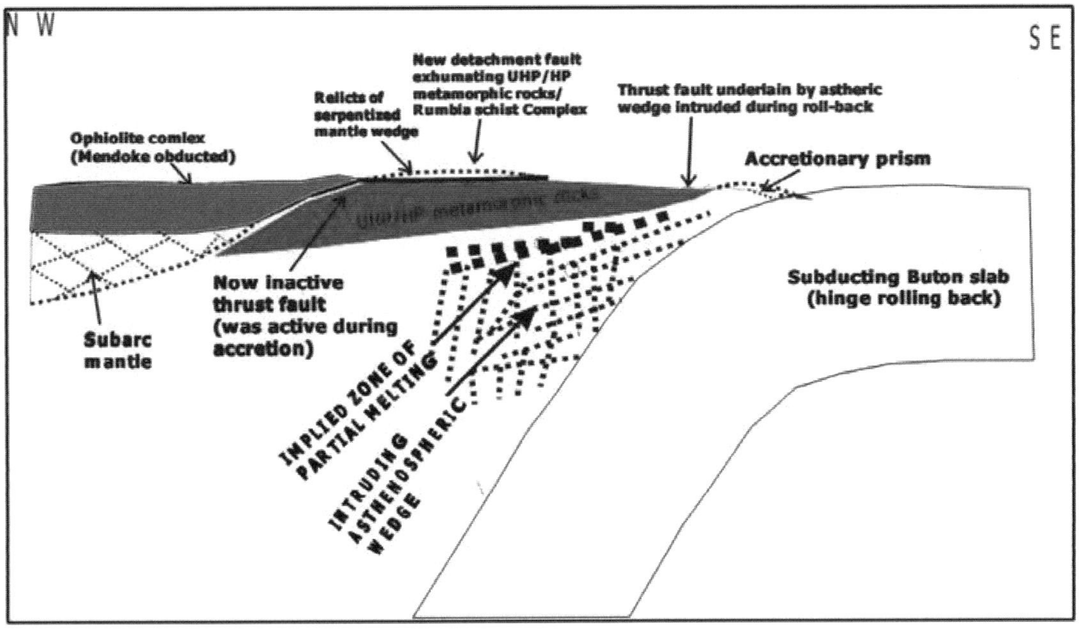

Figure 22 The second scenario of gold deposits in the Rumbia Complex. In this model, the contribution of the subduction of Buton-Tukangbesi platform result of partial melting and intruding asthenospheric wedge (adapted model from Lister and Forster (2009).

6. Conclusion

Rumbia metamorphic rock Complex have been hydrothermal altered, and based on the mineral assemblage are classified as phylic type overprinting propylitic hydrothermal alteration type. Phylic alteration type; the minerals assemblage are: Ser+Cal+Qtz±Oq; while prophylitic alteration type have the minerals assemblage are: (1) Act+Cal+Qzt±Oq; (2) Ep+Chl+Cal±Qzt±Oq. In general, based on $^{40}Ar/^{39}Ar$ age spectrum indicates that the Glaucophane Schist provide age spectrum between 23 Ma - 11 Ma. Mica schist give age spectrum 31 - 11 Ma. Furthermore, chlorite schists giving ages pectrum at 15 Ma - 7 Ma. Referring to the age spectrum $^{40}Ar/^{39}Ar$ against Glaucophane Schist, Mica Schist, and Chlorite schist, integrated with petrographic data, mineragraphy and regional tectonic data, it can be conclude that the tectonic events that occur in Rumbia Complex, is as follow:

Phase 1: There are two micro continent, moving closer to each other. Before each of which has oceanic crust. One of the micro-continent is Proto-Rumbia, whose front section is formed Oceanic Island Basalt (OIB) prior to subduction occurs. Subduction occurs between the Proto-Rumbia and Proto-Mekongga accompanying the oceanic crust. These events resulted of metamorphism of Oceanic Island Basalt (OIB) on 31 Ma-17 Ma, accompanied by emplacement of the oceanic crust. In this phase Glaocophane Schists formed at 23Ma, with protolith of Oceanic Island Basalt (OIB).

Phase 2: Continued subduction events that followed by collisions, which is interpreted occurred on 17 years ago, marking the amalgamation phase of Proto Mekongga, Proto-Rumbia, and Proto-Meluhu, which is now known as the Southeast Arm of Sulawesi. In this phase, interpreted as a first phase of gold mineralization that occurs in the Rumbia Complex, or at least the first phase of mineralization in the Rumbia Complex occurs between 23-11 Ma ago.

Phase 3: Collision peak interpreted occurred in 17 million years ago, the extensional phase followed interpreted taking place since 15 million years ago and lasted up to 7 million years ago. End of this extensional events would be marked by the cessation of the growth of the reef, and is the beginning of Ophiolite complex (Mendoke Mountain).

7. Acknowledgments

Many thanks to Gordon Lister, Oleq Koudashev, Davood, Jessica and Sareh, for discussion and interaction during in the RSES, The Australian National University, Canberra, Australia. My colleagues at Hasanuddin University, Makassar and ITB, Bandung, Indonesia for discussion and interaction during the research and preparing this paper.

8. References

[1] De Roever, W P., 1950: Preliminary notes on glaucophane-bearing and other crystalline schists from Southeast Celebes and on the origin of glaucophane-bearing rocks, Proc.K.ned.Akad. Wet.LIII-9 2-12.

[2] De Roever, W P., 1956: Some additional data on the crystalline schists of the Rumbia and Mendoke mountains, Southeast Celebes, *Verh. K. Ned. Geol.-mijnb. Genoot*, XVI:385-393.

[3] Forster, M.A. And Lister, G S., 2004: The interpretation of $^{40}Ar/^{39}Ar$ apparent age spectra produced by mixing: application of the method of asymptotes and limits. *Journal of Structural Geology*.

[4] Idrus A, Warmada I W, Nur I, Sufriadin, Imai A, Widasaputra S, Marlia S I, Fadlin and Kamarullah, 2010: Metamorphic rock-hosted orogenic gold deposit as a source of Langkowala palcer gold, Bombana, Souteast Sulawesi, Indonesia *Proceedings PIT IAGI Lombok 2010 The* ^{39th} *IAGI Annual Convention and Exhibition*, p7.

[5] Lister, G S., and Forster, M., 2009: Tectonic mode switches and the nature of orogenesis, *Lithos*.

[6] Musri, Suparka E, Abdullah C I, 2011: Preliminary study of sulfide mineralization in limestone hosted, Bombana Area, Southeast Sulawesi, Indonesia.

[7] Musri, 2015: Metamorphic Rocks Evolution of Rumbia Complex, Southeast Arm of Sulawesi and its relation to gold deposit, Dissertation of ITB (unpublished).

[8] Parkinson C D., 1998: An Outline of the petrology, structure and age of the Pompangeo Schist Complex of Central Sulawesi, Indonesia,*The Island Arc* **7**, 231-245.

[9] Simandjuntak T O, Surono dan Sukido 1993: Peta geologi Lembar Kolaka, Sulawesi, sekala 1 : 250.000, Pusat Penelitian dan Pengembangan Geologi, Departemen Pertambangan dan Energi RI.

[10] Surono, 1994: Stratigraphy of the Southeast Sulawesi continental terrane, Eastern Indonesia, *Journal of Geology and Mineral Resources*, No **31**, **IV**, p12.

[11] Surono, 1998: Geology and origin of the Southeast Sulawesi continental terrane, Indonesia, *Media Teknik* **XX(3)**, 33-42.

[12] Surono, Tang H A 2009 Batuan pembawa emas primer dari endapan emas sekunder di Kabupaten Bombana, Sulawesi Tenggara, berdasarkan interpretasi indera jauh, *Prosidings PIT IAGI Semarang, 2009*, Indonesia 11h.

[13] Map of Sulawesi. GoogleEarth. https//earth.com: Download on Januari 9, 2012.

IOP Conf. Series: Earth and Environmental Science **71** (2017) 012031 doi:10.1088/1755-1315/71/1/012031

Late Miocene Molluscan Stage of Jawa Insight from New Field Studies

Aswan[1], Elina Sufiati[2], Desty Kistiani[2], Irman Yudi Abdurrahman[2], Wahyu Dwijo Santoso[1], Alfend Rudyawan[1], and Thaw Zin Oo[3]

[1] Geology Department, Faculty of earth Sciences and Technology, Institute of Technology Bandung, Jalan Ganesa 10 Bandung, Indonesia
[2] Geological Museum Bandung, Jalan Diponegoro No.57, Bandung, Indonesia
[3] Geology Department, East Yangon University, Thanlyin Township, Yangon Division, Myanmar
E-mail: aswan_gl@gc.itb.ac.id

Abstract. Neogene stratigraphic stages of Jawa based on molluscan index fossils firstly compiled by Oostingh in 1938. This concept is widely used by palaeontologists, both for application in and outside Jawa. Based on the estimated ages of previous researchers the Neogene stages from old to young are: Rembangian stage considered equivalent to Early Miocene, Preangerian stage (Middle Miocene), Odengian stage (Upper Miocene), Cheribonian stage (Early Pliocene), Sondian stage (Late Pliocene) and Bantamian stage (Early Pleistocene). Further studies to determine a more precise age of each stage are often difficult due to lack of planktonic foraminifera fossils and other age index fossils within the sedimentary deposits that are mostly shallow marine sediments. This study focuses to re-examine the age of sediments of Preangerian and Odengian Stages on new field studies. The type locality and rock units that are correlated with these stages are part of the well-exposed Nyalindung formation around Sukabumi area. The studies were carried out in four main fosill-rich sedimentary sequences that are exposed along Cijarian, Citalahab, Ciangsana and Ciodeng rivers. More detailed ages were obtained from the sediments. The exposed sequences along the Ciangsana and Citalahab rivers are Middle Miocene (N9-N14) while the Cijarian and Ciodeng rivers sequences were deposited between Middle to early Late Miocene (N9-N16). The Nyalindung formation in the Cijarian river also contain *Vicarya sp*, an index fossil that marks a rise in sea level in the Miocene (12 Ma), which previously only been reported from the Ciangsana river sediments.
Keywords: Late Miocene, Molluscan, Jawa, Age, Neogene

1. Introduction

Neogene stratigraphy in Jawa was divided into 6 stages based on the index molluscs fossil [1, 2]. These stages distinguish the Neogene from Early Miocene to Pleistocene, namely: Rembangian (equivalent to Early Miocene), Preangerian (Middle Miocene), Odengian (Late Miocene), Cheribonian (Early Pliocene), Sondian (Late Pliocene) and Bantamian (Pleistocene). Subsequent studies on these stages are often unfruitful due to difficulties in finding planktonic foraminifera within the type section of each stage which are mostly shallow marine deposits. This study focuses to re-examine the age of sediments from the type formation or type locality based on [1], that is believed to be the equivalent to the Preangerian and Odengian Stage using mollusc and foraminifera fossil association,. The type locality/formation sections of these sedimentary rocks are stratigraphically included in the Nyalindung Formation that exposes around the city of Sukabumi (Figure 1). Four locations of mollusc-rich sedimentary sequence are exposed in Cijarian River, Citalahab River, Ciangsana River and Ciodeng River. [3], [4], [5], and [6] previously conducted research on the Nyalindung Formation, however there are some key river traverse that were overlooked until now. Two important traverses were visited; they are Ciangsana River and Cijarian River next to Ciodeng Village. The results suggest that the age obtained from these two sections are in higher resolution, moreover some locations suggest a completely different age compared to the previous results.

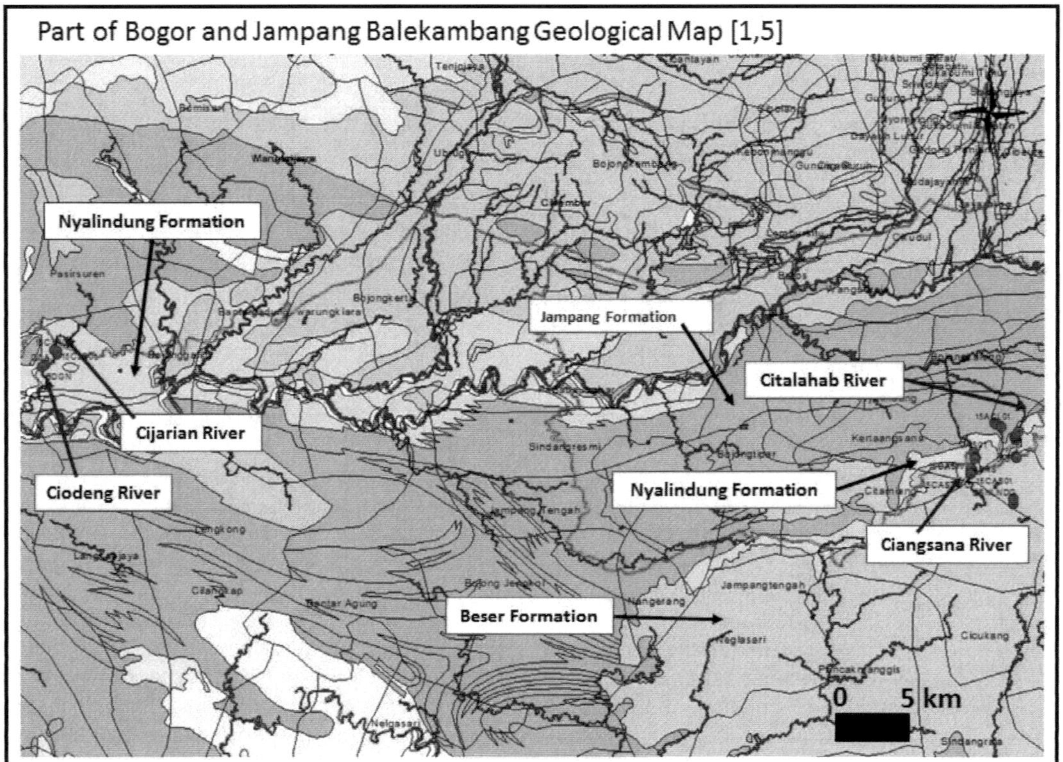

Figure 1. Research locations map shows sampling location along: Cijarian, Ciodeng, Citalahab and Ciangsana rivers.

2. Geological Setting

The study area is located within the Southern Mountain physiography zone according to [2] on the west side of the Bogor Basin [7]. The Nyalindung Formation is the main focus of the study which was found in stratigraphy contact with the Jampang Formation in the north and the Beser Formation in the south [8, 9].

The Jampang Formation has three members [8, 9], namely: (1) Flow breccia of andesitic pyroxene composition that overlies conformably the (2) Tuff and Breccia member. The Jampang member is interpreted to be Early Miocene. The Nyalindung Formation consists of green calcareous glauconitic sandstone, claystone, marl, sandy marl, tuffaceous marl, conglomerate, breccia and limestone. Tuffaceous marl along Cijarian, Citalahab, Ciangsana Rivers and Ciodeng Village are rich in molluscs (Figure 2a-d). The unit is estimated of Middle Miocene (N 13) in age and conformably overlies the Lengkong Formation. Beser Formation consists of two members, they are: (1) Volcaniclastic member and (2) Lava member. The main part of this formation comprises of volcanic breccia, laharic breccia, tuff breccia, tuffs and conglomerate. Rough terrain often represents the Lava Member (or also known as Cikondang Member). The Beser Formation overlies the Cimandiri and Jampang Formation unconformably [8, 9]. It is interpreted to be deposited in the terrestrial to beach environment.

IOP Conf. Series: Earth and Environmental Science **71** (2017) 012031 doi:10.1088/1755-1315/71/1/012031

3. Methods

Representative samples were collected from the field especially from the fossil-rich beds of the Nyalindung Formation. Locations were selected from previous type sections of Neogene stages of [1]. Preangerian stage is represented by exposures along Cijarian (Figure 2a), Citalahab (Figure 2b), Ciangsana (Figure 2c), and Ciodeng (Figure 2d) rivers; represents the Odengian stage. Samplings were designed to stratigraphycaly represent each part of the lower, middle and upper part of the formation in each transects. Monotonous lithology association is observed in the study area. It consists of bedded calcareous mudstone and sandstone with abundant molluscan fossils.

The samples were washed using H_2O_2 for a full day to separate the finer clastics fragments from the planktonic forams specimens. The samples then dried in an oven for 10 minutes. Once the samples are dry, they are ready to be analysed the foraminifera fossils content of each sample from each location under the binocular microscope. Age range of each sample will be determined from foraminifera association according to [10] age zonation or foraminifera age index fossil content. Foram fossil occurrence in the sample from each particular location is arranged stratigraphically from the oldest (bottom row in the table) to the youngest (top row in the table) as shown in Table 1-4.

Figure 2. Nyalindung outcrops that rich of mollusc fossils along: Cijarian River (2a), Citalahab River (2b), Ciangsana River (2c), and Ciodeng River (2d).

4. Results

The following results are the age of the Nyalindung Formation determined from the planktonic foraminifera association in each location (Table 1-4). Age index fossils can be seen in Figure 3.

1. Cijarian River transect: Middle Miocene – early Late Miocene (N9 – N16) based on the *Orbulina universa* (3a), *Globoquadrina altispira* (3b) and *Globigerinoides obliquus obliquus* (3c) fossil association.
2. Citalahab River transect: Middle Miocene - late Middle Miocene (N9 – N14) based on the *Orbulina universa, Globorotalia fohsi fohsi* (3d) and *Globorotalia siakensis* (3e) fossil association.
3. Ciangsana River transect: Middle Miocene - late Middle Miocene (N9 – N14) based on the *Orbulina universa*, *Globorotalia siakensis*, dan *Globigerinoides subquadratus* (3f) fossil association.

4. Ciodeng Village: Middle Miocene - middle Late Miocene (N12 – N17) based on the *Hastigerina siphonifera* (3g), *Sphaeroidinellopsis seminulina* (3h) dan *Globigerina praebulloides* (3i) fossil association.

Table 1. Age analysis of Nyalindung Formation along Cijarian river based on planktonic foraminifera fossils. Sequence from bottom to top in the table represent old to young stratigraphy.

No	Sample No	AGE	Planktonic Zonation [10]	Planktonic Foraminifera Content
1	15 B CJR	Middle - Late Miocene	N9 - N16	*Orbulina universa, Globoquadrina altispira*
2	15 A CJR	Middle Miocene	older than N16 and younger than N9	*Orbulina universa, Globigerinoides obliquus*

Table 2. Age analysis of Nyalindung Formation along Citalahab river based on planktonic foraminifera fossils. Sequence from bottom to top in the table represent old to young stratigraphy.

No	Sample No	AGE	Planktonic Zonation [10]	Planktonic Foraminifera Content
1	15 CTL 05 BF	Middle Miocene	N12 - N14	*Orbulina universa, Globorotalia siakensis*
2	15 CTL 05 AF	Middle Miocene	N12 - N14	*Orbulina universa, Globorotalia siakensis*
3	15 CTL 04 CF	Middle Miocene	N12 - N14	*Orbulina universa, Globorotalia siakensis*
4	15 CTL 04 BF	Middle Miocene	N12 - N14	*Orbulina universa, Globorotalia siakensis*
5	15 CTL 03 BF	Middle Miocene	N12 - N14	*Orbulina universa, Globorotalia siakensis*
6	15 CTL 02 B F	Middle Miocene	N10 - N11	*Orbulina universa, Globorotalia fohsi fohsi*
7	15 CTL 02A F	Middle Miocene	N10 - N11	*Orbulina universa, Globorotalia fohsi fohsi*
8	15 CTL 01 B FA	Middle Miocene	N9 - N10	*Orbulina universa*
9	15 CTL01 A FA	Middle Miocene	N9 - N10	*Orbulina universa*

Table 3. Age analysis of Nyalindung formation along Ciangsana river based on planktonic foraminifera fossils. Sequence from bottom to top in the table represent old to young stratigraphy.

No	Sample No	AGE	Planktonic Zonation [10]	Planktonic Foraminifera Content
1	IV B 15 CAS 9 BF	Middle Miocene	N14	*Orbulina universa, Globorotalia siakensis*
2	IV B 15 CAS 8 BF	Middle Miocene	N14	*Orbulina universa, Globorotalia siakensis*
3	IV B 15 CAS 7 BF	Middle Miocene	N9 - N13	*Orbulina universa, Globigerinoides subquadratus*
4	IV B 15 CAS 6-2F	Middle Miocene	N9 - N13	*Orbulina universa, Globigerinoides subquadratus*
5	IV B 15 CAS 6-1 F	Middle Miocene	N9 - N13	*Orbulina universa, Globigerinoides subquadratus*
6	IV 15 CAS 5F	Middle Miocene	N9 - N13	*Orbulina universa, Globigerinoides subquadratus*
7	15 CAS 04 F	Middle Miocene	N9 - N13	*Orbulina universa*
8	15 CAS 03 F	Middle Miocene	N9 - N13	*Orbulina universa*
9	15 CAS 02 F	Middle Miocene	N9 - N13	*Orbulina universa*
10	15 CAS 01 FB	Middle Miocene	N9 - N13	*Orbulina universa*
11	15 CAS 01 FA	Middle Miocene	N9 - N13	*Orbulina universa*

5. Discussion

The Prangerian stage mentioned in [1] covers a wide range of Middle Miocene age, however based on the analyses results from the sample in Cijarian river suggest an even wider age range to reach early Late Miocene. The new age results of the Odengian stage from the Cijarian river around Ciodeng village suggest an older age range of late Middle Miocene – middle Late Miocene compared to Late Miocene [1].

The occurrence of *Vicarya sp.* (Figure 4) as a Middle Miocene fossil index along the Cijarian river section is another new finding in the study. Previous studies [1, 3-4] were successfully recognised this fossil along the Ciangsana River (Figure 2c). *Vicarya verneulli callosa* (Figure 5) fossil marked the global sea level rise during the Middle Miocene [11] (N13; based on [10]). This recent finding suggests that the sea level rise was recorded by the Nyalindung formation.

Table 4. Age analysis of Nyalindung formation along Ciodeng river based on planktonic foraminifera fossils. Sequence from bottom to top in the table represent old to young stratigraphy.

No	Sample No	AGE	Planktonic Zonation [10]	Planktonic Foraminifera Content
1	Ci VII A	Middle Miocene - Not Older Than Late Miocene	N12 - N23	*Hastigerina siphonifera*
2	Ci VIII A	Early - Middle Miocene	N6 - N17	*Sphaerodinellopsis seminulina, Globigerina praebulloides*

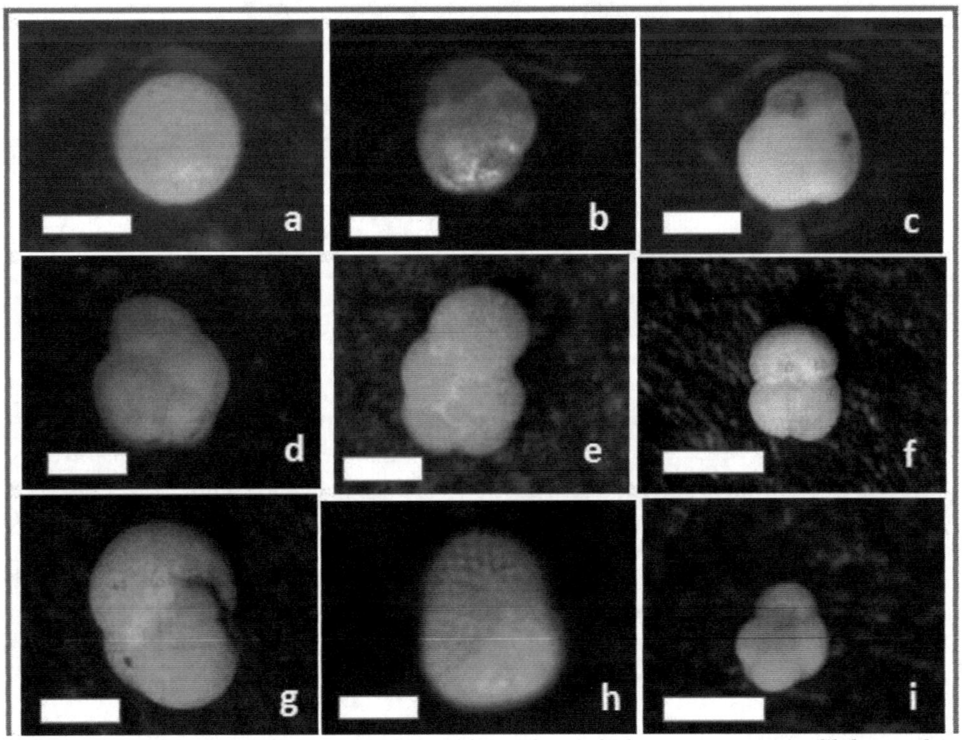

Figure 3. Age index fossils determined from this study: *Orbulina universa* (a), *Globoquadrina altispira* (b), *Globigerinoides obliquus obliquus* (c), *Globorotalia fohsi fohsi* (d), *Globorotalia siakensis* (e), *Globigerinoides subquadratus* (f), *Hastigerina siphonifera* (g), *Sphaeroidinellopsis seminulina* (h), *Globigerina praebulloides* (i). White scale bar: 0.25 mm.

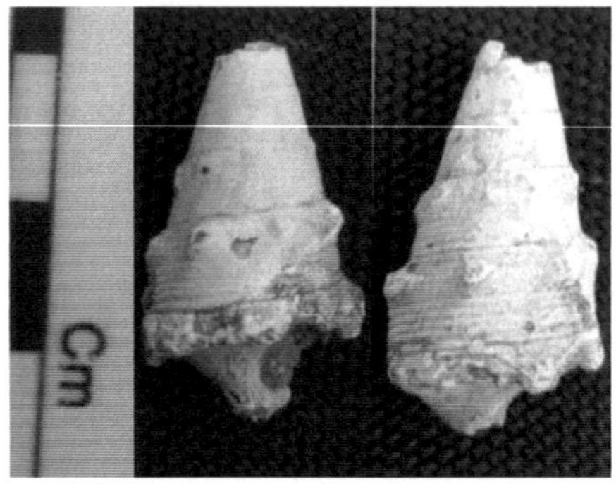

Figure 4. *Vicarya sp*. as a Middle Miocene fossil index found along the Cijarian river.

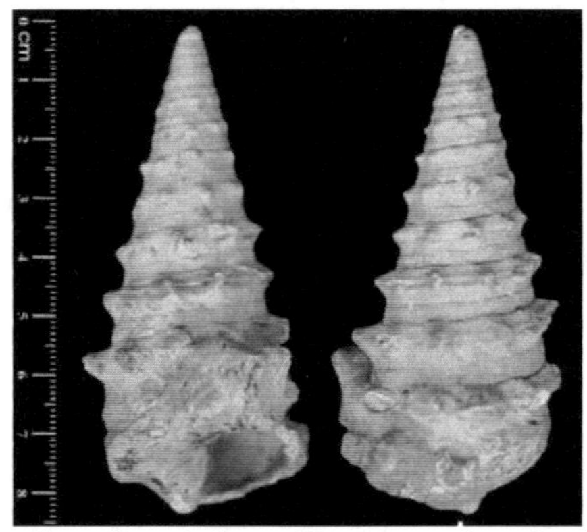

Figure 5. *Vicarya verneulli callosa* found along Ciangsana river fossil marked the global sea level rise during the Middle Miocene

6. Conclusions

The Nyalindung formation age is different in three river sections; (1) Citalahab River (Preangerian: early Middle Miocene~late Middle Miocene), (2) Cijarian River (Preangerian: early Middle Miocene~early Late Miocene), (3) Ciangsana River (Preangerian: early Middle Miocene ~late Middle Miocene). These three new ages from each representative Preangerian stage are results of a more detailed determination from the Middle Miocene age assigned by [1].

The Nyalindung formation age in Ciodeng River (Odengian) is late Middle Miocene - middle Late Miocene compared to Late Miocene suggested by [1].

Vicarya verneulli callosa fossil occurrence in Ciangsana and Cijarian rivers marks the global sea level rise in the Middle Miocene.

7. Acknowledgments

We would like to express our gratitude to the Geology Study Program at ITB for the support and opportunity to use the facility during the work of this study. We would also like to thanks to the Dean and all staff of Faculty of Earth Sciences and Technology (FITB) – ITB and Head of Geology Expertise Group and staff, who helped the author by gave a permission to attend the conference. The Bandung Geological Museum is thanked for the permission to compile and publish the manuscript. Lastly we are thankful to the TREPSEA II 2016 committee for the efforts putting together a proceeding for the conference.

8. References

[1] Oostingh C H 1938. *Mollusken Als Gidsfossielen Voor Het Neogeen in Nederlandsh Indie* (Surabaya: Handelingen Van Het Achtste Nederlandsch – Indisch Natuur Wetenschappelijk Congres)

[2] van Bemmelen R W 1949 *The Geology of Indonesia* (Martinus Nijhoff, the Hague, **IA**)
[3] Shuto T 1974: Notes on Indonesian Tertiary and Quaternary Gastropods, Mainly Described by the Late Professor K. Martin, I. Turritellidae and Mathildidae. Tokyo: Geology and Palaeontology of Southeast Asia, **XIV**, p 135–160
[4] Shuto T 1975: Preliminary Correlation of The Neogene Molluscan Faunas in Southeast Asia. Tokyo: Geology and Palaeontology of Southeast Asia, **XV**, p 289–301
[5] Aswan and Ozawa T 2006: Milankovitch 41000-year Cycles in Lithofacies and Molluscan Content in The Tropical Middle Miocene Nyalindung Formation, Jawa, Indonesia. *Palaeogeography, Palaeoclimatology, Palaeoecology* **235,** p 382 – 405
[6] Aswan 2009: System Tracts Determination Based on Molluscan Shell Association of The Nyalindung Formation (Middle Miocene), Sukabumi, West Jawa; In Terms of Sequence Stratigraphy. *Buletin Geologi* **39** p 147-166
[7] Martodjojo S 2003 *Evolusi Cekungan Bogor* (Bandung: Penerbit ITB) p 238
[8] Sukamto 1975 *Geological Map of The Jampang and Balekambang Quadrangles, Jawa* (Bandung: Department of Mines and Energy – Directorate General of Geology and Mineral Resources – Geological Research and Development Centre)
[9] Effendi A C, Kusnama and Hermanto B 1998 *Geological Map of The Bogor Quadrangles, Jawa* (Bandung: Department of Mines and Energy – Directorate General of Geology and Mineral Resources – Geological Research and Development Centre)
[10] Blow W H 1969: Late Middle Miocene to Recent Planktonik Microfossils. 1st, Geneva Proceeding Leiden, F.J. Brill **1**, p 199-422
[11] Janson X, Van Buchem F S P, Dromart G, Eichenseer H T, Dellamonica X, Boichard R, Bonnaffe F and Eberli G 2010: Architecture and Facies Differentiation within a Middle Miocene Carbonate Platform, Ermeneek, Mut Basin, Southern Turkey London. Geological Society, Special Publications, **329**, p 265-290

Turbidite Facies Study of Halang Formation on Pangkalan River, Karang Duren – Dermaji Village, Banyumas District, Central Java - Indonesia

Yan Rizal[1], Raymond Lagona[2] and Wahyu Dwijo Santoso[1]

[1] Geology Department, Institute of Technology Bandung (ITB), Jl. Ganesha 10 Bandung, West Java- Indonesia (40132)
[2] Alumni Geology Department, Institute of Technology Bandung (ITB), Jl. Ganesha 10 Bandung, West Java – Indonesia (40132)
Email: yan@gc.itb.ac.id

Abstract. The Halang formation had been exposed continuously in Pangkalan river, Banyumas Basin. Detailed stratigraphic studies by which using measuring section method was performed on Halang formation which found along Pangkalan river. The facies of Halang formation based on sedimentological study can be divided into five facies association, namely with: proximal channel, distal levee, frontal splay 1, crevasse splay, and frontal splay 2. The thickness of distal levees deposit indicates that Halang formation is confirmed as mud rich system turbidities. Fasies association of Halang formation is controlled by several factors. Changes in sediment supply are that one of factor which created a change in fasies association. Sediment supply has changes caused by sea level changes which are resulting of global climatic changes or tectonic factor. Moreover, volcanism process during Late Miocene leads the sediment supply changes.
Keywords: Pangkalan, Halang formation, facies association, mud – rich system turbidite.

1. Introduction

Halang formation is one of the formations in the Banyumas basin area, with a good and completely exposure of the series of turbidite sequences. According to Mulhadiyono [1], there are some hydrocarbon traces in the sandstone of this formation, this statement are additional factor to do some study for this formation included also the facies associations study of the Formation Halang due to they environment deposition.

In general, the sedimentation sequences of Halang Formation consist of alternating turbidite sandstone and claystone [1]. The Halang Formation it self can be divided in two groups, Upper Halang Formation and Lower Halang Formation. The lower Halang Formation is characterized by association of mudstone with overbank channel and crevasse - splay sediment [2]; [3]; [4]. The channel deposits are generally not continuous and only localized. The upper Halang Formation is dominated by mudstone with thin sandstone intercalation [1]; [5]. It is a character of turbidite sediment on the submarine distal fan [6].

To know in more detail the characteristics of Upper- and lower Halang formations, then do a sedimentation study of Halang Formation at the River Pangkalan. This location was chosen because at this river found a well and continue exposure of Halang Formation, so that the availability of data is considered good to do a study facies sedimentation of this formation.

Administratively, the River Pangkalan areas belong to village Dermaji, Region Lumbir, District Banyumas Central Java Province (Figure. 1).

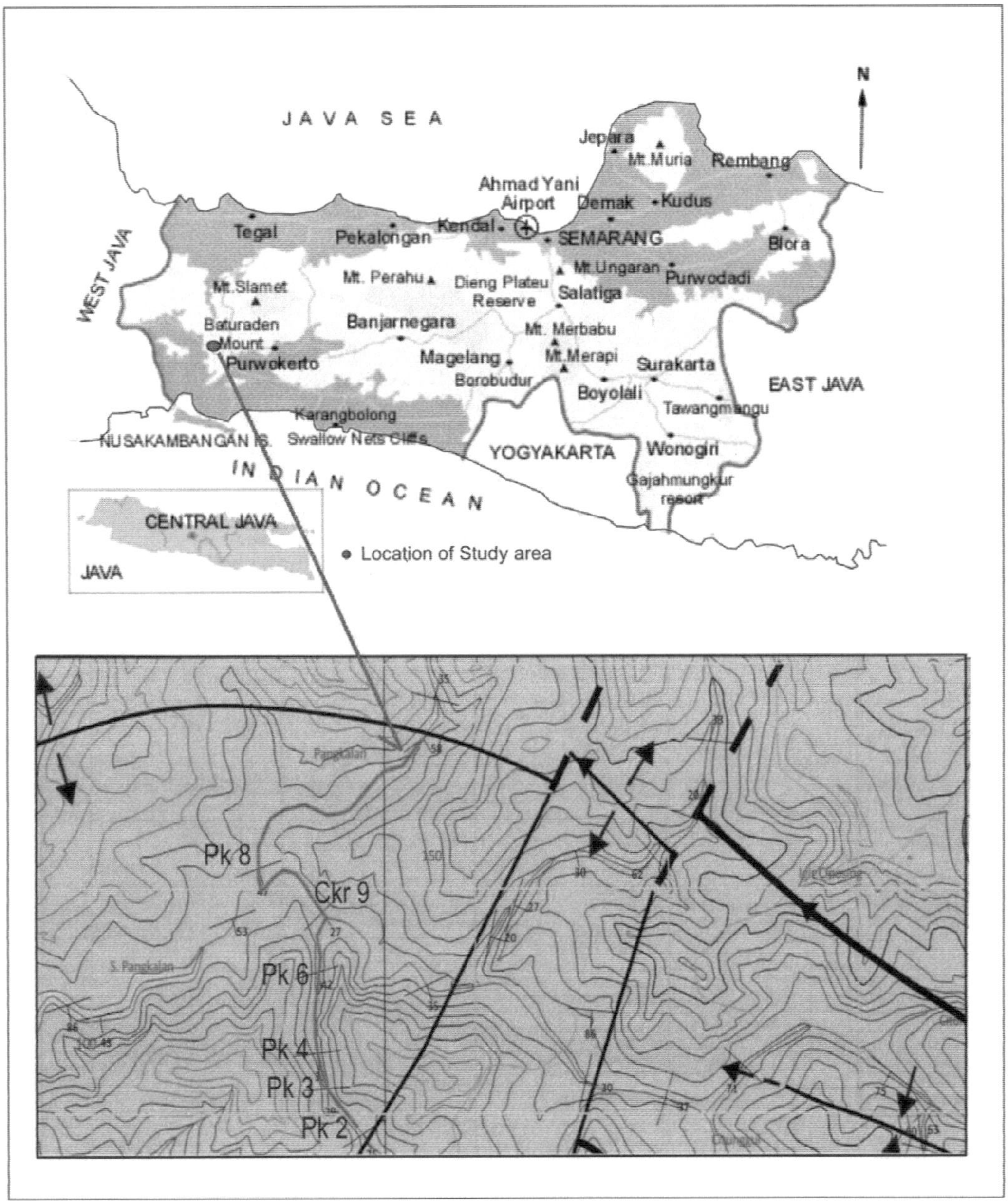

Figure 1. Above: Research Location area [7]. Below: Location map of stratigraphic measuring section at Pangkalan River.

2. Methods

Sedimentological study divided in two stages, field work and studio. The fieldwork was performed on outcrops of Halang Formation which exposed along of selected traverse in River Pangkalan. Detail measuring section was revealed on two kilometers traverse to obtain tipically lithology changes and sedimentary structure, that consist of sedimentary textures, sequence order pattern, and variation of rock layer thickness as well as rock sampling for paleontology analysis. All of field datas were gathered as primary input on studio stage.

The studio work included paleontology analysis and lithofacies analysis. Paleontology analysis has done to identify the fossil content due to the age of the rock and environment as well as lithofacies analysis a due to environment deposition. Litofasies grouped using a parameter of grain size, sedimentary structures, the order of vertical succession and thickness of sediment, where the grain size is a parameter litofasies the most easy to use as the basis for grouping [8-9]. In addition to vertical sequence, geometry layers were also observed, such as the pinch out layer, to know and understand the geometry of sediments during the deposition [8]. Therefore, integrated analysis from field data was required to get the best interpretation of depositional history of the development of the turbidite system in Halang Formation.

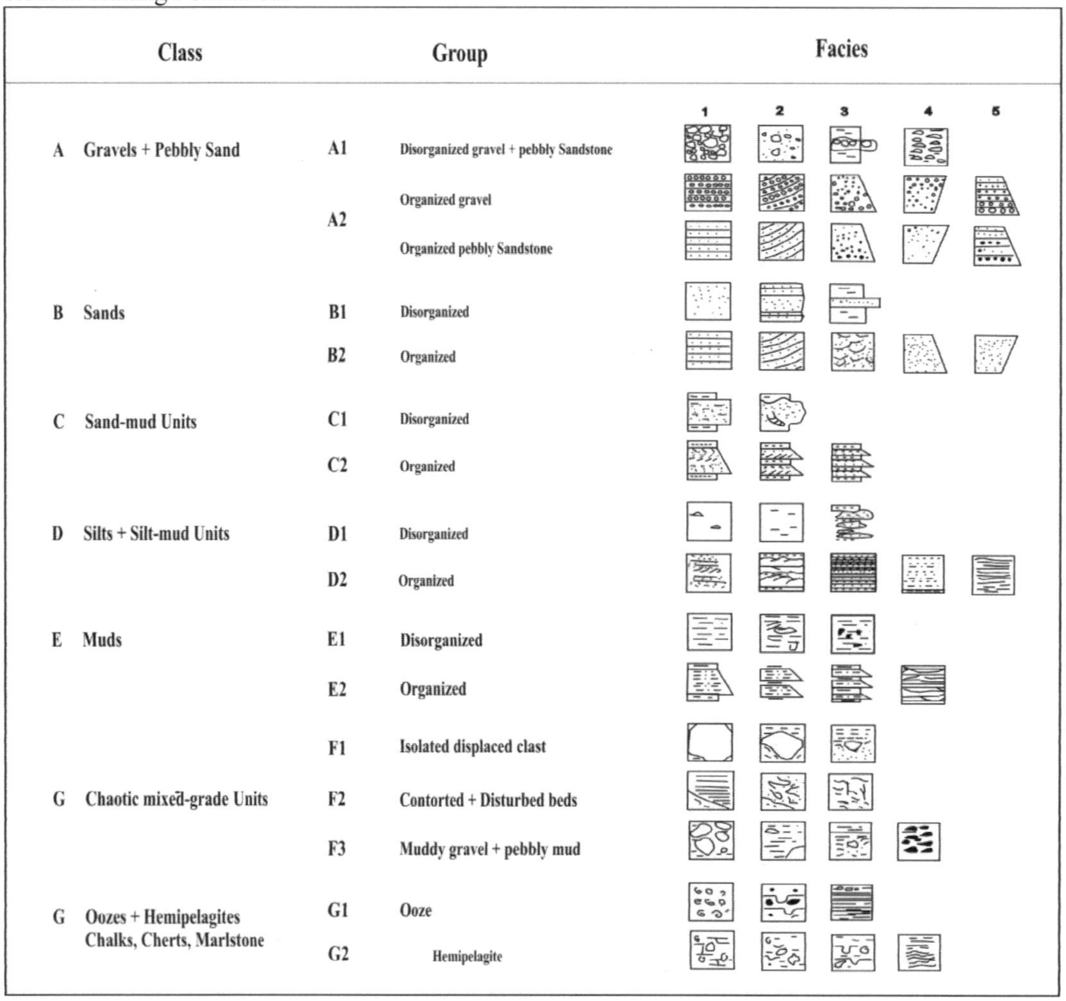

Figure 2. Stow Facies Classification [8].

3. Regional Geology

Halang formation is one of the formations were deposited in Banyumas basin [10]. This formation fills Banyumas back - arc basin in the Late Miocene age [11-12].

The sediments provenace of Halang formation allegedly came from the volcanic arc Southern Mountains uplifted during the Oligocene – Miocene. A volcanic material eroded and transported to the north past a faults that make up the steep slope and deposited in the deep marine environment in the gravity current mechanism [2,4,8].

Based on their sedimentation process and the distance from the source of sediments, Halang Formations can be divided into two parts, Lower Halang Formation and Upper Halang Formation. Lower Halang formation is considered closer to the source of sediment, and deposited in turbidite channel system [5], but the deposition mechanism of Upper Halang Formation, is more dominant a suspention, which resulted a deposition of claystone with thin intercalation of sandstones [1,5].

Stratigraphically, Halang Formation deposited interfingering with a Kumbang Formation [1]; Figure 3). Both formations deposited in the same environment and at the same time, deep marine on Late Miocene age [1,3]. The difference of these two formations lies in their grain size and rock composition. The Kumbang Formation consists of volcanic breccias with andesite fragments, lava flows and tuffs, while Halang formations consists of silisiclastic turbidite deposition in the form of alternation of sandstone and mudstone [3].

Geological mapping in the area Dermaji and Pangkalan river by Lagona [13] proved that a sediment of Kumbang Formation presenced between a sediment of Lower- and Upper Halang Formation (Figure 4).

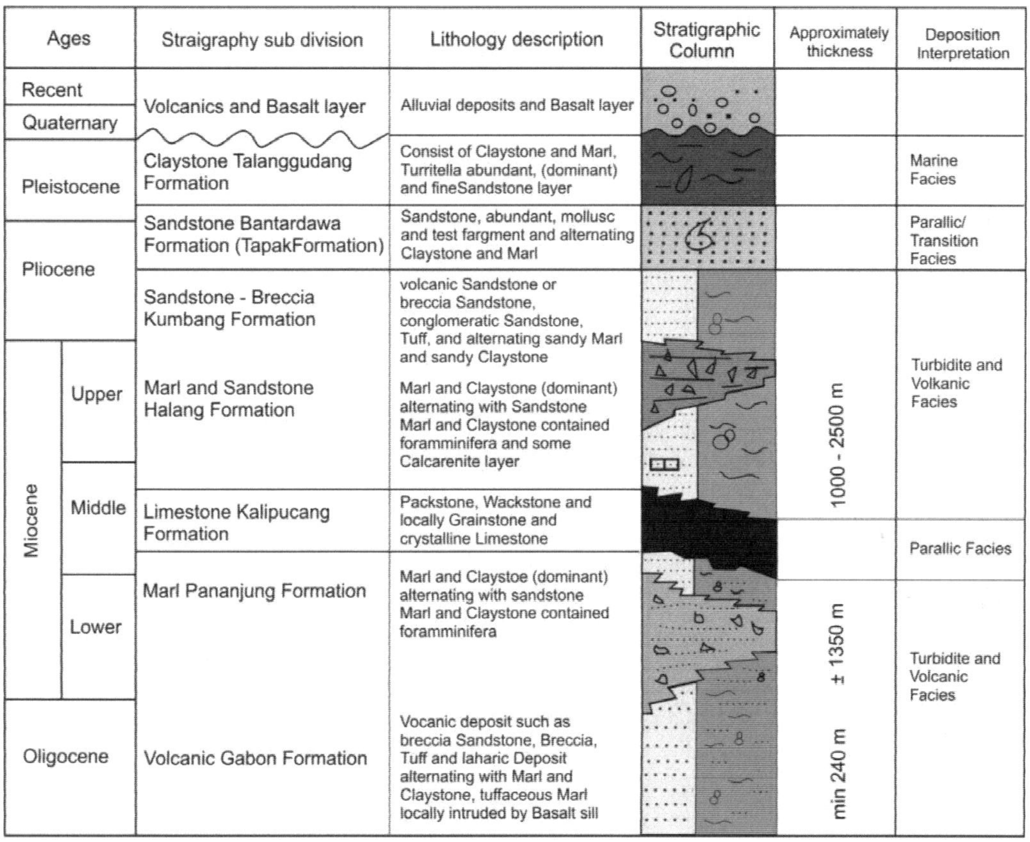

Figure 3. Regional Stratigraphic of Central Java (modified from [1]).

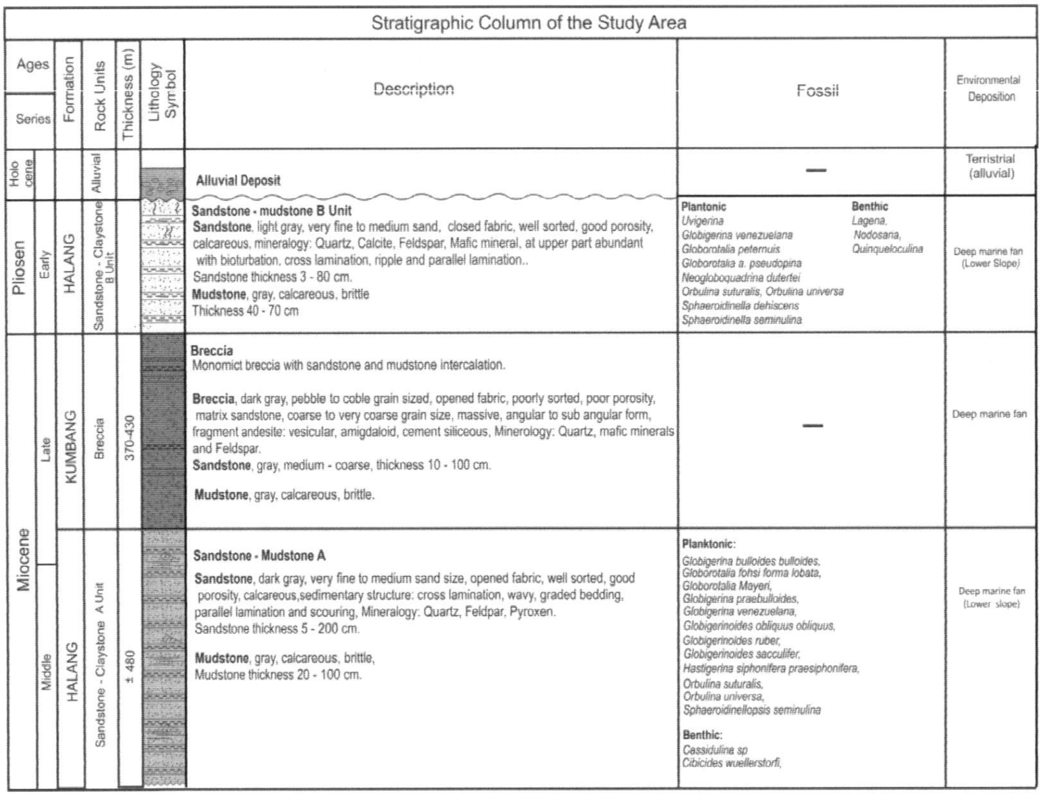

Figure 4. Stratigraphy of Pangkalan Area [13].

4. Result of Observation and Interpretation

The measurement results of stratigraphic cross sections on the Halang formation indicate that an exposure of Halang formation has a vertical succession thickness about 156 meters and layers notch N229^0 / 35^{0}NW. Based on the depositional environmental models of Mutti et al [14] and Posamantier and Kolla [15] in the SEPM Strata [16], a facies which found in outcrop in the Pangkalan river can be divided into five facies associations, namely: Proximal Channel Facies Association, Channel - levees facies Association, Frontal - splay Facies Association 1, Crevasse - splay Facies Association and Frontal - splay Facies Asosiaciaton 2

4.1. Proximal Channel Facies Association

This facies association composed of thick layering of fining upwards Sandstone, overall the thickness of this succession is 11.4 meters (Figure 4). At the lower part of the section, sandstones consist of coarse sand to very coarse grain size contain some fragments of igneous rock basaltic with small amount of calcareous sandstones fragments and coral in size of granule to pebble. Sandstone color is blackish-gray, poorly sorting, poor porosity, closed in fabric, massive compact, graded bedding sedimentary structure, cement silica / noncalcareous, with 1.2 – 1.5 m thick. To the upper part, grain size become finer. To the upper part of the section, sandstone has medium grain size, gray, well sorting, good porosity, closed fabric, mineralogy mostly quartz, calcareous, ripped-up clay clasts sedimentary structure (Figure 5). The main characteristics in this facies association which different with another facies association is thick sandstone with intensive sedimentary structre at the bottom and gradually disorganized and rip – up clay sandstone at the top [9].

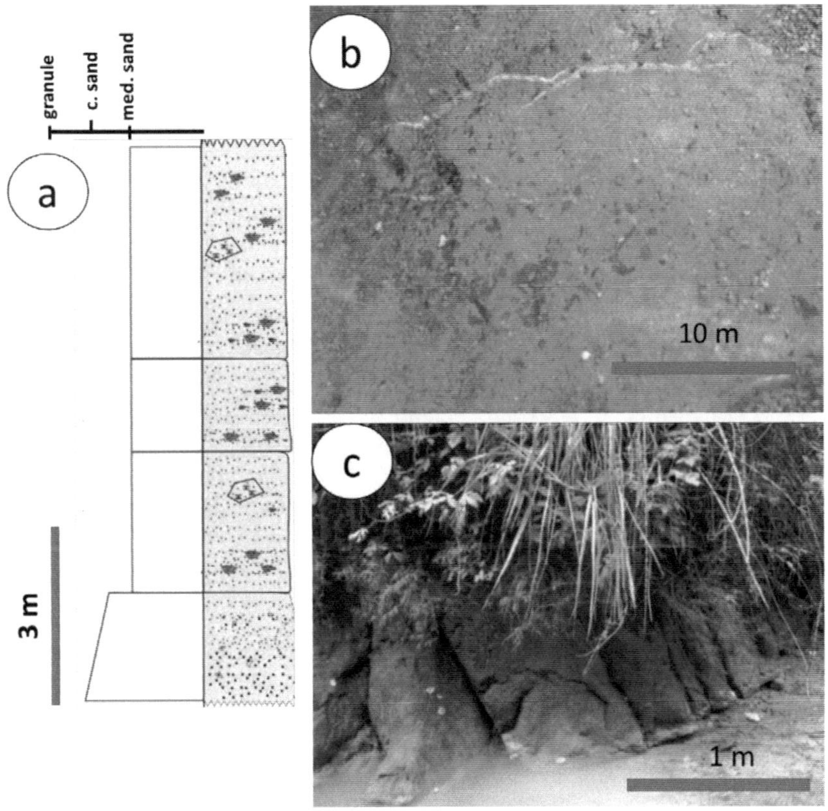

Figure 5. (a) Vertical succession of channel fasies association at sta. PK8 in Pangkalan River. (b) Rip up-clast at the top of channel fasies association. (c) The outcrop of bottom of channel fasies association shows coarse sandstone with granule gradded bedding.

Base on lithology, this sandstone can be divided in two fasies. According to Stow classification [8] in Reading [9], the granule-pebble sandstones are belongs to the Organized Pebbly Sandstone facies (A2-3) and Disorganized Sandstone facies (B1-1).

Organized Pebbly Sandstone facies (A2-3) is characterized by the presence of granule to pebble sized fragments, which shown graded bedding sedimentary structure. Disorganized Sandstone facies (B1-1) is characterized by massive sandstone with anorganized clastic clay fragment without specific sedimentary structures. The clastic clay fragments spread and float on the entire body sandstones (Figure 6).

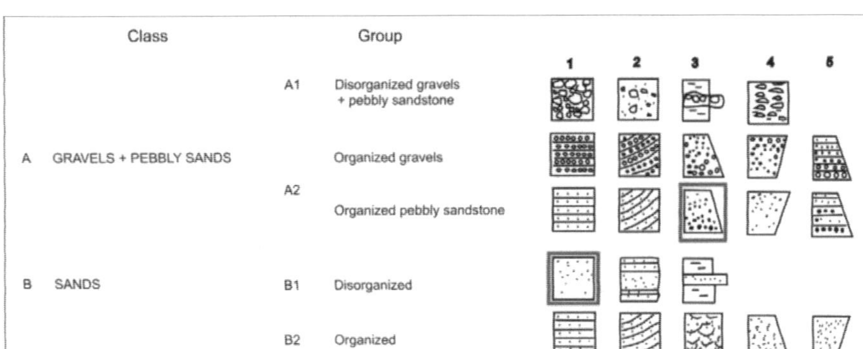

Figure 6. Facies classification of *Proximal Channel* fasies asosiation signed by red boxes [8,9].

In the vertical sucession of the facies, lenses geometry of a channel is absence, also at the contact of coarse sandstone and medium sandstone did not show an erotional surface. This case can be occur, when a mechanism of transportation as a traction carpet where a sediment flows with high concentrations covered by a turbulent flow above it [14]. Gradded bedding sandstone with a granule - pebble size is the result of gravity settling during sedimentation. A very strong and turbulences currents lead up the clay particles into a fluid and forming clay fragments which floating in sand as ripped-up clasts. So that, it can be interpreted that the coarse sandstone and medium sandstone being formed by a turbulent flow mechanism and deposited as F3 and F4 in lower slope and basin as described in the Figure 7 below.

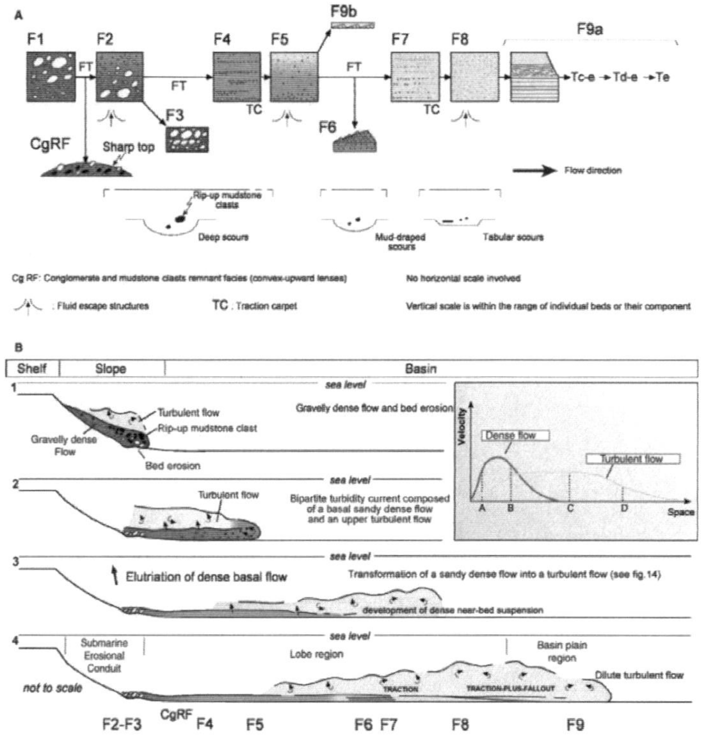

Figure 7. (A) Framework for a predictive classification scheme of turbidite facies. (B) Main erosional and depositional process associated with the downslope evolution of a turbidity current [17].

4.2. Distal-Levees Facies Association

Distal - Leeves facies associations composed of alternating of calcareous sandstone and thin layer claystone with a total thickness of 41.7 meters. Sandstone has a 5-10 cm thickness, gray light color, very fine to fine sand in grain size, calcareous, good sorting, good porosity, fabric is closed, sedimentary structures such as parallel lamination are abundant at the lower part of the section (Figure 8) and a ripple sedimentary structure appears at the top of the succession (Figure 9). The characteristic of the calcareous sandstones is similar to other sandstones, but it has high strong carbonate content. The 20-100 cm thick mudstone has the characteristic of gray color, brittle, calcareous, partly has scaly clay structure. The main characteristic of this facies association is thick interbedded sandstone and clyastone with intensive sedimentary structure. Sedimentary structures that can be found consist of Ta, Tb, Tc, and Td in bouma sequence.

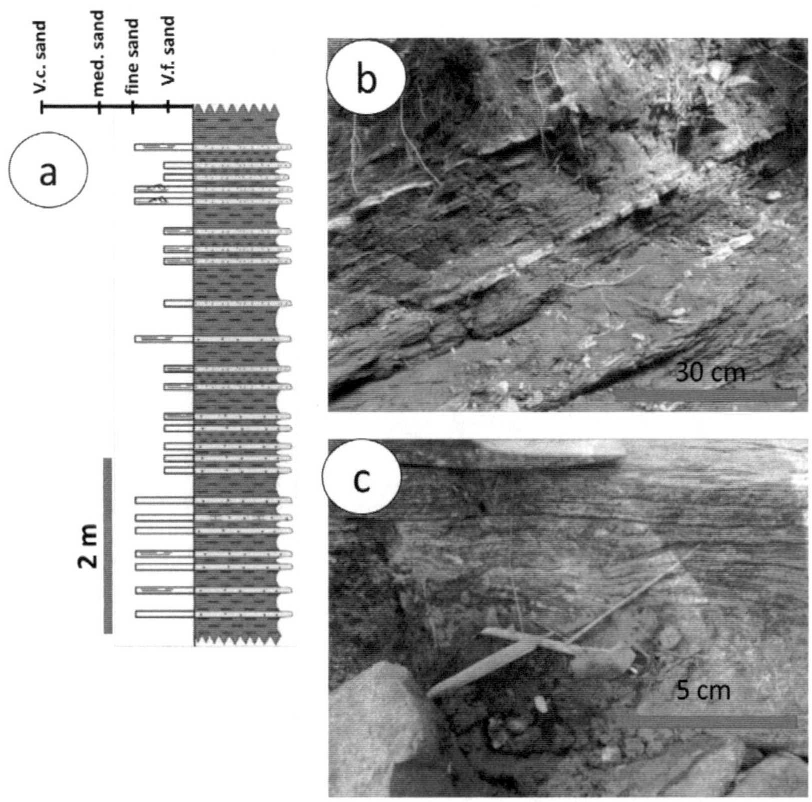

Figure 8. (a) Verical succession of bottom part of distal – leeves association facies at Sta. Ckr 9. (b) The outcrop of distal – leeves association facies shows parallel lamination structure. (c) Ripple structure at distal – leeves association facies.

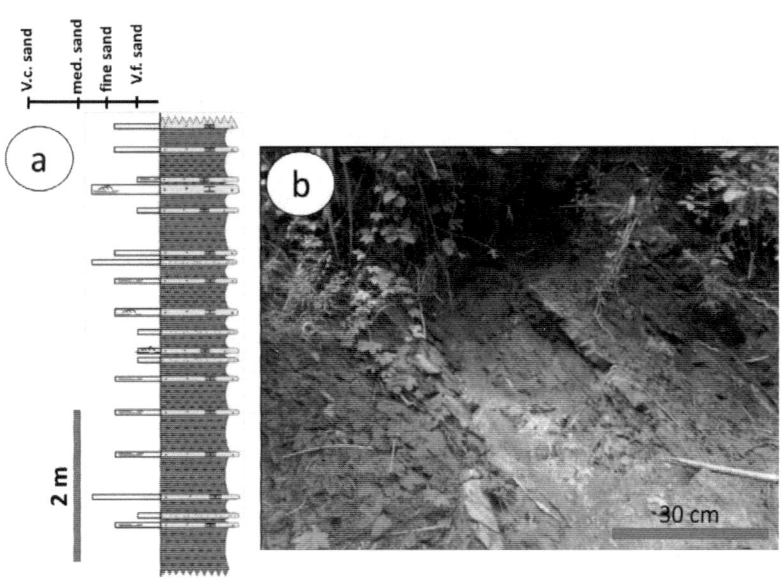

Figure 9. (a) Vertical succession of top of distal – leeves association facies at Sta. Ckr 9. (b) The outcrop of top of distal – leeves association facies shows interbedded sandstone – claystone.

Base on Stow facies classification [8], this alternating mudstone and sandstone succession can be categorized in to Organized Sandstone - mudstone facies (C2-3) because it has an arrangement alternating with parallel laminated and ripple sedimentary structures on sandstone and some of sand layer shown thinning and fining upwards (Figure 10) and grouped in to distal levees facies association.

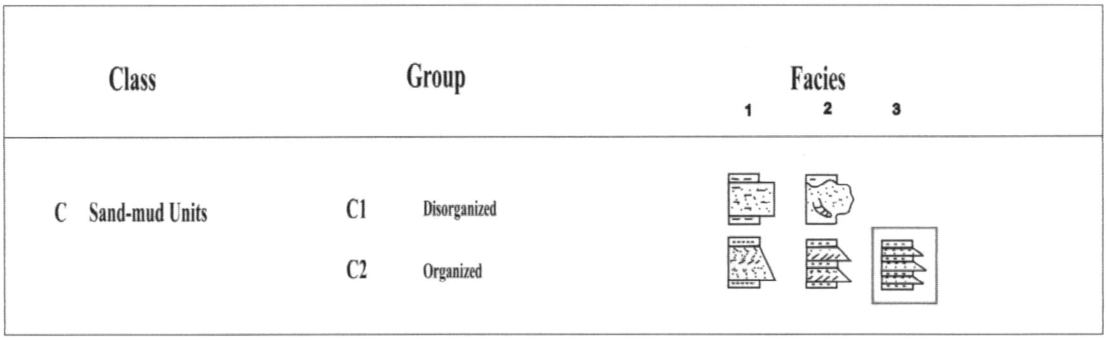

Figure 10. Facies classification. *Distal Levees* Facies Association which signed by red box [8,9].

Distal - levees is an architectural element that is usually located on the outside of the sinusoidal or buckling of channel. A coarser grain is difficult to transport and stuck on the main channel, therefor in the facies associations a graded bedding sedimentary structure not appear. Sediment flow due to gravity mass flow has a flowing rate relative high to down slope. When it crossed the line of the channel edge, the flow passed out of a channel and cross a proximal levee, flowing down a slope and then deposited on the distal levee (Figure 11). At the distal levee, sediments deposited with parallel lamination sedimentary structure and followed by a ripple sedimentary structure. Sedimentary structure such as parallel lamination is a product of *upper flow regime* as traction current, which flowing down slope toward the distal levee. A ripple sedimentary structure indicates a decrease of flowing rate becomes lower flow regime on the transport of fine sand-sized material. A mudstone layer is deposited in the suspension mechanism. Based on it lithology vertical sucession, grainsize and sedimentary structure, so this facies association interpreted as deposit of outer fan environment [18].

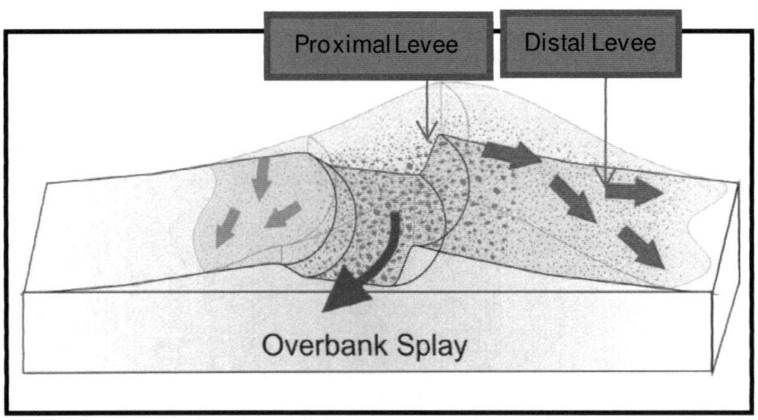

Figure 11. A model of sedimentation system on channel-leeve [18].

4.3. Frontal Splay 1 Facies Association

This facies association composed of sandstones and mudstone alternation, a total thickness of this succession reaches 49.9 meters. The main characteristic of this facies association is thick interbedded of sandstone and claystone, that sandstone can be found in two different characters, a massive sandstone and thin layered sandstone (Figure 12 and Figure 13). The character of massive sandstone is blackish-gray, medium - fine grain size , well sorting, good porosity, closed fabric, siliceous cement, abundant mafic minerals, showing graded bedding sedimentary structures, with the thickness around 1 meter (Figure 12). The thin layered Sandstone has light gray color, fine – very fine grain size, calcareous, well sorting, good porosity, closed fabric, slightly mineral: quartz and mafic minerals, well developing parallel lamination and ripple sedimentary structures, a a small portion shows the sedimentary structure of graded bedding, with the thickness around 10 cm. The specific characteristic of this facies that a slump deposits with alternating of thin layered sandstone and claystone at the top of facies association (Figure 13).

Massive sandstones associated with alternating thin layered sandstone - mudstone is interpreted as channel deposit that associated with basin-plain sediment [15]. Depositional environment that allows these associations can be formed is a fan area at frontal splay that is part of channel mouth on the outer fan toward a basin plain. In basin plain, sandstones with parallel lamination sedimentary structure indicate a system of upper flow regime and a ripple sedimentary structure as an indication of lower flow regimes. Thin sandstones with both the sedimentary structures in separate sandstones body interpreted that the current strength changes from time to time. Coarser grain size in the massive sandstones indicates that a massive sandstone as a product of stong current deposition. The presence of thin layered sandstone-mudstone slump is indicated that shows a resedimentation of semi consolidated sediment which originating from the channel - margin or levee. According to Stow [8] in Reading [9], this facies associations contains Organized Sandstone (B2-4), Organized Sandstone - mudstone facies (C2-3), Disorganized Sandstone facies (B1-1), and contorted + Disturbed Beds / Slump facies (F2-2)(Figure 13). Facies is formed by sediment failure near to the lower slope or along levees , so this facies association interpreted as deposit of basin along a levee.

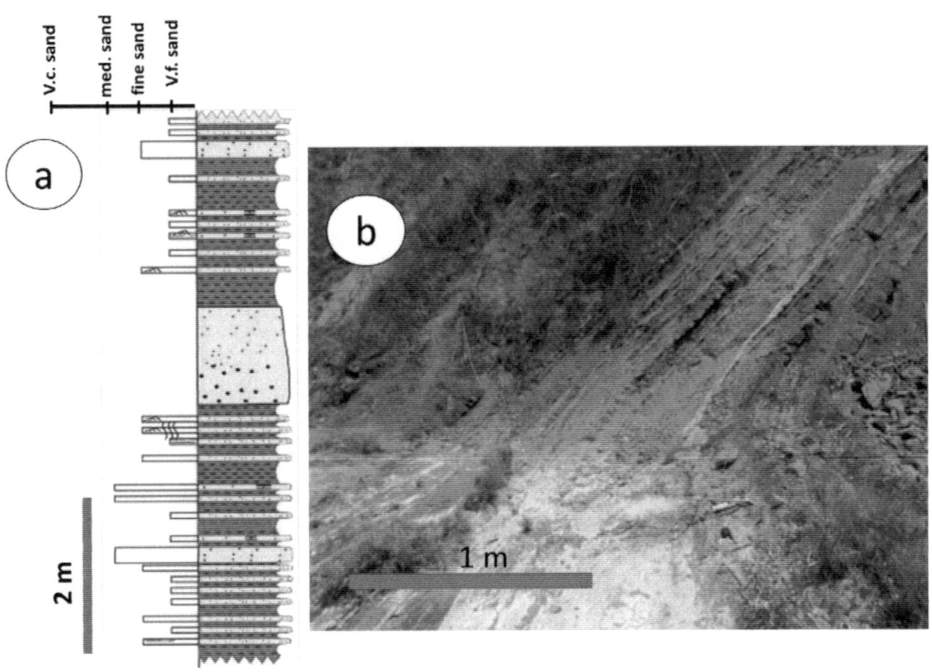

Figure 12. (a) Stratigraphy succession of bottom part of frontal splay association facies at Sta. Pk 4. (b) The Photograph of bottom part of frontal splay association facies shows interbedded thin sandstone and claystone.

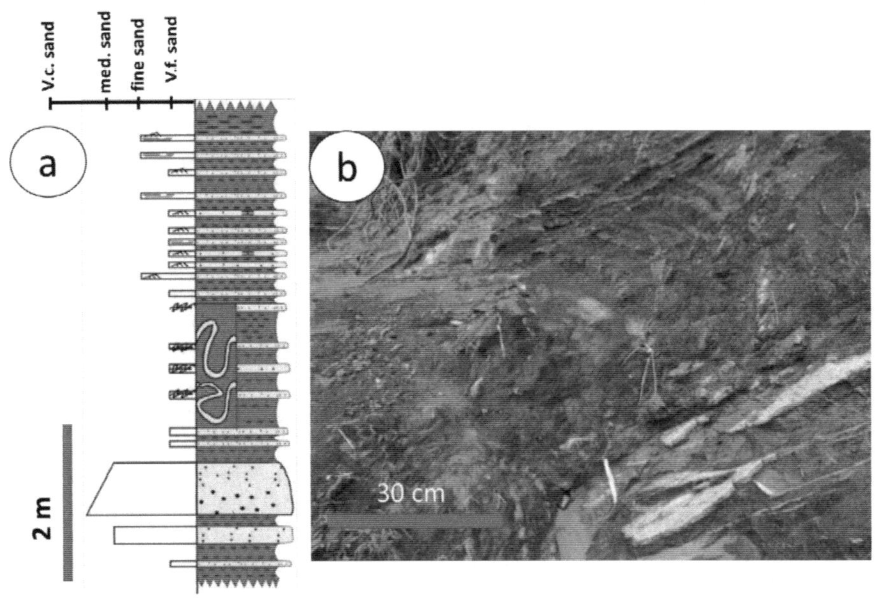

Figure 13. (a) Stratigraphy succession of top of frontal splay association facies at Sta. Pk 4. (b) The Photograph of top of frontal splay association facies shows slump structure.

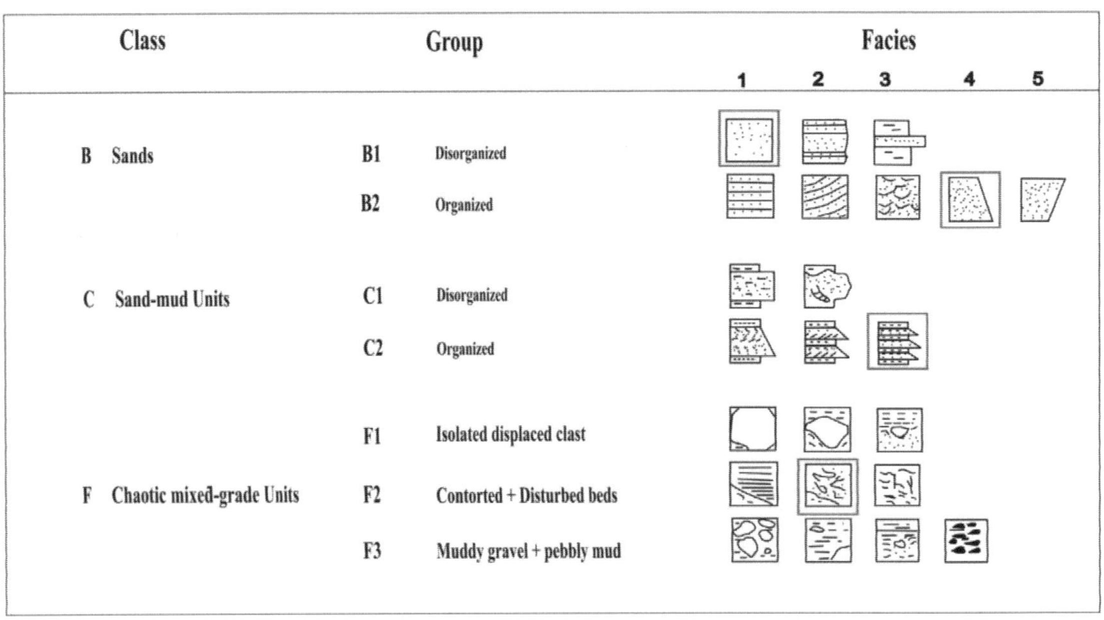

Figure 14. Facies classification. *Frontal Splay 1* facies Association which signed by red boxes [8,9].

4.4. Crevasse Splay Facies Association

This facies association consists of interbedded sandstone with claystone intercalation, with the thickness of succession is 7.3 meters. Sandstone has gray, slightly calcareous, well sorting, good porosity, closed fabric. Claystone has gray color, concoidal, calcareous and brittle. Sedimentary structure of Ta, Tb, Tc, well developed in sandstone. Scour marks structure are also abundant in the succession of sandstones (Figure 15).

Base on it grain size, according to Stow [8] in Reading [9], a lithology, which are found in this succession are grouped in to *Organized Sandstone-Mudstone facies (C2-3)*. Crevasse splay facies association has a similarity to frontal splay facies association, but crevasse splay is finer in grain size than frontal splay [15].

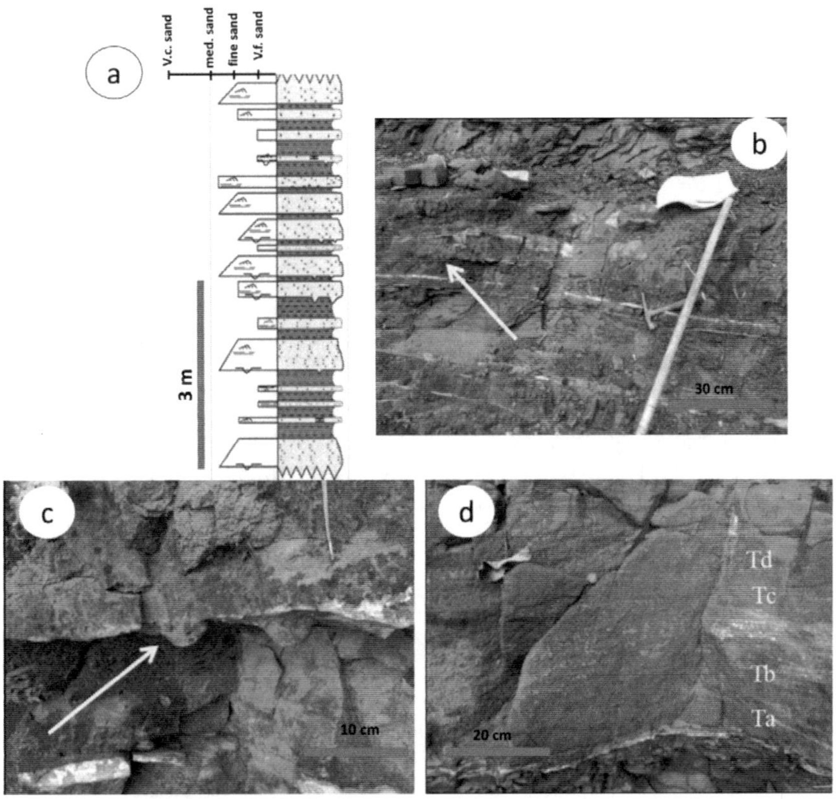

Figure 15. (a) Vertical succession of a crevasse splay facies association at station Pk 3 Pangkalan River. (b) Thinning geometry of sandstone layer (yellow arrow). (c) Scour mark at the bottom of sandstone (yellow arrow). (d) Sedimentary structure of crevasse splay facies association shows Ta, Tb, Tc, and Td of Bouma sequence.

In general, a vertical succession of sandstone and mudstone alternation has coarsening- and thickening upward pattern (Figure 15). A coarsening upwards succession to indicate a breakthrough on the path towards channel - levees that formed as a branch of the channel. Scour mark found at the bottom of sandstone layer (Figure 15c), this structure shows that this facies association deposited in large and high rate of sediment supply, which is one of the characteristics of the crevasse splay deposits [7]. The flow of sediment supplay, which cut a natural levee of main channel and flowing down and eroded a distal part of levee slope, this process is caused of *hydraulic jump* (a phenomenon of transition from the thin, supercritical flow into thicker, subcritical flow), so on the part of the slope are scoured then filled again by a sediment product of rapid flow and turbulent flow occurs in the distal part of levee (Figure 16). Turbulence also occurs when the sediment flow down the slope. Ta, Tb, Tc, Td structure of Bouma Sequences indicated that the traction current in the aftermath of the turbidite flows during the sedimentation process (Figure 15d).

According to Shanmugam and Moiola model [19], the succession of sandstone and mudstone alternation is belongs to facies association C, which deposited in lower fan of deep marine fan (Figure 17). Geometry of sandstones in form of pinch out or thinning laterally, which indicates the geometry of the crevasse splay edge [5].

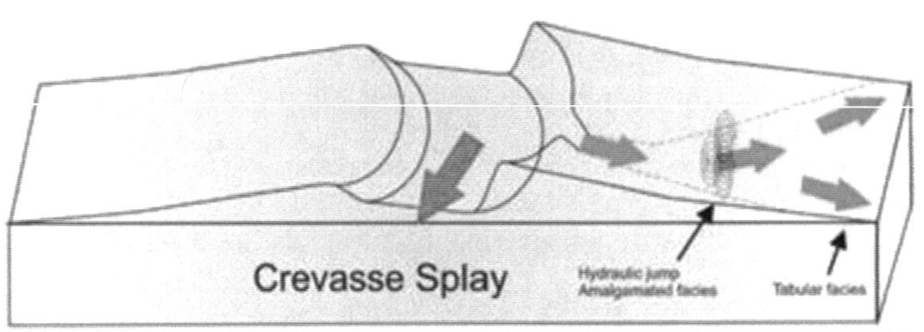

Figure 16. Model of environmental deposition system on crevasse splay [18].

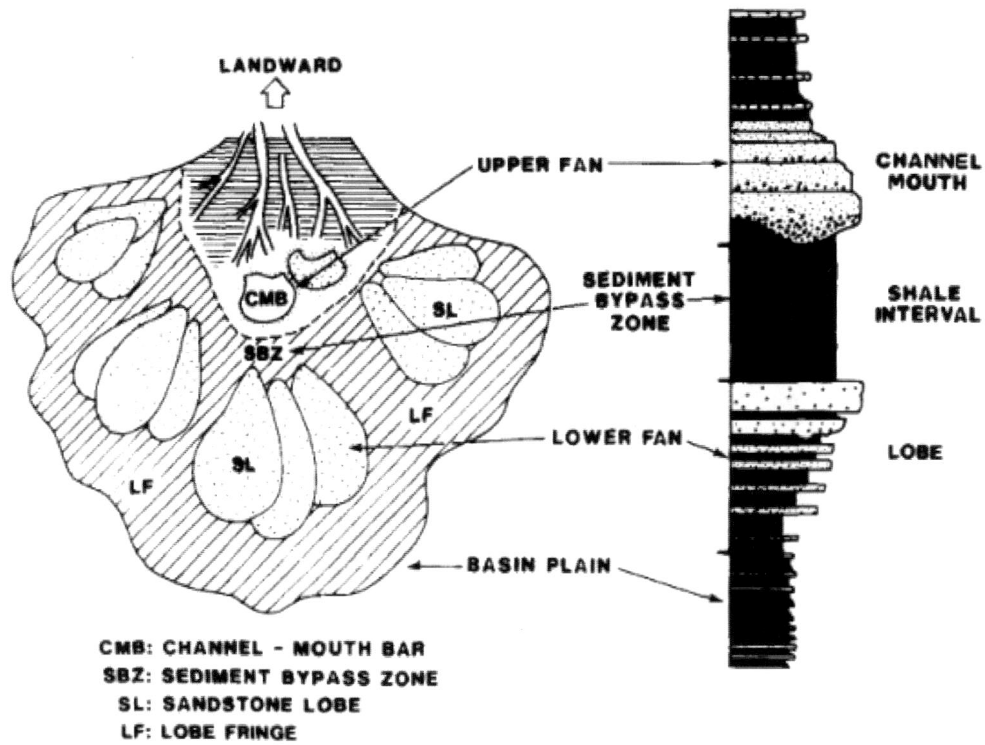

Figure 17. Fan model of detached submarine fan lobe [19].

4.5. Frontal Splay 2 Facies Association

The sediment succession consists of alternating sandstone and claystone, with total thickness is about 46 meter. Base on it thickness, Sandstone layer in this succession can be divided into two kinds of sandstone, massive sandstone and thin layered sandstones (Figure 18).

Massive sandstone characterized by a thickness between 1-1.8 m, grainsized varied from very coarse to medium sand, compact and hard, sedimentary structure at the lower part mostly graded bedding and gradually changed to upperpart be parallel- and ripple lamination with erotional contact at the bottom. The thin layered sandstone characterized by thin layer with a thickness varied from 5-15 cm, alternating with mudstone, sedimentary structure in one sandbody or layer is parallel- and ripple lamination, some of them is dominated by ripple. A mudstone has a thickness about 40 – 60 cm, dark gray, brittle, calcareous.

According to Stow facies classification [8] in Reading [9], a facies which present in the succession is Organized Sandstone facies (B2-4), Organized Sandstone - mudstone facies (C2-3), and Disorganized Sandstone fasies (B1-1). Massive sandstones associated with alternating thin sandstone –

mudstone and interpreted as channel sediment deposition associated in basin plain. This facies association pattern can be formed in invironment deposition of frontal splay area of the mid fan system. At the end of the channel, a deposition of the massive sandstones is following by sedimentation of sandstones with parallel lamination- and ripple sedimentary structure. The structure of graded bedding in the massive sandstones indicates the occurrence of settling when the current energy is decreasing. The summary of facies association changes can be seen in Figure 19.

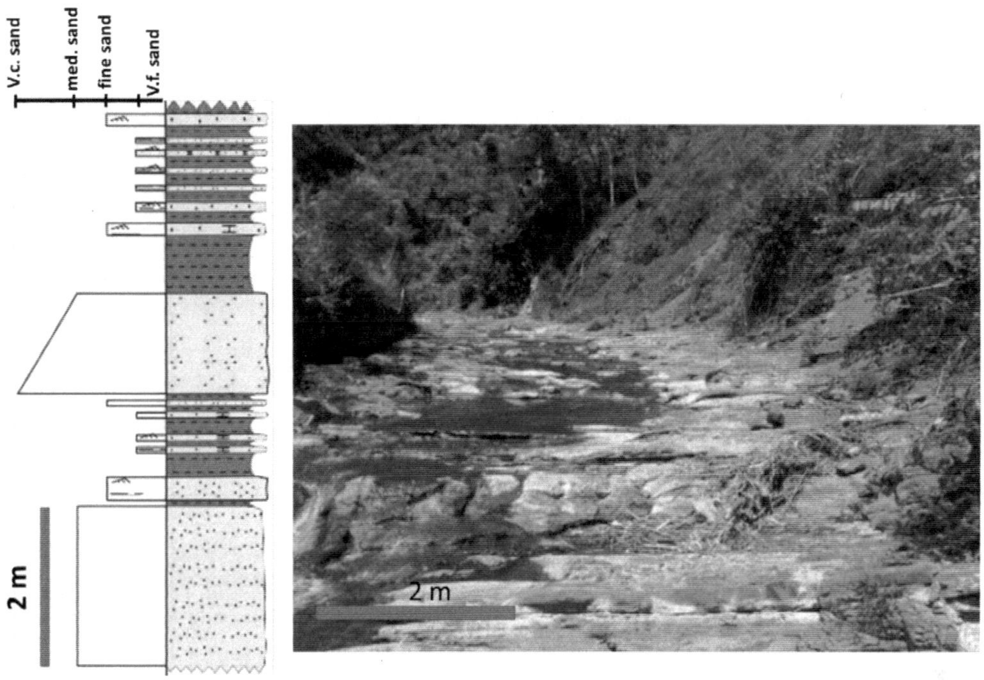

Figure 18. Stratigraphy section at Station Pk 2 in Pangkalan River.

5. Discussion

Stratigraphicly the facies association from bottom to the top in Pangkalan River consists of proximal channel facies association, distal-levees facies association, frontal splay 1 facies association, crevasse splay facies association and frontal splay 2 facies association. The proximal channel facies association deposited in lower fan – basin plain of submarine fan system, and distal-levees facies association deposited in outer fan as mud rich system, the frontal splay as deposit of basin plain area [14,15] (Figure 20).

Turbidite deposition system can be good observed by sedimentary structures slump and Ta, Tb, and Tc on Bouma cycles [20]. Beside of them, the dense alternating layer of sandstone and mudstone indicate a rapid change of current which is caused of the turbulences. A thick of distal leeves sediment indicate that a turbidite process is turbidite mud - rich system [14]. A locally pinch out geometry and layer depletion indicates that the turbidite deposite channel at River Pangkalan is sinuos channel deposit [21].

A change in the thickness of sandstone is also a noteworthy thing, because the sandstone plays an important role as reservoir for oil industry. An oldest facies association in the stratigraphic vertical succession at River Pangkalan is a proximal channel, showing their thick massive sandstone without mudstone intercalation. To the top of sucession, the thickness of sandstone gradually decreased and shows a facies changes toward a distal fan area. This change is controlled by paleomorphologi which indicates paleomorphology in form of to low gradient morphology [5,21].

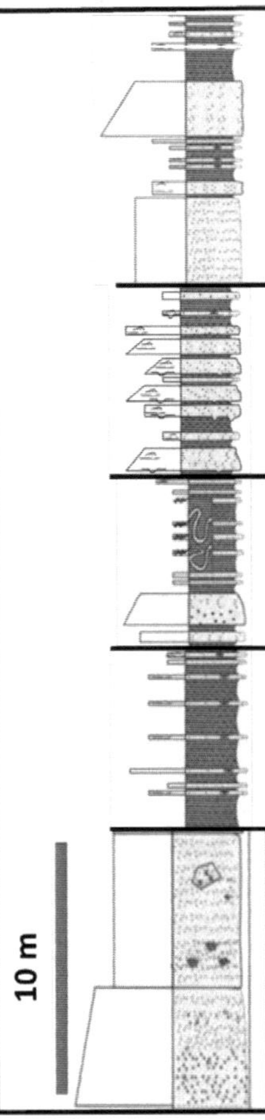

Facies association	Main characteristic	Sedimentary structures
Frontal Splay 2	Massive sandstone at the bottom and interbedded sandstone and claystone at the top	Erosional contact at the bottom of thick sandstone; Ta, Tb, and Tc Bouma sequence
Crevasse Splay	Interbedded sandstone with claystone intercalation	Ta, Tb, and Tc Bouma sequence
Frontal Splay 1	Massive sandstone at the bottom and interbedded sandstone and claystone at the top	Graded bedding at massive sandstone; slump structure
Distal leeves	Interbedded thin sandstone and claystone	Tb and Tc Bouma sequence
Proximal Channel	Thick layering fining upward sandstone	Intensive graded bedding at the bottom and rip – up clay at the top

Figure 19. The summary of facies association changes at Pangkalan River.

Here we provide an illustration of block diagram model for turbidite deposition on River Pangkalan (Figure 21). A change of deposit association from proximal channel to the distal leeves indicates a change in flow velocity from a strong current to weak current, a current becomes weaker, when the channel is full, occurred slumping that produced a slump deposit of thin layered sandstone - mudstone on the leeve side of channel, this is indicated by frontal splay facies associations. Erosion is intensified on the side of the channel and and caused the levee broke, a sediment flew out from a channel. This process produced crevasse splay facies associations deposit, when a sediment supply decrease, the material that passes through leeve diminishing so that the sediment of frontal splay facies deposited again.

Changes in sediment supply and deposition systems is strongly influenced by sea level changes that may be caused by a fall or rise in sea level or tectonic activity during the deposition process takes place [5]. Additionally, volcanism in the Late Miocene [22] causing fluctuations in the supply of sediment forming Halang Formation.

A detail sedimentology studies and correlation with other turbidite deposit of Halang Formation that exposed in the other area in west Java, is very helpful to understand the development and the other factors that influenced a turbidite systems of the Halang Formation in Banyumas basin.

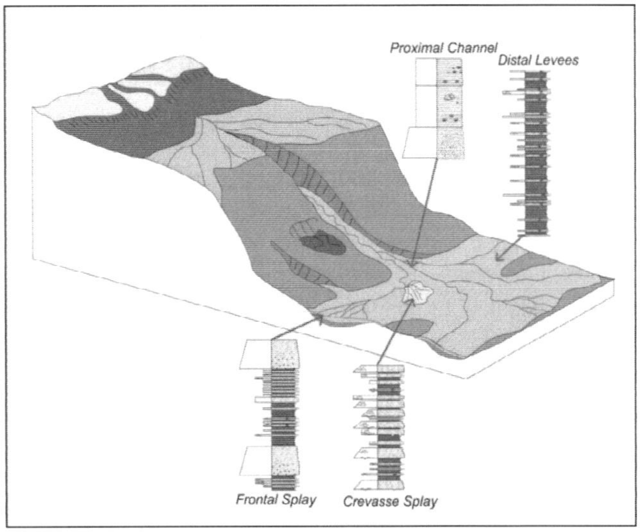

Figure 20. Model of submarine fan and its facies association (modified from [15] in [16]).

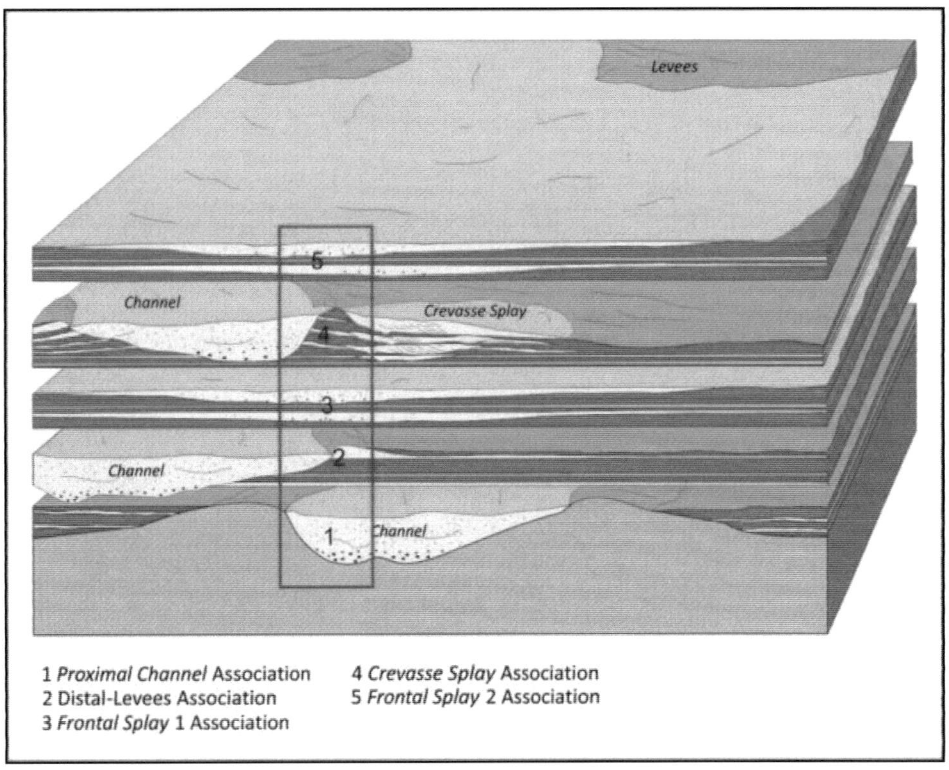

Figure 21. Block Diagram of facies association of stratigraphic section of Pangkalan River traverse (red box).

6. Conclusion

Halang formations at the river Pangkalan can be grouped into: Proximal channel facies association, the association of distal levees facies, frontal splay 1 facies association, crevasse splay facies Association, the Association of frontal splay facies 2.

Leeves distal sediment thickness indicated that the formation Halang turbidite deposition is mud - rich system. In addition, the pinch out geometry of Sandstone and locally observed layer depletion indicates that the turbidite deposit in the area River Pangkalan is sinuos channel deposit. The gradually reduced of the sandstones thickness and changes of facies toward a distal position, shows the change in paleomorphology becomes slighter. A changes of sediment from proximal channel deposition be distal leeves indicates a flow velocity turned out to be decrease.

When the channel is full, ongoing sedimention resulted in an avalanche of sediment on top of it and form a slump deposit of thin layered sandstone - mudstone on the side leeve of the channel which is known as the frontal splay facies associations. Erosion is intensified on the side of the channel which caused sediment flowing out the channel, break through levee resulted a deposition of crevasse splay facies associations, when returning a decline in sediment supply, the material that passes through leeve diminishing, so the sediment of frontal splay facies deposited again.

7. References

[1] Mulhadiyono, 2006, Proceedings Indonesian Petroleum Association 2 nd Annual Convention, Jakarta.
[2] Djuhaeni and Martodjojo, S., 1989, Bulletin Geology of Indonesia No. 12, p. 227–252.
[3] Kastowo and Suwarna, 1996, Geological Research and Development Center, 2nd Edition, Bandung.
[4] Martodjojo, S., 2003, ITB Press, Indonesia. 238 pp.
[5] Mukti, M. M., Ito, M., and Armandita, C., 2009, Proceedings Indonesian Petroleum Association 33th Annual Convention & Exhibition, Jakarta .
[6] Walker, R. G. and N.P. James, 1992, Geological Association of Canada Publication, Bussiness and Economic Service, Canada, 239-263.
[7] Bakosurtanal, 2008, *Administrative map of Central Java.*
[8] Stow, D.A.V., 1985, GSA Special Bulletin Vol 18, Geological Society of London, page 67-93.
[9] Reading, H. G., 2001, Bulletin of the Geological Society of Denmark Vol. 48, Copenhagen, 101-115.
[10] Praptisih and Kamtono, 2011, Indonesia Journal of Geoscience, Vol. 6, No. 1, p. 13-27.
[11] Sribudiyani, Muchsin, N., Ryacudu, R., Kunto, T., Astono, P., Prasetya, I., Sapiie, B., Asikin, S., Harsolumakso, A. H., and Yulianto, I., 2003, Indonesian Petroleum Association, Proceeding 29th Annual Convention, Jakarta, p. 335–346.
[12] Mukti, M. M., and Ito, M., 2010, Sedimentary Geology, Elsevier.
[13] Lagona, R., 2014, Institut Teknologi Bandung.
[14] Mutti, E., Bernoulli, D., Lucchi, F. R., and Tinterri, R., 2009, Journal compilation, International Association of Sedimentologists, 267-318.
[15] Posamantier, H. W., and Kolla, V., 2003, Journal of Sedimentary Research, SEPM.
[16] SEPM Strata, 2014, http://www.sepmstrata.org/page.aspx?pageid=39.
[17] Mutti, E., Tinterri, R., Magalhaes, P. M., and Basta, G., 2007, AAPG annual convention, Long Beach.
[18] Davis, L., Khan, Z., Altosaar, A., dan Arnott, B., 2010, Geocanada- Working with the Earth.
[19] Shanmugam, G. and Moiola, R.J., 1985, Proceeding Annual Conference Nigerian Association of Petroleum Exploration, 1, 18-39.
[20] Bouma, A.H., and Stone, C.G., 2000, Fine-Grained Turbidite Systems systems AAPG Memoir book 72, AAPG, 352 p.
[21] Dykstra, M., and Kneller, B., 2008, Sedimentology, International Association of Sedimentologists
[22] Clements, B., and Hall, R., 2007, Proc. Ann. Conv. Of The Indonesian Petroleum Association, Jakarta, no. 31.

IOP Conf. Series: Earth and Environmental Science **71** (2017) 012033 doi:10.1088/1755-1315/71/1/012033

Paleoenvironmental Study of Miocene Sediments from JTB-1 and NRM-1 wells, in West Ogan Komering Block, Meraksa Area, South Sumatra Basin

Aswan[1], Mirzam Abdurrachman[1], Bayu Sapta Fitriana[2], Mohamad Fery Mustofa[2], Wahyu Dwijo Santoso[1], Alfend Rudyawan[1], Windy Dwi. Rahayu[1], Ahmad Hamdani[1], Yepi Rohiman[1] and Thaw Zin Oo[3]

[1] Geology Department, Faculty of Earth Sciences and Technology, Institute of Technology Bandung, Jalan Ganesa 10 Bandung, Indonesia (40132)
[2] JOB Pertamina – Talisman (Ogan Komering) Limited, Gedung Setiabudi Atrium, Lantai 7 Suite 706, Jalan HR Rasuna Said, Kavling 62, Jakarta, Indonesia (12920)
[3] Geology Department, East Yangon University, Thanlyin Township, Yangon Division, Myanmar
E-mail: aswan_gl@gc.itb.ac.id

Abstract. Paleodepositional study based on microfossil has been done on Miocene sediments from JTB-1 and NRM-1 wells, West Ogan Komering Block, Meraksa area, South Sumatra Basin. This study was carried out to test the different depositional environment of the well-known stratigraphic units in South Sumatra Basin, which was assumed to occur due to different structural position. In this study, stratigraphic unit classification was determined by their lithology characteristics, they are: Lahat Formation, Talangakar Formation, Baturaja Formation, Gumai Formation, and Airbenakat Formation. This study reveals that the shelf deposits, such as The Baturaja and Gumai Formation of NRM-1 well have thicker interval and indicate marine depositional environment compared to the same rock unit of JTB -1 well. The Baturaja Formation in JTB-1 well was deposited in a shallow neritic environment while at the NRM-1 well, it was deposited on a shallow neritic to middle neritic interval. The Gumai Formation interval at the JTB-1 well was deposited in middle neritic to outer neritic bathymetric zone while at the NRM-1 well; the entire interval indicates outer neritic sediment. Integrated study of microfossils analysis, stratigraphic and seismic correlation shows that the position of JTB-1 well is structurally higher than the NRM-1. It is consistent with the interval sediment in JTB-1 well which are generally thinner and interpreted as shallower marine deposit. The results from this study will give significant contributions for the hydrocarbon exploration and development in terms of improving knowledge of paleodepositional environment. Therefore, distribution, characteristics and hydrocarbon potential could be estimated more accurately as well as optimising the hydrocarbon recovery.
Keywords: paleoenvironmental, South Sumatra Basin, Baturaja, Airbenakat, Miocene

1. Introduction

The stratigraphy of the South Sumatera Basin has been established including its paleontology aspect by [1-3]. However, their studies are more focused on the regional development of the basin in terms of age determination and depositional environment. This study focuses on a more detail approach using biostratigraphy method that has been overlooked previously. Some preliminary results suggest there are different aspects to the stratigraphy compared to the regional studies, in particular the depositional environment of each rock formation. Therefore a comprehensive biostratigraphy study is needed.

This study is mainly focuses on two wells in West Ogan Komering Block, Meraksa Area, South Sumatera (Figure 1) where there are indications of different depositional environments for the same formation due to the different structural level. Paleontological analyses have been performed to several samples representing the Baturaja and Gumai Formation from NRM-1 and JTB-1 wells which are only 4 kilometres apart. Previously, the Baturaja Formation is interpreted to be of Early Miocene age and the Gumai Formation age ranges from late Early Miocene to Middle Miocene. This study analyse the depositional environment by determining the benthonic forams association.

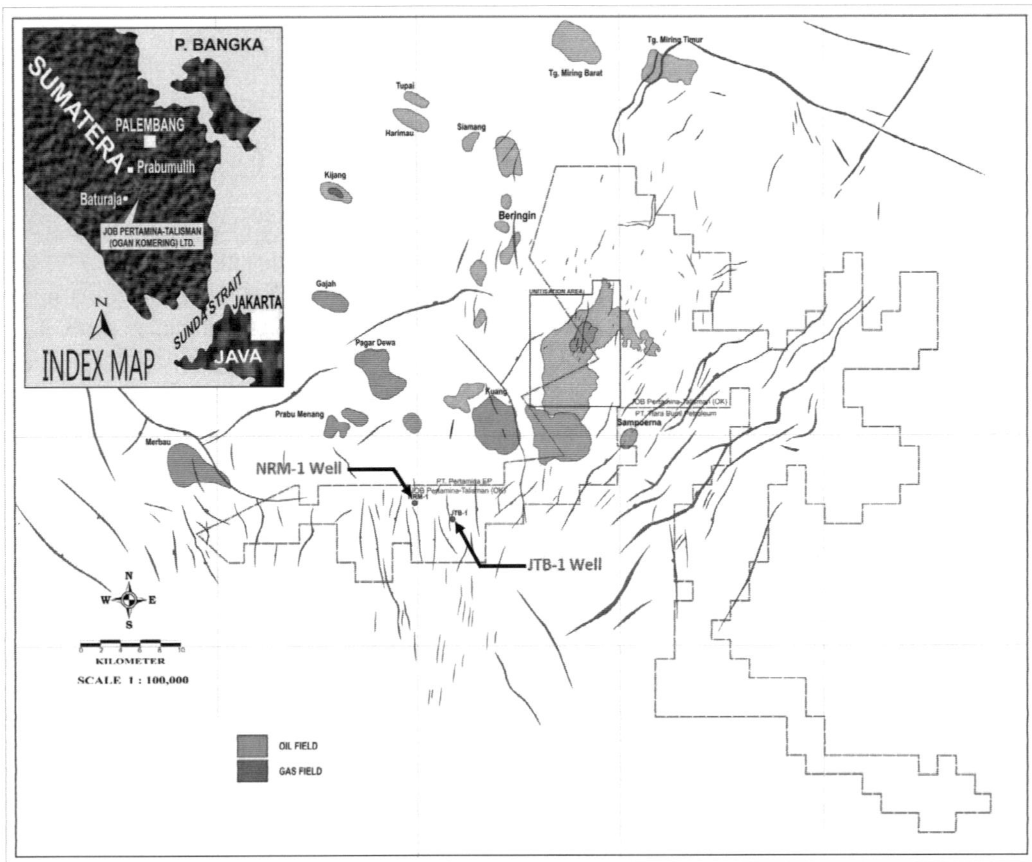

Figure 1. NRM-1 and JTB-1 wells location map in the West Ogan Komering Block, Meraksa Area, South Sumatra Basin.

2. Geological Setting

According to [1,4-6] the South Sumatera Basin is a back-arc basin as a result of the Sundaland and Australian plate interaction. The basin is underlain by the pre-Cenozoic metamorphic and igneous rocks basement that is cross-cut by extensional faults that formed horst and grabens to accommodate thick sequence of Cenozoic sediments from Eocene to Quaternary [1].

2.1. South Sumatera Basin Stratigraphy

The following stratigraphy is summarised from [1] from old to young. :

- Lemat or Lahat Formation (Paleocene – Early Oligocene). It consists of mainly shale with tuff intercalation and occasional coal layers, agglomerates and various coarse volcanic fragments.

IOP Conf. Series: Earth and Environmental Science **71** (2017) 012033 doi:10.1088/1755-1315/71/1/012033

- Talang Akar Formation (Late Oligocene – Early Miocene). The base of this formation is dominated by fluvio-deltaic sediments which consist of thinly layered shales and coals. The upper part consists of sandstrone, shale with occasional coal layers that represent paralic – shallow marine sediments. Baturaja Formation (Early Miocene) is known to be dominated by reefal limestone that was developed in local highs and found to be more silty and muddy in the deeper facies.
- Baturaja Formation (late Early Miocene – early Middle Miocene) is dominated by reef and clastic limestone.
- Gumai Formation (late Early Miocene) is mainly consist of thick muddy and silty shale succession that indicate the increase the maximum sea level in the South Sumatera Basin.
- Air Benakat Formation (Middle Miocene) characterised by thick mudstone layers at the base and more sandstones layers towards the top of the formation.
- Muara Enim Formation (Early Miocene – Late Miocene) is dominated by mudstone and shale with occasional sandstones and coals layers.
- Kasai Formation (Late Miocene – Pliocene) consists of pumice, tuffs, tuffaceous sandstones and mudstones.

3. Data

Selected intervals from JTB-1 and NRM-1 wells that represent the Baturaja (16 samples) and Gumai Formations (7 samples) were acquired. In JTB-1 well the Baturaja Formation were acquired between 1028.5 m and 902 m and the Gumai Formation were acquired between 813 m and 766 m depth. In NRM-1 well the Baturaja Formation were acquired at depth between 1070.5 and 1007 m while the Gumai Formation was acquired from 950 m to 726 m depth. The average sampling interval is 2 metres. In this study, stratigraphic unit classification for Baturaja Formation and Gumai Formation were determined by their lithologic characteristics based on previous study such as [1] and [2].

4. Methods

Samples from cutting and cores were cleaned using H_2O_2 solution so that they are free from fine clastic fragments and the foraminifera fossil can be seen clearly under the microscope. The benthic foram association are determined referring the catalogue published by [7]. In each sample the benthic forams association is utilised to determine the depth of the depositional environment based on the Bathymetric Zone of [8]. This study also determined the age of each rock unit based on [9] age zonation. Biostratigraphic correlation and interpretation then carried out to determine the cause of the difference in thickness based on isopach maps data, regional log correlation of Ogan Komering Block and depositional environment. Isopach maps and regional well correlation of Ogan Komering Block presented in this study are part of a larger research project that will be discussed in another paper.

5. Data Analysis

The analyses results are presented below; benthic foram association is utilized to determine a particular depositional environment that can also be the marker of the depositional environment changes (Table 1 and Table 2).

5.1. JTB-1 well

1. 1030 – 955 m: inner neritic (shallow neritic) characterized by few diversity of planktonic and benthic foraminifers which dominated by *Ammonia umbonata* (Figure 2a) and *Elphidium crispum* (Figure 2b). *Ammonia umbonata* and *Elphidium crispum* are typical fossil found in the high-energy sandy environment.
2. 902 – 813 m: middle neritic characterized by moderate abundance of benthonic and planktonic foraminifers. Association of *Bulimina* sp. (Figure 2c), *Virgulina* sp. (Figure 2d), and *Lenticulina thalmani* (Figure 2e) that usually confirms the middle neritic environment.

3. 778 – 766 m: outer neritic based on the maximum abundant of planktonic foraminifers. Benthonic foraminifers such as: *Uvigerina peregrina* (Figure 2f), *Lagena* sp. (Figure 2g), and increasing amount of *Bulimina striata* confirms the outer neritic environment.

Table 1. Depositional Environment of Baturaja and Gumai Formation in JTB-1 well.

No	Depth (m)	FORMATION	AGE	Planktonic Foraminifera Zone (Blow,1979)	BIOZONATION MARKER: Foraminifera Datum	PALEOBATHIMETRY (Tipsword et al., 1966): Supralitoral (terrestrial)	Littoral (tidal)	Shallow Neritic (0-20m)	Middle Neritic (20-100m)	Outer Neritic (100-200m)	Upper Bathial (200-500m)
1	352	Airbenakat Fm.	Middle Miosen younger	N11 or younger					■		
2	360-362							■			
3	362-364							■			
4	364-366							■			
5	366-368							■			
6	372-374							■			
7	374-376							■			
8	500							■			
9	596-598		Middle Miocene	N9-N11	→ *Globorotalia peripheroronda*			■			
10	600-602							■			
11	626.5							■			
12	626-628							■			
13	628-630							■			
14	630-632							■			
15	632-634							■			
16	634-636							■			
17	636-638								■		
18	638-640								■		
19	766-768	Gumai								■	
20	768-770									■	
21	770-772									■	
22	772-774									■	
23	774-776				→ *Orbulina suturalis*					■	
24	776-778			N5-N9						■	
25	813		Early Miosen						■		
26	902.5	Baturaja Fm.							■		
27	955							■			
28	956							■			
29	957							■			
30	962							■			
31	970							■			
32	971.5							■			
33	980-982							■			
34	985							■			
35	986.5							■			
36	1002							■			
37	1012.5				→ *Globigerinoides subquadratus*			■			
38	1014-1016			N4-N5				■			
39	1020-1022							■			
40	1022-1024							■			
41	1028.5							■			
42	1028-1030							■			
43	1162-1164				→ *Globigerinoides quadrilobatus*			■			
44	1164-1166		Oligo cene	N4 or older				■			
45	1071.5							■			

Table 2. Depositional Environment of Baturaja and Gumai Formation in NRM-1 well.

No	Depth (m)	FORMATION	AGE	Planktonic Foraminifera Zonation (Blow,1979)	BIOZONATION MARKER: Foraminifera Datum	PALEOBATHIMETRY (Tipsword et al., 1966): Supralitoral (terestrial)	Littoral (tidal)	Shallow Neritic (0-20m)	Middle Neritic (20-100m)	Outer Neritic (100-200m)	Upper Bathyal (200-500m)
1	628-630	Air Benakat Fm.	Middle Miocene	N13 - N?				■			
2	630-632							■			
3	632-634							■			
4	634-636							■			
5	638-640							■			
6	726	Gumai Fm.		N9 - N13	*Globorotalia mayeri*					■	
7	770									■	
8	771.5									■	
9	774									■	
10	774-776									■	
11	778-780									■	
12	942-946									■	
13	946-948				*Orbulina universa*					■	
14	948-950									■	
15	1007	Baturaja Formation	Early Miocene	N6 - N8					■		
16	1009.5								■		
17	1010.5								■		
18	1013								■		
19	1015								■		
20	1017.5								■		
21	1020			N4-N5	*Globigerinoides primordius*				■		
22	1022.5								■		
23	1027							■			
24	1029							■			
25	1030							■			
26	1032							■			
27	1034.5							■			
28	1038							■			
29	1039							■			
30	1044								■		
31	1048.5								■		
32	1056								■		
33	1062.5								■		
34	1064								■		
35	1068								■		
36	1070.5								■		

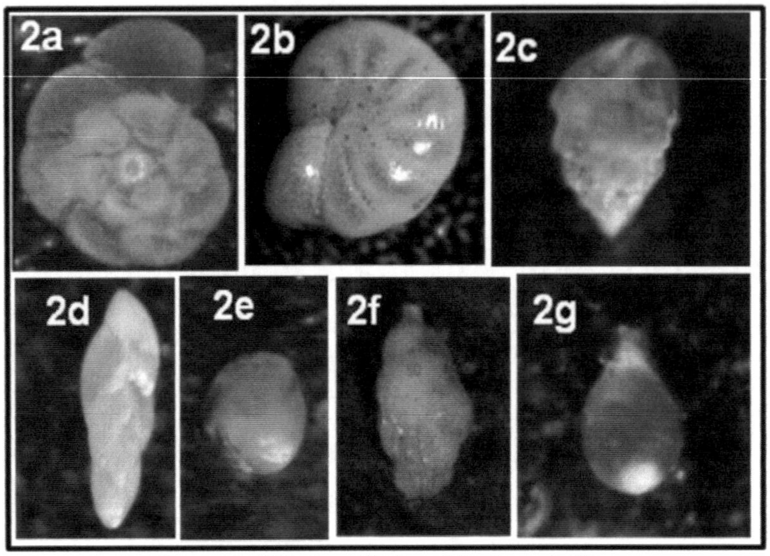

Figure 2. JTB-1 well's benthic foraminifera.

5.2. NRM-1 well

1. 1.070.5 – 1.044 m (BRF): middle neritic based on moderately abundant of planktonic and benthic foraminifers. Benthonic foraminifers association that consists of *Bolivina* sp. (Figure 3a) and *Cibicides* sp. (Figure 3b) confirms the middle neritic environment.
2. 1.039 – 1.027 m (BRF): inner neritic (shallow neritic) characterized by less diverse of planktonic and benthic foraminifers which is dominated by *Ammonia umbonata* (Figure 3c).
3. 1.022.5 – 1.007 m (BRF): middle neritic based on moderately abundant of planktonic and benthic foraminifers. *Bolivina sp.* and *Cibicides* sp. association marked the middle neritic environment.
4. 950 – 726 m (GUF): outer neritic indicated by optimum abundance of planktonic foraminifers. Benthic foraminifers index in this interval are *Uvigerina peregrina* (Figure 3d) and *Bulimina striata* (Figure 3e).

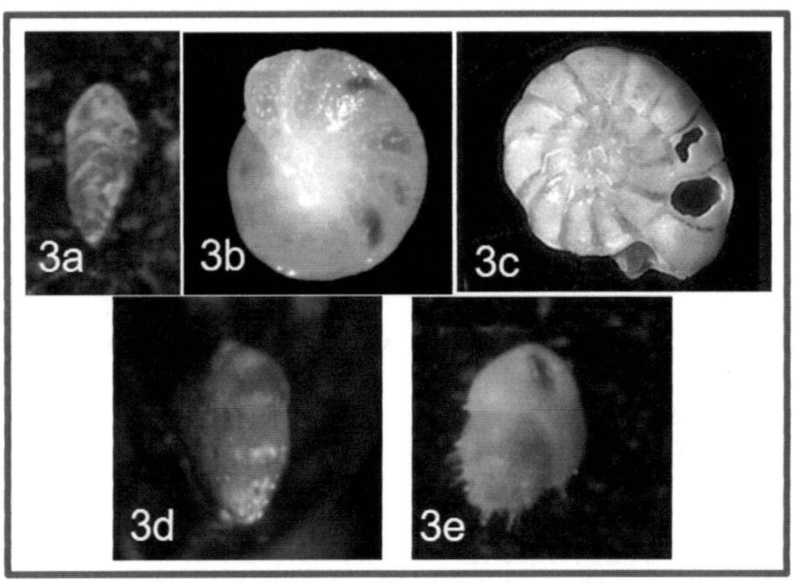

Figure 3. NRM-1 well's benthic foraminifera.

6. Discussion

1. The stratigraphic correlation suggests that the Baturaja and Gumai Formation in the NRM-1 well that were deposited in the deeper shelf environment have thicker sequence compared to the same formation in JTB-1 well.
2. The Baturaja Formation in JTB-1 well was deposited in a shallow neritic environment while at the NRM-1 well the fossil association suggest that it was deposited in a shallow neritic to middle neritic interval.
3. The Gumai Formation interval at the JTB-1 well was deposited in middle neritic to outer neritic environment while at the NRM-1 well the entire interval indicates outer neritic sediment.
4. Integrated study of microfossils analysis, stratigraphic, isopach thickness maps (Figure 4) and regional log correlation (Figure 5) shows that the position of JTB-1 well is structurally higher than the NRM-1. They are most likely to be deposited on the remnant of local highs that was formed by the extensional faults. It is consistent with the interval sediment in JTB-1 well which are generally thinner and interpreted as shallower marine deposit. The isochron and isopach maps of the Baturaja and Gumai Formations supports the interpretation by showing a generally thicker Baturaja and Gumai Formation around the NRM-1 well compared to the JTB-1 well. As written in the methods chapter, "isopach maps" and "regional well correlation of Ogan Komering Block" presented in this study are part of a larger research project that will be discussed in another paper.

The results from this study will give significant contributions for the hydrocarbon exploration and development in terms of improving knowledge of paleodepositional environment. Therefore, distribution, characteristics and hydrocarbon potential could be estimated more accurately as well as optimising the hydrocarbon recovery. Therefore, distribution, characteristics and hydrocarbon potential could be estimated more accurately as well as optimising the hydrocarbon recovery in order to maximize produceable hydrocarbon.

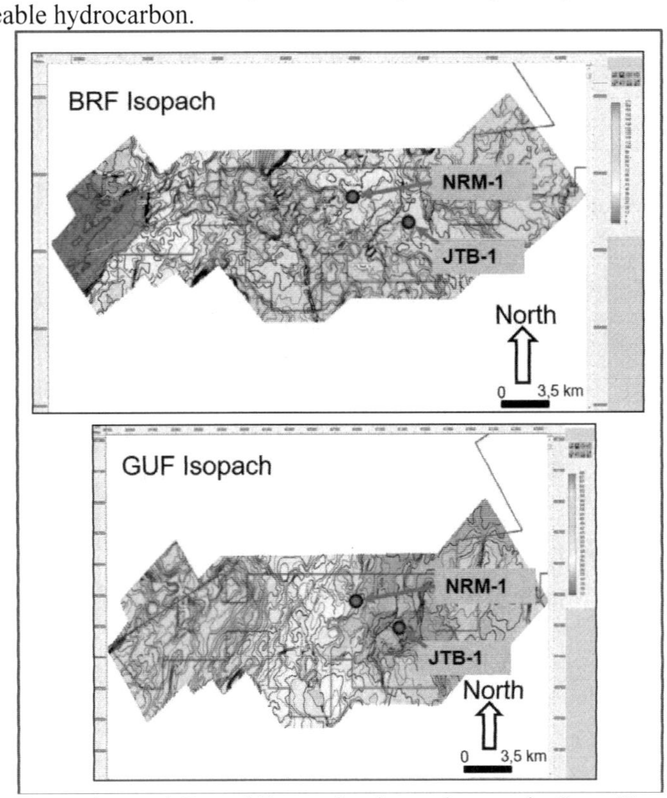

Figure 4. Baturaja Formation and Gumai Formation isopach maps in Ogan Komering Block. The color bar (on the upper right corner of the map) indicate sediment thickness. Blue and red represent thick and thin sediment thickness respectively.

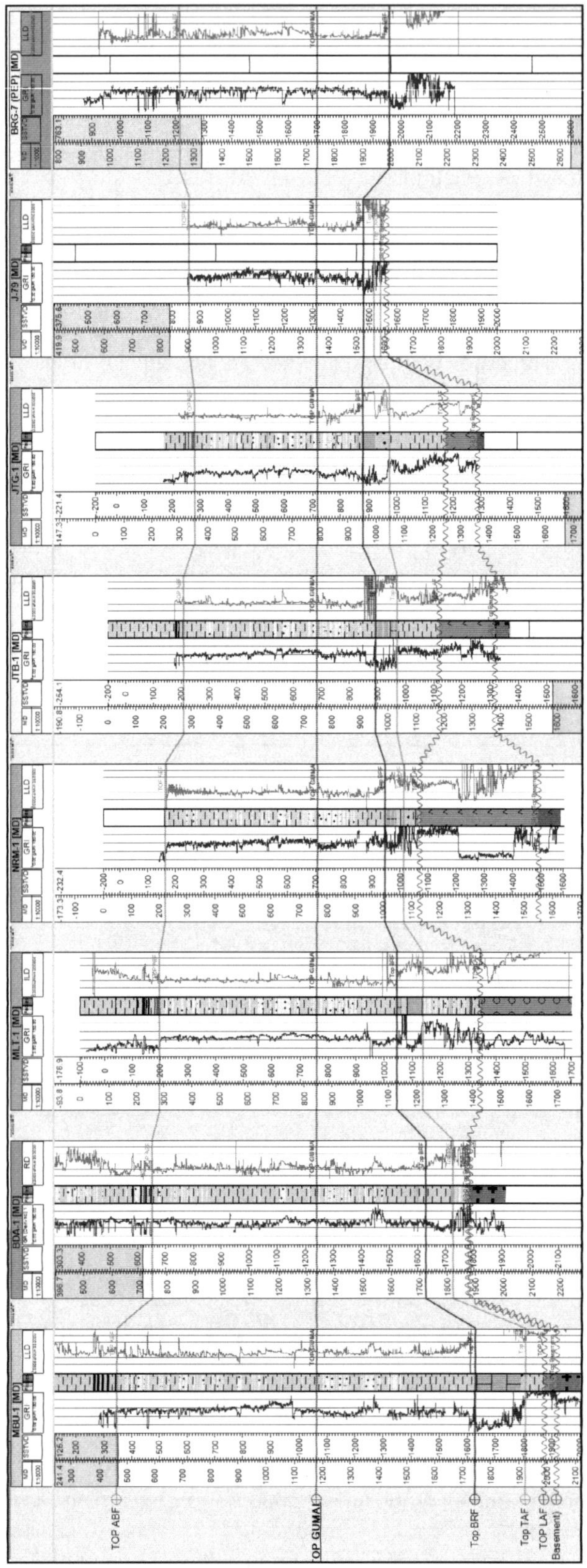

Figure 5. Regional log correlation of Ogan Komering Block.

7. Conclusions

The Baturaja Formation was deposited in a shallow neritic environment and shallow – middle neritic environment in JTB-1 well and NRM-1 well respectively. The Gumai Formation was deposited in middle - outer neritic and outer neritic environment in JTB-1 and NRM-1 well respectively. The Baturaja and Gumai Formation in the NRM-1 well are thicker and located in a deeper marine shelf environment compared to JTB-1 well. Future exploration should consider these new findings in order to better constrain the paleodepositional environment of the two formations.

8. Acknowledgments

We would like to thank the Geology Study Program at ITB for their help in providing the facility for the laboratory analyses in this study. A huge appreciation is addressed tp PT. LAPI ITB and the JOB Management of Pertamina-Talisman (OK) for their permission to publish this paper. Lastly, we would like to congratulate and appreciate the effort carried out by the TREPSEA II 2016 committee that has successfully carried out the conference and made this proceeding materialized.

9. References

[1] Pulunggono A, Suparman A, Assegaf A and Purwanto T 1990 Geology of Garba Area and surrounding, South Sumatra (Jakarta: Universitas Trisakti)
[2] Barber A J, Crow M J and Milsom J S 2005: Sumatra: Geology, Resources and Tectonic Evolution. *Geological Society Memoir* **No. 31** (London: The Geological Society), p 86-97
[3] Wibowo S S, 2015: Studies of geochemistry and maturity modeling on the Talangakar source rock at Tungkal block, South Sumatra Basin. *Buletin Geologi* **42,** p 199-215
[4] Katili J.A 1973 Geochronology of West Indonesia and Its Implication on Plate Tectonics, *Tectonophysics*
[5] Katili J.A 1975 Volcanism and Plate Tectonics in The Indonesia Island Acr. *Tectonophysics*
[6] Pulunggono A and Cameron N.R 1984 Sumatran Microlpates, Their Characteristics and Their Role in The Evolution of The Central and South Sumatra Basin, *Proc 13th Ann. Conc. IPA*, Jakarta, Mei
[7] Rauwenda P J, Morley R J and Toelstra, S R 1984 Assessment of Depositional Environment and Stratigraphy on the Basis of Foraminifera Paleoecology (Singapore: Robertson Research International Limited)
[8] Tipsword H L, Setzer F M and Smith L F 1966 Interpretation of Depositional Environment In Gulf Coast Petroleum Exploration From Paleoecology and Related Stratigraphy (Houston: *Gulf Coast Association of Geological Societies **XVI***)
[9] Blow W H 1969: Late Middle Miocene to Recent Planktonik Microfossils. *1st, Geneva Proceeding Leiden*, F.J. Brill **1**, p 199-422

Geochemical study of pyroclastic rocks in Maninjau Lake, West Sumatra

Endang Wiwik Dyah Hastuti

Geology Study Program, Faculty Engineering, Sriwijaya University, Srijaya Negara Street, Bukit Besar, Palembang, Indonesia
E-mail: endanghastuti59@gmail.com

Abstract. Maninjau area situated in Barisan Mountain range was a central of volcanic eruption during Plistocene-Holocene. Eruption of Maninjau volcano resulted in andesitic lava flows, and pyroclastic flows. Maninjau deposites accummulated around Maninjau Lake are covered by andesitic rocks. This Plistocene - Holocene volcano might have a close link with Semangko fault (Sumatra Fault). Pyroclastic rocks in the study area predominantly consist of pumice deposites with some crystals of plagioclase ± quartz ± pyroxene + hornblende + andesitic, lithics and glass shards. These deposites are slightly unwelded and mostly weathered. On the basis of its chemical composistion, Maninjau deposite can be divided into Ryolitic Pumiceous Tuff (Ignembrite Maninjau) and Andesitic Pumiceous Tuff (Tephra) deposites. Ignembrite Maninjau deposite is characterized by the occurrence of high-K rhyolite, whereas Tephra deposite is able to be grouped into calc-alkaline andesite. In general, these deposites have major elements with various silisic contents; this sugests that Maninjau deposites originated from heterogeneous magmas, and was erupted from a magma chamber. The relationship between SiO2 and another oxide shows negative correlation except Na2O. Pumice deposites contain various high trace elements, suggesting that the source of magma was likely from the subducted slab of oceanic plate underneath an active continental margin.
Keywords: ignimbrite, tephra, pumice; tuff; eruption; Maninjau Lake, West Sumatra

1. Introduction

The study area is geographycally located about 15 km from Bukittinggi to the west (Figure 1). It is a product of centre and fissure eruption of Maninjau Volcano. This fissure eruption resulted in cone volcano collepse and subsequently formed Maninjau Lake. The Maninjau eruption produced volcanic materials, such as andesitic rocks and pyroclastic flows dominated by pumice deposits.

Pumice deposits are resulted from pyroclastic flow which contain crystal, pumice, lithic, and glass shard originated from magma [1, 2]. Pumice deposits generated from Maninjau Volcano are characterized by unwelded rocks, dominated by pumice fragment (90%), and vary in grain size about 0.2 cm up to 20 cm.

Tectonically, Maninjau Lake is located in an active volcano tectonic belts (present-day volcanic arc), extending along the geanticline of Bukit Barisan in the western section of Sumatra (Figure 1). The lake was formed due to the Maninjau volcanic activity. With respect to the Sumatra Fault System, the region is situated at the southern segment of the zone. This regional strike fault is considered to have accommodated an oblique converging plate interaction between the Indo-Australian plate to the south and the Eurasian plate to the north. In the study area, normal faults seem to have been responsible for exposures of the pre Tertiary crystalline limestone.

Maninjau volcano is a stratovolcano type constituted by materials eruption of lava flows, pyroclastic flows, and a number of small intrusive bodies. Maninjau volcanic eruptions during Tertiary produced an elongated caldera forming Maninjau Lake [1]. Maninjau caldera is relatively large, indicating that the eruption lasts longer, and might have to some extent related to the movement of the Sumatran fault system [2].

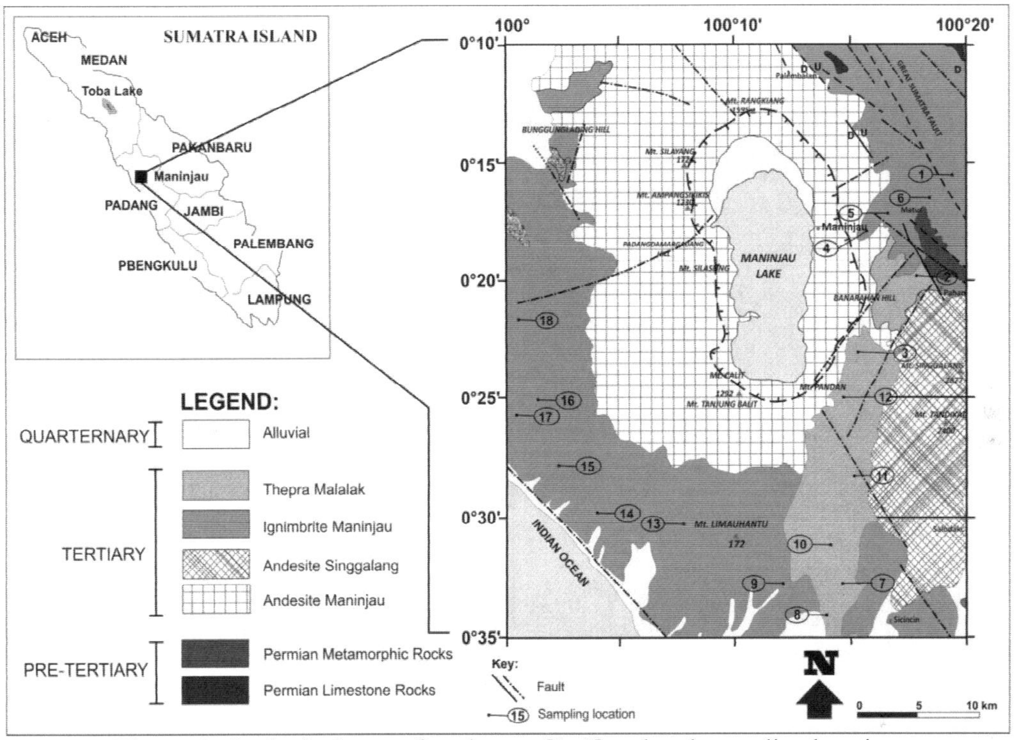

Figure 1. Geological map of study area [1, 3] and rock sampling location.

The oldest rocks exposed in the study area consist of rhyolite, limestone, granodiorite, and diabase. The rocks formed in the Paleozoic until Tertiary [3,4]. While volcaic rocks are formed by Maninjau Volcano activity and consisting of basaltic rocks, tuffaceous breccia, pyroxene andesite, lava and pumiceous tuff [3].

The results of Maninjau volcanic eruption are widespread along the coast of West Sumatra with the general composition of andesite. The oldest volcanic rocks exposed around the study area is andesite lava flows originating from Maninjau Caldera [1, 3] (Figure 1). Pyroclastic flow exposed at around the Maninjau Lake and expand westward. Andesite lava from the eruption of Maninjau formed in the Tertiary to Early Quaternary

2. Data and Method

For this reseach about 18 pyroclastic rocks samples were collected from the Maninjau area. Having completed petrographical studies and x-ray fluorescence spectroscopy (XRF). About 15 samples from pumice fragments were analyzed for major and trace elements. The samples labeled Qpt belong to Ignimbrite Maninjau about 8 samples, while samples labeled Qhpt were collected from Tephra Malalak about 7 samples (Figure 1). The petrography and chemical analysis was done by Activation Laboratories in tekMIRA Bandung.

3. Result and Discussion

3.1. Petrology

IOP Conf. Series: Earth and Environmental Science **71** (2017) 012034 doi:10.1088/1755-1315/71/1/012034

In the field, the Ignimbrite Maninjau is unwelded, greyish white, sand to pabble in grainsize with pumice fragments about 20 cm in grainsize. Pumice is dominant material in Ignimbrite Maninjau (about 75% of volume). Under microscope polarisation, the Ignimbrite Maninjau consists of feldspars, plagioclases, hornblende, biotite, magnetite, quartz crystal, pumice and lithic fragments and laid down on ground mass of glass shards (Figure 2). The crystals content is about 10%, with grainsize about 0.1-2.5 mm. Beside that, in some other samples are contain limonite and serisite which are formed due to alteration of feldspars and glass shards. General textures are vitrophiric with visicle structures and some of them show glomeroporphyritic.

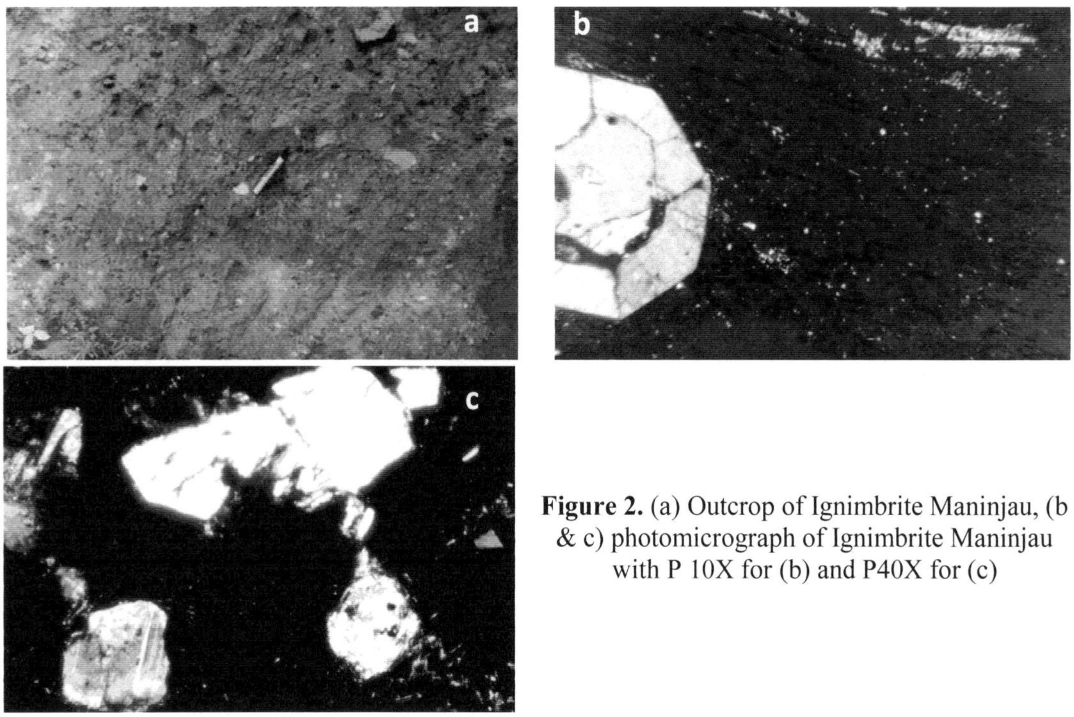

Figure 2. (a) Outcrop of Ignimbrite Maninjau, (b & c) photomicrograph of Ignimbrite Maninjau with P 10X for (b) and P40X for (c)

The Tephra deposit is found in the east to south part ofthe Maninjau Lake (Figure 1). Base on stratigraphic position, it overlay andesite and Ignimbrite Maninjau. In the field Tephra deposit is characterised by granule to pabble in grainsize, redish white, and unwelded. Pumice fragments are dominant (about 85%) and characterised by pink color, unwelded, vasicular structures and glassy. Under the microscope pollarization, Tephra deposit consist of feldspars, hyperstene, hornblende, biotite and magnetite with grainsize about 0.1- 1.2 mm and constitutes about 10% of rock volume, also it contains of pumice and lithic fragments; and they are laid down on ground mass of glass shards. This deposit shows vitrophyric textures and vesicle structures (Figure 3).

IOP Conf. Series: Earth and Environmental Science **71** (2017) 012034 doi:10.1088/1755-1315/71/1/012034

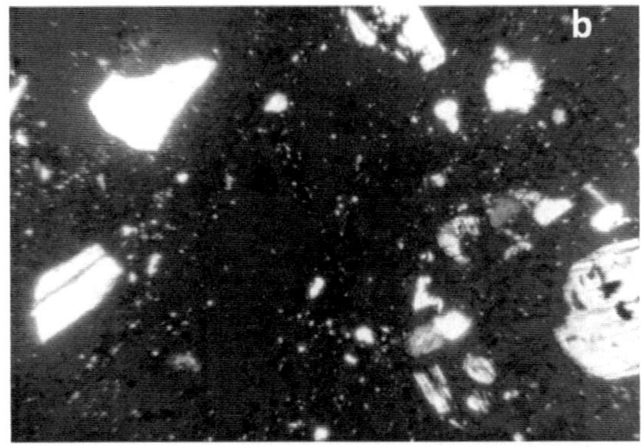

Figure 3. (a) Outcrop of Tephra Malalak, (b) photomicrograph of Tephra Malalak

Base on microscophic analysis, Tephra Malalak deposit is dominated by plagioclases of albite – andesine. The origin of plagioclases are coused by fractional crystallisation selective [5, 6]. So that the magma which is contain pyroxene and hornblende set in the basement of magma chember. Then this magma erupted and solidified to form plagioclase crystals. After that, they are cooling rapidly to produce volcanic ash. The last of this processes, the basement of magma chamber also had undergone eruption and produced deposits with composition of pyroxene and plagioclase which had formed Tephra deposit.

3.2. Geochemistry

3.2.1. Major Elements

Silica contents in the Ignimbrite Maninjau deposit range from 69.50-75 wt % and all samples fall in the high K-rhyolite composition (Figure 4). The relationship between SiO_2 with other major elements shows a negative trend, except Na_2O which shows no correlation (Figure 5). Al_2O_3 and CaO which are controlled by plagioclase and feldspar forming have high value. Whereas, Fe_2O_3, TiO_2 and MgO are controlled by pyroxene, hornblende, biotite and magnetite forming have low value (Table 1).

Table 1. Major elements (wt%, after normalized to 100%) analysis of pumice fragments from pyroclastic rocks in Maninjau Lake.

Samples No	Qpt 1	Qpt 4	Qpt 5	Qpt 9	Qpt 13	Qpt 15	Qpt 6	Qpt 18	Qhpt 2	Qhpt 3	Qhpt 7	Qhpt 8	Qhpt 10	Qhpt 11	Qhpt 12
SiO_2	75.50	77.30	78.01	77.39	72.81	76.37	75.12	73.50	62.72	61.35	63.50	62.50	62.43	61.63	62.22
Al_2O_3	13.21	12.77	12.5	12.75	14.51	12.91	13.80	14.33	18.12	18.43	18.00	18.21	18.00	18.21	18.38
Fe_2O_3	1.,73	1.32	1.12	1.33	2.24	1.47	1.83	2.19	7.27	7.71	7.30	7.35	7.44	7.33	7.41
TiO_2	0.20	0.15	0.15	0.17	0.28	0.18	0.27	0.25	0.81	0.83	0.80	0.88	0.81	0.84	0.82
CaO	0.96	0.83	0.66	0.76	1.24	0.73	1.21	1.18	5.14	5.06	5.11	5.17	5.13	5.20	5.20
MgO	0.25	0.12	0.07	0.10	0.32	0.12	0.21	0.32	1.81	2.19	1.25	1.68	1.80	1.96	1.62
K_2O	5.10	4.24	4.43	4.16	5.65	5.06	4.65	5.21	1.48	2.22	1.34	1.63	1.79	2.23	1.86
Na_2O	3.05	3.27	3.03	3.34	2.95	3.16	2.90	3.02	2.65	2.29	2.70	2.58	2.60	2.60	2.49
Total	**100**	**100**	**100**	**100**	**100**	**100**	**100**	**100**	**100**	**100**	**100**	**100**	**100**	**100**	**100**
LOI	2.80	2.83	3.24	3.43	3.49	2.95	2.90	3.40	3.30	2.50	3.43	3.25	3.12	2.87	2.98

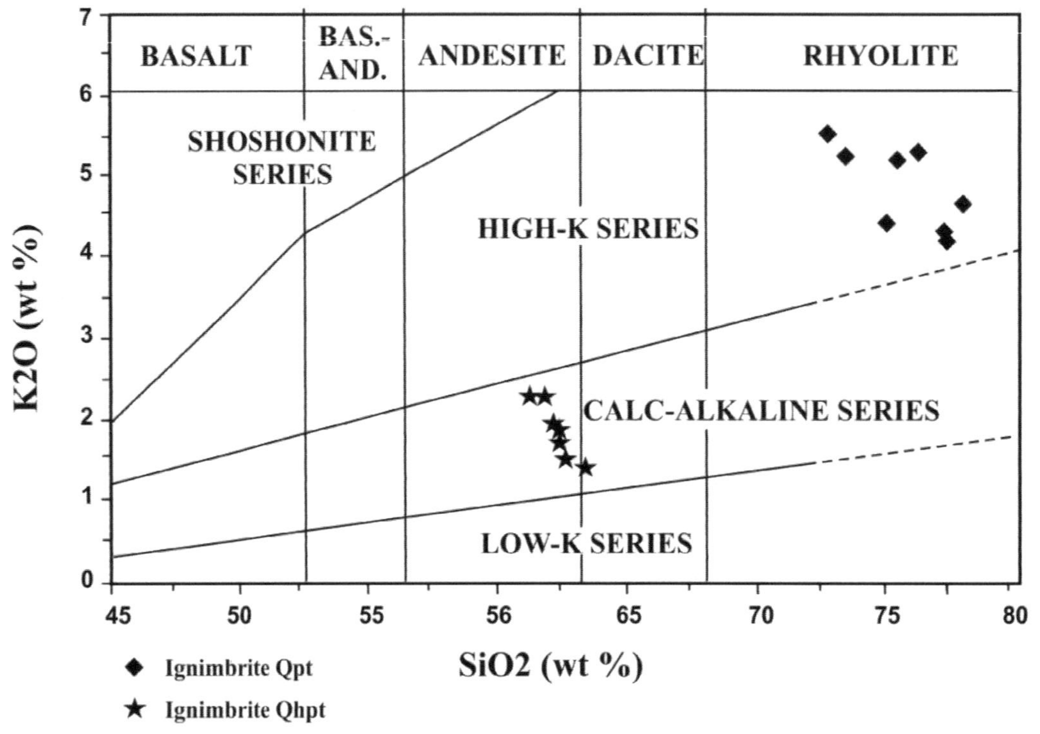

Figure 4. Diagram of K_2O vs SiO_2 (wt.%) from study area show that Ignimbrite Maninjau fall in high-K Rhyolite whereas Thepra Malalak fall in Calk-alkaline Andesite (the line margin based on [7]).

Silica content in the Tephra deposit fall in calc-alkaline andesite (Figure 4) and is characterised by high content of Al_2O_3 and low CaO. They reflect fractionation of plagioclase. Whereas, Fe_2O_3 is more dominant then TiO_2 and MgO are controlled by fractionation of pyroxene and magnetite. They show that magma of Tephra Malalak deposit contain a lot of plagioclase and pyroxene (Table1; Figure 5). They originated from fractional crystallisation selective processes [6, 8]. The relationship between SiO_2 which other major elements show negative trend, except Na_2O which shows positive trend (Figure 5).

Base on chemical compositions (major elements), the magma of Tephra Malalak and Ignimbrite Maninjau deposits are origin from the Active Continental Margin in which magma contains much of silica [9, 10, 11]. So that alot of deposit in this area has acid composition such as pumice deposits. Generally, this magma is present to associate with a thick continental crust [11]. This condition can form to the location of the Maninjau Lake which laid in the Barisan Mountain zone.

IOP Conf. Series: Earth and Environmental Science **71** (2017) 012034 doi:10.1088/1755-1315/71/1/012034

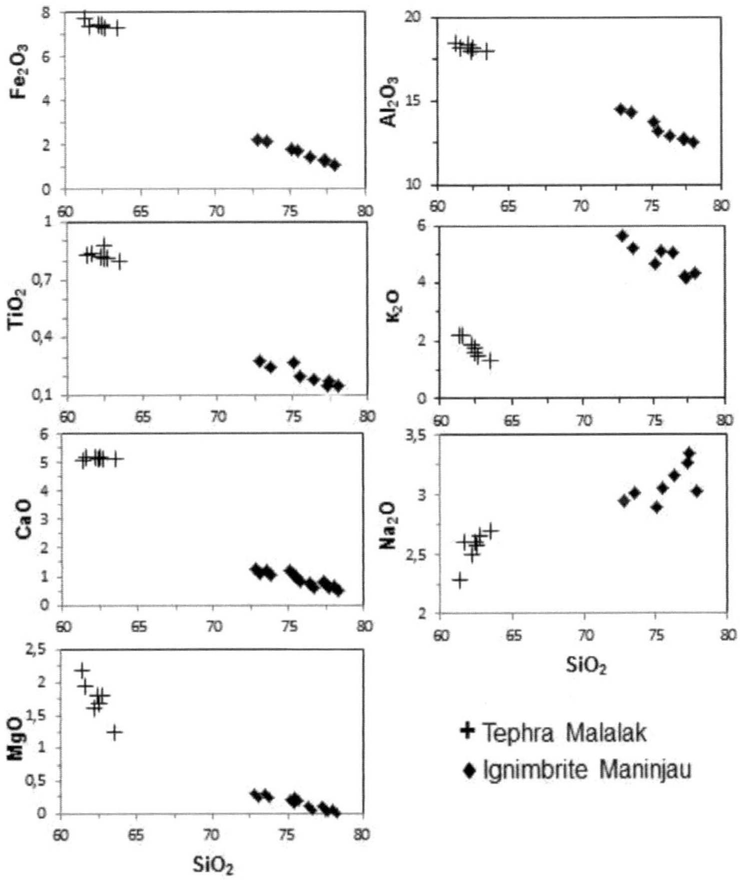

Figure 5. Diagram of major element content (weight %) of study area.

3.2.2 Trace Elements

All trace elements value of the Ignimbrite Maninjau deposit are small varies in composition. In the variation of hacker diagram, they show systematic correlation to silica except Sr which has no correlation if compared to Tepra deposit (Figure 6). The trace elements of Ignimbrite Maninjaudeposit show positive trend, exclude Zr which shows negative trend (Figure 6). All trace elements are high value, except Sc (Table 2). Whereas trace elements of Tephra Malalak deposit are slightly low value than Ignimbrite Maninjau deposit, except Sr which is slightly higher value (Table 2).

Table 2. Trace element (ppm) analysis of pumice fragment from study area included trace element ratio.

Samples No	Qpt 1	Qpt 4	Qpt 5	Qpt 9	Qpt 13	Qpt 15	Qpt 6	Qpt 18	Qhpt 2	Qhpt 3	Qhpt 7	Qhpt 8	Qhpt 10	Qhpt 11	Qhpt 12
Rb	365	364	367	366	330	335	340	336	216	215	211	220	226	219	222
Ba	538	540	524	523	491	495	503	520	442	433	418	429	423	412	394
Sr	355	357	343	351	374	365	358	246	365	369	378	380	386	375	376
La	530	535	524	523	491	496	510	517	438	425	418	417	410	415	399
Ce	636	639	644	635	610	616	620	635	517	512	505	507	504	498	493
Y	245	249	246	248	235	245	245	240	185	182	186	187	192	190	184

IOP Conf. Series: Earth and Environmental Science **71** (2017) 012034 doi:10.1088/1755-1315/71/1/012034

Samples No	Qpt 1	Qpt 4	Qpt 5	Qpt 9	Qpt 13	Qpt 15	Qpt 6	Qpt 18	Qhpt 2	Qhpt 3	Qhpt 7	Qhpt 8	Qhpt 10	Qhpt 11	Qhpt 12
Sc	33	34	31	32	30	31	35	28	23	24	26	29	27	28	26
V	420	422	417	404	396	398	410	415	386	383	379	375	372	368	379
Ni	275	277	271	260	250	255	261	267	220	215	213	211	216	208	214
Nb	217	216	218	216	210	215	213	211	175	175	168	165	172	163	160
Zr	289	281	280	285	305	304	297	292	276	280	279	275	282	279	278
Zn	325	327	324	321	325	323	321	320	283	280	279	275	282	279	278
Cu	401	404	378	377	367	369	368	375	320	323	324	362	331	329	318
Ba/Sr	1.51	1.51	1.53	1.50	1.30	1.36	1.40	2.11	1.21	1.80	1.10	1.13	1.10	1.10	1.05
K/Rb	116	97	109	94	142	125	114	129	57	86	53	61	66	84	70
Rb/Sr	1.03	1.02	1.07	1.04	0.88	0.92	0.95	0.97	0.59	0.58	0.56	0.58	0.59	0.58	0.59
Ba/La	1.01	1.01	1.00	1.00	1.00	0.99	0.99	1.00	1.01	1.02	1.00	1.03	1.03	0.99	0.99

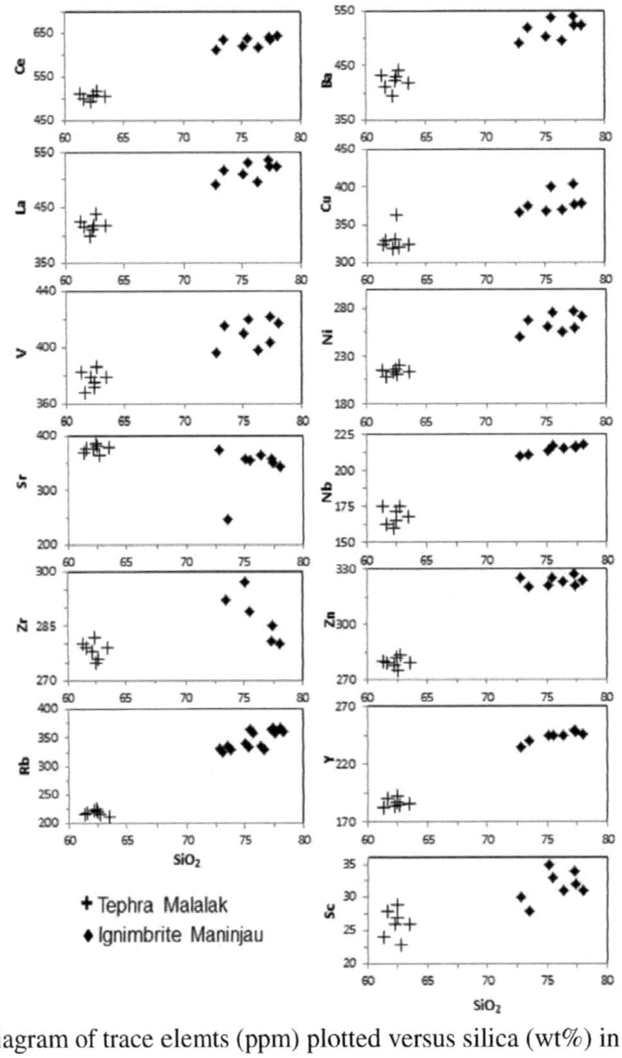

Figure 6. Diagram of trace elemts (ppm) plotted versus silica (wt%) in the study area

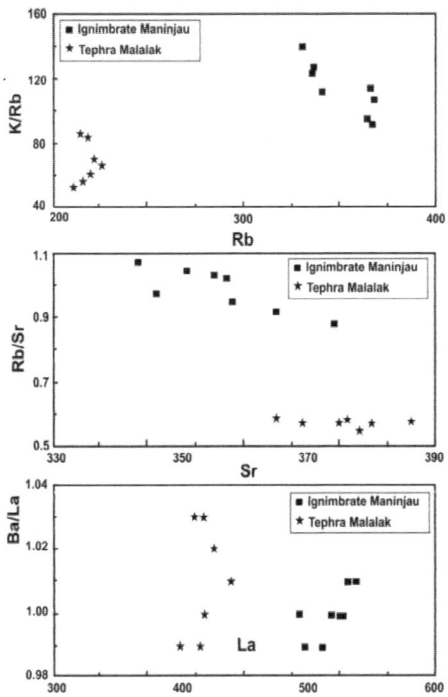

Figure 7. Diagram of trace element ratio versus trace element (ppm) in the study area.

Rb is incompatible element forming crystal in the first phase and control by crystallization of amphibole and biotite. Ba and Sr indicate crystallization of feldspar. Ba/Sr ratio has a tendency increasing with plagioclases crystallization and decreasing of K-feldspar when begin crystallization [12]. Rb and Sr content are high, show that magma origin from crustal with granitic composition. This magma is formed in the first level of magma chamber [11, 13].

Zr and Nb content of Ignimbrite Maninjau deposits are slightly higher then Tephra deposit. Both Zr and Nb are concentrated in liquid residue. La and Ce are rare earth elements and have the same chemical characterization (Figure 4). They have high value in Ignimbrite Maninjau deposit, whereas in TephraMalalak deposite is slightly lower then Ignimbrite Maninjau deposit. La and Ce are controlled by fractionation of pyroxene. But under microscope polarization show that pyroxene only found in few amount in Tephra Malalak deposit. This show that both La and Ce are concentrated in liquid. The study area contains Zn and Cu in high value in Ignimbrite Maninjau deposit, where as in Tephra Malalak deposit has slightly lower.

Ni, V and Sc are compatible elements with ferromagnesian minerals. V and Sc are found in Fe-Ti oxide and hornblende. Ni content of Ignimbrite Maninjau and Tephra deposits are variation and show high value (Table 2 and Figure 6). The high value of Ni content indicate that parental magma origin from peridotitic mantle. V content in Ignimbrite Maninjau deposit is higher then Tephra Malalak deposite. V content controlled by Fe-Ti oxide crystallization. Sc value in both Ignimbrite Maninjau and Tephra deposits are low, about 23-34 ppm.

In the Ignimbrite Maninjau deposit, Ba/Sr ratio show no variation and has value the same as Tephra Malalak deposit (Table 2). They show that Ba and Sr generally, are controlled by crystallization of plagioclase. K/Rb ratios and diagram of K/Rb versus Rb of Ignimbrite Maninjau deposit show negative trend. Whereas Tephra Malalak deposit has lower K/Rb ratios than Ignimbrite Maninjau deposit (Figure 7). Negative trend of K/Rb versus Rb is controlled amphibole crystallization.

Rb/Sr ratios of Ignimbrite Maninjau deposit show a little variation when compared to Tephra deposit. Rb/Sr versus Rb Ignimbrite Maninjau deposit shows negative trend and is controlled by feldspars crystallization, whereas Tephra deposit has no trend (Figure 7). Beside that it is controlled by feldspars crystallization and also it is controlled by alteration process of feldspars and glass shards.

Ba/La ratios of Ignimbrite Maninjau and Tephra Malalak deposits show the same ratio values. When Ba/La ratios are plot versus La, Ignimbrite Maninjau pumice deposits show positive correlation where as Tephra deposit has no correlation (Figure 6).

Base on Sr, Rb, and Ba content in Ignimbrite Maninjau and Tephra Malalak deposits that the magma origin from metasomatism of arc basalt which caused by come out of liquid on subduction zone. Figure 7 show that Ignimbrite Maninjau and Tephra Malalak deposits which contain Ba, Sr, Rb and K are separated. These expresses that the pumice deposits of Ignimbrite Maninjau and Tephra Malalak in the Maninjau Lake origin from from zone magma

4. Conclusions

Base on laboratorium analysis and field study can be concluded that:

- Pyroclastic deposits in the study area have composition rangging from high-K rhyolite to calc-alkaline andesite. This show that magma originated from magma chamber zone.
- The trace elements content in Pyroclastic deposits show that the magma of the study area may be the deposit origin from partial melting due to subduction plate in which produce magma from continetal crust.
- To know the model of magma petrogeneses and their origin is proposed todo detail analyses for trace element especially rare earth elements and isotop elements beside that sampling will good if collected from statigrphic position and the deposit which analyse not only from vocanic deposits but also sedimen around study area should be analysed.

5. Acknowledgements

The author is grateful to the University of Sriwijaya for research fund. We also thank our colleagues Edy Sutriyono and Steve, who have been involved in full paper done.

6. References

[1] Pribadi A, Mulyadi E, and Pratomo I 2007 *Mekanisme Erupsi Ignimbrit Kaldera Maninjau, Sumatera Barat* Jurnal Geologi Indonesia, Vol. 2 No. 1: 31-41.
[2] Smith R I 1960 *Zone and zonal variations in welded ash flo*ws, US.: Geological Survey Profesional Paper, vol. 354-F.
[3] Kastowo, Leo, G W, Gafoer S. and Amin T C 1996 *Geological map of the Padang quadrangle, Sumatra* Geological Research and Development Centre.
[4] Barber A J, Crow M J and Milson J S 2005 *Sumatra: Geology, Resources and Tectonic Evolution* The Geological Sociaty, London
[5] Cox K G and Pankhurst R J 1987 *The interpretation of igneous rocks* George Allen &Unwin, London.
[6] Hildreth W 1981 *Gradients in silisic magma chamber: Implication for lithospheric magmatism* Journal of geophysical Research 86: 10153 – 10192.
[7] Edwart A 1982 *The mineralogy of Tertiary – Recent orogenic volcanic rocks*. In Thorpe, RS. (Ed): Andesite, p. 25 – 98, Wiley, New York.
[8] Walker G P L 1972 *Crystal concentration in ignimbrite* Controbution of mineralogy and Petrology, vol. 36: p 135- 146.
[9] Nicholls, J and Gordon, T.M., 1994. Procedures for The Calculation of Axial Ratios on Pearce Elment-Ratio Diagrams. The Canadian Mineralogist. Vol.32, pp. 969-977.
[10] Wilson M 1986 *Igneous Petrogenesis* Unwin Hyman, London.
[11] Le Bas M J, Le Maitre R W, Streckeisen A and Zanettin B 1986 *A chemical classification of volcanic rocks based on the total alkali–silica diagram* Journal of Petrology 27,745–750.
[12] McBirney A R 1984 *Igneous petrology*. Freeman, Cooper & Company, California.
[13] Cas R A F and Wright J V 1987 *Volcanic succession: Modern and ancient*. Unwin Hyman, London.